MATRIX MANAGEMENT SYSTEMS HANDBOOK

MATRIX MANAGEMENT SYSTEMS HANDBOOK

Edited by

David I. Cleland

Professor, Engineering Management
School of Engineering
University of Pittsburgh
Pittsburgh, Pennsylvania

VNR VAN NOSTRAND REINHOLD COMPANY
NEW YORK   CINCINNATI   TORONTO   LONDON   MELBOURNE

Library of Congress Catalog Card Number: 82-24827
ISBN: 0-442-21448-0

Manufactured in the United States of America

Published by Van Nostrand Reinhold Company Inc.
135 West 50th Street
New York, New York 10020

Van Nostrand Reinhold Company Limited
Molly Millars Lane
Wokingham, Berkshire RG11 2PY, England

Van Nostrand Reinhold
480 Latrobe Street
Melbourne, Victoria 3000, Australia

Macmillan of Canada
Division of Gage Publishing Limited
164 Commander Boulevard
Agincourt, Ontario M1S 3C7, Canada

15 14 13 12 11 10 9 8 7 6 5 4 3 2 1

Library of Congress Cataloging in Publication Data
Main entry under title:

Matrix management systems handbook.

   Includes index.
   1. Matrix organization—Addresses, essays,
lectures.   I. Cleland, David L.
HD58.5.M39   1983   658.4′02   82-24827
ISBN 0-442-21448-0

Preface

The purpose of this handbook is to provide managers and professionals with a reference guide for the design and implementation of matrix management systems in their organizations. The book will be useful for supervisors of matrix operations in contemporary organizations, and for those professionals responsible for applying matrix implementation techniques in such organizations. In addition, the book will help all persons—students, managers, professionals—who require concise reference material on the various matters related to matrix management systems.

Managers and professionals alike need to keep abreast of changes in the structure and management of contemporary organizations. A key characteristic of today's organization is the growing use of some variation of "matrix management," which emphasizes a team approach to the management of organizational activities. The project management context of matrix management may include a project as large as the Alaska oil pipeline. At the other extreme, managing a small research project in a laboratory may involve project management techniques. Project management is firmly established today in corporations and in governmental and educational organizations. Team management is also found in other applications such as task forces, product management, product development teams, people involvement teams, Quality Circles, task teams, plural executives, satellite teams, project center management, new venture management, etc. Current management literature groups all these team-related efforts under the generic title of "matrix management." This handbook provides a pragmatic explanation of what matrix management is all about.

Like any new concept, matrix management continues to emerge. While many authors have published in the field, a reference book has not been made available. Thus this handbook satisfies a need for a reference source that combines synergistically the many theoretical and practical ideas about matrix management that have appeared in recent years.

The handbook provides both summary and detailed guidance for the many varieties of matrix management that have become popular in industrial, governmental, educational, and nonprofit organizations. Many of the authors have provided related bibliography for further reference. Senior executives will find this book useful in determining the potential opportunities for altering their organization toward a matrix form. Once matrix has been implemented or broadened, senior executives will find this book essential for improving the management of the matrix process. For many managers, matrix management is either unknown or surrounded with a mystique that discourages experimentation in this modern management approach. This book should help these executives understand what matrix management can and cannot do for them.

The handbook provides information on both the theory and practice of matrix management. Primary emphasis is on the practical aspects of managing matrix systems. This pragmatism is based on a sound theoretical framework of management experience and thought.

Many people cooperated in producing this book. Their qualifications are stated in the biographical sketches that accompany the chapters.

Seven interdependent areas of matrix management systems are presented.

1. *Introduction to Matrix Management* provides a general framework for what matrix management is and how it evolved.

2. *Matrix Management Applications* deals with the different matrix forms in use today.

3. *The Matrix Culture* examines how values, beliefs, and attitudes of people are shaped by matrix management.

4. *Implementing Matrix Management* describes the problems and opportunities encountered, as well as some strategies to use, when starting up matrix systems.

5. *Matrix Support Systems* describes some of the more important supporting systems required to make matrix effective.

6. *Matrix Management Techniques* deals with several key techniques and approaches for making matrix work.

7. *Organizational Strategy in Matrix Management* provides guidance on how to vary organizational design and development to accommodate matrix management.

I thank the contributing authors for their practical presentations of what matrix management is all about. I am deeply indebted to Claire Zubritzky, who expertly administered the handbook. I also thank Dr. Al Holzman and Dr. Max Williams, who in the School of Engineering at the University of Pittsburgh continue to provide an environment in which the faculty can pursue—in their own way—the generation of knowledge.

DAVID I. CLELAND
Editor

Contents

Start pg 226

SECTION V / MATRIX SUPPORT SYSTEMS

SECTION VI / MATRIX MANAGEMENT TECHNIQUES

SECTION VII / ORGANIZATIONAL STRATEGY IN MATRIX MANAGEMENT

MATRIX MANAGEMENT SYSTEMS HANDBOOK

Section I
Introduction to Matrix Management

Clearly, matrix management is an idea whose time has come. The particular matrix form that can be used by today's managers depends on the circumstances in the environment that stimulate the need for some alternative organizational form. Section I of this handbook presents an overview of contemporary matrix management systems. Matrix management is viewed as a complex of alternative management and organizational forms designed to deal with the interdependencies, complexities, and change of current organizations.

In Chapter 1, David I. Cleland presents an overview of matrix management as a kaleidoscope of organizational systems. He discusses briefly the alternative forms of matrix management in use today. An operational definition of matrix management is provided in terms of structure, process relationships, and patterns of organizational behavior criteria which indicate that some form of matrix management exists.

In Chapter 2 John R. Adams and Nicki S. Kirchof discuss the practice of matrix management. The authors use the typical evolution of the matrix in an organization as a way to identify those conditions in an organization that require some organizational change. By examining the roles of the managerial participants in the matrix, Adams and Kirchof establish the differences between the requirement of the matrix and those of the more traditional organizations.

In Chapter 3 Milan Moravec uses the dual reporting aspect of matrix management as a focal point to examine the sum of the behavioral and organizational factors that contribute to, or detract from, high performance in making the matrix work. He points out the challenge of ensuring high productivity in the simple one-boss, one-worker situation that is compounded when the dual reporting relationship of matrix management is introduced.

1. A Kaleidoscope of Matrix Management Systems[1]

David I. Cleland*

A kaleidoscope of matrix management systems is emerging in the theory and practice of management today. These systems appear to have one overriding characteristic—a departure from the classical model of management in favor of a multidimensional system of sharing decisions, results, and rewards in an organizational culture characterized by multiple authority-responsibility-accountability relationships.

This chapter provides an overview of the various management systems that have evolved in modern organizations, and will help illuminate the value of the matrix approach and demonstrate its flexibility in meeting the needs of different organizations.

PROJECT MANAGEMENT

Project management emerged in an unobtrusive manner starting in the early 1960s. No one can claim to have invented project management; its beginnings are often cited in the ballistic missile program or the space program of the United States. The origins of project management can be found in the management of large scale ad hoc endeavors such as the Manhattan Project, and—on a smaller scale—in the practical model provided by project engineering.

In 1961 Gerald Fisch, writing in the *Harvard Business Review,* spoke of the obsolescence of the line-staff concept and heralded a growing trend toward "functional-teamwork" approaches to organization. Also in 1961, IBM established systems managers with overall responsibility for various computer models across functional division lines. In the 1960s and 1970s a wide variety of organizations experimented with alternative project management organiza-

[1]This chapter is adapted from the article by David I. Cleland, "A Kaleidoscope of Organizational Systems." *Management Review,* December, 1981. Used by permission.
*Dr. Cleland is a Professor of Engineering Management in the Industrial Engineering Department at the University of Pittsburgh. He is author or coauthor of nine books and has published numerous articles in leading national and internationally distributed technological, business-management, and educational periodicals. He has had extensive experience in management consultation, lecturing, seminars, and research.

tional forms. At present, project management has reached a high degree of maturity and is widely used in industrial, educational, governmental, and military circles. A distinct literature has emerged dealing with the management of ad hoc projects in contemporary organizations.

PROJECT ENGINEERING

Project engineers are usually responsible for directing and integrating all technical aspects of the design/development process. Typically, a project engineer manages a product throughout the engineering process from initial design through the service life of a product. When a product design/development problem exists in a functional department, a project engineer is responsible for working with the functional department to correct the problem. In some companies the project engineering function is, in effect, the project management arm of the engineering department. In another context, project engineering involves the building of plants from the preliminary study through the design, procurement, erection, and start-up operation.

PRODUCT MANAGEMENT

Product management in one of its earliest forms appeared in the early 1930s when Procter & Gamble inaugurated brand management. Thus this form of matrix management has been with us for some time. Briefly, it assigns responsibility for a given product or brand to an individual. When product managers are so appointed, a matrix organization is created with a resulting product-functional interface.

In some cases these product managers are basically information gatherers on product performance and have little or no authority over the heads of the functional departments. In other situations, a product manager develops a product plan covering such matters as advertising, use of field sales force, research support, packaging, and manufacturing programs. He is then responsible for negotiating with the suitable functional departments for the support and costs of the product.

Such a plan, once approved by top management, forms the basis of the product manager's authority. Product managers typically represent top management on assigned products. In carrying out their responsibilities, they work across traditional functional lines to bring together an organizational focus, so that product objectives can be achieved.

INTERNATIONAL MANAGEMENT

Multinational companies are usually organized to do business globally on a matrix system of management. In these companies, responsibility for strategic and key operating activities is divided among organizational elements as follows:

- *Product.* Responsibility concentrated in product or product line management with worldwide perspective.
- *Geography.* Responsibility concentrated within a specific territory such as a country.
- *Function.* Responsibility concentrated in an organization's functional specialty such as finance, production, marketing, or research and development.

In the international company there are usually two coordinated avenues of strategic planning: product and geography. Since decisions are shared, accountability for results is also shared in terms of product and geographic profitability through profit centers. Financial visibility by product, function, and geography is the norm in the multinational company.

A basic factor in international management that is significantly affected by matrix management is the traditional concept of the profit center, with its delegation of authority to one manager who is responsible for profitable results. To him, everything counts at the profit-center level, everything is measured there, and people are rewarded accordingly. Certain key decisions, such as product pricing, product sourcing, product discrimination, human resources, facility management, and cash management, are traditionally considered the profit-center manager's prerogative.

But in international matrix management, the profit-center manager will share key decision making with others. Some managers, accustomed to making these decisions on their own, find that sharing decision making with some other manager outside the parent hierarchy can be a "culture shock." For example, in product pricing in the international market, the profit-center manager will find it necessary to work with an "in-country" manager to establish price. Product-sourcing decisions may be made by senior marketing executives at corporate headquarters rather than by the profit-center manager. In practice, decision authority should be complementary. If the manager cannot reach agreement on these decisions, it may be necessary to refer the conflict to a common line supervisor for resolution.

TASK FORCE MANAGEMENT

Task forces are used by companies to deal with problems and opportunities that cannot easily be handled by the regular organization. Usually these problems or opportunities cut across organizational boundaries. A task force can be a powerful mechanism for bringing talent to focus on complex matters. When the objective for which the task force was organized is attained, the group disbands.

A task force composed of persons drawn from appropriate elements of an organization is used to deal with a short-term problem or situation. People are usually assigned to work on a task force on a part-time basis and find that they have to satisfy two bosses during this period—their regular supervisor and the task force leader. Since each member usually represents a different part of the organization and brings different viewpoints, goals, loyalties, and attitudes to the group, the job of integrating individual efforts is no small challenge for the leader.

PRODUCT TEAM MANAGEMENT

Product team management is a generic phrase that describes a relatively permanent product-functional matrix in which business-results managers overlay a functional resource organization. A team of people is organized and charged with managing a product or product line serving specific market segments. Other names used to describe this form include "business boards," "business committees," and "business heads." Product team management is a form of permanent matrix where certain key managers on one part of the matrix structure are held responsible for the *results* of a product or product line. On the other side of the matrix, managers who head up specialized functional activities are responsible for *facilitating* the use of resources within the organization so that organizational goals are accomplished.

For example, a chemical company with sales of $2.3 billion had 24,000 employees in 26 countries. The company was organized into seven operating groups, each of which functioned like a separate company. But corporate controls were ineffective. Corporate management found that it lacked visibility in reviewing major projects, strategic programs, and other investment issues. Major expenditures contemplated by the group profit-center managers were rarely challenged. To bring corporate visibility to the decentralized operations, several management improvements were undertaken:

- A financial manager was placed at each plant and was assigned to report to both the plant manager and to the corporate controller—a form of matrix management.
- A computer terminal, placed near the chief executive's office, gave daily, monthly, and quarterly sales data.
- Sixteen executives were identified as "business heads," each with responsibility for a group of products that included product planning, investment, and product sourcing. These executives had no staff of their own; they functioned as the "eyes, ears, and legs" of the chief operating officer and reported directly to him. Profit/loss responsibility rested with the business heads; functional managers retained cost-center responsibility and provided services and technical support to the business heads.

The use of business head–functional support at this company has provided greater flexibility in establishing or discontinuing products. Strategic programs are given a more thorough corporate review; senior management has been freed from involvement in short-range operations that had taken up much of the chief executive's time. Truly strategic issues are now resolved from a corporate portfolio business viewpoint.

PRODUCTION TEAM MANAGEMENT

At the production level, teams of workers do their own work planning and control. In such a setting, the supervisor becomes a facilitator who helps the teams work out the details of assuming responsibility for the manufacture of the entire product.

The production team is used in many different situations in industry. For example, General Electric uses production teams of some 5 to 15 people to handle a particular responsibility of welding in a fabricating plant. The welders in the team are responsible for scheduling and planning their work load.

General Foods Corporation's Topeka pet-food plant has assigned production tasks to teams of 7 to 17 members. Each worker learns every job performed by the team, and receives pay based on the rate at which each job is completed by the team. There are no conventional departments, no time clocks, and no supervisors—just team leaders who work on equal terms with other team members.

TRW Systems Corporation has created semiautonomous work teams in one of its manufacturing plants. The workers assemble a product on a team basis rather than performing assembly-line tasks separately. Teams are allowed to schedule their own time, as long as they do the job.

Volvo auto has one auto assembly plant at Kalmar, Sweden, where workers are organized into groups of six or more persons working as teams. For example, one team of workers injects a sealing compound in all seams and installs sound-dampening insulation. They rotate the team's 15 functions so that no one gets fatigued or bored performing the same job. As Pehr Gustaf Gyllenhammer, Volvo's chief executive, contends, "We have to change the organization so the job itself provides more for the individual. We will never build another production line as long as I am in command at Volvo."

At GM's Fisher Body Plant No. 2 in Grand Rapids, Michigan, the 2000 employees have been organized into six business teams, each team being essentially a business unto itself with its own maintenance, scheduling, and engineering personnel. Currently, salaried employees at all levels participate in deciding how to meet their team's objectives, but team members paid by the hour are included in the decision-making process.

"The plant is much more effective now than before 1973, when the process began," says a GM spokesman in Detroit. But he adds: "The point isn't to improve productivity. It's to improve the quality of work."

NEW BUSINESS DEVELOPMENT TEAMS

Sometimes teams are used to develop new business opportunities. Occasionally they are organized on a permanent basis to conceptualize, develop, and provide an overview for new businesses. One company stated its policy on new venture development in the following manner:

> . . . new business venture teams will be organized within the company. Each team will be asked to identify problems, devise plans to improve the company's performance and implement those plans. By this process, management and professionals at all levels will become active participants in our new Business Systems. . . . our concept is to have key company personnel organized into teams to conceive, develop, and implement methods of improving our business systems.

In another situation a team approach was used to evaluate worldwide business opportunities in polypropylene. Through the efforts of the team a global, long-range plan for an emerging polypropylene business was created.

In the Microwave Cooking Division of Litton Industries, the organization has developed a task-team approach for new product development. Each team has a manager with representatives from several functional departments and top management participation, as required. For new product development within the engineering department, a team includes a design engineer, a stylist, technicians, drafting personnel, and quality and manufacturing engineers. A representative from marketing, buyers, and a home economist are also included. Specialists from industrial engineering, cost accounting, production management, and other specialties are added as needed during the life cycle of the project. The manufacturing function is organized into operating units as a self-contained task team responsible for all aspects of manufacturing a product series.

Litton Industries claims significant benefits from the task-team organization: increases in sales, market share, and profits. For outsiders and newcomers, the most striking feature of the company is the openness of the organization. The firm's receptiveness to new ideas and people, and its attitude toward sharing information, problems, and opportunities, make this apparent at all levels in the organization.

The General Motors Company adopted a form of matrix management in its engineering development community in 1974. Called a "project center," it represents one of the most significant changes in organizational approaches at General Motors since the profit-center decentralization concept in the 1920s and 1930s. The motivation for realigning the engineering divisions into a project center was a strategic change in the marketplace—in this case, the trend toward smaller automobiles. Project centers are used to coordinate the efforts of the five automobile engineering divisions in General Motors. A project center consists of engineers temporarily assigned from the profit-center division to work on a new engineering design. If a major new effort such as a body changeover is planned, a project center that operates for the duration of the change is formed across the automobile division. Project centers work on engineering problems that are common to all divisions, such as frames, steering gear, electrical systems, and so on. The profit center complements but does not replace the lead-division concept where one division has primary responsibility for taking innovation into production.

New venture development inherently involves some degree of innovation. Emerging evidence suggests that a team approach improves the chances for innovation in the business organization—certainly a prerequisite for developing future competitive strategies!

QUALITY CIRCLES

The Quality Circle (QC) is a productivity and quality-improvement technique that was implemented in Japan in the early 1960s. QCs are composed of small groups of employees who participate in improving their work and work environment. The formation of QCs in Japan grew out of the need to change the inferior image of the Japanese product. The Japanese introduced the concept of making each member of the organization responsible for quality. Training in quality control was therefore extended to both the supervisory and the worker level in the organization.

A Quality Circle is a group of four to ten people with a common interest who meet regularly to participate in the solution of job-related problems and opportunities. It is an ongoing group operating in the work environment which performs "opportunistic surveillance" for the organization: searching for opportunities, defining problems by applying formal data collection and analysis, and arriving at solutions that are presented for acceptance and implementation by management.

The QC working techniques include problem selection, brain-storming, cause-and-effect diagrams, pareto diagrams, data collection, histograms, check sheets, and graphs. More advnced QC techniques include sampling, data collection and arrangement, control charts, scatter diagrams, and other statistical techniques. The services of professionals in the organization such as statisticians, industrial engineers, systems engineers, and other staff services are made available to the QC groups. The use of statistical methods was originally highlighted in QCs. Today the use of QCs has extended to participation in a broad range of work improvement.

The Quality-Circle approach strongly endorses participation in management techniques. Decisions on matters in the Quality Circle domain are shared, and results are usually shared either directly through incentive awards or indirectly through improved productivity. Multiple authority-responsibility-accountability influences are characteristic of the quality-circle community.

THE PLURAL EXECUTIVE

The size and complexity of many contemporary organizations have led to a new phenomenon termed the "plural executive," a permanent, formally established office composed of several individuals who perform the functions of top management as a team. The plural executive is a reasonable alternative to the traditional single chief executive and performs just as well. This collective style of top management has many designations—"Office of the President," "Corporate Executive Office," "Management Committee"—and is used by such firms as the Bounty Savings Bank, Sears, General Electric, IBM, Bendix, Westinghouse, Dow Corning, and Dupont. Benefits realized from the use of the plural executive include the strength and security of group decisions, more objectivity, continuity of management, and development of personnel. There are problems as well, as for instance frustration over the time required to make decisions, potential compromise to satisfy the team, and conflict arising out of the chemistry of the individual members.

MULTIORGANIZATIONAL ENTERPRISE MANAGEMENT

The management of "super projects" such as the space shuttle program or the Alaska oil pipeline requires the amalgamation of many organizations to support a common goal. Known as multiorganizational enterprises (MOEs), these ventures usually contain many participating organizations or groups that have different objectives and exhibit different cultures. This diversity—along with the size, complexity, and interdependence that characterize MOEs—is best managed under a matrix system because of the latter's emphasis on multiple authority and the sharing of key strategic and operational decisions, responsibility, and accountability.

JOINT VENTURES

In a joint venture, consensus among the owners serves as the basis for management of the enterprise. The degree of sharing in overall management varies. At the very least, the partners in a joint venture share in setting of long-term objectives, in approving short-term strategies, and in evaluating results. Sharing comes about through teams composed of a few key representatives of each venture partner. The absence of a single boss, however, may create problems in resolving issues that arise.

A TYPICAL ORGANIZATION

An appreciation of the nature of these various management systems can be gained by looking at one large corporation, representative of today's business community. This transnational corporation uses many forms of matrix management. In its aerospace business, project teams are used extensively in the management of high technology products. Another part of the corporation is involved in the design and development of a variety of industrial systems for handling steel during the fabrication process. Project teams are used to design, develop, and produce these systems. Energy management systems for the utility industry are provided through project management in another part of the corporation. The design and construction of electrical-power-generating plants is a major business activity elsewhere in this corporation where project management is used. Product managers and new business ventures teams are also used in many of the consumer goods activities of this corporation.

Recently the company concluded an extensive study of its international markets through the use of several high-level task forces that conducted an analysis of the firm's competitive position in world markets. A key conclusion reached through this analysis was that the corporation needed to realign its corporate structure along an international matrix management system.

At the top, the chief operating officer heads a management council composed of senior executives who have jurisdiction over the major operating elements of the corporation. This committee deals as a team with key strategic and operational issues in the corporation; the executive incentive system rewards the members of the management council on a team basis. Product management teams are used to market certain key products and product lines. Production teams are a way of life in most manufacturing plants. The corporation has also instituted a company-wide effort to organize and utilize quality circles at both the worker and the professional level. In the engineering departments, project engineers manage the design and development process. The corporation is actively involved in establishing joint ventures overseas and in Mexico. Finally, it is working on several petrochemical projects with other corporations as a multiorganizational enterprise.

Matrix management has become so widespread in this corporation that all managers are required to attend special executive development courses that teach them to build the teamwork—in terms of attitudes, knowledge, and skills—that is the heart of a matrix management system. Executive promotions are based in part on the candidate's demonstrated ability to operate within a matrix. The top 15 managers in the corporation have all had substantial experience in some form of matrix management.

A company which truly believes that productivity can be improved through participative management, consensus decision making, and other approaches emphasizing the sharing of decisions, results, and rewards, needs a basic strategy to change the organization's culture. Such cultural changes come about slowly, but using the matrix management approaches suggested in this book can facilitate them.

REFERENCES

Cleland, David I. "A Kaleidoscope of Organizational Systems," *Management Review*, Dec. 1981. Used by permission.

Janger, Allen R. *Matrix Organization of Complex Businesses.* New York: The Conference Board, Inc., 1979.

Jermakowicz, Wladyslaw. "Organizational structures in the R&D Sphere." *R & D Management,* No. 3, Special Issue, 1978.

Kitch, D. N. "Quality Circles in the Westinghouse Electric Corporation Nuclear Technology Division." Research Report, IE 226 Systems Management II, April 1981, School of Engineering, University of Pittsburgh.

Kolodny, Harvey F. "Managing in a Matrix." *Business Horizons,* March–April 1981.

———. "Matrix Organization Designs and New Product Success." *Research Management,* Sept. 1980.

"Management Itself Holds the Key." *Business Week,* Sept. 9, 1972.

Mee, John F. "Matrix Organization." *Business Horizons,* Summer 1964.

Menzies, Hugh D. "Westinghouse Takes Aim at the World." *Fortune,* Jan. 14, 1980.

Oldham, Franklin Jr. *The Plural Executive: An Experiment in Top Management,* Arkansas State University, Business Administration, State University, AR, 1975.

Osgood, William R., and Wetzel, W. E., Jr. "A Systems Approach to Venture Initiation." *Business Horizons,* Oct. 1977.

Prahalad, C. K. "Strategic Choices in Diversified MNCs." *Harvard Business Review,* July–Aug. 1976.

Tinnin, David B. "Why Volvo is Staking Its Future on Norway's Oil." *Fortune,* Feb. 12, 1979.

Ware, James. *Managing a Task Force.* Boston: Harvard Business School, Intercollegiate Case Clearing House, 9-478-002. 1977

Wickesberg, A. K., and Cronin, T. C. "Management by Task Force." *Harvard Business Review,* Vol. 40, No. 6 (Nov.–Dec. 1962).

2. The Practice of Matrix Management

John R. Adams
Nicki S. Kirchof*

THE PRACTICE OF MATRIX MANAGEMENT

Ever since it has been in use, many managers have questioned the need for the matrix, citing numerous problems generated by this form of organizational structure. Like most forms of organization, the matrix has evolved to resolve a particular set of problems in a specified set of circumstances. If applied in those circumstances, the matrix does an excellent job of dealing with the problems. Most failures of the matrix can be traced to its use in an inappropriate environment. This chapter will trace the typical evolution of the matrix in a given organization, emphasizing the conditions that usually require organizational change. The roles of the various managerial participants in the matrix will then be examined, with emphasis on the differences between the requirements of the matrix and those of the more traditional organization. Finally, the advantages and disadvantages of the matrix will be reviewed, again emphasizing that to be effective the matrix must fit the specific set of circumstances facing the organization. In general, the matrix is to be avoided unless its use is demanded by the nature of the environment and the work to be accomplished.

DEVELOPMENT OF MATRIX MANAGEMENT

Functional Organization

Most organizations begin as a "normal," functional hierarchy characterized by the straight-line chain of command, the one-man, one-boss concept, and the division of responsibility and formal authority, first described by Weber and taught in most organization theory courses.[1] The familiar organizational chart found on the wall of most offices depicts the formal organizational relationships and the designated division of labor in such a structure. A generic chart of a typical functional organization's structure is presented in Figure 2-1.

---

*Dr. John R. Adams is an Associate Professor of Management at Western Carolina University and the Director for Educational Services of the Project Management Institute (PMI), the international professional association of project managers. Ms. Nicki S. Kirchof is a Research Associate of the PMI. The authors are coeditors of *A Decade of Project Management,* published by the Project Management Institute in Drexel Hill, Pa., and are well-known and published in the project management field.

This form of organization is ideal for a situation that is stable, repetitive, and expected to continue for an extensive period of time. Each manager specializes within his own area of concern, striving to improve the efficiency with which each particular task is accomplished. Decisions are made on the assumption that the same tasks will be performed for years to come. The drive for efficiency causes the continuing tasks to consume essentially all the time of the management personnel.

Project Organization

Periodically, however, special requirements arise that are not encompassed within one or more of the functions. Perhaps a new product is to be developed and marketed, requiring the cooperative efforts of engineering, marketing, production, and other functions as noted in Figure 2-1, as well as assistance from the accounting, legal, and contracting staffs. Or perhaps the organization has grown and must now construct and move into new facilities. Since the normal functional organization must continue to operate as usual, the functional managers have little time to devote to this unusual activity. Typically, if this is an important new effort, a dynamic and capable person from the upper levels of middle management is selected to take responsibility for this unique activity. A project is organized around this project manager. A few specialized assistants are provided and a project team is formed. Most major organizational functions will be affected by this team, and each will be required to contribute to its activities. The team is typically removed from the functional organization's structure, is provided a special reporting line to the general manager (since this is an *important* project), and is given authority to have functional elements of the organization do work relevant to the project. In many cases the team also is given authority, within budget limitations, to contract work out to other organizations.

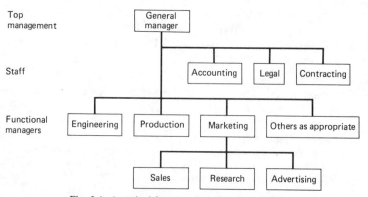

Fig. 2-1. A typical functional organization structure.

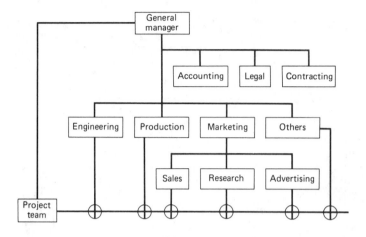

Fig. 2-2. A typical project organization in a traditional organization.

A diagram depicting this structure is presented in Figure 2-2. In this figure the small circles represent the interface between the project team and the functional units, and indicate the authority of the project manager to delegate the work supporting the project.

This project form of organization works extremely well in situations where change occurs relatively slowly and few projects are required at any one time. The functional managers are able to concentrate on their normal functional duties, continuing the usual profit-making operations of the firm. The project team provides the concentrated management attention required for successful completion of the project. Typically, the demands of the project on the functional groups, while sometimes upsetting, seldom constitute a major imposition on any one individual or group, and the total organization's work usually progresses in a reasonably well-orchestrated manner. Note, however, that the project represents the firm's reaction to some needed change (new facilities, a new product), while the functional organization represents stability and a continuation of traditional activities.

Multiproject Traditional Organization

As change becomes more pervasive—that is, as the environment becomes more turbulent—organizations find themselves sponsoring more and more projects. The results are depicted in Figure 2-3.

There is a definite limit to the number of major projects any traditional organization can support. As the number of projects increases, the managerial load on the general manager increases to the point where he can no longer cope. While the work delegated by a single project usually can be readily absorbed

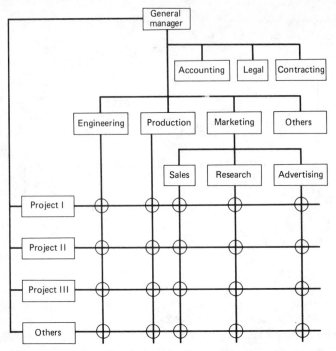

Fig. 2-3. A typical multiproject traditional organization.

by the functional organization, the added work load of multiple projects can quickly overwhelm most efficiency-oriented units. By definition, each project is of major importance to the organization, so none can be slighted. The project managers find themselves in competition with one another and frequently with the functional managers for the resources needed to accomplish the task. This situation can literally tear an organization apart.

The multiproject organization depicted in Figure 2-3 is structurally similar to a matrix organization, and many believe that the pattern described in the previous paragraphs represents the evolution of the matrix. This concept, however, is not correct. At this point, there exists only a functional organization with a number of important projects to be performed. One additional major change is required before this organization can truly be described as a matrix. That change is in the minds and attitudes of the senior managers of the organization.

The multiproject organization still relies primarily on the functional organization for its ongoing, day-to-day operations. The functional organization is the "bread and butter"; it produces the income that makes the projects possible. The projects are there largely as a sideline, and in a sense support the operations of the functional units even while placing additional requirements upon them.

Matrix Organization

As the rate of change in the environment increases and the number and importance of projects continue to grow, a point is reached where management must recognize that the work of the projects is more important than the work of the functional organization. The "profit" to the organization at this point is provided mainly by the project efforts, while the functional units exist chiefly to support the projects. This situation is clearly seen in a construction company that undertakes such projects as building a school, a road, and a dam, with the functional units essentially providing a pool of resources from which the projects can draw as needed. This situation is represented in a simplistic way in Figure 2-4. The same type of situation, however, can occur within a division of a larger functional company—for example, an engineering division where the successful completion of specific research projects that integrate several engineering specialties becomes the driving objective of the division, as opposed to the separate pursuit of electrical, mechanical, or civil engineering activities. In either situation, it is hoped that the various projects are phased so that they do not all require the same resources from the same functional unit at the same time. The ability to conduct multiple projects with limited resources in the matrix structure is dependent on excellent phasing of the projects, so that the functional unit's resources can be shared appropriately.

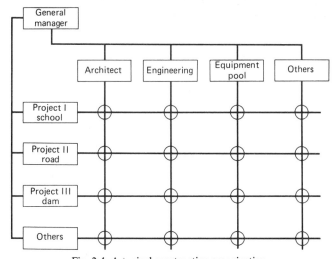

Fig. 2-4. A typical construction organization.

Once top management has recognized the importance of the project work and has convinced their subordinate functional and project managers of the situation, the essence of the matrix has been achieved. It will not be achieved without pain, however, particularly on the part of functional personnel. Those used to operating in a traditional organization understandably resist assuming the support role in a matrix. This change affects both their own and others' views of the relative importance of each unit in the organization.

The process described so far is admittedly simplistic and ignores much of the detail of the transition. Any complete description of a developing matrix in a functional organization would need to exhaustively analyze such issues (to name only a few) as the various structural possibilities for establishing projects in traditional, multiproject, and matrix organizations, and the relative authority structures of the weak versus the strong matrix. Simplistic as this discussion is, however, it serves to illustrate in a general way the typical pattern followed by most traditional organizations—either the organization as a whole or some division of it—as they develop into matrices. The way in which the job requirements of managers change as they move from the traditional to the matrix organization, however, requires further elaboration.

ORGANIZATIONAL ROLES IN MATRIX MANAGEMENT

The specific organizational structure used to implement the matrix concept is unique in its application to each organization. The opportunities and problems that the matrix creates for its organizational members, however, are generic to the matrix and create managerial roles that are amenable to analysis. The marked differences in the roles that managers must perform in the matrix, as opposed to the traditional organization, make the transition to a matrix difficult for most managers.

As the traditional organization evolves into the matrix, the tasks to be performed by managerial personnel change and new management roles are created. Key words used to describe these roles include communication, flexibility, collaboration, negotiation, and trust. It is the manager's task to develop and practice these characteristics in order to improve the performance of the matrix. In this section the organizational roles generic to the matrix are described, analyzed, and, where possible, compared with the comparable manager's traditional role. An organization chart of a typical matrix structure is helpful in defining the basic managerial positions found in the matrix. The positions labeled in Figure 2-5 are discussed, insofar as it is possible in a matrix, in hierarchical order starting at the top.

The General Manager

The formal role of the general manager or top-level management is similar to the top management role in a traditional organization, in that it includes total

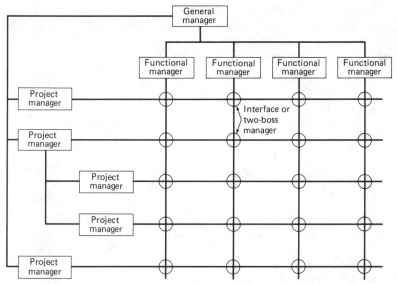

Fig. 2-5. Managerial positions in a typical matrix.

responsibility for those activities subordinate to the position. As seen in the diagram, the general manager is literally on top of or outside the basic matrix structure and therefore should have a clear perspective of all activities and personnel within the matrix. Here, however, the similarity ends. The general manager leads a dual command structure, the functional and the project hierarchies, which must be balanced through a careful blend of autocratic and participative managerial styles. Participation is required, for example, in the arbitration of technical disputes and in resource allocations. Autocracy is essential to the establishment, enforcement, and revision of priorities among and between the functional and project entities within the matrix.

The general manager must personally enjoy the matrix; that is, he must accept wholeheartedly and constantly sell the principles and concepts of the matrix to those required to participate in and work with it. This role involves the three major managerial concerns: balancing power, managing the decision context, and setting standards. The balance of power involves allocating both project and functional budgets, orchestrating personnel assignments, and applying schedule pressures. These are only a few of the general manager's power sources that must be used to establish and maintain priorities among the functional departments and projects. Management of the decision context is accomplished by establishing strategy, policy, and control systems to assure that decisions are made to benefit the overall organization rather than the individual functional department or project. It is essential that the general manager perform this task since the individual subordinate managers will have so much detailed, pertinent information about their specialties that it is unlikely they will be able to take the corporate view. Finally, high performance standards start at the top. They must be established and enforced by the general manager, since his is the only position in the matrix with the perspective and power to require top-quality performance. The general manager must thus remain aloof from the detailed technical disputes and personal quarrels of the project and functional managers. This role demands the skills of the initiator, the coordinator, and the leader.

Functional and Project Managers

It is difficult to discuss the roles of the functional and project managers in a matrix separately. They exist on the same level in the matrix, both reporting to the general manager, and they must interact on a cooperative basis if the matrix is to survive and succeed. Both maintain some degree of authority, responsibility, and accountability with regard to each of the organization's projects, and they must negotiate continuously. The functional manager controls people and is responsible for developing careers within technical specialties. Therefore functional managers tend to practice the classical, empirical, or behavioral school of management thought. Project managers, on the other hand, control project costs and schedules and are responsible for the satisfactory accomplishment of project goals. They tend to practice the decision theory and systems management schools of management thought. To be successful, both functional and project managers must share equally the task of providing leadership within the matrix, and both must be committed to managing in a professional and productive manner the conflict that will inevitably develop.

Functional units in a pure matrix exist primarily to provide support to the various projects. The functional manager in a matrix therefore retains sole responsibility for few of the traditional organization's managerial tasks. Most major decisions are shared with other managers at either the same or a higher level in the matrix. Decisions on performance evaluations, pay adjustments, promotions, and even job assignments for persons who work in the function are frequently influenced by project managers. Functional managers select and hire a wide variety of specialists and provide for easily definable career progression paths. They establish budgets for their functions and, after receiving approval from above, maintain absolute control of functional expenditures. They project work load, define requirements for additional personnel, and establish and maintain standards of technical excellence among their functional subordinates. Since they have manpower flexibility and a broad base of personnel with whom to work, most functional tasks are completed within cost limits. Functional managers, however, must share their subordinates with the project manager, since their function's primary task is to provide skilled personnel to accomplish project tasks. Managers who are good at this job are thus able to take a major load off key top-level management personnel by meeting with project managers face to face and resolving most conflicts at this level, rather than elevating them to the general manager. Their role is broad and all-encompassing. In particular, communication ability and "people" skills are essential to the success of a functional manager in a matrix.

Sometimes functional managers experience the matrix form of organization as a loss of status, authority, and control. For managers used to having total control of their subordinates, this structure frequently comes as a shock and can initially result in frustration, hostility, and resistance to the matrix concept. As the matrix matures, however, functional managers usually adapt to these changes and find the role challenging. They must "serve" as well as "dictate," a task some find difficult. Most find that successfully accomplishing this difficult job is a very gratifying experience.

In the matrix form of organization, project managers are responsible for providing the final end product of the organization's efforts. In a construction company, for example, the project manager is responsible for completing the road, building, or dam. In a research organization, the project manager is responsible for completing the research. Project managers act as unifying agents for controlling the resources and technology allocated to the project. They establish budgets and schedules, and project and levy resource requirements on the functional managers. In addition, they provide the concentrated managerial attention necessary to integrate their projects across the organization's functional lines and to drive them through to completion. They directly supervise all members of the "project team," that small number of specialized people (typically three to five for a $10-to-$20-million project) assigned full time to the project to assist the project manager in integrating the project activities into the matrix. Project managers in a matrix, however, do not have the power to dictate. They must use their knowledge, competence, personality, and group management skills to get their project's tasks accomplished within budget, on time, and within performance standards. They must rely heavily on their interpersonal skills and their ability to persuade, and they must compete with the other projects in the matrix for the critical resources needed for the project.

The project manager in a matrix has the responsibility of accomplishing specific organizational tasks, but does not have the personnel or other resources directly under his control to accomplish the work. In this situation, managers used to the controls and authority of a traditional organization frequently experience frustration, doubt, and a loss of confidence when first placed in the matrix. The lack of hierarchy and formal authority often leaves the new project manager wondering where to start. The sources of influence and power in the matrix are personal in nature, involving the manager's knowledge, charisma, and interpersonal skills, and this concept itself requires a restructuring in the new project manager's thinking. The best project manager is likely to be an individual who is uncomfortable in the confines and controls of the traditional organization, who may well have been known for his dynamic, nontraditional approach to a job, and who has sufficient ego and self-confidence to demand the support needed for the project. This project manager will realize that there are many projects in the matrix, but will sacrifice his project for others only when forced to by circumstances beyond his control or when directed by the general manager.

"Subproject" Managers

Some projects are so large and complex that they must be divided further into "subprojects," each with a subproject manager responsible for a defined portion of the overall project. In the military and aerospace industry the terminology used involves major "programs" divided into "projects," a use of terms that has created much confusion in this field. An example will serve to clarify the concept, with the aerospace industry terminology included in parentheses. If a new combat aircraft is to be developed, a major project (program) is established with a senior project (program) manager in charge. Because of the magnitude and complexity of this effort, it is typically divided into subprojects (projects), each of which is responsible for developing a portion of the new aircraft system. There may be subprojects (projects), large and complex in their own right, responsible for developing the airframe, or the engine, or the armament system, or the integrated avionics systems. The subproject (project) managers are fully responsible project managers in their own right, with their own teams reporting to them, and with full responsibility to schedule and integrate the activities needed to accomplish their portion of the overall project (program). The task of the subproject (project) manager is, if anything, more complex than that of the project manager described earlier, for the subproject manager must be concerned not only with integrating the work of his subproject (project) across the functional hierarchy, but he must also integrate his activities with those of the other subproject (project) managers to accomplish the overall project (program).

When discussing efforts of this magnitude, the military and aerospace terminology of programs and projects is quite useful. However, this terminology is not universally accepted and can lead to confusion when used in other industries. In our current state of the art, then, it is important for the practitioners to understand which terminology is being used in any specific publication or discussion.

"Two-boss" Manager or "Interface Personnel."

One of the most complex positions in the matrix structure is that of the "two-boss" manager or "interface personnel". These positions are represented in Figure 2-5 as circles indicating the interaction of project and functional responsibilities. The career development and progression of these individuals are the responsibility of the functional managers, but they are delegated on a part-time basis, or on a full-time basis for limited periods, to a particular project to accomplish a specified task. The actual work performed, however, is assigned and controlled by the project manager, along with whatever pressure is required to assure the completion of quality work on schedule and within budget. The two-boss manager may also serve a supervisory function, directing the work of others in accomplishing the assigned task. In any event, it is clear that a dual line of authority is imposed on this individual, one emphasizing the short-term task assigned and the other emphasizing long-term career potential and professional development. In such a position, the two-boss manager frequently experiences high levels of stress and anxiety resulting from conflicting directions and competing responsibilities. In many cases there is also a certain level of ambiguity over who this individual's boss really is and who has the right to impose requirements. Communication problems are often experienced in this position, adding to the uncertainty surrounding the job. However, the problems of this role tend to become more manageable as the matrix matures, primarily because all parties become more familiar with the requirements of their roles and through experience learn to cope with and manage the conflict generated by the matrix.

A final point concerning the roles involved in the matrix structure is that interpersonal conflicts can easily occur, because of the initial unfamiliarity of the structure and the need to develop new and somewhat unique working relationships. Davis and Lawrence also believe that the increased communications required of all participants in a matrix creates an entirely new set of problems.[2] More differences are brought to the surface and must be resolved, typically on a one-to-one, face-to-face basis. In the matrix structure these conflicts are valued as leading to the higher quality solutions to technical and managerial problems that are likely to result from an open discussion of differences. The underlying assumptions of the matrix are that its members will work toward the common goal in spite of personal differences, that the general manager is able to define and articulate that goal, and that each member will seek the best solution to problems through personal interaction. Members of the matrix are typically expected to use a confrontation or problem-solving approach to conflict, bringing the causes out in the open for analysis. Smoothing, withdrawing, and other management methods that hide or avoid the conflict are to be shunned. Of course individuals must take certain personal risks in sharing information and must frequently push themselves to "open up" in group discussions and confrontations. This task is difficult for many and requires a high level of interpersonal trust. Without such trust, the organization tends to revert back to reliance on the chain-of-command authority that no longer exists at the center of the matrix, and the structure may be destroyed. Trust and open communications are the basic requirements for success in the matrix organization.

The organizational roles of the matrix are not significantly more complex but are merely different from the roles existing in traditional organizations. When managers have been accustomed to the chain-of-command hierarchy and the formal authority it implies, the "new structure" of the matrix may seem strange. As an efficient matrix matures, however, the new roles become more accepted and a new balance of power is reached, based on relatively high levels of trust and communication. As a result, when the new structure fits the tasks to be accomplished, levels of performance can be achieved far above those of the traditional organization.

CHARACTERISTICS OF THE MATRIX ORGANIZATION

Throughout this chapter, it has been implied that the matrix structure is not appropriate for all organizations or tasks. This structure has distinct advantages and disadvantages that must be considered in deciding whether or not to implement it in a specific set of circumstances.

Kerzner, a well-known author on the topic of project management, cites ten basic advantages of the matrix organizational form.[3] These are paraphrased below:

1. The project manager maintains maximum project control over all resources, including cost and personnel.
2. Policies and procedures can be set up independently for each project, provided that they do not contradict company policies and procedures.
3. The project manager has the authority to commit company resources, provided that scheduling does not cause conflicts with other projects.
4. Rapid responses are possible to changes, conflict resolution, and project needs.
5. The functional organizations exist primarily as support for the project.
6. Each person has a "home" after project completion. People are more susceptible to motivation and end-item identification. Each person can be shown a career path.
7. Because key people can be shared, program cost is minimized. People can work on a variety of problems, which makes possible better control of personnel.
8. A strong technical base can be developed and much more time can be devoted to complex problem solving. Knowledge is available to all projects on an equal basis.
9. Conflicts are minimal, and those requiring hierarchical referral are more easily resolved.
10. Better balance between time, cost, and performance.

Viewing this list, one might think that the matrix is indeed an almost ideal form of organization. All these advantages seem to alleviate major problems found within the traditional functional organization. Davis and Lawrence, the authors of *Matrix,* see the basic advantage of the matrix as that of allowing the organization to take on larger projects, because of increased planning and communication efforts and improved division of labor. In particular, one person can only do so much mentally at one time, and physically that person can certainly not be in two places at once. The matrix form centralizes the resources where they can be shared more easily, and functional units can develop teams to do more complex jobs that a traditional structure would not allow.

The matrix, however, does not always work as planned: some hazards must be taken into account. In fact, each and every advantage in this and all other organizations entails disadvantages that must be known and evaluated. Every organization style is a balance of useful and dysfunctional patterns. Every useful activity, carried to an extreme to develop the "greatest technology," becomes a primary weakness that can cause untold damage to the organizational structure. Thus the "best" organization is the one that is created after a careful analysis of the situation in which it must function, the objective it must achieve, and the resources (especially human resources) available to it.

Kerzner sees the hazards of the matrix as follows:[4]

1. Companywide, the organizational structure is not cost effective because more people than necessary are required, primarily administrative.
2. Each project organization operates independently. Care must be taken that duplication of efforts does not occur.
3. More effort and time are needed initially to define policies and procedures.
4. Functional managers may be biased according to their own set of priorities.
5. Although rapid response time is possible for individual problem resolution, matrix response time is slow, especially on fast-moving projects.
6. Balance of power between functional and project organizations must be watched.
7. Balance of time, cost, and performance must be monitored.

He views the main disadvantage as a necessary increase in administrative personnel to develop policies and procedures, so that both direct and indirect administrative costs increase. Davis and Lawrence view this "excessive overhead" as temporary and explain that, as the matrix matures, these increased costs will diminish. Of course one of the principle reasons for using the matrix is to reduce the "red tape" of policies and procedures and the cost of the overhead. Along with "excessive overhead," Davis and Lawrence develop eight other disadvantages of the matrix in some depth.[5]

- Power struggles
- Anarchy
- Groupitis
- Collapse during an
  economic crunch

- Decision strangulation
- Sinking
- Layering
- Navel gazing

In fact, these "disadvantages" are simply characteristics of the matrix that can be analyzed to determine how best to make the matrix effective.

Power struggles are derived from dual command and the balance of power. In a matrix the balance of power is constantly shifting, usually toward imbalance. The design of the matrix, whereby the boundaries of authority and responsibility constantly change and overlap into other persons' areas of authority, causes individuals to attempt to maximize their own advantage. This situation is otherwise known as building either power or personal empires. Avoiding this situation requires that a fine balance of power be maintained within the organization. This action, of course, is the responsibility of top management. The cure for the problem is to ensure that the key players on the axes of power "understand that to win power absolutely is to lose performance ultimately."[6] Essentially, power in a matrix involves constant negotiation among those in the key power positions, with top management as the formal arbiter when issues cannot be settled locally. Top management must side first with one group and then with another to maintain the balance, so that all ultimately work toward the organization's goals.

Anarchy occurs when explicit arrangements to coordinate critical tasks within the matrix are not made. Those who do the work must know what they need to be doing at each point in time, and under what circumstances they are to perform. The cure for anarchy lies with all levels of management. Middle management must coordinate on a voluntary basis the detailed, day-to-day efforts of shared personnel, leaving top management free do deal with major issues. The Chief Executive Officer (CEO) must be willing to intervene in all major issues, to pull conflicts among all key personnel into the center of attention, and personally to resolve the important decisions which can (and frequently do) disrupt progress toward the organization's goals. Again, the need is to emphasize the power structure and maintain the necessary balance between functional and project personnel.

Groupitis occurs when all business decisions are determined in group meetings. Here the problem generated is one of wasted time and unwillingness to accept responsibility for actions. The key to dealing with this tendency is to frequently re-emphasize what the matrix is all about. Not every decision should be made in groups. Group decision making is definitely not the overall purpose of the matrix. Once again, when this situation occurs, the failure lies with management. In the matrix, the authority for individual decision making typically lies with the project manager. Group activities become necessary when those decisions involve use of the functional manager's resources. Such issues ideally are resolved through informal coordination between the two managers involved. Teamwork is of key importance here and needs to be emphasized at all levels, particularly by top management, to keep problem resolution at the one- and two-person levels. Groups do and should become involved, when top management is drawn into the decisions. This action, however, should generally be construed as a failure of middle management personnel.

Collapse during an Economic Crunch occurs to the matrix during periods of economic decline in the organization. Matrix forms of organization, like everything else, seem to blossom during periods of economic growth and prosperity and tend to wilt during difficult times. Of course this is just the opposite of the matrix organization's purpose. The typical purpose of the matrix is to stretch the human resources—to make more efficient use of the limited resources at hand.

The cost of the matrix should be viewed in terms of conflict and organizational frustrations, not economic costs. The matrix should be generated in periods of economic downturn and predicted economic problems, for it should be a more efficient way of pursuing organizational goals. When economic times are good, matrices should normally be considered unnecessary and *organizationally* too costly and disruptive to consider. There will always be economic cycles. Matrices started in economic upturns are likely to fail when the going gets tough, because they have probably generated an increase in administrative personnel to develop policies and procedures. The very policies and procedures developed in times of economic plenty frequently strangle the matrix in an economic downturn. Matrices born in periods of economic trouble tend to remain "lean and mean" and frequently are the only reason why the organization as a whole is able to survive the economic trouble. The prevention of "failure to survive during an economic crunch" then truly lies in good planning.

Decision strangulation occurs when too much time is spent making decisions and getting these decisions cleared by the many managers in the organization. This problem is typical of the functionally oriented company that has adopted the matrix form but whose top management has refused to change from giving directions from a centralized decision-making hierarchy to negotiating priorities among functional and project concerns. The matrix is *not* an organization where decision authority rests with top management. Rather, in a properly run matrix the project manager makes most of the decisions for his project and requires approval of others only when the decision involves increased use of their resources, or when the priority of his project relative to others is in question. The matrix requires *decentralized decision making*. Requiring approval for routine decisions from many managers, particularly at several levels of the organization, indicates a poorly managed organization in which teamwork and authority delegation have been sadly ignored.

Sinking occurs when the matrix sinks down slowly into the lower levels of the organization. Davis and Lawrence believe that sinking is necessary, that it takes time for the matrix to settle in and become accepted. This is true of all organizations, not just the matrix. "Sinking" actually refers to the overcoming of organizational inertia. It takes time for organization members to learn that the old ways no longer work. They must learn the new power structure; indeed, that structure itself takes time to develop and stabilize. As with any major organizational change, even if top management accepts the new methods wholeheartedly, changes its own ways of working, and fully supports the change, it must allow an extended period during which first one level, then the next, and then the next will begin to function in the newly appropriate manner. The matrix must "sink" into the organization.

Layering occurs when matrices are founded within matrices down through several levels of the organization, which is a normal tendency. Junior managers who see senior managers apparently resolving certain difficulties by reorganizing into a matrix naturally assume that this is an accepted technique within the organization and so attempt to deal with their own difficulties through a similar method of matricing. This technique has sometimes proven successful in very large organizations (such as the U.S. Air Force!) down through the second level of the matrix and beyond. The problem is simply one of tremendously increased informal communication channel requirements and an almost total dilution of an individual's authority. Remember that the matrix, properly used, is designed to delegate specific authority down to a specific project manager, holding him personally responsible for accomplishing his project's work (subject only to the limits of the priorities assigned to that work). When this project manager can in turn delegate his responsibilities to others below him who may or may not actually work for him, the major advantage of the matrix may be totally lost. When this occurs twice or more in a single organization, creating three or more levels of the matrix, then difficulties of enforcing schedules, priorities, and responsibilities become readily apparent. Layering leads to many of the previously noted difficulties, including power struggles, anarchy, groupitis, and decision strangulation. In general, layering should be avoided at all costs.

Navel gazing is evident when the organization becomes too involved in internal relations at the expense of paying attention to the outside world, the clients, and the projects' ultimate goals. This situation usually occurs in the early stages of the matrix. It seldom occurs later, for when it lasts very long the entire organization typically fails. The whole purpose of the matrix is to accomplish the projects' goals rapidly and efficiently in times of economic difficulty. Of course there are internal organizational problems, as in any form of organization. These problems, however, can be dealt with effectively with a minimum of impact on the projects' goals, if management has sufficient foresight to recognize the difficulties likely to occur with this organizational style, and capable enough to plan methods for dealing with them effectively. The methods exist and are available to those willing to study the field prior to committing their resources.

The disadvantages listed here and noted by Kerzner and by Davis and Lawrence must be considered when one is thinking of implementing a matrix form of organization. This decision in particular affects the very survival of the firm. It is critical that management evaluate its own and its subordinates' ability to operate in this new environment prior to instigating such a major and traumatic organizational change.

CONCLUSION

This chapter has reviewed the evolution and development of the matrix structure seen as an outgrowth of changing environmental conditions and of the demands these changes make on the traditional organization. The chapter has also reviewed the changes needed in the organization's managerial roles when progressing from the traditional to the matrix structure, and analyzes many of the hazards that may develop as the matrix is instituted and matures. The matrix is an effective and efficient form of organization, capable of accomplishing feats impossible for other organizational structures under conditions that would confound the best traditional management personnel. These feats can be achieved, however, only when the matrix is applied in the appropriate environmental situation, with the appropriate personnel and the appropriate management style. Most reported failures of the matrix organization occur because it has been applied indiscriminately in situations where the matrix is really not appropriate.

The roles that change the most when implementing a matrix structure are those of the project, the functional, and the two-boss manager. Each of these managers must practice greater interpersonal communications skills, rely more on collaborative techniques, and develop more trusting personal relationships than are needed in traditional organizations, if they are to survive and function well in the matrix. These qualities are difficult for some managers to develop, but the ability to develop them may well identify those who can satisfactorily manage in the matrix.

The advantages of the matrix are many, if this organization structure is used appropriately. The disadvantages or hazards typically result from either improper and ill-prepared management, or from an application of the matrix form of organization in an inappropriate situation. The matrix is certainly no universal solution for an organization's woes. It is a complex structure, difficult to control, and unforgiving of poor management. When appropriately and skillfully applied, however, it is an extremely useful and feasible addition to the techniques available to the competent manager who is willing to monitor, analyze, and carefully control a dynamic organization oriented toward high rates of change.

REFERENCES

1. Peter M. Blau and Marshall W. Meyer, *Bureaucracy in Modern Society,* 2nd ed. (New York: Random House, 1971), pp. 17–56.
2. Stanley M. Davis and Paul R. Lawrence, *Matrix* (Reading, Mass.: Addison-Wesley Publishing Company, 1977), pp. 129–144.
3. Harold Kerzner, *Project Management: A Systems Approach to Planning, Scheduling, and Controlling* (New York: Van Nostrand Reinhold Co., 1979), p. 55.
4. *Ibid.,* 56.
5. Davis and Lawrence, pp. 129–144.
6. Davis and Lawrence, p. 130.

3. Ensuring High Performance in a Matrix Organization

Milan Moravec*

INTRODUCTION

The dual reporting aspect of a matrix organization is one of the shoals upon which effective management is most likely to founder. The person who reports to two bosses has to deal with two sets of expectations, two personalities, and two kinds of impact on his own priorities. How can effective performance be ensured under these conditions, when it is difficult enough to ensure effective performance under *any* conditions?

This problem comes into sharper focus if you visualize a matrix as a diamond-shaped entity with the general executive, such as an operations manager or possibly a division head, at the top; matrix managers who share common subordinates on the sides of the diamond; and the subordinate at the bottom (see Figure 3-1). As the balancing of this diamond on its tip suggests, the top must ensure a balance of the often conflicting objectives of the managers at the sides; otherwise, the balanced mix of decisions relating to the subordinate at the bottom is impossible.

Let us call the *f*unctional manager with *f*ormal organizational responsibility the F-manager and the *p*rogram, *p*roduct, or *p*roject manager the P-manager. Generally, the F-manager is the one with whom the subordinate has the formal, solid-line relationship, and the P-manager the one with whom the subordinate has the dotted-line relationship. Despite the appearance of that broken line, the dotted-line relationship is just as legitimate as the solid-line one.

Both the F-manager and the P-manager are concerned with maintaining high performance—high productivity—on the part of the subordinate who reports to them. And in almost every matrix organization, both managers are obliged to participate in coaching and evaluating that person's performance. This evaluation can be the key to *future* performance.

If they are typical of their peers, both these managers would just as soon not face the onerous task of conducting a performance review. Here is the kind of

*Milan Moravec is Personnel Manager, Arabian Bechtel Company Limited, P.O. Box 121, Al-Jubail, Saudi Arabia. Formerly he was Program Manager of Development for the Bechtel Group of Companies in San Francisco. Milan has worked with matrix organizations in North and South America, Australia, Europe and the Middle East.

31

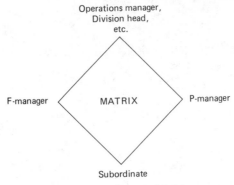

Fig. 3-1. The Matrix Model

comment most supervisors make, openly or to themselves, regarding performance appraisals: "Thank goodness, another one is over. Only two more to go and that will be it for this year!" While managers suppose that performance appraisals may be worthwhile—for record-keeping purposes, or for reminding employees that performance is important—these evaluation sessions are still carried out for one main reason: company policy requires them. Very few managers view performance appraisal as a managing tool to improve bottom-line results for their project or unit.

In a sense, performance evaluation adds yet another dimension to the two-channeled matrix organization. Normally designed and administered by the personnel department, performance evaluation is carried out by line management. The latter tend to view it as, at best, a necessary evil, as they do not feel it is a "managing" tool. As a result, the process often degenerates into a record-keeping system and fails to achieve even a small portion of its potential.

Performance evaluation can be a very valuable management tool that can bring about real improvements in productivity—*if* the process is properly managed. In a matrix organization, that can be a big "if." But it is worth pursuing, because few elements of managerial strategy have as direct an impact on productivity as effective performance planning and appraisal. Human (as opposed to mechanical or technological) productivity comes primarily through

- management's setting clear job standards and communicating its expectations and assessments of progress to employees;
- management's providing and communicating opportunities to improve performance; and
- employees' working toward this improvement.

These three contributors to productivity merge in the performance appraisal process.

PERFORMANCE PLANNING

There are two keys to an effective performance appraisal system: *structure* and *communication*. These elements are interdependent.

Often, a very narrow view is taken of performance evaluation. The process is seen as a once- or twice-a-year task that includes completing a form and telling the employee about past performance. A truly effective system is much more than that. It is not just appraisal but a continuous performance *planning* system. In a matrix organization it begins with an agreement on the appraisal responsibilities of both the P-manager and the F-manager; involves the employee from the day of hiring; builds on early communications with the employee regarding job duties, tasks, and standards; and includes the participation of higher management.

Communication becomes an issue right away. The person atop that matrix diamond (the operations manager or equivalent) has the responsibility, with the concurrence of the P-manager and the F-manager, of establishing the ground rules for the performance evaluation of all people who report to both managers. Both managers need to provide their input individually into the employee's performance review. The following conditions need to be met:

1. Only a cumulative performance review, per performance period, should be presented to the employee. The employee needs to understand that the organization has an integrated overall view of performance. Coaching, however, should occur as necessary.
2. Before the performance evaluation is presented to the employee, both the P-manager and the F-manager should review it and concur.
3. Either the F-manager or the P-manager can present the evaluation to the employee. However, management should be consistent as to which one presents the review to employees throughout the organization.
4. The responsibilities, tasks, and standards of the job should be determined individually by the P-manager and the F-manager. However, only one set of such responsibilities, tasks, and standards should be presented to the employee. The F-manager and the P-manager should agree on one composite set of these to be included in the job description. The manager who presents the performance evaluation should also present the composite set of responsibilities, tasks, and standards to the employee. The employee should have the freedom to obtain detailed clarifications from whichever manager will be assessing the employee on particular elements of that set. (This condition recognizes that each manager is more technically able than the other to explain the parameters on which he/she will be assessing the employee.)
5. The employee needs to be continuously assured of being evaluated on performance to meet job standards, rather than on "pleasing" or "being

nice" to the F-manager or the P-manager. The issue is getting the job accomplished according to (or better than) the standards, rather than loyalty to a person.

6. Salary administration should be handled by the manager who presents the performance evaluation to the employee. Specific merit increases should be based on the employee's performance review and other factors the management of the organization chooses to incorporate. Salary plans should be concurred by both managers.

7. Career counseling and planning should be conducted by the manager who represents the function that the *employee* chooses to progress through. In some cases it will be the P-manager and in other cases the F-manager. A career is an individual interest and the employee's motivation should be embraced by management and guided by projected organization needs.

8. Periodic and day-to-day coaching of the employee should be given by that manager who is evaluating the employee along a specific responsibility, task, or standard. Both managers, however, should periodically share with each other what they have individually communicated to the employee and what they plan to communicate.

9. The supervisor of the P-manager and the F-manager, and possibly higher management as well, should be involved in the planning of the system and also in the reviewing of the evaluations. Performance evaluation is— or should be—a key element in the management of an organization's resources. And the management of human resources should not be isolated from other "business" planning that goes on in the division or unit.

THE COMMUNICATION ELEMENT IN PLANNING

The word "communicate" comes up often in the description of these requirements. Before any communicating is done with the employee, the top-of-the-diamond manager, who supervises the P-manager and the F-manager, has to communicate his/her goals and expectations to them. Then those two have to define objectives, activities, tasks, and standards of performance in support of the goals, and to communicate with each other continually throughout the performance planning and appraisal process.

Ensuring that quality communications take place is the responsibility of the top manager. Because the priorities of the F-manager and P-manager may at times conflict, and because they may present what can be perceived as conflicting requests by the employee, ensuring smooth communication may call for some conflict resolution skills. It also calls for setting some specific times for communication to take place. For example, a meeting of three managers should take place as soon as it is clear that the two of them will have a subordinate in common; at this meeting, the top manager should obtain agreement on a set of conditions like the nine just listed. Also at this time, specific dates should be set on which the F-manager and the P-manager will meet to discuss the employee's performance, responsibilities, and coaching requirements. These dates should probably precede, by an established length of time, the dates on which the performance evaluations will be presented to the employee.

During this initial discussion, the manager should clarify the responsibilities, tasks, and standards on which each will evaluate the employee. This clarification is useful for more than performance evaluation purposes; it also enhances each manager's understanding of the other's priorities, and thus decreases the possibilities for conflict based on misunderstanding.

Such clarification also sets the stage for cooperation in human resources planning. Like other human resource development techniques, performance evaluation is even more effective when integrated with the other management tools used to plan, direct, and process work through employee performance. Various organizations have different ways of integrating performance evaluation with other systems and programs. Some require that career development discussions be tacked on to appraisal sessions, for example. In other organizations, career discussions are separated from reviews of current performance. Still other organizations use performance evaluations only to determine salary increases, while a fourth group uses them as a reference for training, education, promotions, and transfer decisions. Whichever of these is the case, clarification of the managers' roles should be an important element in the discussion.

KEY ELEMENTS IN PERFORMANCE EVALUATION

Once the structure, standards, and responsibilities are understood, the evaluating manager (P or F) must attend to the other elements in the performance planning system: communicating with the employee, evaluating performance, and documenting progress against goals. The following scenario, illustrating these elements, is based largely on the performance planning system used worldwide by Bechtel.

1. As soon as the new or reassigned employee arrives, the manager responsible for presenting the evaluation to the employee tells the employee what is expected on the job. The manager is clear about which job priorities are most important and what performance levels meet job requirements. He or she provides examples of "meets job requirements" levels of performance.

Since communication is not complete until the message is accurately *received,* the manager then hands the employee a worksheet on which to write down his or her understanding of the job. "Come in after you have been on the job for a few weeks and we'll talk about it," the manager says. This sets the employee to thinking and to testing standards and expected results, and encourages commitment to achieving these results. It also allows the manager to make sure, early on, that the communication has gone through—that there is no misunderstanding about job responsibilities and levels of performance.

2. The worksheet is kept by the employee, who later (10 to 15 days before a performance evaluation is due) fills in sections relating to areas of proficiency, job accomplishments, performance difficulties, and suggestions for improvement. This process of reflection, in addition to restimulating a feeling of participation on the part of the employee, provides valuable information for the evaluator and indicates what the supervisor (evaluator) can do to eliminate the roadblocks to meeting or exceeding job standards.

3. Before the supervisor and the employee meet in the evaluation session, the supervisor and the other manager to whom the employee reports get together to review the worksheet and the employee's performance. They agree on a performance evaluation drafted by the presenting supervisor. (Several employees can be dealt with in one meeting, if the timing is appropriate.)

4. The evaluation form completed by these supervisors is reviewed by *their* superior before it is shown to or discussed with the employee. This ensures that evaluations are uniform and timely, and that everyone is operating on the same assumption regarding the performance planning process. It also gives the top manager an opportunity to analyze the range of ratings given by various evaluators so as to spot abnormal distributions indicating unrealistic standards or a "halo effect" in ratings. The manager can review certain employees' previous evaluations by the same supervisor as a check on how well that evaluator is progressing—which comes in handy when the manager evaluates the *evaluator*'s performance.

5. On the evaluation form the managers have listed "action items"—development plans and goals for the employee that will improve performance on the job. This ensures that the manager(s), as well as the employee, take responsibility for performance improvement. Often the action items are within the control of both the supervisor and employee. However, the supervisor can provide the means and create the environment for achievement. Significant action plans, complete with goals and deadline dates where appropriate, are written and scheduled on the appraisal form, so that progress can be measured or observed by both the employee and the manager.

6. There is also a section on the form that lists "improvements over last evaluation period/opportunities for further development." This section has been filled in and is now discussed, even if the employee had previously received a "meets" or "exceeds" rating. The employee worksheet is referred to, as appropriate. "Opportunities for further development" include those additional goals and activities that can further enhance performance ratings.

7. The discussion with the employee is conducted according to the guidelines suggested later in this chapter.

8. Once the employee has begun thinking about current performance objectives and needed development, the next step is to think about future goals. At the end of the evaluation session, the supervisor asks the employee whether he or she wants a career discussion. If so, this discussion is scheduled for a later

date, either with the supervisor or with the other manager (depending on the employee's career preferences), or possibly with a more senior manager.

This part of the appraisal form can be used by higher levels of management in succession and replacement planning. To ensure this, it is wise to build in a mechanism for senior management's regular review of groups of forms. By tying performance appraisal to career planning, the organization can make managers aware of performance evaluation as part of the total human resource planning system. And employees will understand that progression along career paths can occur only following the successful mastery of a job, as measured by performance reviews.

9. At the end of the review session, the supervisor obtains acknowledgment from the employee. "Acknowledgment" means neither agreement with the rating nor "confessing" or pleading guilty. An effective performance appraisal should result in higher motivation, and confessions and guilt feelings are not effective motivators. Acknowledgment means that the employee signs the evaluation form to indicate that he or she has seen the appraisal and discussed it with the evaluator. In other words, the employee simply acknowledges that the discussion took place. During the acknowledgment part of the review, however, the supervisor does try to obtain comprehension from the employee that the issues raised are important enough to work on.

10. Between review sessions the employee is coached by both the F-manager and the P-manager, if both have noted areas for improvement or if the employee requests coaching in their areas of supervision.

PERFORMANCE RESULTS

The above scenario contains built-in assurances that the evaluation process benefits both the organization and the employee:

- It integrates management's performance expectations/standards for positions in a matrix relationship.
- It lets the employee know immediately what is expected on the new assignment, and it ensures that the employee has understood.
- It requires participation of both employee and supervisor(s) in devising ways to meet and exceed job standards.
- It enables higher management to participate in the system and use it for better personnel planning.
- It fosters cooperation between the two managers to whom the employee reports.
- It motivates the employee to improve productivity in specific key areas.
- It provides for fair and objective measurement of performance against communicated standards.
- It relates performance accomplishments to the employee's career goals.

- It provides adequate documentation to eliminate a great part of the misunderstanding and miscommunication that often exist in a matrix organization.
- It provides an objective reference for personnel actions.

SKILLS OF THE EVALUATOR

The achievement of these results, however, can be severely jeopardized by an inept evaluator. Performance appraisal is too important—too closely tied to productivity—to be conducted by people who are uncomfortable with it and/ or not skilled in the techniques involved.

For example, many managers find it difficult to be candid with employees in discussing weaknesses in performance. These managers end up being so "tactful" that the employees are inadequately informed about their performance and are left to draw their own conclusions. This does an injustice not only to the company and the manager but also to the employee. If the manager points out that the employee is working below standard, both know that they have to work to raise the employee's productivity. The employee also has a chance to demonstrate an ability to meet the job requirements. If the manager fudges the issue, however, the employee is denied the opportunity to draw on the manager's supervisory skills in improving performance.

BEHAVIOR IN THE SESSION

In addition to leveling with the employee, there are a number of other activities an evaluator should be advised to do in a performance appraisal discussion:

- *Allot enough time without interruption*, so that the employee realizes that this discussion is the most important matter on your mind. Answering the telephone, looking at your watch, or muttering that you have to be at the airport in two hours will guarantee a less-than-positive response to whatever you want to communicate.
- *Put the employee at ease;* maintain a nonthreatening, frank atmosphere. Offering a cup of coffee or commenting on a matter of common interest can raise the initial comfort level.
- *Be prepared with substantiating evidence.* This can include agreed-upon goals documented on a worksheet, written comments of the other supervisor, and—most important—examples of the employee's work. Remember that your objective is not to pin the employee to the wall, but to clarify your position and provide useful illustrations. Remember also that some employees require more supporting evidence than others.
- *Focus on proficiencies and deficiencies related to the job's crucial aspects,* and not on personality traits. "You don't have the right attitude" is not

job-specific enough. Tell employees what is wrong not with themselves but their work.

- *Listen actively.* This means testing to see whether you and the employee have understood each other, giving feedback, checking out your impressions, and hearing the employee out without interrupting. Remember that "understanding" does not mean agreeing.
- *Allow face-saving and avoid arguing in self-defense.* Your aim is to strive for improvements in performance, not to extract pleadings of guilt or to punish the employee for past inadequacies. By the same token, the employee should not be able to put *you* on the defensive.
- *Discuss proficiencies first, if possible.* This sets a more positive tone for the session.
- *If the overall rating is unfavorable, however, face facts early in the session.* There is no point in lulling the employee into thinking that his or her performance has been good, and then springing a surprise.
- *Avoid comparing the employee with others.* Statements such as "Mark turned in a $20,000 record for this period, and you did only $14,000" foster envy and endanger a unit's spirit of cooperativeness. The focus should be on comparing this particular individual's performance with the standards of the job.
- *Emphasize future improvement in performance.* Explain ways to improve where performance does not meet standards, and point out development opportunities for those areas where performance meets or exceeds job requirements. This may require some work before the performance appraisal interview. For example, if you feel that a certain deficiency could be remedied if the employee learned more about a particular subject area, you may want to arrange in advance for that employee to talk with an expert. If the next step for an employee is increased responsibility, you may have to check with your manager or the training and development department to see what the educational or experience requirements are for broadened accountabilities. Developmental activities should be one of the major areas of discussion between the F-manager and the P-manager prior to the evaluation session with the employee. Remember that identifying performance improvement goals is a responsibility of the supervisor as well as the employee.
- *Be receptive to the employee's input.* Often the employee will have sound suggestions for improving his or her own performance and for learning more about professional development. If the employee does not volunteer any suggestions, ask for them.
- *Identify areas of mutual understanding.* Sum up the discussion with a recap such as the following: "Okay, we've agreed that as far as this project is concerned, you've met all the job requirements for time management, cooperation, and client relations, but that you didn't quite meet require-

ments on dependability because you didn't follow through on one assignment, which was to develop a method to allocate man-hour charges. And
you have an appointment with the planning and control rep next Tuesday
to see if he can provide some suggestions. You plan to improve your job
performance by setting up a flow chart to monitor direct and indirect man-
hour requests and so forth, and I've agreed to provide you with additional
information on how to do that. Then, since Jim Scarlatti reported that
there have been some problems in keeping up with your regular responsibilities since this project entered Phase 2, you need to document your
time schedule so that the three of us can perhaps reallocate tasks and work
something out. Aside from being late on two of Jim's deadlines, your performance in the department met all job requirements, and you exceeded
requirements in policies and procedures and cooperation. Right?"

- *Acknowledge disagreements, but make sure the employee understands
that, willingly or not, he or she is expected to perform the work according
to acceptable standards.* Your task as a supervisor is to get the work
accomplished in the right quality and quantity and on schedule, not to win
a popularity contest with your employee. Human relations techniques are
means toward ends, not ends in themselves.
- *Show all the documentation to the employee* before requesting the
acknowledgment signature.

PERFORMANCE EVALUATION OBJECTIVES

During the session, the evaluator should keep in mind the four key objectives:

1. *Substantiating:* providing evidence of the employee's job performance to
support the assessment.

2. *Developing understanding:* seeing that the employee understands precisely
the assessment and what is expected on the job: duties, responsibilities, priorities, and measurement standards.

3. *Gaining acknowledgment:* obtaining the employee's comprehension that
the discussion has taken place and that certain performance issues and development opportunities have been discussed.

4. *Action planning:* defining the steps that will be taken to improve
performance.

EVALUATOR TRAINING AND FOLLOW-UP

As we have seen, a well-thought-out performance planning and appraisal system can be an effective tool in improving productivity. It serves as a management-by-objectives mechanism and as a device for involving several people in
the effort to upgrade an individual's performance. Since these people come
from different parts of the matrix, performance planning can also be an integrating force for the organization.

However, as also noted, the skill of the evaluator is a critical ingredient in this scenario of success. Yet few evaluators are naturally skilled. Therefore an organization—and particularly a matrix organization, with all its complex personnel issues—ought to have a formal program for ensuring that all managers who serve as evaluators are up to standard in discharging this important aspect of their job. Unequal skill levels invite inconsistency, which is detrimental to career planning, succession planning, and morale.

All supervisors (before they become supervisors, if possible) should be required to participate in a short workshop on conducting performance appraisals. This workshop should provide some, if not all, of the following:

- A picture of the way performance planning and appraisal fits into the goals (P-goals and F-goals) of the organization, accompanied by a message from the top executive reinforcing the importance of performance planning and how it is used as a management tool for productivity and human resources planning and development.
- Printed instructions for planning and conducting appraisal sessions.
- Role plays to give the participants some experience and a feeling of familiarity.
- Videotapes that illustrate what to do and possibly what not to do (there is disagreement among trainers over the desirability of illustrating negative behavior).

The performance appraisal form itself should be well designed to reflect the personality of management, the nature of the business, and the breadth of the employees along the two dimensions of the matrix. Periodic updating may be necessary if the organization's management or business objectives or competitive practices change.

Clearing performance reports with higher-level management provides a way of monitoring evaluators' performance and ensuring consistency across the department or organization. It also aids in monitoring the effectiveness of the performance appraisal form in providing the kind of information needed by higher management in career planning, rotation, salary, and succession planning.

LONG-RANGE BENEFITS

Implementation of both the structure and the communication aspects of the performance planning system will benefit employees, managers, and the organization. For the employees, there is a chance to obtain personal satisfaction and professional pride. Communications between employees and managers in the matrix will be consistent and of the necessary quality. And the employees will have frequent opportunities to develop and apply various skills, thus improving their career prospects and worth to the organization.

The managers, on the other hand, will be able to secure employee commitment and motivation and to develop the employees' potential, which of course makes it easier for the managers to achieve the goals of their units. In addition, the opportunity to communicate purposefully with other managers in the matrix will enhance their comprehensive knowledge of organizational priorities.

Finally, the organization will realize high overall productivity from its human resources and a greater stability in the matrix diamond.

Section II
Matrix Management Applications

Section II contains the core theme of the book—the different ways in which matrix management is used today. Throughout this section the reader should recognize the sharing context of matrix management: from the top level of the organization through the use of the plural executive, down to the worker level where that individual has an opportunity to share in organizational decisions through Quality Circles, autonomous production teams, etc.

The reader may sense that there is nothing new about the sharing context of matrix, that such sharing has characterized organizations for a long time. Indeed, some form of matrix has always been with us. What should become clear, however, is the emerging formal "real-time" application of matrix management in an effective manner in today's management. Matrix management is a form of organizing resources that can be used at many different levels.

In Chapter 4 John E. Fogarty describes the use of "integrative management" in a steel company. With the U.S. steel industry in economic difficulty, matrix management's contribution in this company is noteworthy, and should encourage managers concerned about productivity in their own organizations to see what matrix management might do for them. Fogarty identifies both the positive and negative aspects of management matrix, noting that it is not a panacea for organizational difficulties.

In Chapter 5 Victor G. Hajek discusses matrix management in engineering projects, looking at organizational roles as well as executive qualities related to the planning, organizing, and control of projects.

Chapter 6 deals with matrix management of international projects. H. A. Williams briefly describes certain characteristics of project management and then discusses some of the ways in which leading international corporations use project management. He goes on to show a product line field sales organization organized on a matrix basis, and ends by noting the unusual care to be taken in setting up the proper support and reporting relationships in the international matrix field sales organization.

In Chapter 7 J. F. Ricketts shows how one large transnational corporation operates in the international environment. He describes how the matrix approach is useful in reconciling the needs of managing several dimensions of the international matrix: the different demands of each foreign country; the worldwide product disciplines; and the efficiency of key centralized staff functions. Those persons interested in gaining an appreciation of the macro-organizational reporting relationships found in the international matrix will find Rickett's article particularly useful.

Each manager is dependent on the people that work for him or her. If managed properly, this dependency can be used to more effectively improve oranizational productivity. In Chapter 8 Edmund C. Mechler, Jr., explains how this can be done through the use of Quality Circles. Mechler describes the *what, why,* and *how* of using Quality Circles. He offers some pragmatic guidance on how Quality Circles, with proper management support, can help modern managers deal with the enormous change occurring in their organizations.

Chapter 9 presents a "hands-on" philosophy and techniques for using task forces to solve a wide range of complex organizational problems and opportunities. James P. Ware believes the task force can deal effectively with operating or strategic problems that the existing organization alignment is unequipped to handle. Problems that don't fit the existing organizational structure, or that are too big or complex or require a fresh approach, can be dealt with through an ad hoc task force. Yet according to Ware, many task forces fail. Consequently, he prescribes concepts and techniques that can improve their use.

In Chapter 10 Lynn W. Ellis writes about temporary groups as an alternative form of matrix management to reduce the inadequacies of the hierarchical form of organization. He describes several situations in which a temporary group has been used to deal with organizations' built-in lethargy and resistance to innovation and change. Ellis takes the position of the research manager who attempts to stimulate innovation; he addresses the use of functional authority, the conflicts and needs of informal groups in large organizations, and the special characteristics of informal groups in large organizations. He then describes the use of a formal temporary group in the large organization to deal with ad hoc problems. He concludes by noting the contribution that formal and informal groups make to intra- and cross-functional coordination and communication.

Frank Oldham, Jr., in Chapter 11 provides insight into the use of the plural executive to integrate top management decision making and implementation. According to Oldham advances in technology, the movement toward global management, and the growing product service base have created top managerial responsibilities in large organizations that are beyond the capabilities of a single individual. He discusses the advantages and disadvantages of the plural executive approach, and decides that it is premature to say that the plural executive is here to stay. He ends on the positive thought that if the plural executive is as effective as some proponents claim, then such a matrix management form should help the corporation outperform traditionally organized competitors.

Chapter 12 treats one of the oldest forms of matrix management in existence. M. Dean Martin and Penny Cavendish present product management as a variant form of the emerging discipline of project management. They trace the evolution of the product manager and describe his or her role in facilitating the marketing function in organizations. Their description of the project manager's job, what it entails and how the job can be approached, is particularly useful in establishing the legitimacy of matrix management. In concluding, they remind us of the increasing complexity and uncertainty of the environment, which will cause more and more firms to seek ways to better serve markets and enhance profitability.

Chapter 13 deals with the emergence of multiorganization enterprises (MOEs). According to Mel Horwitch MOEs are made up of many organizations; they are generally both public and private, are usually large in terms of funding and people, have components that maintain their institutional identity, and are organized to accomplish a specific mission. Horwitch sees the management of these institutions as requiring new perspectives that are largely absent from traditional thinking in organizational strategy. This management must give more attention to the potentially radical transformation of such enterprises over time, and to the need for effective "feedback" from the external environment, for champions, for an "overadaptive" organizational structure, and for multiple information channels.

4. Integrative Management At Standard Steel

John E. Fogarty*

Integrative management, an advanced form of matrix, has been alive and well at Standard Steel in Burnham and Latrobe, Pennsylvania, for four years. Its contribution to the revitalization of this 170-year-old diversified company of 2500 people has been formidable.

The first portion of this chapter defines this method, which provides the permanent, formal foundation for managing Standard Steel. The second portion follows the evolution of integrative management into its present state of development. My perception of its remaining shortfalls is discussed in the final section.

THE FORMAL FOUNDATION

Successful management systems have adjusted to the infinite complications of our ever growing, dynamic society. A significant point on that evolving curve appeared a few years ago: it was called matrix management.

In the simplest sense, matrix supplements the normal *vertical* functional management team (Figures 4-1, and 4-2) with a *horizontal* business committee team that is often composed of a separate group of individuals representing various business disciplines (engineering, finance, operations, sales, etc.). The purpose of the horizontal team is to foster greater conceptual understanding of each business segment of a multifaceted enterprise.

In 1977 Standard Steel instituted its own form of matrix management. The matrix approach seemed appropriate because the company is composed of four clearly defined business segments, each with a different manufacturing system and marketing arena. Little or no conflict emerged and the participants continue to enthusiastically endorse the system. The net result, which provides the formal foundation for managing Standard Steel, has been most productive.

---

*John E. Fogarty is President of Standard Steel in Burnham, Pennsylvania. Mr. Fogarty received his undergraduate engineering degree from the University of Massachusetts in 1951; undertook postgraduate studies at the Massachusetts Institute of Technology in 1958; and received executive-management training at Penn State University in 1972. He joined Standard Steel in 1977, after serving in senior technical, operation, and general management positions in the steel industry.

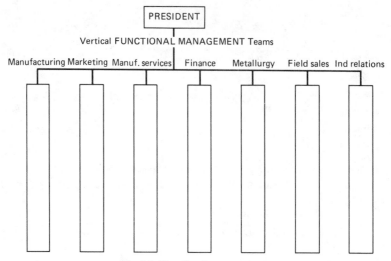

Fig. 4-1. Functional management.

Matrix, however, is known to have occasionally performed with only marginal success, perhaps as a result of some natural inclination toward an adversary relationship between the vertical and horizontal teams. Because of this reputation, Standard Steel has often found itself defending the matrix concept. Its defense as always been the same. It stresses that matrix is not universally

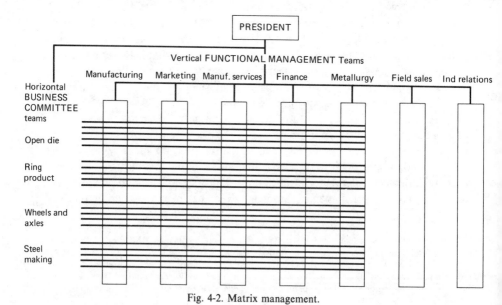

Fig. 4-2. Matrix management.

applicable; that matrix is too complex a concept for implementation based on generalized description; and that Standard Steel's approach is unique in being specifically formulated to avoid vertical/horizontal conflict. Though some reluctance accompanied the company's early use of the word "matrix" to describe its system, no other terminology was available.

This led to Standard Steel's coinage of the phrase *integrative management,* the prime justification for which centered around its implicit exclusion of the vertical/horizontal adversary coefficient so often associated with matrix. More specifically, whereas both the vertical and horizontal management teams of matrix and integrative management organizations can report directly to the President, integrative management differs in that the horizontal team is not constituted as a separate, independent group of managers. With integrative management, horizontal team members also maintain functional responsibility and report to a functional executive, as depicted in Figure 4-3. Consequently, as each horizontal team member is also a vertical team member, adversary relationships and conflict are minimized. Functional executives are also more at ease with this system because they have representation on the horizontal teams. The horizontal teams, nevertheless, do report directly to the President. As such, they constitute a formidable and accountable multifunctional yet unitarian influence within the decision-making processes of the company.

This system of management, which incidentally was incorporated by Stan-

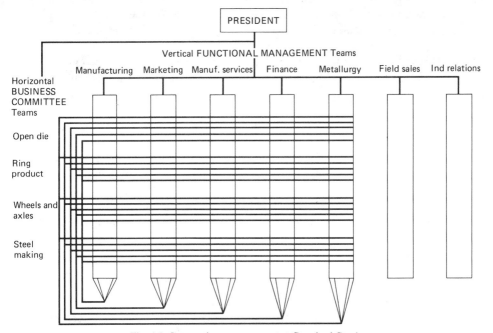

Fig. 4-3. Integrative management at Standard Steel.

dard Steel with no significant increase in staff size, has proven a credible base for innovative and responsible thought generation, the results of which have significantly contributed to the company's growth, diversity, and profitability.

THE EVOLUTION OF INTEGRATIVE-MANAGEMENT

Standard Steel began operations in 1811. After successfully accommodating the economic and political challenges 150 years, its sales volume reached $20 million in 1960. Since then it has added a second division, diversified, and multiplied its sales volume more than ten times. It employes 2500 people and produces 300,000 tons of steel annually, primarily for use in its own manufacturing operations.

However, 1977 was not a good year for Standard Steel. If market prices for its products had been a little lower, or costs a little higher, it would have lost money. In response to that challenge, Standard Steel's labor-and-management team resorted to the same old Yankee "bag of tricks" it has relied upon for so long; increased productivity, process and product innovation, and the investment of a lot of money. *All things being equal,* the net effect would have reduced costs during 1978 by $7 million—a laudable achievement to which all Standard Steelers could point with pride.

However, all things were *not* equal. During 1978, even with the added advantage of overall price increases of $3 million, the company made little headway. This was because of inflation. More specifically, the productivity and pricing benefits achieved were wiped out by labor-cost, material-cost, and energy-cost increases.

This suggested a difficult future for Standard Steel because, while opportunities for further cost reduction were diminishing, inflation was continuing to spiral. It therefore became apparent that the company could no longer totally depend on that *old* bag of tricks. It needed a *new* one—one that would not only help to improve its processes and productivity, but also make it generally "work smarter."

Another way to describe the magnitude of the problem facing Standard Steel in 1977 is to state that it was able to deliver only 50% of its products on time. The immensity of the problem implicit therein suggests a more general source of difficulty than simple individual competence. Fundamental strategic deficiencies had to be involved. Certainly the ability of the company's management structure to cope with its continuing diversification and growth was suspect.

A review of the structure of the entire organization revealed, for example, that all the Vice-President of Manufacturing had time for was the crisis of the minute. He had fourteen functions reporting to him. Accordingly, a structural change was affected whereby that number was reduced to six, and one layer of operations management was removed. In addition, a new department under a Vice-President of Manufacturing Services was created to provide separate executive leadership and accountability for all manufacturing materials flow and engineering staff services.

Even with such organizational changes in place, the company was still structured along simple vertical lines (manufacturing, marketing, finance, etc.). Concern remained that the company's diversity and growth had created sufficient complication to still overwhelm executives with day-to-day detail. It was therefore decided that its conventional management structure could no longer effectively manage the business.

Prof. John Slocum of Penn State helped tackle the problem. He agreed that Standard Steel's functional management system should be abolished and that a relatively new concept, matrix management, should be installed in its place. The matrix system maintains the vertical responsibilities of the functional organization, but supplements it with a separate horizontal group of managers whose responsibility extends across all functional aspects of the various individual business segments. The result projects horizontal as well as vertical responsibilities on an organizational chart, hence the term "matrix." The system is unique because it can focus collective expertise across the spectrum of management disciplines on individual business problems.

It was agreed that each of Standard Steel's product lines—wheels and axles, open-die forgings, ring-mill products, and steelmaking—would be horizontally managed by a business committee normally composed of five middle-management representatives, one each from marketing, manufacturing, manufacturing services, metallurgy, and finance. These committee assignments would be in addition to the representatives' functional assignments. It was perceived that this structural feature would nullify the tendency of such cross-purpose organizational structures to generate counterproductive horizontal/vertical conflict.

In retrospect, the net contribution of that feature probably constitutes a key reason for the success of integrative management at Standard Steel. As previously stated, it also justifies our coinage of that term.

It was also resolved that, although the individual members would continue to report to their functional department heads, each committee *as a body* would report directly to the President. The purpose was to demonstrate the full support of top management for this managing concept, and to provide for direct access by the committees to the President. Interestingly, this latter feature has seldom been utilized. Its incorporation into the system and the perceptions that it creates, however, constitute another formidable contribution to the success of integrative management at Standard Steel.

The responsibilities of the committees were precisely defined. They involved preparation of the annual business plan for each product line, including targets for individual factors such as production costs, product inventory, shipping volume, and gross profit. The committees were also to monitor the performance of each of these factors against plan each month, and to effect appropriate interim changes in policy, or to recommend them to the senior functional executives, so as to accommodate changes in manufacturing and marketing environment. Importantly, functional executives also agreed to reorder the structures of their functional organizations by product line, so as to provide congruence with committee structures.

It was further agreed that though the business committees would meet as required to address themselves to developments within their business spheres, they would hold formal monthly meetings to consider and adjust manufacturing and marketing strategies, with the President joining them informally at the end of each meeting to answer questions or provide guidance. Formal quarterly meetings were also scheduled with the President and the functional executives, during which all pertinent business variances were to be addressed.

Since this approach constituted a major departure from the company's wholly functional management system, it was recognized that it would have to be implemented with great care, communication, and understanding. Hence, it was decided that all committee members should attend a formal, one-day-a-week, two-month course whose purpose was to both introduce integrative management and explain how it could enhance the success of the company. The program recognized that a manager on a business committee is a team member—a colleague—and must be able to persuade people from other disciplines to work with him in peer-group situations, rather than through the disciplines imposed by hierarchy. Accordingly, the course provided instruction in group dynamics, decision making, conflict resolution, and leadership.

Acceptance of this concept during the training period was not immediate. Within two weeks, committee members began to complain: "Why are we doing this? It's disturbing everything we've always done. It won't work. How will we have time for our functional duties and our committee responsibilities, too?" Things got so bad that Dr. Slocum and I had to submit to an extensive question-and-answer session to get things back on the right track. What pulled the idea through was our absolute conviction that this was the right way to go, and our assurance that the managers would be allowed to manage freely but accountably within the new framework. Four years of very successful implementation have since provided an even more credible answer: participation in business committee activities so effectively eases and expedites the implementation of the functional responsibilities, that sufficient time is saved to adequately accommodate those business committee activities. A most positive "vicious circle" is thereby generated, whereby integration breeds proficiency.

Another interesting aspect of committee membership that surfaced rather quickly was an implicit need for reasonably strong personalities and commitment among the members. The committees themselves began to project disenchantment through members who failed to provide balanced input. Such inclinations to optimize effectiveness were quietly corrected via reassignment.

Occasions have arisen, however, when it became prudent for the horizontal middle-management business committees to refer one of their tactical decisions "upstairs" for functional executive approval. The integrative nature of the teams—integrative in the sense that each member has both functional responsibility and business committee accountability—has seldom failed to provide the insight required to judge when this should be done. The net effect has been more competent decision making at the middle-management level, with little or no executive backlash.

The understanding and support of top management has also proven crucial. Early in the game, for example, I became disenchanted with the performance of one of the committees. A more perceptual analysis suggested that I had actually, though inadvertently, been expecting that committee to assume responsibility that belonged to senior executives. The matter involved discontinuing operations at a major manufacturing center. To reiterate, I am convinced that successful implementation of such management systems requires the allocation of a great deal of attention and thought by senior executives during its entire first year.

This newly applied managing strategy also created a much more critical self-appraisal of what we were doing and how we were doing it. Not only did we find more productive ways of doing the old things, but also ways to further diversify into markets that Standard Steel had never before pursued.

Integrative management has even affected the thought processes of the horizontal middle-management team members. They have learned to think more like generalists than engineers, accountants, operators, or salesmen. This contribution provides a personal sense of achievement for participants and an obvious strategic reward for the company. It also enables the senior functional executives more time to formulate strategic business policy. In today's highly complex business world, top managers must have time to recognize change and deal with it; time to deal with the colossus of Washington; and time to achieve a far more credible level of communication with their employees.

Testifying to the benefits provided by integrative management is the fact that, though Standard Steel now ships substantially more tonnage than it did before these changes were made, its on-time delivery performance has increased from the 50% level of 1977 to over 90% today. Improvement in overall proficiency is defined by Figure 4-4.

REMAINING DEFICIENCIES

But nothing is perfect. As pleased as we are with integrative management, important problems remain. Significantly, they do not seriously involve the innovative effectiveness of the committees, the relationship between the committees and the functional executives, or the morale of the participants. This view was supported by David I. Cleland, Professor of Engineering Management at the University of Pittsburgh, after visiting with Standard Steel's executive staff and committees for several days.

The main problem lies not with the participants, but rather with the non-

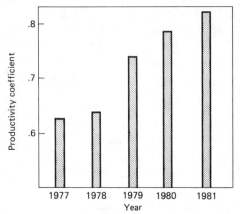

Fig. 4-4. Productivity Improvement

participants. Whereas the middle managers on the committees gain personal recognition from both peers and superiors for their accountable participation in running the business in a general management sense, their counterparts in departments not represented on the committtees (maintenance, plant engineering, industrial relations, field sales) do not share equally in that limelight. Nor do they enjoy the unique opportunities for personal growth that committee membership is perceived to afford.

This constitutes a matter of growing concern, though it is partially being accommodated by the business committees themselves. For example, with increasing frequency the committees are inviting others to participate in certain of their meetings for specific reasons. They have also established both permanent and temporary subcommittees involving members of other departments. These subcommittees deal in an even more "hands-on" mode with quality, operations, development, field sales, and competitive issues. These measures provide only partial accommodation, however.

One fuller measure would be to expand committee constitution and thereby incorporate membership from other departments. The net affect, however, would be counterproductive: committee size would become unwieldy and focus would fade. Implicit too in this scenario would be a certain waste of time by the new participants. Nor, for the same reasons, would it be appropriate to establish new committees just for the sake of providing "committee" participation.

No credible resolution to this problem is presently apparent. Perhaps the continually expanding influence that the committees are exerting will one day find a solution. Some resolution must be provided since businesses cannot afford, however unintentionally, to "turn on" one group of essential disciplinarians while ignoring another.

CONCLUSION

The technical, operational, marketing, and systems innovation generated within these committees have allowed Standard Steel to "work smarter" than ever. Integrative management has created a "new company" atmosphere while developing aggressive, confident, and conceptual-thinking managers. Morale at all levels has improved, and our security as a viable business enterprise has been reestablished.

More conceptually, while recognizing the enormous complexity imposed upon business today by social, inflationary, international, and regulatory forces—all of which make expeditious coping with change the primary task of today's manager—it is great to observe how Americans, by whatever management means, adjust to such challenges with confidence, enthusiasm, and pride. Though I believe that the time has come for our government to help business rekindle these qualities with government appreciation, government understanding, and administrative support, the challenging role of the business policy maker continues to provide a unique opportunity for great fun and great satisfaction.

5. Matrix Management for Engineering Projects

Victor G. Hajek*

PROJECT TEAMS IN LINE ORGANIZATION

In order to provide the necessary surveillance and control of the various systems under contract, management officials usually establish teams of engineers, production personnel, planners, contract administrators, etc., and charge the team with the responsibility of carrying out the objectives for the project. In the traditional pyramid or line organization, the responsible head of each project team reports directly to a top administrator of the parent organization and is responsible for achieving the project objectives with the resources that the parent organization allocates to the project team. The head of the team is usually an engineer with proven managerial abilities and is identified with such titles as project engineer, project director, project manager, etc. Since the responsibility of the head of the project team includes many managerial disciplines in addition to engineering, the commonest title of the project head is project manager.

The typical line organization of a project is illustrated in Figure 5-1.

The basic structure of the project team in the line organization of Figure 5-1 can also be depicted as the triangle of Figure 5-2, wherein each of the groups headed by the different group leaders is directly responsible to the project manager. The resources in each of the groups are assigned to the team for the duration of the project.

The project manager exercises direct and autonomous control over the various discipline groups (such as purchasing, engineering, etc., each headed by a

---

*Upon completion of his naval service in World War II, Victor C. Hajek received his MBA from the University of Michigan and worked for the Westinghouse and Worthington corporations as a sales and applications engineer. He then returned to the Navy Department and served as a project engineer on aircraft antisubmarine warfare (ASW) simulators, advancing to the position of Division Head over branches of project engineers responsible for air warfare simulators. In addition to his regular functions, Mr. Hajek served as advisor to NATO on ASW simulators and headed an engineering team on a flight simulator program in the United Kingdom. Prior to his retirement, he served as the head of a division comprising branches responsible for special engineering, computer applications, and advance planning functions. Mr. Hajek is the author of two books on project management and numerous articles on simulators. Currently he is writing a book on project management in matrix organizations and serves as a consultant on simulator applications and project management.

Fig. 5-1. Typical project in line organization.

supervisor) and is responsible for the coordination and monitoring of the effort of the team. When the project cost, schedule, and equipment performance objectives are realistic and major problems or revisions are not encountered, the resources generally prove adequate and the project is brought to a successful conclusion.

However since most systems handled by project teams possess developmental features and other characteristics requiring adjustments of objectives, the project manager rarely enjoys a smooth-running operation. Most projects have one or both of the following characteristics, which at some point in the project's life cycle require modification of the originally established cost, schedule, or performance objectives.

1. Unique systems requiring engineering whose budget for resources and planned schedule must often be adjusted because of unanticipated complications or revisions to the engineering design, fabrication, or testing effort.
2. Changes precipitated by revisions to performance objectives, schedules, or other reasons that necessitate adjustments to the project plans.

When complications such as these occur, the project manager is expected to implement remedial actions by utilizing the assigned resources. If, for instance,

Fig. 5-2. Project team organization.

the computer programming man-hours that are required have been underestimated, or additional effort is necessary to correct errors, the work load for resolving such problems would fall on the programming personnel assigned to the project and would require a concentrated effort in that specialized area. The hectic effort or "fire drill" that falls on the shoulders of those few responsible for an area of effort usually results in fatigue, excessive errors, and inefficiencies over a relatively short time.

The demands for the various disciplines necessary for the design, programming, fabrication testing, etc., of equipment that is to be created by the project team occurs as different points of time during the project's life cycle, and the magnitude of the demand varies. As a result, different project resources can experience excessive peaks and valleys in their utilization profile. One of the difficult tasks of the project manager in a line organization is to plan and utilize the resources so as to create an even work-load distribution. However, since the resources are generally committed to the one project, the line project manager often has to live with a situation whereby certain resources are idly waiting for other prerequisite functions to be completed, while other resources are concentrating maximum effort to complete milestones on schedule.

In addition to the inflexibility of line organization project teams noted above, isolation among projects often exists. Project managers naturally tend to prioritize their assigned projects over and above other projects in the parent organization. As a result, the competitive environment often frustrates the company-wide cooperation that is so highly desirable. In emergency situations the detailing of personnel with specialized talents from other projects to render assistance to a particularly troubled project is usually resisted, and any arbitrary reassignments dictated by higher management lead to resentment among project managers.

In addition, the basic feature of the self-sufficient project team organization throttles communication among teams. Successes and failures in resolving problems, technical breakthroughs, and the multitude of valuable experiences that occur on various projects are generally not shared. Inhibited communication in the parent organization often leads to expensive redundant design effort and repetition of mistakes otherwise avoidable, and in general creates an undesirable company environment.

With the increase in the size and complexity of projects, the procedures and structures of line organization project teams are frequently modified to prevent possible limitations of the line organization operation. For example, to relieve work-load distribution or unexpected resource problems, subcontractors and job shop personnel are retained to supplement the working force in different areas of effort. Staff personnel and assistants to the project manager are often added to relieve span-of-control problems. To facilitate interproject communication, liaison personnel are retained as members of the project manager's staff. Such patch-work arrangements usually prove helpful, but complicate the project team structure and make the management of the project more complex. Adopting the matrix types of organization for engineering projects solved many problems that handicapped the efficient execution of large, complex projects in the line organization. The matrix concept is not restricted to project applications but is used in a variety of forms for different types of enterprises. Matrix structures are based on geographic areas, product lines, customer groupings, and varieties of forms that are designed to serve the operational characteristics and objectives of the organizations involved. The discussion in this chapter will be restricted to the two-boss matrix system commonly used for the management of technical projects.

MATRIX ORGANIZATION

In the matrix organization, the personnel and other resources that a project manager requires are not permanently assigned to the project, but are obtained from a pool controlled and monitored by a functional manager. Personnel required to perform specific functions in a particular project are detailed for the period necessary, and are then returned to the control of the functional manager for reassignment. Discipline supervisors are responsible for the efforts of the groups constituting assigned project personnel and for other required resources. The members of the groups and their supervisors are charged with the timely completion of the different tasks and are responsible to the project manager and the functional manager. For example, an engineer assigned for a specific period to design a subsystem of a project is responsible (through the discipline supervisor) to the functional manager for completing the task as scheduled, and to the project manager for providing an acceptable design. The two managers report to a matrix executive.

The structure of the matrix is based on the concept shown in Figure 5-3.

An expansion of Figure 5-3 to indicate a matrix organization of two projects and two types of resources is depicted in Figure 5-4.

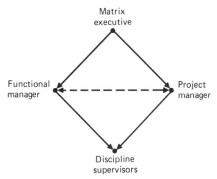

Fig. 5-3. Basic matrix structure.

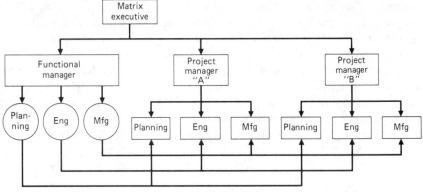

Fig. 5-4. Matrix organization

The triangular configuration of the organization noted in Figure 5-3 for one project can be noted in Figure 5-4. As in the line organization, the project manager is responsible for cost, schedule, and equipment meeting the performance of the project, except as previously noted; specific resources required during specific periods of time are detailed to the project by the functional manager. One of the major responsibilities of the project manager in the matrix is to work with the functional manager to establish the resource requirements and the timetable for their utilization on the project, and to work out the revisions required as the project effort proceeds. On the other hand, the functional manager is responsible for assuring that resources are utilized in the manner best serving the interests of the parent organization.

To function effectively in the matrix, the project manager must understand the role of the three primary parties with whom he interfaces. Therefore brief discussions of the functions of the matrix executive, the functional manager, and the discipline supervisors follow.

THE MATRIX EXECUTIVE

The major areas of responsibility of the matrix executive include the following.

1. To maintain an environment of open communication and candid discussion of problems. The inherent characteristics of the matrix organization serve to surface problems that then raise questions and require discussion of the remedies. The matrix executive must assure that problems are promptly surfaced so that constructive dialogue can lead to expeditious solutions.

2. To crystallize issues and moderate discussions in a "controlled conflict" environment, when problems are not resolved at lower levels. When parties to an issue are not able to agree on a course of action, the conflict is referred to the matrix executive, who is responsible for effecting a compromise agreeable

to all parties to the conflict. Even when completely objective parties are involved in a conflict, sincere differences of opinion can frustrate progress. However, the completely objective consideration of the best interests of the parent organization will establish the proper course of action. The matrix executive is responsible for conveying whatever relevant additional information should be considered regarding the parent organization's best interests, and for promoting an environment of cooperation and compromise until an agreement is reached. Once the parties agree to the compromise, they can be held responsible for implementing to the letter.

3. To maintain a balance of power in the matrix. Assuring the balance of power between the functional and the project legs of the matrix triangle, shown in Figure 5-3, is one of the responsibilities of the matrix executive. Any tendency of either manager to dominate the other requires corrective action. If, for instance, the functional manager dominates the matrix, the compromise features essential to resolving conflicts may be damaged, and the allocation of resources so implemented as to produce results undesirable for the organization.

4. To establish objectives and the scope of resource expenditures for development. A major responsibility of the matrix executive is to translate the negotiated customer requirements of the equipment to be produced into project objectives. If developmental efforts are required, the executive would establish the type and amount of resource expenditures to achieve the objective. The direction would be rendered after discussions and review of the requirements. The proper implementation of this responsibility is particularly important for requirements involving creative engineering and research efforts.

THE FUNCTIONAL MANAGER

Whereas the project manager is responsible for achieving the performance cost and schedule objectives of the assigned project, the functional manager is responsible for providing the resources required to meet the objectives of all projects in the matrix organization. His functions include the following.

1. To maintain records of existing man-hour utilization among different disciplines and projects.
2. To establish work-load projection for future utilization and to update projections for anticipated changes.
3. To maintain a balance of work loads so as to avoid excessive peaks and valleys in the utilization of personnel and other resources.
4. To negotiate with project managers regarding revisions to resource utilization schedules, in order to accommodate necessary personnel reassignments due to problems or changes, and to maintain a balance in work load and other resource utilization.

5. To coordinate with the project manager the supervision, evaluations, assignments, etc. of the discipline supervisors and the personnel assigned to each discipline group.

6. To handle personnel matters such as union negotiations, training, and grievances.

Because the functional manager's primary responsibility is to control the assignments of personnel and other resources that are in demand for different projects, pressures from project managers are common. The functional manager must objectively appraise the relative urgency of resource requests for different projects and must frequently draw conclusions regarding the priorities of allocations. Also, the functional manager must be able to defend his position when conflicts are precipitated, and negotiate solutions that all parties to a conflict, hopefully, can accept as being in the best overall interests. In the event the conflict cannot be resolved, the functional manager must be able to objectively present a position that will facilitate a resolution by the matrix executive.

THE DISCIPLINE SUPERVISORS

The discipline supervisors, often referred to as "two-boss managers," represent the third echelon of the matrix organization. Each discipline supervisor heads up specific areas of effort such as engineering and production as noted in Figure 5-4, and is responsible for meeting his or her assigned areas's objectives of cost, schedule, and performance, which constitute subsets of the overall project objectives.

Although assigned to a specific project team, the discipline supervisor is responsible to both the project and the functional managers. As a member of the project team, he or she is responsible to the project manager for meeting the objectives established for his or her area of effort. Also, the discipline supervisor is responsible to the functional manager for effective utilization of the assigned resources, maintaining smooth personnel relationships among the discipline team members and in general meeting the objectives of the second boss.

The ability to function effectively in situations where two competing demands must be resolved requires that the discipline supervisor possess special qualities of logic, communication talents, flexibility, and negotiation abilities. The resolution of conflicts usually involves compromises that require adjustments in the work plans of the group. The discipline supervisor then must convey the revised plans to the subordinates and obtain acceptance of the revised plan and unreserved cooperation in implementing the new direction.

Some of the tactics and functions that the discipline supervisor must exercise in the two-boss matrix relationship include the following:

1. To analyze the prime issues of a conflict and establish a viable course of action that can be promoted and negotiated with the project and the functional managers.

2. To actively communicate with each of the managers and other individuals that are affected by a conflict, and to promote the desired course of action prior to the negotiation meeting.

3. To obtain an insight regarding points that represent the firm positions of the other parties, and to select feasible trade-off points that can be easily negotiated.

4. To cultivate traits of sincerity, competence, and cooperativeness that will encourage other members of the matrix to discuss issues and problems with candor and trust. The possession of such personal characteristics will promote open discussion and candor by others.

5. To function as the mediator between each of the individuals in the group and the two bosses as far as performance, discipline, and other personnel matters are concerned.

THE PROJECT MANAGER

The project manager must understand the roles of the other matrix managers and be able to effectively interface with all the parties mentioned in order to function successfully in the matrix.

The manager who previously operated in a line organization will have to adjust to a radically new world. If the individual lacks the flexibility to make this adjustment, difficult situations will generally occur, with resultant failures. Major functions for which the matrix project manager is responsible include the following.

1. To carry out the objectives established by the matrix executive. The objectives and directions that the project manager receives from the matrix executive are shared with the functional manager, and the implementation requires a cooperative effort. The project manager is expected to resolve issues and establish plans of action that can conform to the parameters of the approved objectives without involving the participation of the matrix executive.

2. To effectively negotiate from a position of knowledge, when conflicts are surfaced. In discussing and negotiating a particular issue, the project manager must exhibit a persuasive personality, articulation, and knowledge of the issues and of the technical, financial, schedule, and other effects on the project that are at stake.

If a conflict must be referred up the line for resolution, the project manager will find that the executive assumes the role of a strategy manager and consultant, guiding the discussion and providing additional pertinent information to promote understanding of the larger issues coming into play. Once a solution or course of action has been established, the role of benevolence that the matrix executive has assumed is transformed into an autocratic posture. The matrix executive then holds the project and functional managers directly responsible for carrying out the actions that have been mutually agreed upon.

3. To perform the project team management functions in cooperation with

the functional manager and other key members. The echelon directly under the matrix executive consists of the managers responsible for the project—as depicted in Figure 5-3, the two bosses that characterize the matrix.

The fact that the project manager must depend upon the functional manager for the resources required by the project represents the most radical change, when the transition from the line to the matrix organization is implemented. To be effective in the matrix, the project manager must prepare realistic requirements for essential resources and then successfully negotiate for those resources. Also, the project manager should have a knowledge of the status of other projects with which resources must be shared, since in many cases the managers of different projects will compete for special resources. Thus the matrix forces the managers to take a more global view of the company's activity.

The third echelon of the matrix organization with which the project managers must interface, consists of the different discipline supervisors and their subordinates. Although the discipline supervisor is a member of one project team, two managers must be satisfied and two managers will render performance evaluations of his effectiveness. The interface between the project manager and each of the discipline supervisors goes beyond the boss-subordinate relationship. Directions rendered by the project manager that would be rigorously followed in the line organization, may require modification because of considerations that the functional manager may raise.

4. To recognize the priority of the assigned project as secondary to the general welfare of the parent organization. In the matrix, the project manager must recognize and accept the fact that other projects may be assigned higher priorities and impact his project in such areas as available resources and schedules. In such cases the matrix executive will expect the project and functional managers to negotiate a compromise solution and get on with the job. Directions from the matrix executive will be given primarily to break deadlocks. Even then, directions would generally be a formality after compromises have been mutually agreed upon.

In the final analysis, the objectives of the project manager in the matrix are not only to meet the performance, schedule, and cost objectives of the project, but to achieve the objectives in the manner best serving the overall welfare of the corporation or activity. In a matrix organization managers compete for resources; however, when a decision is reached, all project managers must give unqualified support to the decision, even though the directed course of action may require adjustments in projects that did not benefit from the decision.

MATRIX MECHANICS

The performance requirements for a project are stated in the specification portion of the contract, and in most cases the project manager has had a role in establishing design approaches, etc., during the proposal preparation stage.

The details for carrying out the functions in the matrix that are necessary for meeting the contract and specification requirement vary among organizations. However, the basic procedures are similar. Work packages are created by those individuals assigned to discipline supervisors who are responsible for the different areas of the project. The work packages are derived from the work breakdown structure of the project and contain information regarding the technical approaches, the performance objectives of each subsystem covered, and the schedules, budgets, and other related information. The generalized work packages reflect the resources to be used, the performance targets to be achieved, and the time frames for the effort upon which the functional and project managers have agreed.

The matrix executive crystallizes the contract objectives and directs the matrix managers to initiate the project effort. The key matrix team members usually participate in the proposal preparation and are therefore familiar with the details of the contract requirements. However, the technical negotiations and clarification process usually result in revisions to the proposal, so that the differences between the initial proposal and the final contract require review and amendments to the various work packages affected.

The project manager has the major responsibility for having the work packages revised to assure consistency with the finalized requirements and the directions of the matrix executive. This responsibility is shared and discussed with the functional manager in order to reach mutual agreement on resources and other affected areas. The detailed effort on individual work package revisions is accomplished by members of the discipline supervisor teams, subject to ultimate acceptance by the project and functional managers.

The master project plan is also one of the basic documents used in the matrix organization. The plan identifies the major milestones to be achieved in order to meet the contract delivery schedule. The project manager is responsible for coordinating the plan with the functional manager and assuring that details are consistent with the work package. Preparing the master project plan involves allocating different types of manpower disciplines, detailing projected costs of labor and material, establishing schedules of project personnel; in effect, the plan presents what is required in order to meet the performance, schedule, and cost objectives of the project.

Once the project master plan and the work packages for the project are in place, the initial phase of the procurement cycle usually runs smoothly. The teams under the discipline supervisors who are responsible to the project manager cover such areas as contracts, engineering, quality control, planning, administration, and finance. Ideally, if everything proceeded as planned, the unique matrix forces relating to conflicts would not come into play. However, the nature of most projects, particularly if engineering development is involved, precipitates problems that must be resolved through negotiations by the matrix managers.

The project manager's main concern is to assure that the effort necessary to satisfy the specification requirements is proceeding within the cost and schedule parameters. The achievement of those project objectives requires that responsibilities of each of the discipline supervisors be implemented as established in the work packages. The tools that the project manager uses are essentially the same as those used in the more traditional line organization. A PERT (Program Evaluation Review Technique) system designed for the project provides essential status information and indicates developing problem areas. PERT reporting documents such as the Management Summary Report give the project manager information relating to the schedule and cost status, trouble areas, and cost predictions. The Manpower Loading Display reports would be of particular value to the functional manager, who has a primary interest in the control and distribution of manpower resources. The LOB (Line of Balance) monitoring report, which is a simpler version of the PERT system, provides current project status without the predictive features of PERT and is used extensively for project monitoring. One of the discipline supervisors who is generally given the title of planning supervisor is responsible for tracking the status of the project and providing reports to the project and functional managers.

After the project is launched, the basic plans for the performance schedule or cost of a project rarely, if ever, remain in force without further change throughout the procurement cycle. The need to revise the work packages and project plans can be precipitated by a variety of internal or external causes. External causes include revisions of the performance requirements or delivery schedules. Internal causes include engineering problems requiring additional manpower resources, time, and funding; design approach revision to permit adopting new technology; manufacturing problems; delays in material deliveries; and other complications that can plague a project.

When problems arise in the line organization, the project manager is expected to implement solutions utilizing the assigned project resources. When such resources are available, the project manager can unilaterally shift the necessary manpower as required. However, when external resources are required, the inherent inflexibility of the line organization does not permit assistance that may be available in other parts of the company, without petitioning the hierarchy of the company to direct reassignment of personnel. Such direction often generates hostility from other project managers and jeopardizes the smooth functioning of the company.

Project changes precipitated by external or internal forces almost invariably affect the project cost, schedule, and in many cases the design approach for the product. To accommodate the changes, the reallocation and rescheduling of personnel and the revision of project milestones are required. (As contrasted with the line organization, the matrix possesses the flexibility to provide available resources to overcome an unexpected problem.) The project manager is responsible for preparing the petition for relief by detailing the areas to be revised, identifying the required resources, establishing schedule revisions, and in general providing the information necessary to the discussion and negotiations with the functional manager.

The functional manager in turn would review and establish the status of resource assignments, including schedules and other information to be considered in the discussions and negotiations with the project manager. The success of the negotiations, which is the essence of the effectiveness of the matrix organization, rests upon the flexibility of both negotiating parties, and the acceptance of the precept that the parent organization must be a priority consideration over any single project.

In entering into discussion with the functional manager, it is incumbent upon the project manager to prepare adequately so as to sell the plan of action considered necessary for achieving the revised project objectives. However, fallback positions representing reasonable compromises in schedule, cost, and even performance parameters must be prepared, to assure that the project will not be compromised to a degree that might prove untenable.

The functional manager is concerned with the distribution of resources not only for the project under discussion, but for all other company efforts that come under the matrix organization. If, for example, the project experienced problems requiring supplemental engineering hours, the project manager would prepare the case covering the following major points:

1. Type of engineers required
2. Number of man-hours necessary
3. Period of time for their utilization

In response, the functional manager would include the following for discussion:

1. Availability of required personnel
2. Costs involved
3. Periods of availability
4. Alternative proposals to be considered

The functional manager attempts to match specific resource requirements with their availability for the periods identified. If the match can be made, the plans, personnel assignments, cost allocations, and other adjustments of records of both managers would be implemented. However, when a match cannot be accommodated, adjustments in project objectives and/or resource loading plans must be worked out. In essence, the objective of the discussion and negotiation of the two managers would be to minimize the impact on their respective areas of responsibility.

In converging on a compromise, the project manager considers points like the following:

1. Based on the periods of availability, to what degree will the schedule, cost, or performance objectives be affected?
2. To what degree will the compromise affect the project contract objectives and the customer's acceptance?
3. What alternatives to the compromise can be considered?

The functional manager would be responsible for considering the following:

1. Establishment of the impact on the scheduling and man-hour loading of resource allocations.
2. Degree to which other projects may be affected by the revisions to the existing resource allocations.
3. Establishment of a priority judgment regarding the relative importance to the parent organization of the various projects affected by the reallocation of resources required to resolve the problems under consideration.
4. Evaluation of possible alternative solutions involving effort external to the matrix or parent organization.

The various possible solutions to the difficulties confronting a particular project would require objective discussions and consideration not only by the project and functional managers, but also by the different discipline supervisors of the project team. Further, when problems of a particular project are of such magnitude that the solution requires the diversion of resources from other projects, the discussions would require the participation of representatives of other projects that might be affected.

Discussions and compromises will normally result in a course of action acceptable to all parties. However, if compromise to a controlled conflict cannot be achieved, arbitration by the matrix executive becomes necessary. The failure of the participants in a conflict to reach acceptable compromise results in the following undesirable situations: (1) The matrix executive is drawn into the problem and is obliged to assume the responsibility for selling a compromise to which the conflicting parties voice acceptance; but tacit disagreement by one or more of the parties may jeopardize the team concept upon which the matrix is based. (2) At least one of the parties to the conflict may lack the personal trait of objectivity essential for functioning in the matrix.

Once the conflict resolution is achieved, the project manager must revise the work packages to reflect those changes brought about by the compromise. The functional manager is responsible for revising the schedules for resource utilization, man-power loading, and assignments and other revisions.

PROJECT MANAGER QUALITIES

To successfully perform in the environment of the matrix organization, the project manager must develop and exercise many attitudes and behavior modes that, though desirable, may not be essential to managing projects in a traditional line organization. The major points previously discussed will now be highlighted separately for emphasis. Any project manager or other official in the matrix who is deficient in any of the following could detrimentally affect the successful functioning of the matrix operation.

1. *Engineering competence.* All project managers must be technically competent. However, in the matrix the manager should also possess the technical imagination or creativity and judgment needed to adopt alternate approaches that will achieve the permanent objectives with minimum impact on project costs and schedule.

2. *Flexibility.* In a line organization, the project manager must possess sufficient flexibility to be able to effectively respond to dictates imposed by higher authority. The matrix project manager must possess the additional flexibility to participate in compromises of his assigned project objectives and the revisions of plans of action deemed necessary.

3. *Objectivity.* The project manager becomes party to a conflict when circumstances threaten the achievement of his project objectives and solutions are not available from the project resources. In many situations, solutions may exist within the resources of the parent organization, but any shifting will adversely affect other projects. In such circumstances, the project manager must be able to view all projects from the corporate point of view, and accept the conclusion that the interests of the parent organization will be best served by compromise in his assigned project.

4. *Candor.* The matrix project manager must suppress any inclination to be less than candid where the welfare of his project is concerned. Candor is related to objectivity. The effectiveness of the matrix depends to a large degree upon the accuracy and validity of information relating to the status, resource requirements, and other aspects of the project.

5. *Communication abilities.* One of the major functions of the project manager is to establish and maintain open communication channels with the functional manager, the discipline supervisors, the matrix executive, and other parties inside and outside the parent organization who have an actual or potential interest in the project. The project manager must effectively communicate with the members of the discipline supervisor teams, since they are responsible to the functional manager as well as the project manager.

6. *Managerial abilities.* All successful managers must possess the skills of organizing, motivating, directing, etc. Because the matrix project manager functions in an environment demanding cooperation and team-effort, he must show a high degree of sensitivity to other members in the matrix and parent

organization; a thorough understanding of the matrix organization and the various mechanisms necessary for its function; a good talent for coordinating the efforts of subordinates; and the ability to work effectively with all levels of the matrix and parent organizations.

In summary, the matrix represents a major step in organizing working groups of individuals whose personal loyalty to their assigned task or project is secondary only to their concern for the welfare of the parent organization.

6. Matrix Management in International Projects

H. A. Williams*

The first international project of which there is any written historical record was completed in six days; the Project Manager rested on the seventh. Like other, more contemporary project managers, this Manager achieved a specific objective (the creation of the Earth) within the context of a defined starting point (a void) and ending point (a planet inhabited by carbon-based forms). It is possible that this project was only one phase of a large program—the evolution of the universe.

A project is a specific undertaking with a clearly defined beginning and end; it exists to achieve a concrete objective. In contrast, a program is a continuously evolving approach to achieving a broad goal; it is ongoing and its point of completion is indefinite. The first lunar landing, for example, was a *project* that achieved specific objectives of one phase of the NASA space exploration *program*.

Understanding what a classic project organization is and how it works is important in any discussion of management tools and methods, because its singleness of purpose makes it attractive for many applications. To be sure, this also makes it susceptible to disruption by the imposition of management systems that, though meant to improve performance, actually impede it.

THE CLASSIC PROJECT ORGANIZATION

Creating a project organization separate from the parent company, while maintaining the basic characteristics of that company, is a management method that has evolved in the construction industry to handle operations physically remote from the parent company. Figure 6-1, which shows a typical project management organization in the construction industry today, illustrates

---

*H. A. Williams has spent over thirty years in management positions, working in and with projects of all types, including factory-oriented production projects on the B-47 and B-52 programs for the Boeing Company, the NASA and Air Force Titan, Gemini, and classified nuclear space projects for the Martin Marietta Corporation. For the last twelve years he has participated in various aspects of the Westinghouse nuclear power programs, dealing with both domestic and international power plant construction projects.

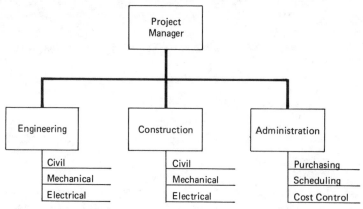

Fig. 6-1 Construction project organizaion.

how this simple project organization functions. This project organization is an entity within itself, needing little or no organizational support to carry out its basic functions. If we substitute the word "manufacturing" for "construction" and add a sales function, this organization chart would resemble that for any medium-sized industrial company.

Classically, a simple project management organization will have one manager in charge, making decisions and taking almost total responsibility for the project. Organizations of this sort became very popular with military procurement agencies in the 1960s because of the efficiency of dealing with one person in the contractor's organization responsible for any operational or functional problem. This method has also been utilized extensively by NASA in the space program and is now a key management tool in many organizations outside the construction and aerospace industries.

THE CLASSIC MATRIX ORGANIZATION

In all probability the first matrix project management organization came about when the owner of a small construction company sent a project manager off with a contract and a bankroll to complete a job. As an afterthought, he sent along a cost-control clerk more loyal to the owner than to the proejct manager. This was also the first matrix management organizational problem.

Formal matrix management, as applied to a project, is a system whereby any given individual or discipline group within a project organization is responsible both to the project organization to complete the project within given project restraints (time, cost, quality, etc.), and also to a functional discipline-control group or individual for the manner in which the functional discipline is performed. While the degree of difficulty in implementing a simple in-house engineering project may vary widely from those involved in implementing a complex, multidiscipline, multifunction project halfway around the world, the methods used in organizing and structuring these two projects are no different. The process is to staff the project with the talent necessary to complete the job, make sure that everyone understands the objective of the project, and furnish necessary tools, materials, and other resources (normally data and logistic support) to accomplish this objective.

Since tools and materials and other resources are not normally *all* supplied at the beginning of the project, a support function that is not a part of the project operates to supply the project as needed with the necessary implementing capability, whether it be design drawings, wrenches, or legal expertise. Unless this implementing capability is under the full control of the project director, an informal matrix organization is established even without a formal matrix chart and defined reporting lines. This process is the same for the simple in-house project and the complex international project. The differences are those of degree rather than of kind. It should be noted, however, that frequently the complexity of implementing an international project tends to increase the degree of difficulty in geometric progression.

The earliest discussion of matrix management that I have found is a June 1954 article published in *Aviation Age,* "New Management Approach at Martin," by William Bergen of the Glen L. Martin Company, now Martin Marietta Corporation. Bergen does not actually employ the terms "matrix" or "matrix management," but speaks of "systems management" in much the way that today's writers speak of matrix management. The article includes a typical project organization chart that looks surprisingly like the classic matrix organization of today.

It is interesting to note that most writers depict the project matrix as it appears in Figure 6-2, with the functional disciplines displayed across the top of the chart and the project arranged vertically on the side. This is the normal progression that results from imposing a project matrix on a classic functional organization, and as a matter of fact is the form depicted in Bergen's 1954 article. On the other hand, if you ask a project manager in the construction industry to depict in chart form how his projects are supported by the functional organizations in the company, he will draw the same chart, but with the projects listed across the top and the functions depicted vertically. While there may be some Freudian overtones in the choice of which organization to designate in the top position, of more importance is how this revision of the classic organization chart reflects the actual functioning of a matrix project organization. This helps us to understand what makes it work and what makes it fail.

All projects matrixed with a functional organization are either project-driven or functionally driven, as graphically presented on the organization charts shown in Figures 6-3 and 6-4. Many factors determine whether a matrix project is functionally or project-driven. Probably the most common factor is

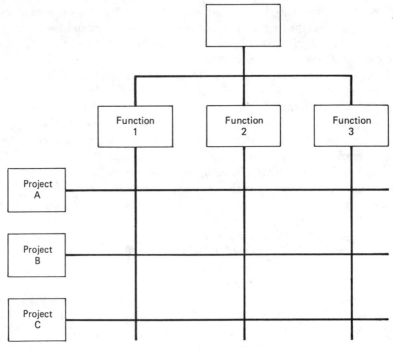

Fig. 6-2 Classic matrix organization.

the relative strengths, weaknesses, and capabilities of the individuals involved, though this should be the factor with the least weight. If company management, in initially setting up a matrixed project, understands the functions and dynamics of the project components and has realistic objectives, it will establish the project initially as either functionally or project-driven and clearly identify it as such. The following items identify the most important factors determining the organizational drive behind a project:

- *Reporting line* (or "Who owns the bodies?"). When a functionally driven organization is decided upon, the project personnel doing the work should report to and be managed directly by the functional manager; when a project-driven organization is decided upon, the project personnel doing the work should report to and be managed by the project manager.
- *Work location.* Grouping all project personnel in one location apart from individual functions develops a cohesiveness and ease of communication within the project that tends to promote a project-driven operation. Conversely, physically locating each functional group of a project within the environs of its own functional department promotes a functionally driven project and makes the project-driven mode difficult to achieve.

- *Geography.* Locating the total project work force halfway around the world from the functional staffs is folly, if the object is to have a strong functionally driven project. This is a corollary of the work location factor but is much more determinative, particularly in international projects.
- *Management desires.* In setting up a matrixed project organization, management should understand whether a functionally driven or project-driven organization will achieve the desired results and should structure the organization accordingly, since if outside forces or a given set of circumstances dictate a number of factors that tend in one direction, management's attempt to produce the opposite result will cause chaos. Conversely, if management structures the organization in line with these forces, the desired results can be achieved almost automatically.
- *Personnel capabilities.* If all other factors are equal, which they never are, or tend to cancel one another out, which they sometimes do, the capability and personality of a few key individuals in the functional and the project groups may well determine whether the project is functionally or project-driven.
- *Objective.* The project objective will most often be project-related (such as, classically, completing the project), yet interim objectives (such as a

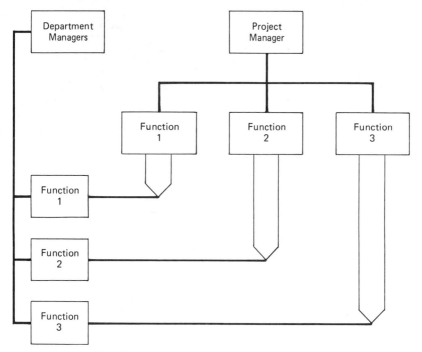

Fig. 6-3 Project-driven matrix project organization.

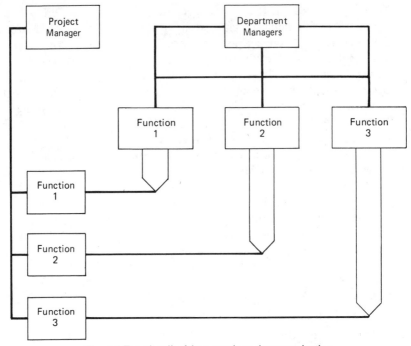

Fig. 6-4 Functionally driven matrix project organization.

cost reduction) may at times override the project objective to the point where the original objective becomes secondary. When this happens, the project drive can change from operational to functional. Too, an objective exclusively the province of a functional department may create a functional drive from the beginning of the project, which occurs most commonly in research projects.

Most of the literature on matrix management would lead one to believe that, to obtain all the advantages of improved communication, motivation, and synergism, it is necessary to achieve what I would describe as a balanced project, or one where the influences of the project drive and the functional drive are in equilibrium. One way of depicting this balance organizationally is shown in Figure 6-5, which graphically represents the difficulty inherent in achieving such balance in an operating project organization.

Since a project is time-limited, the ability to develop an integrated operating balance between functional and operational individuals and organizations is also limited. Seldom is there the luxury of allowing attrition, coercion, training, and management to change the sharp conflicts of organizational and personal interests into the smooth coordination required for a balanced project. Other more permanent organizations can aspire to this objective, but a project, by its nature, must place more emphasis on short-term objectives. This means that a matrix project structure is inherently flawed, being an interim organization designed to lead to the ideal, but with a life so short that the ideal is not achievable.

A way round this dilemma is to properly mix the project objectives, the project structure, and the project environment so that the project dynamics promote rather than hinder the accomplishment of the objectives. Take as an example a small research project to design a sophisticated manufacturing tool. The project manager as well as the dozen or so engineers required to complete the project are expected to spend about half their time on the project during the six months required for completion. Engineering comes from three separate functional disciplines; some of the work can be done in all three disciplines in parallel, but much of it must be done in series.

This is an obvious example of a project that should be set up on a functionally driven basis. As long as the organizational structure is clear to everyone from the outset and the interrelations of the functional and operational entities are apparent, the organization should work smoothly. The managers of the three engineering disciplines should assure the necessary production from the

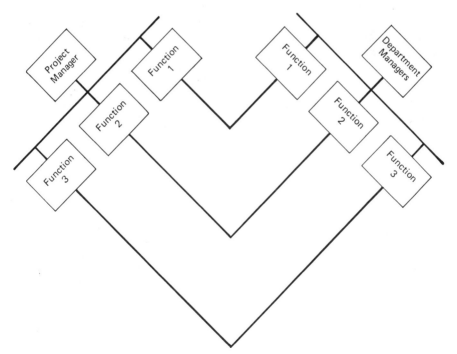

Fig. 6-5 Balanced matrix project organization.

individual groups, while concurrently establishing priorities with other work to keep the engineers productively active for the 50% of the time they are not working on the project. The part-time nature of each individual's assignment to the project and its relatively short time span preclude establishing a separate project office area, and consequently each individual will remain in the location to which he has previously been assigned. The project manager must manage his own time and priorities for the project and for his other work while coordinating, scheduling, and following the flow of the work among the engineering groups.

Now add another element to the situation: the manufacture of the experimental tool. The total project plan, now including delivery of the tool, requires that the design be completed on schedule to allow six months' fabrication time. Material availability accommodates to this schedule, which is critical to the overall success of the project. Because of the tight fabrication schedule, the project manager has recommended (and management has approved) the assignment of a manufacturing manager to the project earlier than normal during the design phase. This manager's function is to review the preliminary design for feasibility of production; to plan the manufacturing effort; and to coordinate operations between engineering and manufacturing to achieve an optimum engineering release schedule. Because the tool is experimental, about half the original engineering force required to design the tool will stay on the project during fabrication to handle design changes necessary to improve the manufacturing process. During fabrication a schedule specialist and a cost-control engineer will be assigned to the project full time. Considering these additional circumstances, it becomes less evident that the project is or should be functionally driven.

There is increased complexity in that the manufacturing department and two administrative functions have been added. Moreover, the objective of the project has broadened from that which is primarily the province of the engineering department to that which includes the manufacturing department. Because of the additional personnel, the increased time required to complete the project, and the greater complexity of the project, it would be prudent to consider establishing a project office to include all project personnel in one location for ease of communication and coordination. If a decision is made in favor of establishing a project office, the operation will, with the broadened objective and increased complexity of the project, tend to become project-driven. The failure of company management (and the management of the functional departments) to recognize, accept, and promote the project as project-driven rather than functionally driven could well result in another failure attributed to the difficulty of matrix management. On the other hand, if all parties recognize and accept the nature of the operation as project-driven, give the project manager direct control over the utilization of project personnel, and assure support of functional disciplines (rather than trying to provide overall project direction), the operation has at least a reasonable chance of succeeding.

AD HOC MATRIX ORGANIZATIONS

Using the simple project organization as a structure for a task force or an ad hoc committee is ideal, since it collects the services of varied disciplines under one leader. The very act of organizing a project task force requires setting its direction and objectives, and establishing lines of accountability. Since the life of a task force is by definition limited, and since participation is usually part-time, all members maintain strong lines of communication and ties of loyalty to their home organizations. Consequently, a project task force is automatically a matrix management organization, even without a formally structured matrix reporting system.

The project as a management tool has evolved a little differently in Europe than in the United States. The European construction industry does not include broad-based architect-engineering and construction companies, which in the States undertake a complete project and handle the project's systems design, architecture, detail design, procurement, construction, erection, and start-up. In Europe there are, of course, many competent engineering firms and construction companies capble of handling such work, but the responsibility for integrating all aspects of a project is normally left to the ultimate owner, to a major equipment supplier, or to an ad hoc joint venture made up of two or more contributing companies. (This business structure, however, seems to be changing, probably as a result of the influence of large-scale projects in the Middle East. Many European engineering firms are broadening their scope of operations to include the other aspects of a complete project. This is particularly true in the design and construction of chemical and petrochemical plants.) Particularly in the electrical equipment business, major suppliers such as Alsthom, Siemens, Brown Bovari and ACEC all have experienced project organizations. These groups vary in organizational make-up but are all structured to support their manufacturing divisions. These groups propose, negotiate, and implement projects, utilizing equipment supplied by their manufacturing divisions. They are typically rather small organizations, with each participating individual chosen for his capability and talent while bringing with him strong ties to his parent organization. These organizations draw on their manufacturing operations for manpower during the proposal and implementation phases of a project. The result is an ad hoc matrix organization, formed in much the same manner as a special-purpose task force, and with much the same result.

In a slightly different manner the large multinational and multiproduct Japanese firms such as Mitsubishi and Hitachi also operate with ad hoc matrix organizations. Rather than maintain a core project organization, a lead manufacturing division (or company) usually manages a given project through the proposal, negotiation, and implementation stages, calling on other divisions for products, services, and support as required. The very nature of the Japanese business culture assures the same degree of coordination, communication, and cross-fertilization of ideas that one would expect in a formalized matrix project reporting system. An idiom frequently heard in Japanese business is *nemawashi,* meaning literally to "dig around the roots"—that is, to dig carefully around the roots of a tree before transplanting it. It is used in Japanese business as an admonition to be sure that everyone involved is informed and concurs before any significant action is taken. Thus the culture, in and of itself, has accomplished one of the major objectives of a formal matrix management organization: to make informed, intelligent decisions by capitalizing on the expertise of internal resources.

MATRIX MANAGEMENT IN NONPROJECT OPERATIONS

This discussion so far has dealt exclusively with matrix management in terms of project organization. For contrast, it is revealing to look briefly at how matrix management is utilized in other, more permanent organizational structures. Figures 6-6 and 6-7 show two field sales organizations, each supporting a number of product lines to be marketed in a particular geographical area. Figure 6-6 shows a product-oriented organization with marketing organized geographically by product line. Figure 6-7 shows the same organization, but with the product-line marketing organization matrixed to the sales organization. This classic matrix organization (note how closely it resembles the project

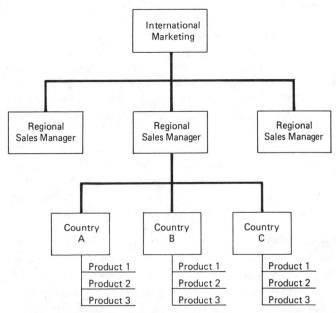

Fig. 6-6 Product line field sales organization.

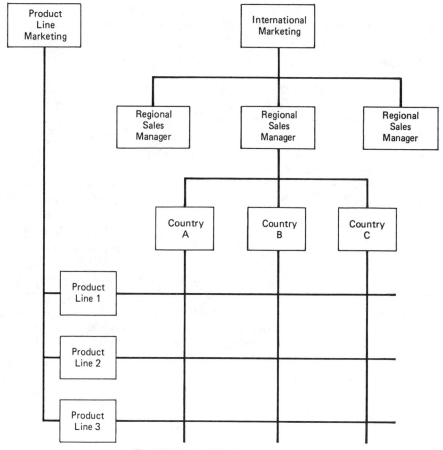

Fig. 6-7 Matrix field sales organization.

matrix chart of Figure 6-2) has a limited number of product lines and a limited number of sales offices.

If the organization actually included a large number of product lines (for purposes of illustration assume 100), it is clear that this organization could become cumbersome with 300 matrix interfaces in one marketing region alone. This potential for excessive numbers of product/function matrix interfaces is a problem for relatively permanent matrixed organizations. The solution is of course to formalize the matrix reporting relationship at a higher level in the two organizations being matrixed. Unfortunately, no one has yet developed a method for determining just the right level at which the matrix should be formed in any given organization. In the simpler project organization, a fixed matrix reporting interface is not normally a problem, since project organizations rarely have more than three levels of management, and usually contain only a limited number of functions that would be practical to matrix.

CONCLUSIONS

In establishing an organization to complete a project in the international arena, more than usual care should be taken in setting up the proper support and reporting relationships. The difficulties of communication, if only resulting from time-zone differentials, make it essential that whatever communication capability exists be utilized to get work done and not to resolve intramural organizational problems. If any support is required from the functional disciplines in the home office, it is important to work out in advance the type and nature of the support required and the reporting relationships of all individuals involved.

A careful analysis of the structure of the project should be made in the beginning to determine what factors constitute the organizational drive behind the project, whether it is a simple in-house project or a complex international project. Periodic reviews of the initial analysis should be made to determine whether or not these factors have changed, and if so whether or not the changes suggest restructuring the project's management.

In the final analysis, organizing work in terms of projects is the best way to accomplish many tasks and the only practical way to accomplish some. It is, by its natural structure, a method of matrix management. If this is recognized and accepted by management, by the functional support groups, and by the project personnel, work can proceed successfully and smoothly, whether functionally or operationally driven.

7. Matrix Management in a Transnational Mode

John F. Ricketts*

This chapter describes the Westinghouse International matrix organization, makes some observations based upon experience in working in an international matrix, and examines some of the accommodations made to support it.

THE WESTINGHOUSE MATRIX ORGANIZATION

The purpose of the matrix structure in an international business is, basically, to reconcile the following needs:

1. The different demands of each country;
2. The disciplines of worldwide product planning—whether and how to serve each market, and where to manufacture; and
3. The efficiencies of some centralized staff functions

Westinghouse formed an international matrix organization that has been operating and evolving for several years. In early 1979 Westinghouse established a major task force to study and make recommendations on the most effective approach to the international marketplace. The task force arrived at a nucleus of fundamental conclusions about international business that formed the basis for its recommendations:

- Real economic growth will be much higher in the developing countries than in the United States and other developed countries.
- Companies around the world are abandoning the concept of a "home market" and are thinking, planning, and acting on a world scale.

---

*John F. Ricketts has worked in an international matrix management organization as a general manager with responsibility for one side of an international matrix interface. He has also had an opportunity as Executive Assistant to the President, International, of Westinghouse Electric Corporation, to observe the development of the global matrix structure initiated in 1979. Mr.Ricketts earned a B.A. in Business Administration and is a member of the Canadian Institute of Chartered Accountants. He has attended the Harvard Advanced Management Program.

- Technology is a key to penetrating world markets.
- Success in the international marketplace will depend largely on the ability to dovetail country-oriented planning and operations with the product-oriented planning and operations.
- The individaul country with its unique cultural, political, and economic characteristics is the main strategic element in the organization of a successful transnational corporation.

As a result, the task force recommended and Westinghouse concluded that a matrix structure would fit best. The matrix suited the Westinghouse concept that the country was the basic building block in the global marketplace. It permitted a flexible approach to each country based on the characteristics of the country's economy, culture, and politics, as well as the nature and extent of the existing Westinghouse presence.

With a matrix organization, Westinghouse could build on existing experience and the strength in its strategic business units. The Westinghouse International matrix organization (see Figure 7-1) is multidimensional, consisting of:

1. Country managers
2. Strategic business-unit managers
3. Centralized functional department managers

The distinctive contribution to the Westinghouse organization design is the addition of the country manager structure. The new organization was positioned to promote a balance of power so as to facilitate the working of the

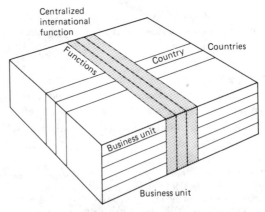

Fig. 7-1. The balanced matrix

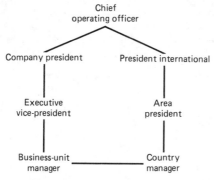

Fig. 7-2. The International matrix

matrix, with decentralized functional department managers reporting to corporate senior officers as well as into the International Organization. Some examples are: Treasury, Law, Controller (see Figure 7-2).

The new structure required a change in the management culture of the corporation. The driving forces in support of that change were

1. A recognition of the quickly evolving global marketplace
2. Opportunities for growth
3. Increased competitive threat

From the outset, all the players had to recognize the time needed for a cultural change to take place, typically some three to five years. Moreover, Westinghouse would have to modify and develop systems and procedures to support the new organization. In looking at how those changes came about, let us first examine the matrix design.

DEFINITION OF ROLES

Early on in the reorganization process, Westinghouse developed a "responsibility chart" (see Figure 7-3). The chart describes lead and shared responsibilities on each side of the matrix. All international decisions made wtihin the matrix management system, however, would be joint decisions. For that reason, effective communication would be essential to the success of the matrix.

The complexity of this Westinghouse matrix design can be better appreciated when one realizes that there are 14 country management structures and over 35 business units. There are varying levels of involvement on both sides. Some country managers have major operating subsidiaries reporting to them in their country. Other country managers have little or no involvement in oper-

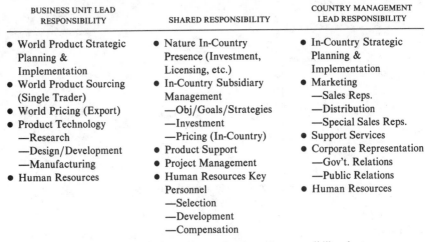

BUSINESS UNIT LEAD RESPONSIBILITY	SHARED RESPONSIBILITY	COUNTRY MANAGEMENT LEAD RESPONSIBILITY
• World Product Strategic Planning & Implementation • World Product Sourcing (Single Trader) • World Pricing (Export) • Product Technology —Research —Design/Development —Manufacturing • Human Resources	• Nature In-Country Presence (Investment, Licensing, etc.) • In-Country Subsidiary Management —Obj/Goals/Strategies —Investment —Pricing (In-Country) • Product Support • Project Management • Human Resources Key Personnel —Selection —Development —Compensation	• In-Country Strategic Planning & Implementation • Marketing —Sales Reps. —Distribution —Special Sales Reps. • Support Services • Corporate Representation —Gov't. Relations —Public Relations • Human Resources

Fig. 7-3. The Westinghouse International matrix responsibility chart.

ating subsidiaries, but play a major commercial and administrative role. On the other side of the matrix, some business units are well established internationally, others have little or no international involvement.

In using this form of matrix organization, Westinghouse was attempting to benefit from the potential for synergy from its existing experience, as well as to encourage more international awareness by those business units lacking broad international experience. The catalyst for that synergy and awareness would be the country management structure.

THE ACTORS

The Westinghouse International matrix involves cultural and behavioral changes on the part of three types of managers:

1. The business-unit manager with international involvement, who must share the experience of his earlier international initiatives and bear some new costs for his share of the country management structure. He must do so with initially limited perceived benefits from the country management structure for his busines unit.
2. The business-unit manager who has limited involvement. He perceives a new country management organization promoting international involvement for his business unit. He sees his business as having little international experience, product suitability problems for the international market, and too many unaddressed domestic market challenges.

3. The country manager, who is newly appointed to his position and rapidly assimilating the culture and commercial practices of his country. The country manager has a background that is rooted in one of the three Westinghouse operating companies. He has launched a program to
   - develop his knowledge and contacts in the business units of the other operating companies.

The challenges involved in operating in this matrix are formidable, and they are increased by the geographic separation. Absence does not always make the heart grow fonder. In the seventeenth century the Duc de La Rochefoucauld wrote, "Absence diminishes commonplace passions and increases great ones, as the wind extinghishes candles and kindles fire." Whether the passions are positive or negative, geographical distance serves to magnify them, while commonplace matters tend to be overlooked. The geographical disbursement intensifies the need for support systems to help to bridge the gulf.

What are the support systems that have been or, in some cases, might be put in place to help this matrix structure achieve its objectives?

PLANNING

In the initial period of the new matrix organization, there was a discontinuity in planning process beween the business units and the countries. As a result, Westinghouse found itself with two major global planning problems:

- Uneven participation of the international organization in the business-unit planning process.
- Inadequate follow-through from the opportunities foreseen in the country planning to the stragegies and actions developed in the business unit plans.

These problems in particular helped to undercut the objective of increasing the awareness and the involvement of those business units which had limited international market participation.

Changes in the planned allocation of business-unit resources to the international market, which sometimes occurred in the interval between country planning and business-unit planning, were frustrating to both sides of the matrix. While that situation was understandable, in many cases it did not contribute to the smooth functioning of the matrix.

To overcome this problem, Westinghouse developed a new comprehensive planning system to integrate the planning of the country and the business unit. The new planning process allocates responsibility for specific tasks, support, and interaction. This increases the cooperative interaction and eliminates a major source of frustration in the matrix.

PRODUCTIVITY IN THE PLANNING PROCESS

The new planning system also better defines responsibilities in planning, and thus disciplines the overall planning process. Formerly there were "great drives" to generate planning information, initiated from both sides of the matrix. This surge in international planning activity was caused by the development of country plans and the increased emphasis of international planning by the business units. In this period there was competition for attention and involvement in the many planning processes. At times this planning activity tended to overload the matrix circuits. In some cases there were not sufficient resources, people, or time available to do justice to the planning need.

When this happened, the business-unit planning need was not well served and/or the country manager was not able to make an optimum contribution. The new integrated planning process defines the scope of planning information and orders its collection. As a result, the business unit now receives the concentrated efforts of the international organization. These efforts are based not only on the market environment, but also on consideration of the potential and the limitations of each business unit. The result will be a significant improvement in international strategic planning and greater appreciation of the value added by the country manager and the international organization. The improved coordination in the new planning system will contribute to the functioning of the matrix.

ACCOUNTING

The management and reward system of the corporation involves performance measurement on specific financial goals. This is part of the culture of the business unit. Westinghouse developed accounting visibility by country, so as to permit performance measurement on specific financial goals for the country manager. The business-unit manager's appreciation that the country manager is operating under a similar financial discipline contributes to better understanding on boths sides of the matrix.

The increased cost of the new international country management organization was absorbed, initially, as a strategic expense at the corporate level. Such an expedient allowed the international organization to "earn its stripes," so to speak—to demonstrate value added before costs were allocated to the business units operating in each country. But that process is now drawing to a close. In the future, an accounting determination will be made for each country, to divide the international organization costs between operational support and strategic contribution. The operational supports costs will then be allocated to the benefiting business units, while the costs of the strategic contribution will continue to be absorbed at the corporate level. One of the purposes of the country management structure is to act as a catalyst for businesses that are not participating internationally up to their potential. Westinghouse established this treatment of the costs of the international organization to assist the necessary cultural change in support of the matrix.

One inhibition to doing business in the international marketplace has been the "double mark-up" effect of a domestic division and an in-country distribution arm. The resulting price may not be competitive, in which event the corporation loses a sale. This situation has inhibited some market penetration where Westinghouse does have a presence. A system of dual credit for the sale is being pioneered, to permit both parties to benefit from the sale. This accounting system should open up many new international opportunities. Conflict between the domestic and the international arm has been changed to cooperative promotion of new business opportunities.

COMMUNICATION

In a geographically dispersed matrix, travel, differences in time zones, differences in the weekends for moslem countries, and holiday differences all compound the problems of establishing effective communication. The introduction of an electronic mail system has greatly improved the communications within the international organization and between the international organization and the business units. Access is so easy that it has overcome any tendency not to communicate because of the difficulty involved.

It is my experience that, in the early stages of a new matrix relationship, it is important to over communicate to build trust. This is not the same as being there, but there are no more excuses for not communicating, at least electronically.

If the matrix does not function effectively below the business-unit manager/country manager level, a bottleneck can quickly develop. The business-unit manager/country manager relationship is the prime interface in the matrix, but it is not the only one that matters. The potential for a bottleneck further down the organization chart is a critical consideration in a multidimensional and geographically dispersed matrix. Every opportunity should be taken to improve the knowledge, understanding, and experience of key people on both sides. No one system will cure this problem. A combination of exchanges of people for assignments, exchange visits, seminars, courses, cultural awareness programs, written communication, video tapes, sensitivity to cultural mores and language needs, and clearly articulated corporate goals all contribute to better interaction.

In my own experience, the strength of the interaction among the people below the official matrix interface is a measure of the strength of the matrix itself. I have seen people working in the best interest of the corporation as a whole and not solely in their own departmental or provincial interests, and I would certainly not want to be without that potent source of ideas and action.

CONFLICT RESOLUTION

The matrix organization is designed in such a way that unresolved conflict can be negotiated at successively higher levels until it reaches a common superior (see Figure 7-2). Yet in practice, unresolved problems do not flow steadily up the dual chain of command. This is so for at least two reasons:

1. People at all levels appreciate that unless the problem is a policy issue that has not been previously addressed, it should be negotiated at the primary matrix interface;
2. A steady flow of unresolved problems would indicate a management failure at the primary matrix level.

Processes for conflict resolution are more important in a matrix organization, because the organization cannot tolerate unilateral decisions. The slow-to-change cultural and behavioral patterns in an evolving matrix organization will undoubtedly lead to testing the matrix process. This testing process is particularly evident where there are real or perceived "turf" issues involved. Managers in different environments will inevitably have legitimate and strongly held differences of opinion. In such cases an objective resoution of the difference may not be possible without the use of a facilitator to help both sides resolve the issues. Such a process should work toward an objective examination of the issues in a manner consistent with the best interests of the corporation— not the parochial interests of either side of the matrix. From a parochial perspective there will always be winners and losers, but the matrix helps to ensure that the corporation wins consistently. Westinghouse and others are developing a facilitated process to anticipate, address, and resolve issues in the international matrix.

Mentor Relationship. In a geographically disbursed and many-faceted matrix organization, it might be helpful to supplement the informal mentor relationships with formal ones. Suppose a country manager has personal ties mainly to one area of the corporation, because that is where his "roots" are. It might help the functioning of the matrix if a formal "buddy" system were established in each of the other operating areas. This relationship among peers might be better named than "mentor" or "buddy," but the purpose is exactly that. To look out for, offer guidance to, and develop a special rapport with each other. The country manager would benefit from such an arrangement in many ways. There would be someone in the domestic operations sensitive to situations that could create opportunities or problems for him. Such a pipeline into those areas of the corporation where he has the least contact could be invaluable in the initial period of his country management responsibility. Such a relationship would be a two-way street and could contribute much to the evolving cultural change.

CAREER PLANNING

As the matrix matures, there will be an exchange of experience through an exchange of people from one side of the matrix to the other. The empathy resulting from such a change will do much to alter the culture. Such moves, then, are a necessary ingredient in the evoution of the matrix. At Westinghouse that aspect has been recognized in the human resource planning. Both sides of the matrix have a vested interest in seeing that this planning is implemented.

In summary, while the literature on the subject tells us that a matrix is a most difficult organization form to implement, I believe that, if the conditions for a matrix do exist and one takes the trouble to make it work, it can yield superior operating results and superior management.

A matrix organization takes time to mature. It must be supported with systems and procedures that recognize its special needs. Those needs in a multi-dimensional and geographically distributed matrix are not all that different from those of a domestic matrix. The needs, however, are magnified many times by the lack of opportunity for day-to-day nuturing and care. The support systems must address these needs or the matrix will stop evolving and begin unraveling.

8. Matrix Management: A Quality Circle Model

Edmund C. Mechler, Jr.*

INTRODUCTION

A Quality Circle Program is a method of increasing organizational efficiency and effectiveness. Organizational efficiency is measured in terms of how the resources—human, financial, equipment, buildings, etc.—are used in achieving the mission, goals, and objectives of an organization. Organizational effectiveness measures the quantitative and qualitative success of resources used in achieving the organization's and the society's growth. Organizational efficiency and effectiveness must substantially increase in the 1980s if society is to continue to advance. These two variables, efficiency and effectiveness, may be the most important factors studied in the last 20 years of the twentieth century. The Quality Circle Program can be viewed as the next step in an evolutionary process from Taylor's scientific management through Hawthorne's Experiments to behavioral science's influence, always attempting to increase organizational efficiency and effectiveness. Quality Circles, if implemented as a function of the organization and its constraints, increases efficiency by eliminating those processes within the organization that increase costs. Additionally, the increasing complexity of society's activities and business's endeavors requires that more people become involved in charting and implementing the organization's growth path successfully. Quality Circles is a popular method of promoting organizational growth. But what does this mean to the two dominant human components within an organization, management and workers, where management is defined as those who accomplish work through others, and workers as those who accomplish the work?

An organization's management can use a Quality Circle Program to release hidden potentials within all individuals and guide these hidden potentials toward the organizational goals. A common manager's comment is, "We must find a simple way of getting the employees more involved in their work." If this is true, why are current entertainment and relaxation activities so complex,

---

*Edmund C. Mechler received a B.S. in mathematics and an M.S. in engineering management from the University of Pittsburgh. He has also received training in management techniques, adult education, interpersonal skills, and group dynamics. He has worked at M.I.T. (Apollo Program), Westinghouse Electric Corporation (Process Control and Human Resources), and Brown Boveri Control Systems, Inc. (Manager, Training and Organizational Development).

creative, and consuming? To relax in modern terms takes work, and to be entertained takes even more work. But this type of work is constantly done: observe a person making arrangements for a weekend outing. Maybe it is not a matter of finding simple ways, but of allowing the employees to become involved in their work and the business. Providing a minor verification to this hypothesis are television interviews in which workers say that they need to be involved and to use their creativeness.[1] If modern managers change their thinking from finding ways to involve employees to allowing their employees find their own ways, then the individual's hidden potential will be released to increase organizational efficiency and effectiveness. If we changed our hiring practices to hiring employees for their minds as well as their physical strengths, we could double a manager's potential for accomplishing work through others.

The workers within an organization can use the Quality Circle Program to release internal forces to create and to acquire control—so elusive—over one's own life. Comments like "I feel a high school graduate could do my job" from a professional,[2] or "I would like to know what this company does" from a blue-collar worker,[3] can be eliminated. Satisfaction in a job well done can also increase with Quality Circles, which results in a revitalized desire to work and to involved; certainly spending approximately 25% of one's life in misery is completely illogical. These conditions can only improve the results of an organization that further develop pride in the individual. Also, if a worker is bored in his or her present position because of industry's method of task separation, the Quality Circle Program opens the doors to new learning and a deeper understanding of the factors affecting the business and the individual's tasks. Working on the factors that affect a job increases the individual's sense of control over those factors that influence not only one's work life but one's personal life as well. Quality Circles offers the individual worker a more creative work life and increased involvement in the running of the company.

The involvement of both the dominant human components of an organization in increasing organizational efficiency and effectiveness is truly an evolutionary step, when compared to the Quality Circle predecessors. In effect, the Circle Program may be the only true efficiency program, because it involves both human factors. Also, the benefits to both managers and workers increases the potential return to the organization since the attitude of the people doing the tasks changes from putting up with the time spent at work to wanting to be there. The Quality Circle Program, in addition, gives the organization a method of growth by using all the brain power of the people involved. If these results intrigue you and you want a Quality Circle Program, you'll be joining the crowd, since many business and corporate leaders want the program as soon as possible. However, as the old saying goes, you don't get something for nothing: a Quality Circle Program has its costs, and they must be paid before results are obtained. In addition, many organizations have Human Resource Programs either existing or starting, which provokes questions concerning program compatibility, remembering that the prime objective of this type of program is to increase organizational efficiency and effectiveness. But before the costs are listed and program compatibility is examined, what is a Quality Circle?

QUALITY CIRCLES DEFINED

A Quality Circle is commonly defined as a small group of workers doing similar work who volunteer to meet for an hour each week, on company time, to identify and solve quality or work-related problems. But this definition is written predominantly for blue-collar workers, not the professional or management circle. This type of Circle orients itself toward increasing the efficiency of the organization. With this distinction made, the blue-collar Quality Circle requires expansion. But before one expands the blue-collar Circle, the Circle types require new names. The blue-collar worker Circle, at least in this article, will be termed a Section Circle, because this type of Circle is normally used within a work section and handles section problems. The professional or management Circle will be called a Company Circle, because it cuts across functional lines and handles company-oriented problems. Figure 8-1 shows the relationships of the Section and the Company Quality Circle.

Section Quality Circle

A Section Quality Circle is a small group; it should not exceed 12 members, including the leader. The dynamics of a small group is vastly different from that of a large group: participation, adjusting unruly behavior, and handling outside influences are easier, etc. This is because the interactions of a group

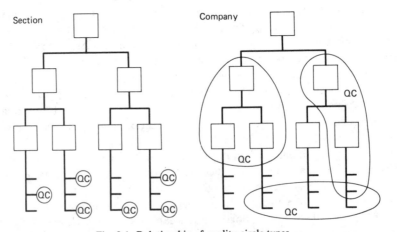

Fig. 8-1. Relationship of quality circle types.

increase exponentially as the number of members increases. To solve the efficiency problems of an organization, it is necessary to get the inputs of the people accomplishing the task not in a bitch session, where all the complaints are listed, but in a solving mode. Large groups have more of a tendency to turn into bitch sessions and therefore should be avoided.

All groups have three characteristics: structure, task, and dynamics. The structure of a Section Quality Circle is partially defined within the definition given: members should be doing similar work, should volunteer to join the Circle, and should meet for one hour a week on company time. In addition, the leader of a Section Circle should be the supervisor/manager of the section where the Circle is initiated—a condition that will at least improve communications within the section. Finally, when a problem is solved, the Circle presents the recommended solution to the management required to authorize its implementation, for it is still management's responsibility to run the company. So the structure dimension of a group is clearly defined within the Quality Circle Program.

The Quality Circle's task is also clearly defined: within a Section Circle the members will work on quality or work-related problems. The problems do not include union, financial, or personal problems. The organization has other structures to deal with these types of problems; these structures may not be efficient or even working, but if such problems enter the Circle meetings, Circle inefficiency must result. Again, another important group dimension is clearly defined within the Quality Circle Program.

The final group dimension, dynamics, should be constantly evaluated. The dynamics of a group can be defined as how a group accomplishes or fails to accomplish the task. How do the members work together? Do members group into subgroups? Do certain members get more "air time" and others very little? Who are the informal authority people within the group and how much do they effect the group? Is there too much or too little social time? These are but a few questions to constantly evaluate within a dimension that only now is being studied and adjusted properly within groups. A helpful question to ask is: "Are all the members participating somewhat equally? If the assumption is made that the person doing the job is the best person to make the tasks more efficient, and the Circle members are doing similar work, then the higher the participation, the better the solution will be. Of course the study of this dimension is still in its infancy and assumptions are not validated, but this dimension has the highest potential for organizational rewards.

Company Quality Circle

The Section Quality Circle is well defined as previously described; its requirements, to implement a successful program, are absolute. But before requirements are examined, it should be stated that another version of the Quality

Circle is developing for professionals and management. The Section Circle version can be used successfully with professionals and managers to increase efficiency, but greater returns can be realized within the Quality Circle Program working on the organizational effectiveness or business decisions.

The Section Circle definition need be changed only minimally to fit the Company Circle: a Company Quality Circle is a small group of workers from the needed specialties who volunteer to meet for a regular period of time to solve a designated problem. So defined, the small group basically functions like a Section Circle; membership should still be voluntary, except that the leader could be requested and appointed. The leader of a Section Circle, the supervisor/manager of the section that the Circle will be in, should be presented with the program and volunteer to receive the training, and only then decide whether or not to lead a Circle. On the other hand, the leader of a Company Circle should be presented with the task and then ask to volunteer. It is the author's personal opinion that all supervisors and managers should have Quality Circle training and experience. As a leader and member of many meetings, the supervisor/manager will receive training and experience that can save an organization enormous amounts of money and time by developing team-building expertise. While this requires more investigation, it is clear that the leaders of a Company Circle do not have the same constraints as a leader of a Section Circle: that is, they do not lead their own subordinates in a new personal interface along with all the other daily interactions. Rather, they are dealing with peers and the levels of management that they deal with daily. Therefore the voluntary characteristics for a Company Circle can be changed to fit the circumstances.

The members of the Section Circle differ from those of the Company Circle, because in both cases the members should be the people required to solve the problem in question. In the Section Circle the problem is predominantly the efficiency of the organization, which implies that the people doing the task or the people interfacing with them are those who will develop the best possible solution to the problem. In the Company Circle, however, the type of problem dealt with involves the organization's effectiveness and so requires a cross-section of specialities to review all aspects of the problem. Since the task of a Company Circle concerns organizational effectiveness, the management, especially top management, are the most likely to discover the problem first and in a timely manner, which is why the Company Circle's tasks are designated. Therefore the type of people within each circle should be those required by the nature of the task.

Finally, a difference between the meeting times of the two types of Circles is dictated by the task. Within the Company Circle, the organizational-effectiveness type of problem may require longer meeting times to clarify the data, and longer intervals between sessions to collect the data. Although the task type dictates the meeting times, it will be helpful to hold all circle meetings for one hour per week over a period of time, if the leader or members are new to the Circle Program, or if the Circle Program is new to the organization. Where the Circle Program is new to the participants or the organization, circle meetings require regularity to develop their importance to both the organization and the participants and to become habitual. After Circle regularity is established, in a Company Circle it is advisable to allow the meeting times to be set by the participants according to the task.

Section and Company Circles are similar in many areas, the major difference being the type of task. Table 8-1, adapted from Edmund J. Metz's article "The Verteam Circle,"[4] further defines the differences and similarities of blue-collar versus professional or management circles.

The requirements for a Circle Program are similar for both types of Circles. Therefore both types of Circles can be started at the same time if the requirements are satisfied, but initially it is advisable to start only one type in organizations that have not had exposure to the Circle experience, so as to build credibility and acceptance. Before listing the steps needed to implement the program, let us examine the Quality Circle Program's needs.

Table 8-1. Section and Company Circle Comparison.

DIFFERENCES		SIMILARITIES
SECTION	COMPANY	
• Homogeneous work group	• Heterogeneous work group	• Participation is voluntary
• Meets for one hour/week	• May meet for extended periods	• Receives training in basic circle problem-solving techniques
• Solves efficiency problems	• Solves effectiveness problems	• May implement solution
• Supervisor is leader	• Leader appointed	• Utilizes facilitator
• Decides problem to solve	• Problem to solve decided by top management	• Steering committee monitors
• Operates under regular organizational rules, procedures and protocols	• Has rules, procedural and protocol freedom	• Requires no change in organization structure
• Solves numerous problems	• Solves a single problem	• Results measured
• Generally blue collar	• Generally white collar	• Members work as a team
• Regular circle policy applies	• Special circle policy applies	• Makes management presentation on recommended solution
		• Has a people-building philosophy

QUALITY CIRCLE PROGRAM REQUIREMENTS

To assure a good chance of success with a Quality Circle Program, considerable work must be done in four major areas: management support, cultural ambiance, facilitation, and training.

Management Support

In any organization, no matter what the organization's prime objective, management is the key to achieving success. Management's role is to run the organization—to make the daily and strategic decisions to achieve the prime objective. Management is in the best organizational position to observe and understand the key factors that can make or break the authority, responsibility, and accountability required to either achieve or miss the prime objective. This does not imply that the organization's professionals or blue-collar community have little effect on an organization's prime objective; but experts have reported that most of the problems of modern organization professionals and blue-collar individuals can be traced to management.[5] With this in mind, it is absolutely imperative that management support be developed first and continued throughout the Circle Program. Management support can be broken down into two subareas, *knowledge* and *resources*.

Knowledge.   All levels of management must learn about the Quality Circle Programs, before the managers/supervisors who will start the Section Circles or the managers who will start the Company Circles are asked to volunteer. One approach would to start with the CEO, then proceed to the CEO's staff and finally to middle management. Any skipping of a management level assures failure.

In developing management's needed knowledge, certain areas are extremely important. Foremost is the fact that the Quality Circle Program is their program, not a product of the Personnel or Human Resources Department. One or both of the latter departments will do the work associated with the program, but management owns the program and eventually will take complete control, usually in the form of a Steering Committee that decides where the Circles will start and eventually decides on the Circle's solution implementation. The organization's management cannot start a Quality Circle Program and let it continue on its own, for in our society's view, authority figures are either gods to adore or devils to rebel against. Because managers are authority figures within the organization, managers must be careful in any conversation with subordinates. One inadvertent negative remark, whether intended or not, can render a hundred positive comments or actions useless. These comments are not offered as a defense or a suggestion to change managerial behavior; they are reality. The inadvertent negative comment can destroy a Circle member's enthusiasm or a Circle's esprit de corps or the entire Circle Program.

Another important point the managers must understand is the return on the investment they will make when implementing the Quality Circle Program. Most American managers are overly oriented toward short-term results, whereas the Quality Circle Program is long-term oriented. The point just made concerning authority figures will require considerable time to change, before unleashing the employees' hidden potentials. Also, building teams of individuals is not an overnight task. A management that expects short-term results where few immediate results will occur, can also destroy the program.

Once an initial set of meetings explaining the Quality Circle Program has been completed and the previously mentioned points explained, a meeting should be scheduled to present and adjust an initial implementation plan. There should be a series of meetings, not just one, for a Quality Circle Program should not be entered into without considerable thought. A Quality Circle Program can cause much damage to an organization, if implemented incorrectly, by building employees' hopes for more involvement and then dashing them. But the last meeting should present a start-up and construction plan for Quality Circles. Management should voice inputs or suggest changes to the plan, and a consensus should be arrived at. Also, periodic update meetings should be held to keep all levels of management informed on the Quality Circle Program and the Circle's activities. Since this is management's program, they should be involved continually right from the beginning.

Resources.   The second subarea within mangement support is resources. Management must understand and authorize the needed facilities, money, and people. An adequate meeting room must be provided. A Circle meeting cannot be displaced from the meeting room, least of all in the start-up phase. The credibility of the program will be damaged: the program is not as important as everyone says. Management must also supply the funds for the program: training materials, promotional materials, data-gathering funds, etc., must be budgeted and accepted as a long-term investment. The necessary people, facilitators, management, and support personnel must also be provided, depending on the magnitude of the program. In addition, the entire organization must understand that if a Circle requests a special expert for data collection, that expert should participate in the meeting with honest answers and no fears. All members of an organization must accept the program and try to work with it.

The last but most important resource that management must supply is time—*their* time. Management must be present at Circle presentations of problem solutions, and must talk to Quality Circle leaders and members about Circle activities and learn Circle procedures. It is important that all management levels actually learn the Circle's techniques for two reasons: first, a much higher understanding of the Circle Program is acquired by actually learning and using the Circle procedures; second, management may be able to make more efficient use of their meetings.

Management support is an important aspect of a Quality Circle Program and is probably the key factor in success or failure. It isn't easy for management to accomplish the tasks described, but if a Circle Program is desired, the changes must occur. And management must want to change not just as a result of higher management orders. But while management support is critical, other needs must be met for a Quality Circle Program.

Cultural Ambiance

Within systems theory various components are defined—environment, boundary, input, output, feedback, etc., as shown in Figure 8-2. If an organization is viewed within this theory and either Circle Program type is being considered, the cultural ambiance of the organization must change. David I. Cleland defines the organization's culture as "the synergistic set of shared ideas and beliefs that are associated with a way of life in an organization."[6] Within any organization there is a set of behaviors that are associated with every task—there is a way of doing things. The implementation of a Quality Circle Program will disrupt this ambiance. For this reason alone, it is important to implement the program slowly. Both management and workers need time to change, and if denied this time, will usually revolt against the program. As an example, the Circle Program commonly enters an organization via the CEO, who usually directs his or her staff to get a program going. A staff member usually hires a facilitator to handle the program. Finally, the Circles start at the lower levels of the organization, and the possibility of failure is probably one or two years away. Sometimes a contest develops between or within large organizations as to who has the most Circles, and the objective of Circles—efficiency/effectiveness—is forgotten. The Circle leaders and members will see management's attitude and say, "Just another program." The program must be implemented slowly to change both management's and employees' attitudes.

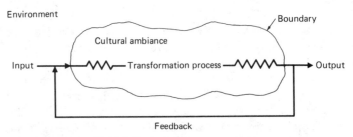

Fig. 8-2. Systems theory components.

Facilitation

The common term for the individual who helps the Quality Circle is "facilitator." The role of the facilitator within an organization is relatively new. The facilitator has been described in many instances as a person who knows the area of work where the Circle will operate. Unfortunately, the individuals who know the work area seldom have the required skills needed to help Circles. The facilitator acts as an outsider to the Circle and in some respects needs to know very little of the actual work being done. In fact, being too close to the problem may even be detrimental. Also, it is rather difficult to acquire a person from a work area, for it is assumed that the person is already busy. Finally, because the facilitating role is new to the organization, the person involved should be able to devote some time to learning and adjusting.

The facilitator has a number of duties. This individual must be an adult educator, organizer, interfacer, and observer. He/she must be skilled in communications—not only in teaching, but in feeding back observed behavior of the leader or members of a Circle. Special training at the National Training Laboratory's Human Interaction Lab and the A. K. Rice Institute's Group Relations Conference is highly recommended. The former develops an understanding of how the individual communicates and what factors affect communication, while the latter promotes group dynamics concepts. Additionally, both courses provide facilitators/consultants as models for developing the Quality Circle facilitator's style. These duties demonstrate the need for a new organizational position. It would be highly advantageous for all managers/ employees to become adept at these functions, but the time involved would seriously affect the prime objective of the organization. Therefore why not train one or two individuals to pass on the skills to others in the organization? In any case, this is a massive change and the time required is enormous.

Training

The training of Quality Circle leaders and members can be broken down into two areas, the logical and the psychological processes. It may seem redundant to teach people to be logical, especially professionals and managers, but most people mix logical and emotional thinking. Also, new and powerful techniques have been developed in the last few years to develop the logical approach and increase creativity. Finally, the Quality Circle Program develops a systematic approach to problem solving.

The logical approach should start by using a logical system like Creative Problem Solving.[7] Here the major steps are defined: where we are and where we want to be, and how to generate solutions, pick the optimum solution, and plan, sell, and implement it, if directed to do so. The logical approach should then develop the techniques to use in each step: brainstorming, selection, cause-

and-effect diagrams, data collection techniques, data analysis techniques, and presentation techniques. The integration of the techniques with a logical system is a recent development in creative problem solving.

The psychological approach should be started along with the logical approach. This type of training is not readily acceptable to most people and must be introduced slowly. The emphasis should be on participation and helping others to participate. The beginnings of certain principles can be laid: for example, listening, which is extremely hard to do. One technique to convey the message of listening is to stop two to five minutes into the discussion of listening and ask how many people have thought of some other topics during the discussion. Sometimes it is useful to use a survival exercise to demonstrate that a group can get better decisions than an individual.[8] The exercise can also show how groups often do not use their total knowledge to solve a problem. Other concepts can be introduced as the group needs arise: subgroups, socializing, outside influence, etc. But it is absolutely necessary to enhance the logical approach by introducing the psychological approach.

The method used to implement the two learning areas, logical and psychological, is as important as the two approaches themselves. Usually the first meeting is used to develop both approaches. The film *The Enchanted Loom*[9] is excellent for this purpose, along with an introduction to the logical problem system. Then have the Circle decide how the secretarial function of the group will be accomplished. This gives the group a chance to use the new techniques. This, indeed, is the general approach to teaching the techniques: one teaches a technique like brainstorming, then has the group use the technique in listing the present problems associated with their jobs, or in selecting one of the problems previously listed to be the first group problem dealt with. When teaching the techniques, it is useful to coteach with the leader. Of course the leader has already been taught the techniques in a similar manner to be discussed shortly. This teaching technique supports the leader's authority and prevents the facilitator from becoming the informal leader. Therefore the training of Quality Circles leaders and members integrates logical, psychological, and adult learning into a uniform attack on the group decision-making process.

After two or three years and two or three problems, other phases of training can be used as the need arises. Group dynamics is a slow, ongoing, teaching function. As an example, most Circles will rebel in various degrees against the management presentation of the solution to their problem, especially if the problem is considered small and falls within the Circle's organizational section. When one talks with the Circle, it becomes evident that there is fear of top management. By having the first Circle problem always presented to the CEO and part of his/her staff, a portion of this fear is usually removed, since this gives both groups an opportunity to interface, explaining their point of view. This fear is due to the authority-figure psychology of most people. To overcome part of this fear, talking and being honest with the leader and members can be extremely helpful. Other areas within group dynamics include: feeding back areas of facilitator observation to the leader, (the facilitator acts as an observer after the initial training period); feeding back to the members about their group behavior; various group dynamics exercises; etc. Group dynamics is an ever growing area of concern because of its newness; the application of its principles will depend, to a large extent, on the group, which is the main reason for the necessary facilitator training.

After the initial training and experience in using the logical systems and techniques, other areas may be introduced as a function of the Circles problems or desires. Such areas could include more complicated logical systems and techniques, and statistical methods. After the initial training period, the group and its problems dictate where additional training is needed.

A final word on training concerns management training in the topics covered by the Circles themselves. All levels of management should take a version of the training that the Circle leaders and members take. This action aids management in understanding Quality Circle activities and can greatly enhance their meetings and their decision-making processes. Quality Circle techniques do not have to be confined to Circle activities alone.

QUALITY CIRCLE PROGRAM STEPS

In any organizational change with the magnitude of a Quality Circle Program, a systematic approach is absolutely necessary. A Quality Circle Program is a change from a hierarchical process to a participative process. The change, to increase either efficiency or effectiveness, will take time and should be planned first and constantly monitored. A recommended decision process that management should follow in determining whether or not a Quality Circle Program should be implemented in their organization is shown in Figure 8-3. Once the decision is made to begin the program, a recommended series of steps might be as follows.

Management Presentations

Starting with a presentation of the Quality Circle Program plan to the CEO, all aspects of the program should be outlined, including the needed resources. The presentation should progress from the CEO to all of his/her staff and then to all of middle management. These presentations should be completed before asking supervisors/managers to volunteer to be trained as Circle leaders. A set of representatives from each higher level, including the CEO, should be present to show management support, as the presentations progress down through the organization.

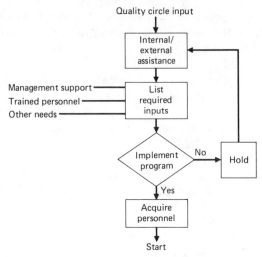

Fig. 8-3. Management's decision process.

Supervisor / Manager Presentations

The last management presentation should be to the management team who are potential leaders of Circles. At this presentation the CEO and a portion of the CEO's staff and middle management should be present to show support. It should be emphasized at this meeting that the supervisors/managers who volunteer are only volunteering to be trained; after the training there will be an opportunity to volunteer to start a Circle in there respective sections. The only difference at this point between a Section Circle and a Company Circle is that most of the organization management team should be involved in the presentations for a Section Circle, whereas for a Company Circle only the areas affected need be involved, the other sections being given a brief overview.

Management / Supervisor Training

Once the managers/supervisors volunteer to be trained in Quality Circles techniques, the course should have all the concepts that will be taught to the members plus exercises on trust theory and group dynamics. The course should be at least two and a half days in duration and off site. If the course is within or near the work site, interruptions usually disrupt the work sessions.

Employee Presentations

If, after training, the supervisor/manager volunteers to form a Circle within his or her section, the manager calls a section meeting where the manager and the facilitator explain the program and ask for volunteers. It is important to tell the section members that every member can be in a Circle, but that if not enough people volunteer, the Circle will be postponed; and that if more than ten people volunteer, the ten needed for the Circle will be drawn at random, while the remaining volunteers will join a Circle at a future date. This prevents the perception that the supervisor/manger will pick favorites. For the Company Quality Circle a posting regarding volunteers within the needed task areas can be used to acquire volunteers. This posting should state that certain types of individuals will be selected, if they volunteer, while the remaining volunteers will be randomly selected.

Circle Start-up

When the Circle Program is in its infancy within an organization, its growth should be slow. I recommend starting two to four Circles at a time, depending on the size of the organization, with intervlas of three to six months between start-ups. The training should present one logical technique at a time, with work on the Circle's first problem as an example of the new technique. The psychological techniques should be interspersed at the facilitator's and leader's joint discretion.

Additional Steps

Management training at all levels in an abbreviated session should begin as the first Circles start. At times a Circle will move rapidly on a problem, and the understanding achieved in the training will be extremely helpful for management at the first Circle's presentation of a solution.

A periodic meeting schedule should be developed for regular feedback sessions to the CEO and his staff plus middle management, so as to keep interest high and update mangement on Circle activities. Publications should be developed to inform the entire organization of the general Circle workings. Figure 8-4 shows the implementation steps to start up a Quality Circle Program. If, at any of the three decision diamonds, not enough employees volunteer in the appropriate categories, the recommendation would be to return to the "hold" position in management's start-up decision process (Figure 8-3). If the decision process has been followed thoroughly, the probability of a negative response is extremely low. The return to the start-up decision process strongly indicates a cultural ambiance problem that has been overlooked.

AREAS OF CONSIDERATION

There are areas of consideration that a Quality Circle Program manager or facilitator should be aware of, so as to increase the chances of success. How-

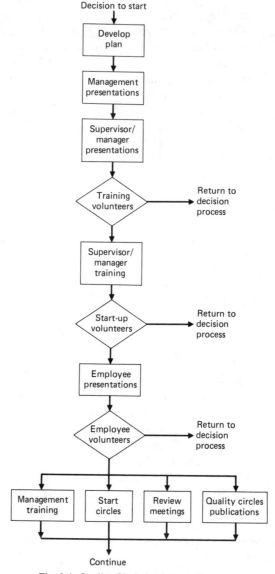

Fig. 8-4. Quality Circle implementation steps.

ever, the appropriate remedies may or may not be usable in a given organization. The decision maker must be in tune with the organization's cultural ambiance to determine if the remedy that is proposed is applicable. The individual must use intuition as defined by Eric Berne: "Intuition is the acquiring of knowledge through sensory contact with the object, without the intuiter being able to explain to himself or to others exactly how he came to his conclusions."[10] In other words, intuition is an unconscious gathering, analysis, and synthesis of data. At times it will be extremely difficult to point to the reasons for picking a prescribed remedy. Emphasis on intuition is necessary because the scientific method has not progressed far enough to supply proven remedies; also, it is philosophically debatable whether in this discipline the scientific method will ever provide proven methods. But let us now examine the areas of concern.

Flip Charts

It is advantageous to have Circle members use flip charts to help record Circle activities. A chalk board can also be used, but if a group wants to refer to a previous point, the chalk board can only hold a limited amount and for a limited time. The flip chart also provides a means of increasing participation by having Circle members rotate as recorders on the chart.

Privacy of Meetings

The Circle meetings should be private, with few or no outside observers. During the initial training period no one should be allowed into the meetings, because the group is learning how to work together. After the initial training, observers may be permitted, as long as the Circle knows about the visit ahead of time and the outsider's role is defined before entering. If a manager wishes to observe a meeting, be careful that he/she understands the Circle Program, the problem being considered, and how far the Circle has progressed in the solution of the problem. The processes can be misleading to an individual who doesn't understand groups and how they operate.

Canceled Meetings

It is extremely dangerous to cancel meetings, especially in the initial phases of a program or any Section Circle start-up. Given the present attitudes of employees about management and the organization itself, cancelation of a meeting may be perceived as a sign that Circle activities are not important.

Members Turn-off

Occasionally a member will change his or her mind about the Circle Program. It is best to discuss the problem openly and honestly to try to change the individual's mind. If an entire Circle decides to quit, the same procedure should work. Pep talks can help, but only for a short amount of time unless the problem clears up. It is usually more effective to address the problem and try to eliminate it. Of course there are times when a Circle has completed its tasks or the dynamics are incorrect. In this case the Circle should indeed be disbanded. Circle members or the leader may also have hidden agendas, other than the Circle's tasks, for the Circle existence. If this occurs and the member or leader tries to use the Circle for other than normal purposes, the Circle should likewise be disbanded. It is imperative to keep the Circle focused on its defined tasks, or the organization will stop the program.

Fear of Management Presentations

Many people actually have a tremendous fear of talking in front of people, esepcially authority figures. However, the leader or facilitator should try to have all members take part in a management presentation of the Circle's problems, since most people are much better in front of others than they think. But if one or two will not help in the presentation, they should play some other part, as for instance remaking the flip charts. Note also that it is usually advantageous for the CEO to be present at a Circle's first presentation, since the Circle will feel important: the CEO takes time to see his Circles at work.

Picking the First Problem

In a Section Circle where the members pick their own problem, the first problem should be an easy one. This facilitates the understanding of the techniques and the problem-solving system. Yet this is not usually the case, even though both the facilitator and the leader urge the group to pick an easy problem. Usually it is the problem that is foremost on the members' minds, or a nagging problem that has long been affecting the section, that is picked. One can only try to convince the group to tackle a small problem first. If a general problem like communications is picked, the facilitator or leader can help the group narrow the problem down. This area of concern is one for which it is extremely difficult to find a workable remedy.

Groupthink

Within the Circles there may be various levels of Groupthink, "a mode of thinking that people engage in when they are deeply involved in a cohesive in-group, when the members' strivings for unanimity override their motivation to realistically appraise alternative courses of action."[11] The usual type exhibited in the beginning of a Circle Program is following the leader. Because of the authority figure associated with a supervisor/manager, some people will not disagree even though they know the manager is wrong. On the other hand, if the manager is a taskmaster, a subgroup might develop in opposition to the manager just to disagree. The facilitator should address the problem honestly with the individuals involved, explaining that the Circle meeting is not the battleground for these problems.

Another type of Groupthink that might occur in the beginning of the Program is the end-of-the-meeting Groupthink. If the meeting's work is being accomplished within the last portion of the meeting, the group should be informed of this and a more detailed agenda written for the next meeting.

The above are a few of the numerous areas of consideration that may develop as the Quality Circle Program grows to adulthood. The manager/facilitator of the program must be constantly on guard for newly developing problems and use training, experience, and intuition to satisfy the constraints of both the organization and the Quality Circle Program. The program manager/facilitator must also be aware of other Human Resources Programs and how the other programs interact with Quality Circle Program, since the prime objective of all these programs is to increase organizational efficiency and effectiveness.

QUALITY CIRCLES AND MATRIX MANAGEMENT

During an initial comparison of matrix management and Quality Circles, the two programs appear to affect the organization in different areas: matrix management affects the organization's structure, and Quality Circles, its processes. Of course no program of the magnitude of these two affects only one characteristic of an organization—structure, task, or process; however, a program's effects are usually felt most strongly in one area. Upon further investigation, though, the Quality Circle Program turns out to complement matrix management.

Matrix management is a restructuring of the organization's human resources from a traditional hierarchy structure to a "multiple command structure."[12] Matrix management is the "system approach to implementation" of management.[13] Because of environmental and internal cultural demands upon organizations, especially high technology organizations, a new structure is developing to share the human resources of the organization. The basic idea behind matrix management is the sharing of human resource expertise within normal organizational functional areas with the newly developing, project-oriented functions within the organization. Quality Circles constitute the delegation of authority and responsibility for problem solving to the organizational level best suited to solve the problem. The program develops teamwork or a sharing of work and ideas to determine the best solution. A Quality Circle Program can aid matrix management by developing concepts of sharing to facilitate the cultural change required by such management.

Compatibility is another important issue to evaluate, when implementing two programs of the magnitude of Quality Circles and matrix management. Does one program work against the desired effects of the other? Does the program work against the prime objective of the organization? Both questions should be answered in the management decision phase, before any resources are dedicated to program implementation. Quality Circles is compatible with

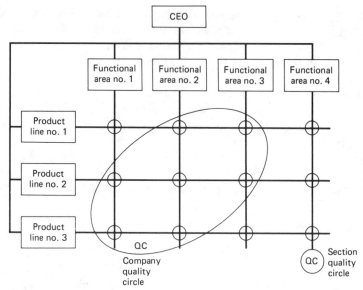

Fig. 8-5. Quality circles-matrix management compatibility.

matrix management because the two programs emphasize team building. In addition, while matrix management is oriented toward horizontal integration of resources, Quality Circles—especially a Company Circle—can emphasize the vertical integration needed within organizations. Figure 8-5 depicts in a simplified form the relationship between matrix management and Quality Circles.

Therefore, upon further comparison of matrix management and Quality Circles the two programs appear to complement each other. But at this point a word of caution is needed: too many programs at one time can be as damaging to an organization as none at all. Although Quality Circles and matrix management form a reasonable partnership, any intervention of the magnitude of these two programs should be entered into slowly and with caution.

SUMMATION

The modern organization is embedded in a time where changes are occurring at enormous rates. Traditional methods of running an organization no longer develop the needed return on investment. New methods must be found to increase the organization's efficiency and effectiveness. The Quality Circle Program is a modern method that can be used to increase either, and to help produce the return on investment. Modern methods require the normal program inputs plus additional inputs to assure their success. Without management support, organizational and Circle facilitation, training, and systematic implementation, the modern program will usually fail, which generally results in increased organizational problems. Programs like Quality Circles and matrix management, if reasonably implemented, are the most likely to increase organizational efficiency and effectiveness for the remainder of this century.

REFERENCES

Notes

1. *NBC White Paper,* "If Japan Can . . . Why Can't We?" June 24, 1980.
2. *Ibid.*
3. *Ibid.*
4. Edmund J. Metz, "The Verteam Circle," *Training and Development Journal,* Dec. 1981, pp. 78–85.
5. *NBC White Paper.*
6. David I. Cleland, "The Cultural Ambience of the Matrix Organization," *Management Review,* Nov. 1981, p. 25. Reprinted by permission of publisher from *Management Review,* Nov. 1981, © 1981 by AMACOM, a division of American Management Associations, New York. All rights reserved.
7. Morris O. Edwards, "Doubling Idea Power," IDEA Development Associates, Pal Alto, Ca. 1975.
8. "The Desert Survival Situation," Survival Exercise, Human Synergistics, Plymouth, Mich., undated.
9. *The Enchanted Loom,* BBC, London, England; Heritage Visual Sales, 505 Consumer Road, Ontario, Canada.
10. Eric Berne, *Beyond Games and Scripts,* ed. Claude M. Steiner and Carmen Kerr (New York: Random House, 1976), p. 29.
11. Irving L. Janis, *Groupthink: Psychological Studies of Policy Decisions and Fiascoes,* 2nd ed. (Boston: Houghton Mifflin, 1982), p. 9.
12. Stanley M. Davis, and Paul R. Lawrence, *Matrix* (Reading, Mass.: Addison-Wesley, 1977), p. 3.
13. David I. Cleland, and Williams R. King, *Systems Analysis and Project Management* (New York: McGraw-Hill, 1975), p. 183.

Bibliography

Barra, Ralph J. *Putting Quality Circles to Work.* New York: McGraw-Hill, 1983.
Dewar, Donald. *The Quality Circle Handbook.* Red Bluff, Calif.: Quality Circle Institute, 1980.
Ingle, Sud. *Quality Circles Master Guide.* Englewood Cliffs, N.J.: Prentice-Hall, 1982.
International Association of Quality Circles, Midwest City, Okla. Journals and annual conference transactions.

9. Making the Matrix Come Alive: Managing a Task Force

James P. Ware*

Matrix organizations exist to deal with uncertainty. Davis and Lawrence, in their classic book, *Matrix,* cite the September 1976 issue of the General Electric Company's *Organization Planning Bulletin,* which contained the following statement:

> . . . all of us are going to have to learn how to utilize organizations to prepare managers to increasingly deal with high levels of complexity and ambiguity in situations where they have to get results from people and components *not* under their direct control.[1]

Matrix organizations, with their dual command structures and heavy reliance on group problem solving, arise in complex, dynamic, and uncertain environments—environments that require rapid processing of high volumes of ambiguous information. And, as Jay Galbraith has pointed out, the only effective means of coordinating diverse, differentiated activities in that kind of uncertain environment is through interdepartmental management groups.[2] But the key to effectiveness is not the *existence* of such managerial groups; what counts is how effectively such groups *work.* Matrix organizations require significant increases in both the volume and the quality of managerial communication. Unfortunately, as Davis and Lawrence stress,

> this increased communication creates a new set of problems. More differences are surfaced and have to be dealt with. . . . In a matrix these differences are resolved with people from different functions who often have different attitudes and orientations. They are resolved without a common boss readily available to arbitrate differences.[3]

---

*Dr. James P. Ware is a principal and Director of Education Programs with Nolan, Norton & Company, Lexington, Mass., a data processing management consulting firm. A specialist in organizational design and change, Dr. Ware served on the faculty of Harvard Business School for five years prior to joining Nolan, Norton in 1981. He is presently engaged in a variety of client projects, with an emphasis on problems related to the implementation of organizational change and the introduction of automation into managerial and professional work.

Thus the critical ingredient in making a matrix work effectively is the ability of the group members to confront and resolve their differences in ways that generate effective, high-value solutions. This is far easier to state than to accomplish, however. One fact we have learned all too painfully in one organization after another is that the greater the differences among individuals, and the more ambiguous the problems they face, the greater will be their difficulty in understanding each other, let alone in resolving their problems in any mutually satisfying way.[4] And matrix organizational settings are guaranteed to create precisely those conditions.

This unavoidable reality places great pressures, and great responsibilities, upon the leaders of management groups in matrix organizations. It is within such groups that the critical work of a matrix organization gets done, and whether that work gets done well depends on the leader's skill at structuring group assignments and managing the relationships among the group's members. The challenge for the group leader, then, is to foster a results-oriented but open, problem-solving culture: to achieve a balance between encouraging exploration of new ideas and requiring implementation of existing commitments.[5]

These issues are particularly critical in the leadership of a task force. While task forces are not unique to matrix organizations, the leadership responsibility is significantly more difficult in matrix environments because of their inherently greater complexity and ambiguity. Effectively managing a matrix task force requires a careful understanding of why task forces are established, how they typically function, how the issues they face evolve over time, and what kinds of leadership tactics are effective. This chapter will deal with these kinds of questions.

THE TASK FORCE APPROACH

The task force approach is basically a simple concept, and hardly a new one. Many companies in a wide variety of industries have found task forces to be highly effective mechanisms for solving a wide range of complex organizational problems. Typically, an organization establishes a task force, task team, or project team when senior management perceives an operating problem or business opportunity that cuts across traditional departmental boundaries, or that the existing organization is simply unequipped to handle. Sometimes the task to be done just doesn't fit the existing structure; sometimes the problem is too big or complex for an already overworked staff; and sometimes the effort requires such a different approach from established procedures that it simply cannot be dealt with by the existing organization.

Consider these examples, which are typical of the organizational problems frequently addressed by task forces:

- A medium-sized textbook publisher established an Inventory Control Task Force to investigate why inventory levels were so high on so many books at the same time that other products were repeatedly out of stock. It quickly became apparent that the problem required assessing procedures being used in product planning, sales forecasting, purchasing, manufacturing, warehousing and distribution, order processing, accounting, and data processing. Every one of these departments had established policies and procedures that affected final inventory levels, yet no one area was fully responsible for the entire inventory system. And the existing matrix, which was organized around products and markets, was inadequate for addressing the totality of the inventory system.
- A large commercial bank recognized the need to engage in a comprehensive, long-range strategic planning process that senior management suspected might lead to a massive reorganization of its major operating divisions. The bank established six focused task forces, each containing three or four levels of management, to develop ten-year plans for each of its principal business areas. The six topic areas did not correspond to the bank's existing divisions, and the members of each task force were drawn from all the operating divisions without regard to their specialized expertise or job experiences.

Ideally, task forces like these bring together individual managers who in combination possess the knowledge and skills needed to construct a high-quality solution, and who have the organizational power and influence to ensure that their solution will be implemented. Then, once the task force's work is completed, the group can be disbanded, freeing its members to return to their regular jobs or to attack new problems in new groups.

Special-purpose task forces like these seem to be especially appropriate when organizations face problem situations containing some or all of the following characteristics:

- Several different functional areas are or will be affected.
- The problem is relatively urgent.
- The existing organization is not equipped to address the problem.
- The nature of the solution is highly uncertain.
- The solution will potentially involve significant changes in organizational structure, responsibilities, or operating systems.

Although task forces are most commonly established in response to some organizational crisis, many matrix organizations have also had great success incorporating task force projects into their regular operating routines. Task teams can be especially effective in companies experiencing rapid growth. For example, William George has reported that Litton's use of task teams was a significant factor in their ability to create a microwave oven division that grew from $13 million to $75 million in revenues in less than five years.[6]

However, it is important to distinguish between task forces and more permanent interdepartmental working groups. For example, task forces are significantly different from standing committees, whose function is to bring together periodically a group of full-time functional managers who must actively coordinate their interdependent activities and responsibilities. In contrast, a task force is formed to work on a specific problem for a specified period of time. Task force members may be assigned on either a part-time or a full-time basis, but generally the assignment is a major one consuming at least 25% of each member's time. In the words of A. K. Wicksberg and T. C. Cronin:

> Membership on the team may be permanent, but it is more likely to consist of a core of permanent individuals with additions of workers with special skills from time to time as required by the particular problem under consideration at the moment. . . .
>
> Structural fluidity is the theme. . . . The heart of the concept rests in the ability to add new members when their potential contribution is high, and to subtract members from the core group when specific contributions have been secured and no further need of these particular talents is foreseen in the immediate future.[7]

These defining characteristics of the task force approach are precisely the factors that make task forces effective and highly productive mechanisms for organizational problem solving. A narrowly defined mission keeps the group's attention focused on the task and reduces time spent on unrelated activities. Since task force members are specifically selected for the project, their skills and knowledge are highly appropriate to the task, and their level of commitment can be very high. The temporary nature of the assignment concentrates members' efforts on ends rather than on means, and there is considerable satisfaction in closing in on a specific, well-defined task. Clearly articulated goals and task deadlines further generate concentration and motivation; and since task forces are usually high-visibility projects, they provide significant opportunities for ambitious, upwardly mobile managers to "show their stuff" to senior executives. In combination, these characteristics can lead to high levels of commitment by task force members, and high potential for success.

Unfortunately, however, several of these very same characteristics are sometimes the source of major operating problems. The limited objectives of a task force can also lead to shallow, pseudo-solutions, if task force members take such a narrow view of their mission that they are unwilling to look at deeper, underlying problems. The temporary nature of the effort sometimes reduces members' commitment levels and willingness to spend personal time and energy on the group: "This too shall pass." Rather than sharpening effort, deadlines can lead to superficial analysis in the rush to complete the job on time. Finally, although the multifunctional, heterogeneous mix of members creates the potential for comprehensive and innovative solutions, it all too often leads to divisive battles, scapegoating, and attempts at forcing pet solutions on the group. Task forces frequently fail because their meetings become public arenas in which members fight out old battles, seeking primarily to win recognition for their own department's views rather than joining together to achieve a more comprehensive common solution. And if ambitious managers "score points" with senior executives, these private agendas can seriously interfere with a group's problem-solving effectiveness.

Thus in spite of their high potential, task forces often fail miserably. When poorly managed, they can be disastrous "black holes" that soak up organizational resources, create more serious operating problems than they solve, and even destroy managerial careers. Consider these examples:

- A national franchising organization created a high-level task force to investigate several new approaches to marketing its products. The group's ultimate recommendations involved major refurbishing of franchise outlets all over the country. The recommendations were accepted, and the reconstruction work was initiated, even though market research data suggested the concept might not be accepted by consumers. (Task force members convinced themselves the data was wrong.) Two years later, after spending over $30 million of a planned $50-million commitment, the entire project was abandoned and several members of the group were forced to resign.

- Metalco, a specialty metals manufacturer, set up a task force to review forecasting procedures that had led to major losses on raw materials inventories because of price fluctuations in the commodity markets. The task force included several line managers and an equal number of corporate staff analysts. John Brown, the project leader, discovered only during the group's final presentation to management that there were three separate factions within the task force itself; furthermore, the company's regional sales managers were adamantly opposed to the recommended changes in forecasting procedures. The proposal meeting was actually canceled halfway through the presentation, because the meeting had degenerated into a shouting match between the two vice-presidents who had joint responsibility for implementing the group's recommendations.

The tragedy is that these problems need not have occurred. Nothing inherent in the issues addressed, or in the structure of the task forces, made these disasters inevitable. Rather, the difficulties arose because of the way the activities of the two task forces were managed.

The key to success with a task force lies precisely in how the task force leader plans for and responds to a number of critical operating dilemmas. Both existing research and personal experience suggest that every task force passes through essentially the same series of stages, and that each stage presents several predictable dilemmas. Understanding those dilemmas, and thinking through the consequences of the choices they require, can make a task force leader's job much simpler. The following specific tactical suggestions, as simple and basic as they are, can have a major impact on the operating effectiveness of any temporary management group.

It is most practical to group these tactical principles into four broad categories, based on the sequence in which the task force will confront the underlying problems:

1. Starting up the task force
2. Conducting the first meeting
3. Running the task force
4. Bringing the project to completion

These guidelines have been written primarily for a matrix manager with task force leadership responsibility. However, both senior executives and task force members should also find the ideas useful. Task force leadership is often used as a developmental experience for promising middle managers, so that senior managers are frequently involved not only as commissioning executives responsible for a task force project as a whole, but also as "coaches" for the individual project leader. Individual task force members should also find these ideas personally useful, both as an aid to understanding the dilemmas faced by the formal leader, and as a source of suggestions for their own work as members of a problem-solving group. One of the clearest research findings on effective managerial groups is that there are several critical leadership roles within a group, and that groups function most effectively when those roles are shared by several group members, either simultaneously or sequentially. Thus these suggestions should prove helpful even to task force members who do not have formal leadership responsibilities.

STARTING UP THE TASK FORCE

The period before the first formal meeting of the task force requires several key decisions that will affect many of the group's later activities. Careful attention to details at the very beginning is one of the best ways to avoid major conflicts later on, when the group confronts the tough issues that constitute its reason for existence. In fact, these "front-end" activities probably represent a task force leader's greatest opportunity for defining the group's working style.

Defining Objectives

Although the specific circumstances surrounding each project will be unique, most task forces are established to accomplish one or more of the following general objectives:

- To investigate a poorly-understood problem
- To recommend and/or implement a high-quality solution to a recognized problem
- To respond to a crisis that results from a sudden change in the organization's business conditions
- To develop managers by providing them with exposure to other functional areas and people
- To gain commitment to a decision by involving the people who will be affected by its implementation

Usually the commissioning executives will have several objectives in mind. (By "commissioning executives" I mean the senior managers who determined that the task force should be established, and to whom the task force will present its findings.) For example, a new product development project can also be an excellent training experience for junior marketing managers. Similarly, a study of excessive inventories might be part of a deliberate strategy to reduce the power of a well-entrenched but ineffective purchasing manager.

Often, however, these multiple objectives are incompatible, in that one goal may be obtainable only at the expense of another. Additionally, various senior executives generally have differing objectives, or differing priorities for conflicting objectives. In many cases these differences will not be openly expressed or even recognized.

Thus one of the leader's first critical tasks will be to meet with the senior executives who have an interest in the project outcomes. In those meetings he or she should explore very explicitly the relevance and relative importance of each possible objective. It will probably be necessary for the leader to define alternative objectives and possible conflicts himself.

The task force leader should also clarify whether the task force is expected to conduct a preliminary investigation, to engage in active problem-solving, or to design and actually carry out a new set of procedures. The nature of the work to be done will influence decisions about who should be on the task force, who should be kept informed about the group's work, and what kinds of operating and decision-making procedures will be appropriate. It is essential to develop an explicit statement about the boundaries of the problem being attacked, and to share that statement with the commissioning executives in order to verify their expectations.

A useful technique for confronting these issues is to write out a formal statement of purpose and fundamental assumptions, and then to engage each of the commissioning executives in an assessment of how well the statement reflects their expectations. This process has the added benefit of getting the senior managers solidly behind the task force effort; their personal involvement in helping to revise and refine the formal statement will develop their commitment to it, and thus to the task force's ultimate success. While this emphasis on a written statement of purpose may seem overly formal, the failure to agree on initial objectives is one of the major causes for task force disasters.

The forecasting task force at Metalco, described earlier, conducted three simultaneous investigations and presented three independent recommendations to senior management, largely because the group never discussed what their mission was. John Brown, the project leader, had simply accepted his own boss's definition of the problem and never even discussed it with other functional officers of the company. As a result, his task force members never came together as one group with a unified sense of their task; instead, the members fell into natural subgroups that merely perpetuated the parochial views of existing functional groups—the very views the task force had been expected to overcome.

Defining General Operating Procedures

These questions should be asked at the outset:

- Will members be assigned to the group on a full-time or part-time basis?
- When should the task be completed?
- What will the group's budget be?
- What organizational reports and other information will be available to the group?
- How much decision-making power is being delegated to the group?
- What information should be reported to functional managers, and how often?

There is no way to anticipate all the procedural issues that the group will face, and many of those that can be anticipated should be worked out by the whole task force group. Again, however, these questions should be discussed in advance with the commissioning executives. Confronting concrete operating details is another way of testing senior management's objectives for the task force and their level of commitment to its work. And it is far better to be told explicitly that some topics and decisions are beyond the group's charter than to discover those boundaries only by crossing over them later.

One of the most important procedural issues to be resolved concerns the way in which the group will make task-related decisions. The more exploratory and open-ended the basic project is, the more open and participative the discussion and decision-making procedures should be. There is considerable research evidence to suggest that task-oriented groups prefer relatively directive leaders, and that decision making is usually more efficient in a structured climate. However, if the problem requires an imaginative or wholly new perspective, then an unstructured climate will generally produce more innovative ideas.

Although leadership and decision-making styles should be discussed with both the commissioning executives and the prospective task force members, this is not a decision to make once for the entire project. The effectiveness of each style depends very much on the nature of the current problem and the people working on it. In fact, it often makes sense to vary decision-making procedures as the project progresses. Nevertheless, explicit attention to these topics at the very beginning is well worth the time and effort involved.

Choosing Members

As much as possible, individuals who are asked to join a task force should

- Possess knowledge and skills relevant to the task
- Be personally interested in the problem
- Have, or be able to get, the time to devote to the task force
- Enjoy working in groups, and be effective in group settings
- Not dominate the meetings or decisions solely on the basis of personality or power

While individual competence is important, however, it is not an adequate basis for constructing the project group. It is equally important to consider the overall composition of the group. Does each member possess organizational credibility and influence relative to the problem? Is each of the functional areas that will be affected by the group's work represented? The exclusion of important departments will not only generate resentment and resistance, but may also reduce both the quality of the task force's recommendations and the probability that the recommendations will eventually be implemented.

A major membership selection dilemma is whether to include persons who are likely to obstruct the group's investigations and to slow down progress toward a consensus solution. Although including such individuals may reduce problem-solving efficiency, it will increase substantially the probability of their later supporting (or at least not actively opposing) the group's recommendations. Lowered efficiency in early deliberations is usually more than compensated for by smoother implementation. In addition, when individual resistance is based on valid information or experience, a solution that ignores the sources of that resistance is likely to be less than optimal or even unworkable.

For example, the Metalco task force excluded the four marketing managers who would ultimately use the new forecasting procedures, on the grounds that their past resistance to change would "unnecessarily complicate" the task force's efforts. For the same reason, the marketing managers were rarely consulted during the group's investigations. As one might expect, the group's recommendations were completely rejected by these four managers: they were being offered a solution to a problem they denied they had.

It is vitally important that the task force leader be personally involved in the membership selection process. He or she may have information about the prospective members that will have a direct bearing on the appropriateness of their involvement. Individual skills and attitudes, as well as existing interpersonal relationships, are often more evident to the task force leader than to senior managers, and this knowledge should not be ignored in composing the group. The leader's involvement in the membership selection process increases the probability of building a team that will work well together and with the leader. It also affords the leader another opportunity to learn more about senior management's expectations about how the task force will operate and what it will accomplish.

Contacting Members

The first contact with each prospective member provides an opportunity to begin defining not only the problem the group will be addressing but also the procedures it will follow. Whenever possible, this contact should be made in person, and the discussion should provide adequate time for responding to questions and exploring initial views on the problem. This process helps bring the task force to its initial meeting already possessing some common sense of why it is there and how it will be working.

It is highly desirable to include the prospective member's senior management in this first meeting, or even to meet with them before talking to the member. These discussions permit the task force leader to clarify the group's purpose, to establish its validity, to enlist future cooperation and support, and to reassure any managers who are initially skeptical or uneasy. The contacts also provide an opportunity to explore each person's current knowledge and feelings about the problem, and to build a productive personal relationship if one does not already exist. This information can prove invaluable as the leader prepares for the first meeting of the full task force.

Preparing for the First Meeting

The first formal meeting of the full task force is a critical event, because it usually sets the tone for all later activities. The task force leader should prepare a careful and complete agenda well in advance. There are two major objectives to be accomplished at the first meeting:

1. To reach some consensus about the group's task.
2. To begin defining working procedures and relationships.

Since the most important function of the first meeting will be to define the problem to be addressed and the organization's expectations for the group's output, the commissioning executives should attend the meeting if at all possible.

A word of caution: only in unusual circumstances will the task force leader have the time and resources to carry out all these start-up activities as thoroughly as described here. Time pressures, the dispersion of group members, and existing formal relationships may well prevent the kind of thorough, rational analysis these suggestions imply. In addition, the commissioning executives often find it very difficult to articulate a clear statement of the problem. After all, most task forces are established because the organization actually does not fully understand the nature of the problem, let alone have a solution in mind.

The task force leader usually plays a major role in defining the problem. But the leader also has to know when to stop *discussing* and start *doing*. That point can be determined only by exercising judgment within the context of a specific situation. These start-up suggestions can get a task force off to a good start, but often cannot be followed in their entirety. The important point is to be aware of the risks incurred by omitting a preparatory step, and thereby become more alert to potential problems as the task force officially begins its work.

CONDUCTING THE FIRST MEETING

This meeting is important not only because it is the first time all task force members will be together, but because interactions begun here can influence all later group activities. The two major objectives for this first meeting have already been suggested; each will be discussed below.

A Common Understanding of the Group's Task

This goal is clearly the most important item on the agenda; yet in most instances it will be the most difficult to accomplish. Few of the other members will have devoted as much time or attention to the task as the leader; yet until they get "up to speed" and share the leader's sense of the problem, they will be a "group" in name only.

Each task force member will typically come into this meeting feeling a responsibility to represent his or her own department's interests. Each will interpret the problem in terms of those interests, and will possess a unique combination of relevant ideas and information. Furthermore, many of the group members may be feeling highly defensive, anticipating that other managers will blame their departments for the problem. These feelings are not only natural, but are very likely based on past personal experiences. If such feelings do exist within the group, the leader should try to find a positive way to bring them to the surface.

There are a number of simple techniques for encouraging an open discussion. For example:

- Encourage everyone to express opinions, but forbid them to evaluate one another's views until all the information is out in the open.
- Ask open-ended questions that focus on facts, rather than seeking to blame anyone.
- Actively elicit ideas and opinions from all group members.

At this point in time the group probably does not possess enough information to achieve a deep understanding of the problem and its likely causes. Indeed, lack of information and understanding is generally one of the major reasons for establishing a task force. Nevertheless, it is essential for the group to achieve at least a general agreement that the problem exists, and to determine where its boundaries lie.

However, one of the most important roles of a task force leader is to prevent premature consensus on a solution to the problem. All too often, experienced managers are "sure" they *know* what the problems are and what actions are required. Thus, even though the leader wants to reach some level of agreement on the nature of the *problem,* at this first meeting it is important to avoid letting the group settle on a *solution.*

Metalco's forecasting task force blew apart after three months of work precisely because John Brown failed from the beginning to recognize the difference between agreeing on the problem and agreeing on a solution. At the group's initial meeting, Brown avoided confronting the obviously different views of Metalco's marketing and operations research managers; worse yet, he let the Operations Research staff go off on their own to begin building a forecasting simulation model—their predetermined solution—independently of the other task force members. It is only a slight exaggeration to suggest that Brown's weak management of the task force's very first meeting lit the fuse on a time bomb that blew up in his face three months later.

An effectively managed first meeting will help the task force members to develop a sense of their joint responsibilities, and of what the appropriate next steps are. An explicit recognition of the areas in which the group members disagree is also highly desirable. The members' participation in a candid, task-focused discussion of this kind serves to generate commitment to the group and its general goals, even if differences of opinion about appropriate strategies are openly recognized.

Working Procedures and Relationships

A second essential topic for the first meeting is the question of how the group will work on its task. Among the issues that require explicit attention are the following:

1. Frequency and nature of full task force meetings.
2. Whether subgroups are required, and if so, how they should be structured.
3. Ground rules for communication and decision making within the task force between meetings.
4. Ground rules or norms for decision making and conflict resolution.
5. Schedules and deadlines for accomplishing subtasks and for completing the final report.
6. Ground rules for dealing with sensitive issues; agreement on which ones require involving other members.
7. Procedures for monitoring and reporting progress, both within the task force and to functional area managers.
8. Explicit processes for criticizing and modifying task force working procedures.

Spending time on these procedural issues serves two primary purposes. First, the discussion will help group members form clear expectations concerning their projected activities and working relationships. These expectations will reduce the tensions inherent in an otherwise very unstructured situation. Second, the process of reaching agreement on procedural matters can become a model of how the task force will resolve other problems.

Resolution of these kinds of issues provides the task force members with a positive experience associated with the group. However, this is obviously a full agenda, and some topics may have to be carried over to subsequent meetings. But the first meeting should end on a note of agreement, if at all possible. Achieving a solid consensus on some portion of these procedural matters serves as a foundation for the way the members will work together, no matter how divided the group is on substantive issues.

RUNNING THE TASK FORCE

Once the front-end work has been completed, the leader's efforts will focus on keeping the project moving and on monitoring and reporting the group's progress. Although specific circumstances will vary, there are several general principles to keep in mind:

Frequent Full Meetings

Though each meeting should have a specific purpose, periodic meetings should be scheduled well in advance, and all members should be required to attend. A meeting can always be canceled if there is nothing substantial to discuss. However, full meetings do have an important symbolic value. They are the only time the entire group is physically together, and anything said there is heard simultaneously by everyone. Very often the most valuable and creative discussions are those that evolve spontaneously in response to someone's raising a nonagenda item. For example, one of the most successful fund-raising projects in the history of public television (the cast party following the final episode of *Upstairs, Downstairs*) grew out of two spontaneous comments during an informal staff meeting at WGBH in Boston.

While spontaneous creativity cannot be purposefully planned in advance, it is certainly possible to create opportunities for unstructured, exploratory discussions. These kinds of discussions provide some of the greatest benefits of using a task force, as they combine the perspectives and ideas of managers with very different experiences, responsibilities, and goals. Yet, given deadline pressures, other work responsibilities, and the typically narrow focus of most task forces, genuinely open-ended, creative problem solving occurs far too infrequently.

Running effective meetings is a critical aspect of a task force leader's job—especially within matrix organizations, where the culture normally creates high expectations for meaningful meetings. However, an extended treatment of meeting leadership skills and tactics is beyond the scope of this chapter. Several useful articles on running meetings have appeared in recent years; the interested reader is encouraged to pursue them.[8]

Subgroups

Unless the task force is very small (fewer than five to seven members), dividing up into subgroups will be virtually mandatory. This process must be managed very carefully. Dividing the project into separate tasks that can be worked on simultaneously can facilitate rapid progress. But remember that one of the virtues of the task force approach is the synergy that results from new combinations of individuals looking at problems in areas they are not overly familiar with. If the task force members end up working only in their own functional areas, or only with persons they already know well, the group will lose one of its major advantages—the creativity generated when "outsiders" bring a fresh perspective to old problems.

Of course, when outsiders start poking around in parts of the organization they are unfamiliar with, they may ask "stupid" or naive questions that insult or unintentionally threaten the operating managers they are talking to. The task force leader must be careful to warn the group members of this risk, and then be prepared to spend time both informing the operating managers about the task force's work and smoothing ruffled feathers.

It is also important to remember that working in subgroups can cause individual managers to lose their overall perspective. If the task force becomes too differentiated, the various subgroups may form their own identities and develop an advocacy style of pushing for their own pet solutions. This danger is another reason for forming subgroups of managers from significantly different parts of the organization; these internal differences can serve as checks and balances that prevent one-sided thinking. Nevertheless, the more the total job is broken down for subgroup work, the more the task force leader must encourage formal intergroup sharing of problems, findings, and ideas as the project moves along.

Interim Deadlines

Interim deadlines are the only way of measuring progress toward the project completion date. Separating a project into definable phases with a firm schedule is a common and well-accepted practice in virtually all formal project management systems,[9] and is especially important in matrix organizations, where the multiple demands on managers make them all too ready to let deadlines slip whenever possible. Scheduling and progress reporting are equally critical for task force projects as well. One reason scheduling is so difficult in exploratory projects is that the final deadline is often set rather arbitrarily; senior management sets a date based on its own desire to solve the problem, rather than on any external requirements or understanding of the magnitude of the work required.

Under those conditions, it is all too easy for the task force leader to assume that time lost along the way can be made up later—or even that the final deadline can easily be shifted. No matter how arbitrary the interim deadlines are, however, if they are missed, the final deadline will be missed also.[10]

Insistence on meeting deadlines is doubly important—and doubly difficult—when task force members are working on the project on only a part-time basis. When they continue to carry out other responsibilities, they inevitably feel powerful pressures to spend as much time as possible on operating tasks that have immediate and visible outcomes. This pressure to accomplish immediate tasks will *always* outweigh the perceived needs of the longer-range task force efforts.

Part-time task force members face significant dilemmas, and will be under continual stress. The task force leader must be prepared to spend a significant amount of time prodding and cajoling group members to complete their tasks on schedule. At the same time, however, it is important to remain sympathetic to the legitimate needs of the functional areas, and to avoid antagonizing either the task force members or their senior managers. This task is especially delicate when the leader has no formal authority over the part-time members, or where the lines of authority have been spelled out only vaguely.

Sensitivity to Conflicting Loyalties

It is not only part-time members who find task forces highly stressful. As all the members work together in group activities they normally (and naturally) begin to develop personal commitments to the project and to one another. These commitments then become a powerful source of stress, as members experience loyalty both to the task force and to their "home" departments.

Full participation in the task force frequently requires group members to share previously confidential information (e.g., department salary costs, sales district profitability statistics, standard cost calculations, etc.). This may be a difficult or stressful action for an individual to take, and it often involves personal risks. An individual who provides the group with confidential or sensitive information has become dependent on the group's integrity in a very fundamental way. Maintaining that integrity is a crucial responsibility for the task force leader.

Such interdepartmental and interpersonal conflicts are of course inherent in any matrix organization. Perhaps the most important challenge facing the task force leader throughout the evolution of the group is maintaining an open, confronting, problem-solving style of confict resolution. Meeting this challenge calls for an exceptional array of interpersonal skills.

Liaison Work

The leader's communications role can be excessively time-consuming, but it is absolutely essential to the success of the project. There are several critical communication tasks:

- Monitoring and reporting progress against schedules
- Bringing appropriate subgroups together to share information and ideas on an interim basis
- Reporting both progress and problems to the commissioning executives, to other affected managers, and to all task force members
- Staying in touch informally with individual group members

This last task is usually the most time-consuming one, but it is also the most important. It involves a lot of careful, concerned listening, passing information from one task force member to another, and bringing together on an ad hoc basis managers who must exchange or share information and ideas. While these activities often seem inordinately time-consuming, they actually form the "glue" that binds the individual task force members together. As individuals and subgroups independently pursue their investigative tasks, the leader is usually the only person who retains an overall understanding of the total effort. This holistic perspective is especially important when most of the task force members are involved on a part-time basis, but it is always a critical responsibility.

BRINGING THE PROJECT TO COMPLETION

The work of an *investigative* task force typically culminates in a written report and a summary presentation of findings and recommendations to upper management. An *implementation* task force will normally have more concrete operating results with which to demonstrate its accomplishments, but even so there is often a formal meeting at which the task force officially relinquishes its responsibilities to an operating group.

The written report documents the work of the task force, but its importance lies in the decision-making process it generates. In fact, the preparation of the final report often provides a useful structure and focus for the task force's concluding activities. In one instance a tentative outline of the final report was circulated among the task force members a full eight months before the scheduled completion date. The outline actually served as a guide to the development of concrete recommendations; the group's need to write the report forced it to reach specific decisions on the issues that the outline identified. In effect, the outline became the group's work plan.

Early drafts of the final report can then become the basis for working out any differences remaining among the task force members. Except in highly charged situations with major organizational stakes, the leader should strive to reach a group consensus before presenting any recommendations to senior management. The importance of this principle cannot be stressed enough; unless the group members are agreed on what actions are needed, there is little chance of obtaining management acceptance or approval of the group's report. Failure to achieve consensus can have disastrous results. Consider these examples:

- As noted earlier, the Metalco task force presented not one but three independent reports to senior management. The reports, representing the work of different subgroups, used different operating assumptions, presented contradictory findings, and reached unrelated conclusions. John Brown had continued the pattern of avoiding conflict that he began in the group's very first meeting; by failing to coordinate and integrate the three subgroups' efforts, he presented senior management with an all-too-accurate picture of chaos. The end result, as mentioned earlier, was the cancelation of the entire study in the midst of the presentation, and the virtual end of Brown's career at Metalco.

- Near the end of another task force effort the leader and an outside consultant who was working full time on the project realized that one component of the group's recommendations was too politically hot to put in writing. They carefully reworked the report and removed the topic from the agenda for the group's final review meeting, prior to presenting their report to senior management. However, they neglected to share these changes with a key subgroup member who had worked on that part of the project. He discovered the change only in the midst of the actual review meeting, and then raised the issue on his own. The subsequent discussion, lasting several hours, very nearly destroyed the entire project. The issue under debate involved a major restructuring of the purchasing function, and the purchasing manager—a powerful figure in the company who was on the task force and present at the meeting—was vehemently opposed to even discussing the topic, which he and several others felt went far beyond the group's original mandate. Fortunately, the group agreed to disagree, and the issue was left out of the final report.

THE FINAL PRESENTATION

The summary presentation of findings and recommendations to senior management is just as important as the task force's first formal meeting. The presentation should be carefully organized, with explicit attention to who will say what, in what sequence, and with what visual aids. The importance of these premeeting preparations varies in direct proportion with the extent to which the recommendations will be surprising, controversial, and/or expensive.

It is a good idea to brief key senior executives informally before the actual presentation, if at all possible. This briefing does not necessarily require their approval or agreement, but their advance understanding can help to prevent defensive reactions or categorical rejections of the group's proposals. This kind of preview is especially important if the recommendations involve major changes in organization structure, budget allocations, or strategic focus for any of the executives who will be present at the formal presentation.

As important as this formal presentation is, however, it rarely constitutes an adequate wrap-up of the task force project. Only if the recommendations are very straightforward and noncontroversial will the senior management group be able to understand and act on them at one sitting. A more effective strategy will be to plan *two* meetings. In the first, the task force can present an overview of the findings and recommendations, as described above, and distribute the formal report. At the end of this presentation, a second decision-making meeting should be scheduled for the near future. The interval between the two meetings gives the executives an opportunity to read the report and consider its implications.

The interim period will also be a busy one for the task force members, who can meet individually and in subgroups with key executives to clarify the report and its underlying assumptions. Only when the report has been formally acted upon can the task force consider its work actually completed.

RISKS AND CHALLENGES

More often than not, task forces serve as initiators of major organizational change. The problems they confront are typically complex, uncertain, and loaded with political implications. The stakes are usually quite high, both for the organization and for individual managers. Given this context, it is tragic that so many task force efforts fail, ending either miserably with a bang or sadly with a whimper. Task force leadership can be a highly stressful experience, and it certainly involves high visibility and high risk. But it also presents developing managers with a major opportunity to practice the fundamental skills essential to general management:

- Planning and leading effective meetings
- Interpersonal communication, negotiation, and persuasion
- Planning and conducting analytical studies
- Planning allocations of time and money for the project
- Developing formal proposals and presentations
- Balancing analytical and political considerations in change proposals
- Meeting budgets and deadlines

The guidelines suggested here can go a long way toward increasing a task force leader's personal effectiveness. In the heat of the battle it is sometimes difficult to remember basic principles, but that is precisely the most important time to return to fundamentals.

REFERENCES

1. Stanley M. Davis and Paul R. Lawrence, *Matrix* (Reading, Mass.: Addison-Wesley, 1977).
2. J. R. Galbraith, *Organization Design: An Information Processing View* (Reading, Mass.: Addison-Wesley, 1977).
3. Davis and Lawrence, p. 104.
4. Paul R. Lawrence and Jay W. Lorsch, *Organization and Environment* (Homewood, Ill.: Richard D. Irwin, 1969).
5. James P. Ware, *Some Aspects of Problem-Solving and Conflict Resolution in Management Groups* (Boston: Intercollegiate Case Clearing House, 9-479-003, 1978).
6. William W. George, "Task Teams for Rapid Growth," *Harvard Business Review,* March–April 1977.

7. A. K. Wicksberg, and T. C. Cronin, "Management by Task Force," *Harvard Business Review*, Nov.–Dec. 1962.

8. Antony Jay, "How to Run a Meeting," *Harvard Business Review*, March–April, 1976; James Ware, *A Note on How to Run a Meeting* (Boston: Intercollegiate Case Clearing House, 9-478-003, 1978.

9. See *Project Management: Techniques, Applications, and Managerial Issues*, Edward W. Davis (Norcross, Ga.: Production Planning and Control Division, American Institute of Industrial Engineers, 1971).

10. Frederick Brooks makes this point very dramatically in his entertaining book on project management, *The Mythical Man-Month* (Reading, Mass.: Addison-Wesley, 1975). Brooks argues persuasively that adding people to a late project simply makes it later, primarily because of the time it takes existing project members to bring new members up to speed.

10. Temporary Groups: An Alternative Form of Matrix Management

INTRODUCTION

That the hierarchical form of organization is inadequate in its primary vertical communications channels has long been known by scholars, managers, and employees, including the anonymous person who penned the ode to the organization chart: "Across this tree / from root to crown / ideas flow up / and vetoes down."[1] To complement the line organization, staff based on functional authority was added, providing a second group of communications channels at least partially orthogonal to the first. Yet these additional channels were still largely vertical in nature. The hierarchical/functional organization has the advantage of stability and provides a clear place for each skill. This stability, however, makes it resistant to change and difficult to focus on the cross-functional activities needed for the overall good.

The evolution of classical team or matrix management resulted from a need to focus on the whole. The building of a complex structure, the design of a new airframe, the execution of a customer's contract, and the carrying out of a complex R & D project are all task-oriented activities where various skills are assembled for the duration of the work. Horizontal communications predominate over vertical ones as the project manager's authority for the here and now becomes more important than the tenuous ties of the professional to his home base functional manager.

The use of classical matrix management requires, first of all, that the task be complex yet cohesive enough to be manageable by a team. Second, it requires that the task be large enough not only to afford the overhead of project management, but also to overcome the indivisibility of key resource people by having enough work to use them full time on the project. Third, the time scale must be sufficiently long to provide some team stability, and time enough to establish team momentum. And finally, the task needs to be one for which a

---

*Dr. Lynn W. Ellis is a consultant, was a vice-president responsible for research, development, and systems engineering at Bristol Babcock Inc. Previously he was Director of Research at International Telephone and Telegraph Corp., following a succession of engineering and general management positions with that company in the United States, Europe, and Australia. An earlier version of this article was presented at the 1978 Annual Meeting of the Industrial Research Institute.

horizontal communications structure is adequate, so that the project manager's authority is not undermined by random and diagonal pulls on team members. Clearly, not all organizational tasks are amenable to the classical matrix management concept. Many are too small, have short time cycles, need some limited effort from some unique resource people, or require a multiplicity of communication channels. Executives in a hierarchical organization, and particularly functional executives, need many of the tools of project management to overcome their stable organization's resistance to change, but through flexible, temporary group structures rather than specific task-oriented matrix management. Such was the experience of E. Maurice Deloraine:

Deloraine, appointed in 1933 as European Technical Director of ITT, found a problem with the operation of manufacturing companies in several countries, each with engineering departments for which he was functionally responsible. One of his solutions was European Technical Committees, each focused on a particular technical area, with a chairman selected for his skills rather than his nationality. By the use of temporary groups he achieved unifying technical recommendations without compromising the country managers' line authority for results.[2]

Since the first appearance of group-action phenomena in the Hawthorne Studies (1924–32), behavioral scientists—and in particular sociologists and social psychologists—have examined group behavior in organizations with a particular zest for blue-collar workers in smaller groups.[3] Yet we are rapidly becoming a nation of white-collar institutions and, because of economies of scale, a nation of large institutions and large factories.

From the viewpoint of the research manager attempting to stimulate innovation, this chapter will address: (1) functional authority, or the delegation by line management of authority to specialists; (2) the conflicts and needs of informal groups in large organizations, and where they parallel small ones; (3) the special characteristics of informal groups as impacted by organization size; (4) stimulation of and barriers to innovation; and (4) the temporary group in the large organization, both as a task and communication resolver and as the mechanism for establishing an informal peer group that outlasts the original temporary purpose.[4] It will be shown that informal or temporary groups that are well structured and have controlled autonomy can be an effective alternative form of matrix management, stimulating innovation by overcoming resistance to change.

FUNCTIONAL AUTHORITY FOR INNOVATION

Functional authority, or the delegation by line management of a portion of its authority to specialists, is an unavoidable necessity in large-scale organizations.

It is unavoidable because individual managers cannot possibly themselves perform all the tasks of their position. As the scale of the organization increases, the price paid is that every action taken depends on every other action taken elsewhere in the organization. To make the organization effective, a division of work must be made by specialty, and these specialists, wherever located hierarchically, need lines of information flow and control with their functional counterparts, both up and down in the organization.

Since so much of any functional activity takes place in groups, either formal or informal, a basic structural need is to nest these groups in a functional sense. At the same time, they must also be nested within the organizational level at which they are located. This dual role creates conflict that must be properly managed. For example:

> The installation of a new telephone exchange requires the involvement of the Buildings & Equipment Engineering group in the telephone company, and both the manufacturing engineering and application engineering groups in the factory which will supply the equipment (the two largest U.S. telephone holding companies also own their principal manufacturing suppliers). The telephone company chief engineer is delegated functional authority over the construction program and is held responsible for the on-time installation of the exchange. At the same time, the factory engineering groups must interface effectively with the factory manager, who may have several exchanges in manufacture for different telephone companies.

The larger the organization, the more tasks are likely to be delegated to functional chiefs. Yet a study of innovation has pointed out that the process involved an interaction between sales, production, and research specialists, who are part of functional departments.[5] To achieve control and coordination, both the medium-sized companies investigated had a central coordinating department and cross-functional coordinating committees. The coordinating department head was then still another functional chief, with limited specialists of his own and multiple interfaces with the other functions.

Looking at a larger-scale organization, one would expect to see such groups both in the operating divisions and at one or more higher hierarchical levels. Functional relationships prevail in innovation, as well as separately in sales, production, and research. As an example, consider the structure in the food ingredients product line of ITT Corporation in the early 1970s:

> The food ingredients division was small and its manager led the innovation effort supported by his functional chiefs. At the food group level, the lead in innovation was assigned to the Vice President, Marketing, and at corporate

level to the Director of Research. This functional line was cross-linked with other functional lines of authority in Marketing, Engineering and Production.

This innovation functional authority is essentially task-oriented and asks the questions what? where? when? and how? Yet Koontz and O'Donnell state that "functional authority is usually limited to the area of 'how' and sometimes 'when' and seldom applies to 'where', 'what' or 'who'."[6] They reason that "functionalization of management, if carried out to extremes, would destroy the manager's job." Avoiding this consequence is the major task of the manager with functional authority in the large-scale organization.

Also to be considered is the viewpoint of the lowest groups in the line. Their orientation is overwhelmingly to the questions of *who?* (which the functional authority does not pre-empt) and *how?* (where there is clearly a divergence of interests). Another area of divergence is on *where?*, which each level in the organization wishes to control.

In the management of innovation in the large organization, conflict is inevitably built into the complex structure outlined above. The spheres of authority are imprecise. Levels have differing objectives and perspectives. Clashes between strong personalities are frequent. The coordinating committee is a necessary mechanism to contain the conflict by bringing together individuals from various constituencies to talk together and bargain.

Harold Geneen, elected Chairman of ITT in 1959, lost little time in establishing a group of task forces to foster innovation in specific products or services he identified as essential to the growth of the company. For example, Task Force P1 was on Transmission Multiplex, P2 on Microwave Radio, P4 on Mobile Radio, etc. Chairmen were selected not based only on skills as under Deloraine, but from senior technical staff with the functional authority of the Technical Director of ITT. In addition, the task forces included staff product line managers and a few division managers and chief engineers to widen the diagonal lines of communication. A broad range of success was achieved, largely related to the effectiveness of the individual task force chairmen.

Obviously, the skill and leadership of the coordinating committee chairman have a major bearing on the success of conflict resolution. His alternatives to group bargaining are few:

1. Referral to a cross-over executive is tantamount to saying that he couldn't handle the delegation of authority.

2. Suggesting that the two protagonists engage in individual bargaining outside the meeting is of uncertain outcome: often the dispute has deeper roots in individual ambitions or in personality conflicts.
3. Writing rules is also of uncertain outcome, unless the dispute is narrow.

One example may serve to illustrate, as recalled by the author who was one of the participants.

A brilliant international executive with another company was recruited as Vice President-European Marketing for ITT. He set out to become and remain influential in the innovation area, in which his predecessor had largely operated by the book, according to the "Product Planning Guide." This guide had been put together several years previously by the World Headquarters Marketing and Research Departments acting in concert. It was, in effect, a top-to-top bottom edict to the divisions: write down the objectives in each innovation, then manage by them. It was moderately successful depending on the commitment of division and group product line management—in fact, there was a strong correlation between both percent new products and return on assets, with the percent of innovation projects covered by product plans. The Marketing VP seized on this point in the coordinating committee and ridiculed the "Product Planning Guide" as too bureaucratic and ineffective. Two of the headquarters members of the committee, one each from Marketing and Research, had been in at the conception of the "Guide" and naturally reacted with pride of authorship. Group bargaining went nowhere because those in functional authority resisted innovation in "how"—how to document an innovation objective. The Chairman suggested individual bargaining outside the committee, with no resolution. Finally, the Chairman proposed an ad-hoc working group to write new rules, with the Marketing VP as Chairman. This master stroke resolved the conflict by (1) creating a mechanism in which he could propose enhancing his constituency; (2) leaving the headquarters Marketing and Research members with their functional authority intact. They could still approve or reject the working group proposals.

In practice the "Product Planning Guide" as rewritten was so little changed in financial approvals that it was readily accepted. And the Marketing VP, by both fighting for his constituency, and involving them in the re-write, added to their commitment. The percent of innovation projects covered by product plans climbed rapidly, followed in time by an increasing percentage of sales based on new products.

In summary, functional authority is an unavoidable necessity of large-scale organizations. The structuring of this authority inevitably creates conflicts. Those assigned to a position of functional authority must work to create impartial mechanisms to get people to talk with each other for resolving the inherent conflict in the organization structure. They will need to sell and resell conflicting (or potentially conflicting) individuals and group leaders so as to get all pulling together. Only by properly managing the conflict inherent in functional authority can they get the parts of the large-scale organization in their particular function working together toward the organization's goals. Properly structured temporary groups such as task forces, committees, and ad hoc working groups can be one such effective mechanism.

LARGE AND SMALL ORGANIZATION PARALLELS

The division of work in a large organization creates efficiency through specialization. It also creates many small groups. For example, the administration department in the small organization may in its large counterpart become a cluster of small functional departments: organization, personnel, office services, transportation, etc. Similarly, a large organization may obtain economies of scale through centralization of some functions such as research and development, but have geographically dispersed small production facilities because transportation costs are a major factor.

In such large organizations, the rules of group behavior are the same in the small departments and facilities as in small organizations. The organization functions best in informal groups with high goals. Management must therefore manage and motivate groups as well as individuals. The individual need most strongly affected by the group is that of affiliation with other people. As a consequence, management must achieve its goals by appealing through the affiliation need. Contingent reinforcement theory stresses that the reward of group acceptance should be contingent on producing more. At the same time management must be aware of effect on output. One researcher cites as "factors conducive to group cohesiveness: homogeneity, easy communication, group isolation, limited size, outside pressure, and formal status."[7]

Another researcher advances the hypothesis that "the guiding perspective is that the culture of informal work groups is a manifestation of autonomy with the confines of the organization," and that "worker-autonomy can be regarded as a part of the barter arrangement between workers and the organization where limited affiliation with the organization is exchanged for a degree of autonomy."[8] Carefully controlled by management, such autonomy of informal groups can be an asset to productivity. Yet by abdicating in this manner a portion of its authority, management often has begotten a group that may limit its effort and output in the interest of self-perpetuation like the famous Bank Wiring Room at Hawthorne.[9] Where management's efforts to raise goals are thwarted by the actions of such autonomous groups, a small organization has limited scope to redesign the job by enrichment, or by retailoring the work station to individual effort. Where this latter approach is utilized, some substitute must be offered to meet the individual's need for affiliation (e.g., congenial rest areas for work breaks).

IMPACTS OF ORGANIZATION SIZE

In really large organizations structured into large subdivisions, dehumanizing factors strongly affect group dynamics as well as individual motivation. Groups (as pointed out above) are more cohesive in limited size. Management's task is thus to fit groups into the organization and to nest groups within groups. Communication is distorted by the number of hierarchical layers through which it must pass. Technicians are often unionized, which creates a parallel communications channel with both good and bad effects. Maintaining free flow of information becomes a dominant function of middle management. Large organization management also has many more restructuring opportunities to correct inefficiencies that have crept into its informal groups.

Even when organization size imposes the need for large departments, management retains the option of so dividing the work that group action is possible. The study by Ford of the service order function in a telephone operating company gives an excellent example.[10] In experimenting with job enrichment he describes a further technique of job nesting that improved both performance and morale by integrating service order representatives, typists, and control clerks into a multiskill team. He concluded that the nesting of related work may be a big step forward from single-job enrichment.

Another example where this has been successful was in ITT's Food Group. A small group of high potential researchers was segregated as an advanced development team and pointedly assigned "special funds" from corporate headquarters. This isolated the group from the swing of the business cycle to where they were able to come up with a number of strategic new product ideas and form them into new products. These included a zero cholesterol egg yolk replacer for baked goods and a truly low caloric weight reducing bread.

Dividing the work so that group action is possible is one management technique to use when the momentum is lost in a larger temporary group, or when such a group's charter is found to be unmanageable.

The success of large technical task forces in the late 1960's at ITT led the product line manager for data peripheral equipment to ask that such a group be organized in his area. It was soon found that the degree of commonality in Task Force No. 7 (as it was known) was limited only to concern peripherals physically connected to a computer: only one West German division and one United Kingdom subsidiary were interested in teleprinters; only another German division and one in Belgium were involved in mail handling, etc. Task Force No. 7 was terminated after a few meetings and replaced by several lesser and more successful narrowly focused groups.

The problem of communications failure through the barrier of hierarchical levels is a common one in large organizations. Often it is caused by too short a span of control, or over-supervision. With today's highly educated work force and proper job-enrichment procedures, less supervision and direct control are required. Simple mathematics shows that a span of managerial responsibility of 10 over 1000 productive workers leads to 3 levels of hierarchy and 111 overhead employees, as against 4 levels and 249 overhead employees when a span of tight control of 5 is utilized. Larger spans of responsibilities help not only with communications but also reinforce the need of the individual supervisor for self-esteem.

Other channels of communications exist in a large organization, and all should be used by management for setting high goals. The channel of communications via the union can be a force for or against management goals, depending on the state of labor relations. In the large organization it can also be a vehicle for information flow on output norms to the detriment of productivity. The company newspaper, public address systems, managers' letters, temporary groups (as discussed below), parallel reporting systems by function, and many other means are available to management to assure the flow of communications across hierarchical levels. When the ITT Corporation was faced with a multinational communications problem in the 1960s, Mr. Geneen added a worldwide product line management function to ensure both-way information flow between parallel business areas in different countries. Communications flow becomes a dominant function of many middle-management employees, to enable the hub of the wheel to operate in proper coordination with the operating activity on the rim.

Still another advantage of the large industrial organization is its ability to resolve impasses from past management mistakes by restructuring. Within an operating sector, an under-productive activity may be moved to a different department or made to fit into the organization structure in a different manner with different leadership, in a flexible manner not open to the small organization. For example, the recalcitrant bank-wiring function at Hawthorne could easily have been divided into the final assembly step on a number of separate product assembly lines, so as to counter the informal group's output-limiting actions. Finally, as the textile mills of New England found to their regret, the large industrial organization can simply close down and move to a new location where the labor force will accept management's high goals. Such an ultimate tactic is not, however, available to the public utility or large government organization that must remain on site to serve the public. Management must then use other methods to change the prevailing group dynamics that offer barriers to innovation and change.

STIMULATION OF AND BARRIERS TO INNOVATION

As Drucker points out, innovation is a condition of social and/or economic change.[11] For innovation to occur, people must change the manner in which they behave.

Stimulating Innovation through Organizational Change

Every text on management focuses, and has for years focused, on the need of innovation for business survival. As a consequence, a generation or two of managers has built into their organizational structure a science and technology function (research, development, and engineering are the usual nomenclature) and a marketing function. Some invention has resulted, a little more innovation has been noted, but by and large the paybacks from these functions have been greater in cost reduction and in market-share enlargement. Thus the technical effort of the past two decades has brought forth principally improvement rather than innovation, with the exception of a few sectors such as plain paper copying—a true innovation in clerical behavior.

The 1952 reorganization of the General Electric Company into product lines was copied by many companies around the world, but neither in G.E. nor in other companies adopting this structure did managers of various product lines innovate, because of the short-range focus of their day-to-day activities. *Business Week* equally attributed the decline of innovation to "a super-cautious, no-risk management less willing to gamble on anything short of a sure thing."[12]

Innovation and its stimulation are clearly tasks charged to the top management of any business. But top management cannot do everything and for only a few chief executive officers is the challenge of economic and social change a dominant factor. Short-range consideration of this quarters' earnings and cash position have become a principal focus of very many chief executives. Thus their organizations must face the reality of delegation—delegation of innovation to a director of research, a manager of market development, and a specialist in organization development. In the absence of crisis, these delegated internal change agents are often given the ball and told to run with it, with the enjoinder: "You're the experts!"

Given the situation described by *Business Week,* such "experts" apply their best knowledge to the situation as they find it, but often on reflection discover that in terms of their assigned task they have accomplished little. There has been a change in the behavior of people neither externally nor internally, the former being a direct consequence of the latter. One such "expert" stated his impact as follows:

> There had been no resistance, in fact his associates were not even conscious of their behavior—they thought they were doing everything he wanted them to do and considered him both productive and a great success. But actually there had been no change in attitudes, policies and procedures whatsoever. The organization worked exactly the way it had been working when he first started out.[13]

Many executives face this problem in their careers. It is a management problem, a behavioral problem, and not a problem of their functional speciality. What can be done, then, by an individual confronted with this problem? What practical strategies for organizational change can stimulate innovation in the individual, the group, and the organization? Again in the 1970s there was a need to analyze the situation at ITT Corporation, whose management perceived it as exhibiting the symptoms described by *Business Week*. For those then working on innovation at ITT, stimulating innovation in such an atmosphere was their most important problem.

Barriers to Innovation

In 1975 the barriers to innovation in ITT were essentially the same as those to any change in any organizational structure in any company. Some selected factors are listed below:

1. "Innovation isn't relative to this product line." Over half of ITT's product lines were mature businesses where the shakeout of weak competitors had been completed and well-situated firms remained. Improvement in features, and in service or customer relations, had become mainstays of business practice. The risk was that some other competitor would innovate with dramatic impact on the product line's future.
2. "The business is going well with present policies." Perhaps, but it may have been living on the innovative capital accumulated in prior years.
3. Division and product line managers had a short-run orientation. As noted before, the profit "in the hand" this quarter was worth two innovations "in the bush."
4. There were vested interests in the status quo. Since converting to a product line structure some 15 years previously, the product line managers in ITT and their technical product line counterparts had held the reins of power, while the more broadly based functions such as research had a lesser position.

5. "Change equates to an admission that something is now wrong." This is closely related to the previous point: taken together, they meant that efforts to stimulate technical and marketing innovation threatened the present establishment.
6. "Central control is likely to be eroded or lost." Change indeed threatened ITT's centralized control mechanism.

Thus the barriers to innovation were principally resistance to change in the individuals in authority, and the demotivating impact of the then centralized structure on the individual engineers and scientists and market and organizational developers who were "turned off" when it came to long-range ideas.

THE TEMPORARY GROUP

Formal and informal groups have a strong role to play in innovation stimulation. In practice, more formal and long-term matrix management has been a major contributor in this area. Lorsch and Lawrence prescribed such a solution, called the Coordinating Department.[14] As a top-down type of solution where only skill-based functions existed, this solution would have had merit; but since the product-line management function had been previously introduced in ITT, a second coordinating function would have been difficult to implement in the mid-1970s as an agent for change.

Yet some method was required to highlight, both to management and to individual technical contributors, that a change in their behavior was being requested; that the generation of innovative ideas was desired; and that such ideas as could be grafted onto the company's mainstream businesses would meet both management and financial support outside local budget limitations. Given ITT's forty-year tradition of temporary groups for such purposes, it was logical that still another such group should be considered.

One management tool of large organizations that is often useful for initiating change is the temporary group. Depending on need, management may use the term "committee," "task force," "working group," etc. In reality this temporary group is usually constituted from several different herarchical levels, contributing to communications flow. In a multilocation large organization, several or all sites may be represented. The group is meant to accomplish some assigned task and then—after weeks, months, or years—it is disbanded. As an indirect effect, its members often tend to constitute a peer group whose impact can be important to management long after the task which brought them together has been accomplished.

Maier has carefully analyzed group problem solving and found that, of the forces operating in them, some are assets and some liabilities, while some can be either, reflecting the skill of the group leaders.[15] He listed as assets (1) greater total of knowledge and information; (2) greater number of approaches to a problem; (3) increased acceptance because of participation in the decision; and (4) better comprehension of the decision. As liabilities he includes (1) social pressure for conformity; (2) tendency to accept the first concensus, not the best; (3) domination of meetings by certain individuals; and (4) the possibility of adversary proceedings. Depending on the leader's skill, forces that could be either assets or liabilities include (1) disagreement and conflict within the group; (2) willingness of the group to take risks; (3) divergence of initial position which affect time needed to obtain results; and (4) the stature of who changes his or her initial position. Given adequate weight to the above factors, the temporary group is often superior to the individual as a task resolver.

In the ITT Corporation the multinational, multilocation, and multiproduct character of its business make temporary groups indispensable. The informal peer groups established by participation in temporary groups create diagonal and random communication channels through the organizational structure that contribute to resolution of the temporary group's task. Many of these informal communications channels are so useful that they are adopted for other purposes than the assigned task while the temporary group is together, and many of these channels remain open long after the temporary group is disbanded.

Many recent studies on the substitution of telecommunications for travel have noted that this substitution is appreciably more effective when the participants are personally acquainted. Here again is an important secondary benefit of temporary groups in the large, geographically dispersed organization.

Looking further ahead, new series of informal groups for stimulating innovation could be started; the combinations are almost endless. Among cross-function groups, McGuire suggests several possibilities: grouping "developments, scientists, marketing specialists, and other product technicians close together in a single facility so that there is continuing informal contact between them"; "new product brainstorming groups—made up of five to ten persons"; and corporate "think tanks."[16]

Intrafunctional groups have long been used in ITT for multinational technical coordination and communication. Certain specialties in the materials area lend themselves to interdivision technical function groups in a single country. These tend to reinforce the esteem needs of individuals in the same way as presenting papers at conferences, but the proceedings may be kept confidential within the company with almost no restriction on the individual's text. The optimum mix of group efforts must continually be determined in accordance with current top-management strategy. It is likely that some new group efforts each year will assist in stimulating innovation, and some older groups must correspondingly end their operating lives with tasks done or recognized as not doable by the group as then organized.

AN EXAMPLE: THE ITT RESEARCH COUNCIL

A new temporary group—the ITT Research Council—was designed in 1976 following these principles. ITT had realized the need to place a premium on growth from within. To accomplish this, more attention had to be paid to research and advanced technology to provide the foundations for new higher-margined products. A separate research and advanced technology (RATEC) program was funded, beginning in 1976, to complement and supplement the in-house R & D programs in the groups and divisions. A new research director was appointed to administer this program, as well as to stimulate innovation throughout the company.

To obtain best value from this program, it appeared essential to obtain the assistance of the best and most creative brains of the company. To do this, an ITT Research Council was set up as a temporary group to advise on this investment in the future via the RATEC program. This permitted taking advantage of a greater total of knowledge and information among the membership, and created the potential for obtaining a greater number of approaches to the program.

The most important factor in structuring the council was the need to obtain acceptance of its decisions. ITT was structured by product groups in the United States and the European Economic Community Common Market. In several nations of the latter, ITT had group R & D laboratories, and was structured by country elsewhere, several countries having R & D laboratories. As a consequence, participation by the existing laboratories was essential to acceptance. The initial membership was therefore appointed as follows from these elements:

- 5 U.S. group laboratory directors
- 2 U.S. group product planning directors
- 3 Common Market group technical directors
- 2 department heads in 2 European country laboratories
- 1 divisional chief engineer (U.S.)

These individuals were divided into three classes, with three-year terms. Subsequent classes appointed in 1977 and 1978 slightly altered the make-up of the Council members from laboratories and divisions, without altering the intent of broad grass-roots membership.

Acceptance was also needed from the headquarters product line managers and their staff technical counterparts, and the following further appointments were made:

- 2 product group managers
- 2 product line managers

- 3 technical directors
- 2 group general managers
- 1 market development director

Rounding out the group were the General Technical Director of ITT, his two deputies, and the Research Director, for a total of 27 people. Because of the need to reinforce self-esteem and recognize the leadership role of head-quarters personnel, these latter 14 were constituted an executive committee, thus creating a two-tier structure.

By providing the line executives an opportunity to participate in the formulation and evaluation of policies for research through the ITT Research Council, an indirect way was provided for stimulating innovation. The Council offered a channel downward for communicating innovation goals. Equally, it became a means of feedback to top technical management as to the needs of the business as seen from the grass roots. Based on this bottom-to-top flow of information, modifications to the research and advanced technology program were formulated to increase its acceptance and value.

All told, some 40 programs were funded under the RATEC program. Not all of these projects were in new research areas, since advancement and acceleration of some earlier programs was also supported by the Council. Several of the programs did not get second-year funding from RATEC and, finding no other funding sources, were terminated. Two projects initiated under RATEC were successful in innovation, and radically changed the manufacturing and marketing outlets: controlled solubility metallic ion glasses, and electromotively stirred thixotropic metal billet casting.

Even more important, however, was the signal for increased innovation given to traditional product lines, where an ever greater acceptance of creative projects occurred through the period of the Council's activity. In addition to word-of-mouth summaries of the Council's activities carried home by the division and laboratory members, Council meetings took place in 1977 and 1978 in European laboratories, further passing the message to the field. By 1978 the Research Council had begun to repeat old arguments, and at the end of that year it ceased meeting, its designed task being as accomplished as its structure permitted.

Factors considered in structuring the Council—besides its formal status, the individual's creativity, acceptance, and communications—were group size and where to meet. Each individual had normal working relationships with one (and in most cases two or more) of the other members. Group size was held to 27, of which only 20 to 24 attended any one meeting, because of schedule conflicts. While this is larger than most behavioral scientists would deem ideal, ITT had a tradition of temporary groups and had found 20 to be quite manageable with its homogeneous executive cadre. The meeting locations were outside headquarters, to get away from external pressures and isolate the group for more effective effort.

As constituted, the group was well structured to take risks in allocating investments in innovation; since research is a long-time constant task, it had ample time to obtain results. There is every indication that this group was superior to any individual member in planning the research and advanced technology program-for ITT during 1976–78.

TEMPORARY GROUPS APPLIED TO FUNCTIONS OTHER THAN INNOVATION

While the past pages have dealt in detail with the innovation function, other areas of functional authority have used temporary groups in the past and can use them to advantage in the future. The principal advantages of such groups are in improved communications flow, a greater knowledge base, and participation's increasing the acceptance of the outcome. In almost all functions these needs can be met by temporary groups as a management tactic. A selection of such approaches from the author's experience follows, with no claim to present a comprehensive look at these other functions, which are not the author's primary background.

Functional Conferences

Each staff function in a large organization needs communications with corresponding functional specialists at lower levels in the organization. The periodic conference of these specialists (the Comptroller's Conference, Treasurer's Conference, etc.) can be a valuable means of improving this communications flow, particularly downward from the functional chief, or diagonally by having presentations made by specialists from other functions on topics of interest. Liabilities of such conferences stem from their size—typically so large as to impede upward communications—and from social pressures for conformity caused by mingling with those who control promotion or are competing for it.

The Quality Assurance Function

The Quality Department at ITT in the 1970s not only used the periodic functional conference, but also established regional Quality Councils as a means of better communications flow and of accessing a greater knowledge base. In this manner each factory or service quality manager met frequently with others nearby for frank discussions of one another's current problem areas. While the products of these factories or services were radically different, their employees' behavior was often very similar.

Another form of temporary group in the quality assurance function is the

Quality Circle employed in many Japanese factories. This is a multilevel group where workers meet together and with supervisors to discuss steps to improve quality. The advantages are the greater knowledge base acquired by involving the workers, and the comprehension and acceptance of the group decision derived from having participated in it.

Productivity Councils and the Manufacturing Function

The basic concept of a productivity council (as practiced in the United Kingdom and Australia 20 years ago) was for manufacturing engineers of noncompeting factories in an industrial district to meet periodically to review their plans for productivity improvement with their peers. The assets were the greater knowledge base, although at that time this often meant a heavier flow out from subsidiaries of U.S. companies than into them. The liability of such groups is now the legal restraints on actions that diminish competition.

Distributors' Councils and the Marketing Function

In geographically large countries, the independent distributor is a national element in marketing. The communications flow between a manufacturer and his distributors can entail conflicts of interest. A distributor's council can help improve communications flow, especially since such councils can be constituted of such a size as to produce good communication, by dividing distributors by region. Being in close contact with customers, distributors bring a greater knowledge base, and benefit from participation in the decisions of such councils. Liabilities of distributors' councils include the tendency of certain distributors to dominate the meetings, and the possibility of developing adversary relationships. Strong leadership skills are needed to keep council meetings focused on constructive topics.

Geographical Councils and the Public Relation Function

Recently, the Eaton Corporation has created State Councils in those states in which it has multiple operations.[17] Each council consists of designated managers of local factories who form a temporary group to work with state and local government officials and legislators. The advantage of such councils is local knowledge of issues, and the greater knowledge base provided by several plants rather than by just one alone.

FUNCTIONAL LEADERSHIP AND COMMUNICATIONS

While change takes place best when pushed by the Chief Executive Officer or when there is a crisis, neither factor was present in the mid-1970s to suggest a change to more innovative organization at ITT. Leadership, however, was required, if the improvements in individual creativity and group effort were to take place. This leadership had to be exercised within the scope of the functional authority of the key innovating executives: research and product line management. Besides the temporary group approach, researchers suggest that classic staff methods, two-way communications, and managed conflict offer possibilities of achieving an understanding and action.

The classic approach for a staff specialist to obtain compliance with an increased innovation program depends on the degree of support from the CEO. Without it, appeal to higher authority will not accomplish the assigned task, and is likely to obtain little allocation of the CEO's time. Technical competence and status will obtain a hearing, but count little against the short-run orientation so broadly established in many companies. Selling, persuasion, and allied approaches often have little effect on the set ways of the top-line executives. Education is equally unlikely to affect those same senior executives. Finally, owing to the long period when short-run "bottom line" philosophies have dominated management policies, often there is no large body of like-minded executives with whom alliances can be formed to bring about change by the "back door."

By providing the line executives an opportunity to participate in the formulation and evaluation of policies through temporary groups (such as a Research Council), an indirect way may be provided for reaching the goal. The temporary group can provide a line of communications downward for the communication of innovation goals. Equally, it can become a means of feedback to top technical management as to the needs of the business as seen from the grass roots. Based on this bottom-to-top flow of information, modifications to the innovation program can be formulated so as to increase its acceptance and value. This change in focus will probably produce desirable actions, but-not lasting understanding.

For change to take place in an enduring way, a broader understanding must be created of the need to stimulate innovation. This entails the risk of conflict between those responsible for the profit-and-loss accounts and those innovators working on the longer-term program. Resolving these conflicts will require mechanisms where people can talk together to resolve conflict through group bargaining. Temporary or formal groups, when established, can serve as this mechanism. Often these disputes will have deeper roots in individual ambitions or personality conflicts, and the skill and leadership of the group chairman will be an important factor in conflict resolution. This skill may in time become one of the necessary skills of future managers.

Still another potentially successful method might be to dramatize the lack of innovation by deliberately creating a conflict, thus showing how top executives must change their ways. To maximize the chances of success, a proper time and place must be chosen, and a topic selected where the potential of innovation is high and a strong opponent is on relatively weak ground. The risks of such a strategy are several: one may lose; one may win the battle but create resistance and so lose the war; multiple conflicts may be required. Given the right opportunity, however, the potential for attitude change by such a strategy is high.

The impact of exerting such leadership from a functional position will take time to be felt. Best results are likely by starting small and building on success. Efforts for change should be directed to tasks where the need is greatest, receptivity is high, and the impact on the company as a whole is likely to be the most significant, owing to the importance of the task.

SUMMARY

Management tasks such as stimulating innovation are often delegated within the organization. Examples of temporary groups give useful insights into practical strategies. A diagnosis of the problem finds that resistance to change in individual executives is a principal barrier to innovation. Individual creativity can be stimulated by easing demotivating factors, and by positive reinforcement and job enrichment. Formal and informal groups contribute to intra- and cross-functional coordination and communication. The key requirement is the exercise of leadership within functional authority so as to have the organization pulling in the same direction. Change is likely to proceed slowly in the absence of a substantial push from top management or a crisis situation. Conflict is inevitable in a situation requiring behavioral change and must be effectively managed for success in stimulating innovation.

"It must be remembered that there is nothing more difficult to plan, more uncertain of success, nor more dangerous to manage than the creation of a new order of things. For the initiator has the enmity of all who would profit by the preservation of the old institutions, and merely lukewarm defenders in those who would gain by the new ones."[18]

REFERENCES

1. Attributed by Peter Drucker, in *Management: Tasks, Responsibilities, Practices* (New York: Harper & Row, 1974), p. 797, to a senior Unilever executive.
2. E. Maurice Deloraine, *When Telecom and ITT Were Young* (New York: Lehigh Books, 1976), pp. 95–96.
3. Donald F. Roy, "Banana Time: Job Satisfaction and Informal Interaction," in David R. Hampton and others, *Organizational Behavior and the Practice of Management,* revised ed. (Glenview, Ill.: Scott, Foresman, 1973), pp. 229–47.
4. An earlier version of this article appeared under the title "Effective Use of Temporary Groups for New Product Development," *Research Management* 22 (Jan. 1979):31–34.
5. Jay W. Lorsch and Paul R. Lawrence, "Organizing for Product Innovation," in Hampton, pp. 381–94.

6. Harold Koontz and Cyril O'Donnell, "Functional Authority," in Hampton, pp. 476–85.
7. Hampton, p. 292.
8. Fred E. Katz, "Informal Work Groups and Autonomy in Structure," in Hampton, pp. 247–55.
9. H. Parsons, "What Happened at Hawthorne?" in *Science*, March 1974.
10. Robert N. Ford, "Job Enrichment Lessons from A.T.&T.," in *Organizational Behavior and Industrial Psychology*, ed. Kenneth A. Wexley and Gary A. Yuhl, (New York: Oxford University Press, 1975), pp. 94–105.
11. Drucker, *Management*, p. 785.
12. "The Breakdown of U.S. Innovation," *Business Week*, Feb. 16, 1976, pp. 56–69.
13. *The Case of the Frustrated Vice President* (New York: New York University, Graduate School of Business Administration, 1954).
14. Jay W. Lorsch, and Paul R. Lawrence, pp. 381–94.
15. Norman R. F. Maier, "Group Problem Solving," in *Organizational Behavior and the Practice of Management*, pp. 316–26.
16. E. Patrick McGuire, *Generating New Product Ideas*, New York, The Conference Board, No. 546, 1972.
17. "The Lobbying Arena Shifts to State Capitols," *Business Week*, Nov. 30, 1981, p. 58.
18. Niccolò Machiavelli, *The Prince* (1513), cited in David S. Hopkins, *Options in New Product Organization*, New York, The Conference Board, No. 613, 1974, p. 2.

11. The Plural Executive: A Unique Top Management Structure

Frank Oldham, Jr.*

The inability of one Chief Executive to possess all the skills necessary and to find the time to deal with all the pressing issues of the office is becoming more and more prevalent in the business world. In the days of the small, uncomplicated business, one man had little trouble dealing with all the management functions, but advances in technology, the movement toward national and international operations, and a widening product service base have created a serious time problem for the Chief Executive. As the pace of business quickens, demands for faster reaction to problems are inevitable. This in turn creates pressures that were probably unimaginable just a few years ago.

The literature indicates that corporate action has fallen short of solving the problem imaginatively. The introduction of assistants to the President to lighten the President's load was attempted with some success, but in the long run that success seems to be only temporary, as the demands of the office continue to grow.

The sharing of the load by the President and the Chairman has been tried by several companies with varying degrees of success. Some companies have formed small groups responsible for particular product divisions or staff functions at various levels of management. One problem with this approach is that it widens the distance between the "firing line" and the board room.

A possible long-term solution to the problem was introduced by the E. I. Du Pont de Nemours Company in 1921, who termed their new set-up a "committee-line" system. In 1921 this appeared to be unique in industry. A management team of ten men devoted full time to company affairs. The du Pont principle that "authority must be commensurate with responsibility" was the basis of operation and decision making. As one executive of du Pont put it, "The executive committee runs the company and the general managers run the business."[1]

---

*Dr. Frank Oldham, Jr., is Professor of Management at Arkansas State University, Arkansas. He also serves as President of Profit Consultants, Inc., a management consulting firm catering to the needs of community banks, and as a director of the Security Bank in Paragould, Arkansas.

After 1921 several other companies adopted a similar management organization. One of these was the General Motors Corporation which had as a major stockholder the du Pont Corporation. Alfred P. Sloan, Jr., a vice-president and later Chief Executive, proposed the reorganization plan, which he said was designed to preserve the advantages of decentralized operations but add a measure of coordinated control.[2]

The idea was slow in catching on, but other pioneering corporations followed, including the Standard Oil Company of New Jersey.[3] Many of the Fortune 500 companies have adopted variations of this concept in their organizations. Some have dropped the structure, while others apparently feel that it works.

ORIGINS

The idea of team management at the top of the business organization instead of a one-man Chief Executive is not confined to the United States. The concept has been prevalent in Europe for a number of years in varying degrees and forms. It appears the plural executive is most commonly used in Germany and Holland. In Germany it is known as the *Vorstand* and in Holland as the *Directie.*

The objective of the plural executive is to increase the capacities of the office of the Chief Executive to deal effectively with the full range of its responsibilities. Various names are given to the plural executive in the United States, including "Office of the President," "Office of the Chairman," "President's Office," "Corporate Executive," "Office of the Chief Executive," and "Plural Executive." Throughout the remainder of this chapter the title "plural executive" is used to indicate the top management team.

The literature suggests that the plural executive performs most effectively those Chief-Executive tasks where the limitation of one man's judgment is most apparent; in setting objectives and policies, in capital and manpower decisions, and where the advantages of a team effort are badly needed. On the other hand, it seems less effective where personal negotiating is important, as in labor negotiations.

MEMBERSHIP REQUIREMENTS

There is some disagreement about the membership requirements of the plural executive. Some writers and businessmen indicate that members must be aggressive, individualistic, and arbitrary, whereas others advocate just the opposite. However, there are some characteristics that most writers and businessmen agree on, including (1) social skills (human relations), (2) maturity, (3) judgment, (4) native intelligence, and (5) diverse backgrounds.

DECISION-MAKING RESPONSIBILITY

In broad general terms the operation of the decision-making process by the plural executive at du Pont is similar to those established later in other companies, but the plural executive in each company appears to vary in response to that particular company's needs. The plural executive is both a policy-making group and a supervisory body. It has control over objectives, strategy, appropriations, selection and promotion of employees, and relations with government, employees, and suppliers. Since it is the top decision-making group, it formulates the basic organization of the company.

In the purest theoretical sense, team members have equality in decision making, but an effective team cannot be leaderless. As a result, it is necessary for some team member to be the *primus inter pares*. This apparently is a very difficult role and requires a great deal of understanding.

The organization chart of a company using the plural executive could be represented by a circle at the top of the organization to differentiate it from the traditional organization (Figure 11-1). From the circle, at least two relationship patterns could exist. First, the members of the plural executive could share the top responsibilities of the Chief Executive and have specified responsibility and accountability for various portions of the organization (Figure 11-2). Or the plural executive could be structured so that there is no direct accountability or responsibility for any specified portions of the company (Figure 11-3). All portions of the company would report to the "office." In some cases it appears difficult to differentiate between these two organization methods.

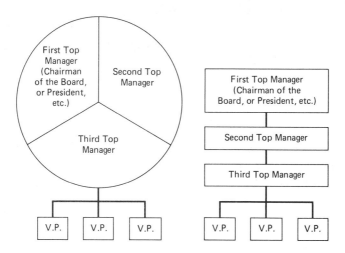

Fig. 11-1. Plural executive and traditional organizations.

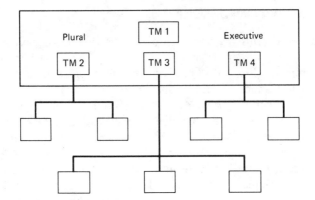

Fig. 11-2. Accountability for specified operations are retained by individuals in the plural executive.

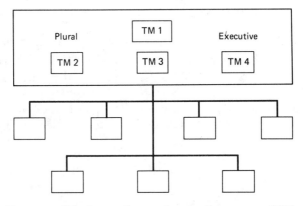

Fig. 11-3. No accountability for specific operations, but liaison responsibilities could exist.

SIZE OF THE PLURAL EXECUTIVE

The size of the plural executive seems to vary considerably. The original plural executive in this country, the du Pont Company, had ten members. The trend seems to be toward small plural executives. According to a study conducted by the author, most companies seriously pursuing this concept have between three and seven members.

Both large and small plural executives have certain advantages and disadvantages. The larger the team, the more difficult it is to find qualified team members and to get agreement among members. On the other hand, the smaller the team, the less diverse the team members' backgrounds, making it more difficult to make intelligent decisions. Regardless of the team size, it is important that the team produce a united front. In other words, once a decision is reached by the plural executive, even though there might be some disagreement, all team members must stand by it. This, of course, could create additional problems.

MERITS OF THE PLURAL EXECUTIVE

A feeling of consensus on the merits of the plural executive seems to pervade the literature. Although various writers and researchers list a few more than others, the majority include those that follow.

Expansion of Capacity to Perform

The most frequently mentioned advantage of the plural executive is that of a "time stretcher." Top executives are required to spend much time traveling and meeting various audiences face to face. With the plural executive, this burden can be spread.

Reduction of Inhibitions Among Executives

The traditional organizational alignment has been found to inhibit senior executives. In this respect the plural executive could contribute to an increased freedom of expression. The plural executive could be successful in promoting a more open and effective set of working relationships without reducing unsoundly the primacy of the Chief Executive.

Development of an International Viewpoint

A company that intends to expand its manufacturing and marketing operations to an international scope needs the diversification of talents that the plural executive offers. It especially needs men knowledgeable in international political and economic areas. In the large corporation, management's perspective must be worldwide. The ability to predict and evaluate the changing political and economic climate of the world is a necessity. A group of well-informed and experienced members of a plural executive are better suited to meet such demands.

Flexibility in Resource Allocation

The need for accurate allocation of millions and sometimes billions of dollars annually requires mature judgment that the plural executive can provide. The pooling of judgment and knowledge in the plural executive adds an invaluable dimension to resource allocation. It provides the flexibility that the top spot in the organization demands: the flexibility to apply expertise in resource allocation wherever needed in the organization.

Integration of Planning

The plural executive can expand and improve corporate planning. It provides the opportunity and scope for free expression. Historically, several companies have indicated the facilitation of divisional inputs into the planning process by the use of the plural executive.

Orderly Succession to the Top

If a company were operating on a single Chief Executive concept, the unexpected loss of that executive could have serious repercussions on the organization. With the plural executive, the loss need not be so severe. The plural executive provides a training and testing ground for successive generations of top management. If a three-man team were at the top, a loss of one of the executives would still leave two for the operation of the company. The needed time and careful study for a replacement would be provided.

Fighting Isolation

Various studies indicate that the isolation of the single person at the top can be a serious problem. With more than one person at the top, this problem can be eliminated to a large degree. The plural executive can make provision for discussing ideas and problems at the top.

DEMERITS OF THE PLURAL EXECUTIVE

As with almost every management concept, the plural executive is not without demerits. Various limitations that have been suggested are as follows.

Qualified Team Members

It is very difficult for some companies to find one qualified Chief Executive, let alone three, four, five, or more. A study by the National Industrial Conference Board indicates that the most perplexing and continuing problem faced by businesses today is, "Identifying and developing those rare individuals who have character, intelligence, motivation, education, stamina, and personality to be top caliber executives." In addition, an executive who functions well as a Chief Executive may not work well as a team member.

Increased Isolation

Just as the plural executive can fight isolation, so it can also work in the opposite direction. The team can unintentionally become so involved with themselves and working together that they lose contact with the day-to-day functioning of the company.

Authority of Individual Members

When a plural executive is formed, the formal or informal relationships that team members previously had with other members of the organization may continue to exist. For example, suppose a former direct subordinate of a team member asks for an answer to a particular question. Is this advice or a decision? Regardless, the team should always present a common front.

Image of Leadership

The image of many large companies today is set by the man at the top. This image is related in some degree to the success or failure of that particular company. What happens to that image when the plural executive, composed of men with varying personalities, traits, and abilities, is established? Some studies show that the *primus inter pares*—the leader—sets the image. Whether or not this is true in all cases is yet to be determined.

Evaluation of Subordinates

How does the plural executive evaluate subordinates? If all members of the team have the same view toward subordinates, the problem is solved, but this appears very unlikely. The best solution is for each team member to constantly keep in mind that he has to balance his judgment against that of his teammates.

IMPLICATIONS FOR FUTURE DEVELOPMENT

If the use of the plural executive does increase, as some predict, what are the implications for further development in the organization of the company? Based on the research to date, two developments are indicated: the emergence of a "super staff," and revision of executive development.

Super Staff

When a company is organized along the single-executive line, the advisory staffs are functionally organized with each one specializing in marketing, finance, or other functional areas. The requirements of team management emphasize that the plural executive is responsible for developing company policy, carrying out company objectives, and efficiently allocating company resources. The present functionally aligned staffs seem poorly equipped to advise in a competent manner. What is needed are staffs capable of coordinating and integrating functional viewpoints.

Executive Development

As already discussed, the plural executive requires a different type of individual, and more of them, than the single Chief Executive organization. This will probably call for a revision of present executive training programs with emphasis on broad-minded, knowledgeable individuals who are not overly aggressive and are adaptable to team work.

CONCLUDING COMMENTS

From the above evidence, it seems premature to conclude that the plural executive is here to stay. Only a minority of management theorists support the team approach to company management, but for these the plural executive seems to be an escape from the increasing pressures on the single Chief Executive.

Any method of top management organization that leads to an interaction of ideas with the possibility of synergy will bring a greater understanding of the modern corporation and sounder management decisions. If the plural executive firms are as efficient and effective as their proponents say, then they should outperform their traditionally organized competitors. Additional data concerning plural executive firms and their effectiveness are needed before any sound conclusions can be made.

REFERENCES

1. William H. Mylander, "Management by Executive Committee," *Harvard Business Review* 33 (May–June 1965): 51.
2. Harlow H. Curtice, "General Motors Organization Philosophy and Structure," in Harold Koontz and Cyril O'Donnell, eds., *Management: A Book of Readings* (New York: McGraw-Hill, 1972), pp. 341–346.
3. D. Ronald Daniel, "Team at the Top," *Harvard Business Review,* Vol. 43, Mar-Apr 1965, p. 75. Also see Harold Koontz and Cyril O'Donnel, *Principles of Management: An Analysis of Managerial Functions* (New York: McGraw-Hill, 1972), p. 379.

12. Product Management in the Matrix Environment

M. Dean Martin and Penny Cavendish*

INTRODUCTION

Product management is generally considered to be a variant form of the emerging discipline of project management. Other forms include projectized management, and new venture and program management. As the environment for marketing has become more and more complex, top management has consistently sought to develop organizational structures that are adaptive and more responsive to environmental change. Firms have grown larger, have developed multiple product lines, and have practiced market segmentation in an effort to cope with the demands of specific groups of consumers and to meet the threats of competitors.

Product management as an organizational approach to the marketing of products is generally considered to have originated at the Procter and Gamble Company in 1927. A new product, Camay soap, was not meeting sales expectations. One individual was assigned as a product manager to focus market efforts for the product. This concentration of management attention was successful, and the approach has since gained adherents with the passage of time. Some companies such as Colgate-Palmolive, Kimberly-Clark, and American Home Products, have successfully implemented the approach. Others such as Purex, Eastman Kodak, and Levi Strauss, did not find the approach as useful as anticipated and have turned to other organizational structures. In some companies, notably Procter and Gamble and Johnson and Johnson, the product manager is referred to as a brand manager. Regardless of title, the basic concerns are the same: one individual is singled out and given responsibility to manage a product or group of products. Depending on the company, the position is usually delegated a range of authority and responsibility, and exists within an organizational format termed the matrix. The purpose of this chapter is to examine the functioning of the product manager in this matrix organization.

---

*Dr. M. Dean Martin is an Associate Professor of Marketing and Management, and Penny Cavendish is a Graduate Research Assistant in the School of Business, at Western Carolina University, Cullowhee, North Carolina.

THE MATRIX ORGANIZATION

The matrix organization is described in other chapters of this handbook; however, a few words are appropriate here to provide a context for the product manager discussion. There are several forms that the matrix may take. In Figure 12-1 a typical structure is illustrated to provide a basis for discussion purposes. Product manager A, for example, reports to the group product manager, who in turn reports to the Vice-President for Marketing. The circles in the figure illustrate the interaction of responsibility areas for the product manager and the functional managers. For example, at node X the joint responsibility of the product manager for product A and the sales manager is emphasized. The product manager must depend on personnel who report directly to the sales manager to market product A. The implications of this type of relationship will be considered later. In like measure, at node Y the joint responsibility of the product manager and the Vice-President of Manufacturing is highlighted. Both individuals must have an understanding of the market and the ability of the firm's products to meet this need. It is a joint responsibility that requires coordination and a sense of dedication and loyalty to the total organization, not merely allegiance to one functional empire.[1] Thus questions of authority, responsibility, and a myriad of other issues are brought into focus in Figure 12-1.

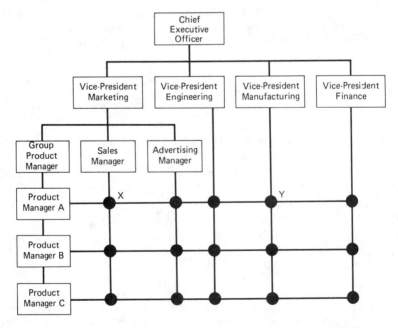

Fig. 12-1. Representative matrix structure.

The Marketing Function

In the organizational setting the marketing function is basically responsible for the marketing of the products manufactured by the firm. This objective involves decisions such as product design, package design, inventory management, distribution channel management, and others. The Vice-President for Marketing will need to coordinate marketing activities with engineering, manufacturing, and finance in order to minimize conflict.

For example, manufacturing will be concerned with production costs and seek large production runs. However, marketing must be sensitive to the location, intensity, and magnitude of demand for any given time period. Since products must be stored if they cannot be sold during a given market period, the marketing staff, including the product manager, may well want to manufacture a smaller volume of the product to meet a sales forecast.

The Marketing Environment

The sales forecast is based on a screening of the environment external to the firm which provides the boundaries for the implementation of the marketing program. This environment has become increasingly complex and uncertain in nature. Top management generally uses forecasts, projections, and other information to assist in formulating the broad strategic objectives for the firm. These inputs relate to political, economic, social, technological, and ecological variables, as illustrated in Figure 12-2. The variables, such as politics, are related, but can be considered separately to illustrate the key points of environmental complexity and uncertainty. For example, a federal law dealing with product safety must be anticipated. If this is anticipated, then perhaps its outcome can be influenced. If not, then the firm must react to it after passage to determine the impact on the design of each product in the product line. The other variables must be considered in like manner. This situation of conflicting needs and environmental complexity and uncertainty illustrates why firms are led to the product manager concept. It permits the firm to assign responsibility for a significant product to an individual who will then be responsible for the product or group of products from birth to death.

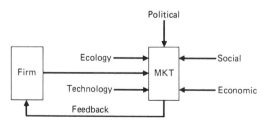

Fig. 12-2. Environmental scanning.

Marketing Strategy

In the multiproduct firm, feedback from the environment forms the basis for planning and control as related to the marketing function. This information is used for both strategic and tactical product planning. In terms of a given strategy, both the target market and the marketing mix come into focus.

The Target Market.   Multiproduct firms often have product lines that feature competing products directed to different market segments. For example, one detergent will appeal to people who are energy conscious and wash their laundry in cold water, whereas a more traditional individual will want a detergent that works best in warm water. The product design decision as well as other decisions are therefore unique; with a multiplicity of products, target market selection and maintenance can become complicated. Each market segment will require a differentiated marketing mix.

The Marketing Mix.   The marketing mix for each product must be tailored to the capabilities of the firm and the demands of the target market. Each element of the marketing mix will be briefly considered.

Product.   Product design involves a delineation of features that will appeal to consumers. For peanut butter, for example, taste and texture may well be the critical factors used by consumers to evaluate the product. If the product is appealing from this perspective, what about the package? Should it be a glass jar with a label or a metal can? Assuming a glass jar, what type of label will attract buyers? What type of structure should the jar possess for ease of handling and shipping, and for placement on display counters? These questions are indicative of those that must be asked for each product—certainly a complex and difficult situation for a manager responsible for many products.

Price.   Obviously, production cost and competitors' prices influence the pricing decision. Also, where is the product in its life cycle, as related to consumer acceptance? Should the manager use a skimming strategy and price the product high, with eventual change to penetration strategy as the product moves through its life cycle? Again, a complex situation for the manager with multiproduct responsibilities.

Place.   A key question for the marketing manager relates to the distribution channel for the product. Should the product receive exclusive, selective, or intense distribution? What type of channel is best? What channel is the one with the lowest cost? Can the traditional channel for a given product be changed to provide a lower overall distribution cost and thereby give the firm a competitive edge in the marketplace, higher profits, or both? These and others are critical questions that the marketing manager must face for each product in the product line. Each question involves decisions fraught with varying levels of uncertainty.

Promotion.   For the marketing manager, promotion becomes a key decision variable. Consumers must know about the product if they are to purchase it. What promotional devices will be best for a given product? Should the manager use a pull strategy with mass advertising, or a push strategy with personal selling? What about the use of coupons in newspapers, displays in stores, special discounts, and other promotional tactics? Once again for the marketing manager with responsibility for multiple products, the marketing job becomes complex and complicated.

Situations such as these have led firms to adopt the product manager concept for the marketing and management of individual products, where the product's success is critical to company success.

The Product Management System

This system as shown in Figure 12-1 is a subsystem of the firm as a whole. In the case at hand it is a component part of the marketing division, yet the concept of product management implies that an individual will be designated as the manager for one or more products. Depending on top management's preferences, the individual can be granted great or very little authority. Regardless, this individual becomes the organization focus of planning and control for a product or products. The individual is visible and thus can be held accountable for the success or failure of a given product.

This accountability places considerable pressure on the product manager in the matrix, who must operate within the constraints and boundaries established by upper management, yet at the same time work with the functional managers in marketing, as well as in other functions such as manufacturing and finance, to develop a marketing plan for each product under his or her purview. A useful tool for management and the product manager is the product charter. This document is essentially the result of strategic planning. Suppose the decision has been made to develop and introduce a new product to a given market segment. The stakes for the company are high, so management decides to assign a product manager to develop and introduce the product. A product charter would essentially include:

- A description of the product and its features or attributes
- Proposed target market for the product
- Organization and authority boundaries
- Authority and responsibilities of the product manager
- Estimate for development cost

- Schedule for introduction of the product
- Suggested marketing-mix considerations
- Identification of other organizational elements and their responsibilities
- Management approval for the product development and approval program

Thus the stage would be delineated for the product manger, whose role is clearly defined. As mentioned previously, charters for product managers can grant broad or narrow belts of authority and responsibility. The position of this chapter is that the product manager should have authority to manage all aspects of the product, including responsibility for at least detecting and reporting the need for new product development. If the individual is to be held accountable for the product's success or failure, then he or she should have the authority to make or at least influence critical decisions. Subsequent sections of this chapter will assume that the product manager is essentially the manager of the marketing channel for a given product—that is, through its complete life cycle from its conceptualization to its demise. Thus the decision to use the product manager approach will be based on these concepts, but will also focus on the product itself and on several other product considerations.

PRODUCT CONSIDERATIONS

As seen from the above discussion, one of the four components of the marketing mix is the product. A product can be defined as a set of tangible and intangible attributes designed to satisfy the wants and needs of a buyer. Being the medium through which organization objectives are met, the product is the fundamental reason for the organization's existence. Therefore some method to manage products is a necessity. Where appropriate, this management can usually be achieved most effectively and efficiently by the product manager approach. This approach is most appropriate in a multiproduct company where a large number of products are produced and marketed. Two prime examples of companies using this approach are General Mills and Procter and Gamble. The activities to be performed will differ, depending on the product classification: consumer or industrial good.

Consumer Goods

The activities of the product manager involved with consumer goods are more impersonal and can be classified according to the shopping behavior of consumers. The classification includes convenience, shopping, and specialty goods. Factors that affect the buying process can be divided into four groups. The first group comprises buyer characteristics, which include cultural and personal determinants, family, social class, motivation, perception, age, and other influences. The second group involves product characteristics such as styling, quality, price, availability, and support services. The third group relates to seller characteristics, which influence the buying outcome based on the image the consumer has of the manufacturer as well as the friendliness, knowledge, and service capabilities of the retailer. The fourth group involves situational characteristics: time pressure, time of year, weather, and the current economic outlook.

The product manager must be aware of all of these factors, but his main focus will be on product characteristics, notably promotion, price, and distribution. Before developing his marketing strategy, the manager must fully understand the role his product plays in the life of the consumer. The meanings assigned to the product must also be understood. For promotion decisions the media, advertising, and appeals must all reflect the attitudes and personality characteristics of the buyer. The product manager must market products that appeal to certain segments of the market in selected types of outlets. For example, designer jeans are not usually sold in discount department stores. Price is the third area with which the manager is concerned. Some market segments are more sensitive to price differences than others, which tends to limit pricing strategy. Consideration of these factors by the product manager is a must, but he or she should also know that these factors are applied with different emphases to the marketing of industrial goods.

Industrial Goods

Activities for industrial goods—products to be used directly or indirectly in the production of other goods—differ from activities involved with consumer goods. This market includes manufacturers, government agencies, mining firms, wholesalers, retailers, utilities, real estate firms, and other institutions. For industrial goods, the relationship between the product manager and the buyer is more personal. As with the consumer goods, the focus of the manager is on the marketing mix, but with a different emphasis. Industrial buyers purchase products that will generally be used to produce other products. In most cases the product is manufactured to a specific set of specifications. The product manager must keep in mind the need to service the product after the sale. An example of this type of relationship can be seen in high-cost projects such as a data-processing system. Promotion, one area of major concern, is again highly personalized. The product manager depends on personal selling with trade shows and other specialized media. Yet industrial products are bought infrequently: for example, a company does not buy a large computer system very often. Small products used daily, such as nuts and bolts, are usually bought under a contract covering a distinct period of time. Even though they do not produce sales daily, these customers cannot be ignored. Providing excellent feedback, sales calls must continue to keep the buyer and manager up to date on needs and developments.

Place, the second area of concern, is more or less fixed. Since there are fewer potential buyers, direct buying is more common in the industrial market. For example, buyers of expensive machinery often go directly to the manufacturer to buy. The third area of consideration, price, is usually negotiated between the buyer and seller. These are the major characteristics with which the product manager is concerned when focusing on the industrial buyer. As with the consumer goods, these factors are not independent of one another. The manager also must remember that the demand for industrial goods is *derived,* meaning that the demand is based upon the demand for consumer goods. Thus the activities of the product manager depend in large measure on the type of good involved—consumer or industrial.

Product Mix

The company also generally has different product mixes and product lines within each classification, with which the product manager must be concerned. The product mix includes *all* products offered by the particular organization. General Mills is a good example of a company with a broad product mix that includes cereal, pizza, cake mixes, icing, flour, instant potatoes, and granola. A product mix is usually described as having width and depth. Width refers to the number of product lines, and depth, to the number of products in the product line. These dimensions help determine whether or not the company should increase the width of the mix and capitalize upon the company's reputation and skills. Also, depth may be increased, enticing buyers with different needs and tastes. With the aid of feedback and periodic evaluations, the product manager can determine what lines to add, strengthen, or delete. Product managers must be concerned with optimizing the product mix, making additions and deletions to the mix in a timely manner in response to opportunities and resources. The product mix is further broken down into product lines.

Product Lines

When a group of products are related by customer, marketing or production considerations, it is considered a product line. For example, Procter and Gamble's detergent product line is composed of Cheer, Dash, Duz, and Gain. The product manager must be familiar with the sales and profits for each product in the product line, as well as the product's position compared to competitors' product lines. There are a number of issues that the product manager must face, such as product line length, line stretching, line modernization, line pricing, and line pruning. In some cases a product line can be too large for a single person to coordinate effectively, so many organizations are turning to the product manager concept where an individual will be responsible for one or more brands in the product line.

Product Life Cycle

The product manager must manage all aspects of the product throughout the product life cycle. The marketing mix of the product changes over time as the product moves through this cycle. It changes because of competitive environment, buyer behavior, market composition, and other influences. The four stages of the product life cycle are illustrated in Figure 12-3.

During the introduction phase the product is placed on the market. Sales are slow while demand is being developed. During this stage technical improvements are made. If people respond favorably, the product will move into the growth stage. Sales expand rapidly, increasing the profits of the firm. Attracted by the market opportunity, competitors enter the market, each trying to develop a competitive product. This stage marks the peak of profitability. Sales continue to expand, moving the product into the maturity stage. At this point competition is even more aggressive, causing profits to decline. After the maturity stage, sales of products decrease in the decline stage. The product is no longer a profit generator and may even be a burden. Hence either the product is dropped, or a way is found to extend its life.

Product life cycles do exist, but they may vary in length. The importance of the various aspects of the product mix will vary, depending upon the stage the product is in. Since sales partly reflect the market effort expended, such efforts have some impact on the product life cycle. Knowledge of the stage the product is in, how long it has been there, and when it is likely to move to the next stage, can be quite useful to the product manager, affecting the decisions to be made. Next, the impact of the marketing mix on the life cycle, and how the product manager changes his focus during each of the stages, will be considered.

Introduction Stage. During this stage the objective of the product manager is to gain acceptance and initial distribution of the product. Competing prod-

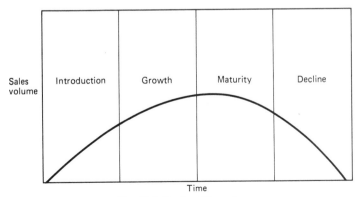

Fig. 12-3. Product life cycle.

ucts may or may not exist. Promotion is needed to create awareness of the product's nature, uses, and availability. There must be a fairly heavy level of informative advertising. Product managers can be selective in their distribution of the product and may well use an exclusive distribution technique. The market price will depend on the product, competition, and other environmental variables. The manager may want to use trade discounts. In this stage the product manager must detect product weaknesses and consumer attitudes.

Growth Stage. The objective of the product manager is to establish brand position with the distributors and users. With rapid growth and competition, the product manager must use the marketing mix to his advantage. The product strategy needs to be flexible, so as to adapt to developing segments. While demand is still increasing, prices tend to remain the same or decrease only slightly. Distribution is intensive as well as extensive. Service is provided to dealers. To meet competition, promotion expenditures are about the same or somewhat higher. To create brand awareness and preference, mass media are used extensively.

Maturity Stage. This third stage poses the most significant challenges for the product manager. His objectives are to maintain loyalty, defend market position, and strengthen distribution relationships. At the beginning of this stage, competition is still intense, but later on competitors tend to drop out. The product manager would be well advised to lighten product lines by improving or differentiating the product. This is the time to attract new customers. Price may be cut in order to extend market share. Another way to increase the number of customers is to increase the promotion. To attract attention and interest, advertising efforts need to be revised. More aggressive promotion such as cents-off coupons and gifts may be developed as a more direct way to attract users. Distribution should continue to be extensive and intensive.

Decline Stage. This stage is usually characterized by rapidly falling sales. There are few competitors. Trying to maintain profit levels is a losing battle. The product manager is selective in his choice of distribution channels; marginal channels are phased out. Promotion is decreased but sustained in order to sell as much of the product as possible. During this stage the product manager needs to determine if the product should be eliminated. Such a decision should not be made in a hasty manner. Rather, the market should be examined to see if opportunities exist for product extension or renewal. A classic example is du Ponts decision after World War II to enter the consumer textile market as an alternative to deleting nylon from its product line.[2]

The product life cycle concept is used by the product manager to interpret product dynamics. It is useful as a planning tool to help identify conditions of the market during the different stages of the life cycle. The product life cycle can also be used as a control tool, allowing the company to compare how well their product is doing in relation to comparable, successful products launched in the past. Overall, the concept can aid the product manager in formulating a specific marketing strategy at each stage of the life cycle. Responsibilities such as these place a premium on the type of individual selected to fill the position of product manager.

THE PRODUCT MANAGER

Typically, the product manager is assigned one or more products for which he or she is responsible. For the product manager concept to work, outstanding individuals must be selected for the job. Being fully committed to the product, this individual greatly influences its destiny. Companies have different job requirements, depending on the authority delegated to the product manager. Different product assignments also require different skills and characteristics. Before management chooses a product manager, the desired skills and characteristics for the job should be detailed in a written format.

Selection Characteristics

To find the right individual for the job, top management should identify the characteristics that the product manager must possess. Most important, the product manager must be able to get the job done by working through other people. He must be sensitive to the cultural ambiance of the organization. This sensitivity involves a knowledge of both the formal and informal aspects of the firm's employees and their value systems, working group norms, organizational objectives and structure, and policies, processes, procedures, and rules characteristic of the matrix.[3] This involves a number of other characteristics. The person must have great strength in human relations skills—in other words, in communication, motivation, and team building. Communication is of importance when articulating views and persuading others. Since the product manager will be working across functional lines, he must be able to open up two-way communication with functional personnel. In many ways the product manager will be a communication link. The product manager will be working not only within the boundaries of the firm, but also with individuals of other firms—for example, industrial buyers. The product manager must also be multidisciplinary oriented. Dealing with a range of business problems, he or she will need sound business judgment. Working with a number of other people, the manager must have coordination and negotiation skills as well as the ability to intergrate activities. Within the bounds of corporate policy, the product manager must be innovative—open to suggestions, while encouraging others to be innovative also. The manager should be able to cope with pressure, handle conflict, and adapt to change.

These skills should be written in a checklist form for use when screening and selecting individuals to fill the job of product manager. It may be difficult to find a person with such a range of skills, but the choice is still critical.

Product Manager Choice

After setting standards, top management must select the individual to fill the product manager's job. Top management must take care in choosing the right person. For product development it is most appropriate for the individual to be chosen at the very beginning of the process. It is also important that the company provide a thorough orientation of those chosen.

There are two approaches top management can use in selecting the product manager. The first is to look within the organization at proven and successful line managers. The second is to go outside the organization and hire someone with experience. There are advantages and disadvantages in both cases. Choosing someone within the organization is easier. Although this experienced line manager knows the organization, he or she may not be a good product manager, since the two jobs are quite different. The line manager may be comfortable in managing specialists, but ineffective in the high-pressure, multidisciplinary product manager position. By going outside the organization to find a qualified experienced product manager, top management is taking more risk. If an individual is brought in from outside, the organization may experience some adverse effects on morale. Being in a new organization, the outsider must be granted time for adjustment and orientation.

The importance of the choice by top management cannot be stressed enough. The decision to hire within or outside the organization must be made while taking all advantages and disadvantages into consideration. To attract qualified individuals, top management must offer rewards commensurate with the job.

Evaluation and Rewards

The criteria to be used for evaluation must also be determined. As a consequence of the qualitative factors existing in product manager assignments, it is hard to develop criteria providing a fair basis for evaluation. Both product profitability and share of market have been identified as the focus of the product manager, but on a short-term basis it would not be fair to hold the individual accountable only to these factors. A number of other factors beyond the product manager's control contribute to profit. The evaluation criteria as a minimum should consider the following:

1. Is the individual aware of market conditions and how they affect the product?

2. Does the individual plan the marketing mix appropriately for the product?
3. Does the individual know where the product is now and where it is going? Does his or her planning cover how to get there?
4. Does the individual follow up on the product, and use implemented control measures to modify the product, if necessary?
5. Is the individual knowledgeable about all aspects of the product?

Some companies will be more precise in their measures of performance. Also, since product management functions vary from company to company, there may be considerable variations in criteria. Top management must next decide whether to reward the product manager, and how.

Rewards should not be based only upon failure or success of the product. To attract outstanding individuals to the position, rewards must be evident and of substantial importance. These rewards can be both intrinsic and extrinsic.

Intrinsic rewards are those generated within the individual. In the case of the product manager, intrinsic rewards must be highly perceived. This type of reward motivates the product manager to take a position of such ambiguity and high risk. One advantage of this position is the ability to see the product through from beginning to end. It is also rewarding to control the activities involved with the product. The product manager's opportunity to influence other people based on expert and referent power is another intrinsic reward. Intrinsic rewards are not controlled directly by any one person, but environmental characteristics can inhibit or enhance the opportunity to obtain this type of reward.

Extrinsic rewards include above all money: for an individual to be attracted to this position, he or she must be well paid. Another such reward is promotion. The product manager, as seen in Figure 12-1, is in the prime position to move up in the hierarchy to positions such as group product manager, Vice-President of Marketing, and Chief Executive Officer. For example, in January 1981 General Mills announced the promotion of three men to its top posts of Chairman of the Board and Vice-Chairmen, respectively. These men had risen through the ranks primarily as product managers.[4] The possibility of recognition and praise are two other rewards strongly associated with the product manager position. Other extrinsic rewards include fringe benefits and social acceptance.

Training

Product management functions amid rapid change; to avoid obsolescence, further development of skills is necessary. Skills can be developed by training programs, by experience, or both. Training programs might include a combination of university courses, professional seminars, presentations by consultants, and in-house presentations. Care must be taken that the training is transferable back to the product management environment. Training is necessary, but the real essence of this job can be learned only through experience. A comprehensive training program will benefit the company by increasing the performance of the product manager.

THE PRODUCT MANAGEMENT JOB

Assuming that the product manager is fully trained and experienced, we need next to examine the product management job. The product manager exists to manage a product. Redundant as that thought may seem, it is nevertheless critical to this approach. By virtue of his or her position in the matrix, the product manager has access to many specialized skills, and so becomes in large measure a generalist whose key job is to manage the assigned product or products by integrating the activity of needed and available resources. The normal management functions of planning, controlling, staffing, directing, and organizing are involved, as well as certain specialized skills. The basic management functions will be applied in the context of the chosen marketing strategy. These relationships are outlined in Figure 12-4. For example, each product manager will have staffing responsibility as related to his or her full range of duties. In Figure 12-4 the most significant responsibilities are highlighted at the intersection of the market factors and functions. These are specified by an "X" with subscript.

Planning

The primary function of the product manager is planning. Three basic planning levels are germane:

1. Input to the overall strategic corporate plan
2. Development of operational plans covering a one-year period
3. Activity planning for the day-to-day work effort

Market factors \ Functions	Planning	Organizing	Staffing	Directing	Control
Target market	X_1	X_6		X_9	
Product	X_2				X_{11}
Price	X_3				X_{12}
Promotion	X_4		X_8	X_{10}	X_{13}
Place	X_5	X_7			X_{14}

Fig. 12-4. Significant responsibility matrix.

These planning levels are illustrated in Figure 12-5. At point To the product manager is working in the context of the strategic plan (A), which is being implemented in terms of its first year. The plan was generated at time T-1. This overall plan provided objectives and goals for the development of the annual operating plan (B), which laid out the milestones for the period. At the same time activity planning (C) is proceeding on a daily or other basis such as weekly, monthly, or quarterly. Concurrent with these planning activities, the product manager is responsible to input information for the strategic plan that will begin at time T+1. Additionally he or she will be responsible for developing the operational plan for year 2.

As disclosed in Figure 12-4, planning becomes a most significant responsibility of the product manager. This fact is evident by the intersections, X_1–X_5. The manager's responsibility for input to future strategic and operational plans requires screening the environment for threats and opportunities relative to the assigned product or products. The exact contribution to strategic planning will be determined and communicated by top management. Thus the focus in this section will be on operational planning. A suggested format for an operational plan is included in Figure 12-6. The idea of a suggested format is just that. Each product manager will have to determine the best format for the product plan based on the planning requirements of the corporation and the demands of the market. A plan should be prepared for each product. Each part of the plan will now be discussed.

Product Description.   Quite often the name of the product will evoke a mental picture—for example, a brand name in the consumer goods market. However, for an industrial product the description might well include specifications, operating characteristics, and a physical description of the product, all relating to its form, fit, and function. The essential concern is to describe the product so as to differentiate it from other products in the product line.

Target Market(s).   Developing products in terms of the marketing concept envisions that products are designed to meet the needs of a specific market

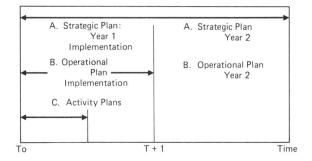

Fig. 12-5. Planning levels.

Product description: ————————————————————————

——

Target market(s): ——————————————————————————

Competing products: (include market share) ——————————

——

Life cycle stage: ————————————————————————————

Product strategy: ————————————————————————————

Pricing strategy: ————————————————————————————

Promotion strategy: ————————————————————————————

Distribution channel: ——————————————————————————

——

Financial information: ——————————————————————————

——

Schedule of activities to support marketing strategy:

——

Other information:

Fig. 12-6. Suggested plan format.

segment. In this part of the plan, this market segment should be described in detail. The location, intensity, and distribution of the demand for the product should be spelled out. Perhaps the product is being offered to different markets, in one form to the general public and in another to hospitals, schools, and the like. Maps showing the location of distributors also help to visualize the distribution of existing demand and to identify areas where future demand might be stimulated by some form of promotion.

Competing Products.   If there are competing products, they should be identified and listed with a statement as to their market share. Each competing product should be described and compared to the product manager's product.

It is also useful to evaluate the competing products in terms of the marketing mix for each. What is its design? How is it packaged? What is the price? What stores are selling it? This information can then be used in formulating the marketing mix for the manager's own products.

Life Cycle State. The stage that the product is in affects the planning effort. Whether it is in the introduction, growth, maturity, or decline stage will aid in identifying and formulating strategies relative to price, promotion, place, and product. If, for example, the product is in decline, a decision would have to be made about extending its product life with "a new, improved" version.

Product Strategy. For the next marketing period the product should be evaluated to determine if there is a need to change it in any way. Should the package be changed? Should the product be offered in a larger, economy format? If the product is in the growth stage and consumer acceptance is strong, it should probably be left as is. However, a product in the maturity stage with fluctuating sales might benefit from some design change.

Pricing Strategy. The price for the product should be examined in the context of change. Should the price be lowered to undersell competition? If the price is lowered, will the competition also lower their price? If the competition lowers their price, what will be the impact on market share for each product? Also, market information may have revealed that the firm has an opportunity to practice price discrimination. If so, what price should be charged in what markets? Over time, can these markets be kept segregated? If so, what actions will have to be taken to ensure that the variable pricing structure will persist?

Promotion Strategy. Again, where is the product in its life cycle? Does promotion need to be increased? To counter a threat against its biggest product, Woolite, American Home Products beefed up its promotion. This tactic was successful in beating out the Bristol-Myers Company's new product entry.[5] Is there a better way to inform consumers about the product? If the product has been promoted in the past by mass advertising, is this still a good way to go, or should the product manager change to direct mail coupons to stimulate customer demand? If the product is in the decline stage, perhaps promotion should be decreased and the product permitted to die a natural death. Whatever the case, a decision needs to be made during the planning period as to the promotion approach.

Distribution Channel. The effectiveness and efficiency of the product distribution policy must be examined. The cost of distribution is high, so any change that would lower this cost should be considered. Inventory levels need to be evaluated in terms of shortages and surpluses, and adjusted for the future based on the effectiveness of past plans. Transportation or shipping mode must be evaluated. Have new developments in the industry occurred? If so, can they be used to reduce costs and increase profits? The channel options may be many or limited in number, but there is still a planning responsibility to examine alternative distribution channels during the planning process.

Financial Information. Once the elements of the marketing mix have been specified, the product manager is in a position to develop estimates of revenues, costs, and profit based on the specified plan. Several configurations might be detailed, each one assuming a different level of demand. Then the most likely one could be adopted, depending on the impact of all qualitative and quantitative factors.

Schedule of Activities. This part of the plan deals with the time element. For example, demand should be delineated for the year based on whether it is fairly constant or seasonal. A product manager for automobile antifreeze would schedule sales visits so that the product is in the sales outlet by July, to be in place for the peak fall buying activity. Promotion would have to be scheduled to inform consumers of product availability and remind them that winter is coming. Cash inflows and outlays need to be scheduled to reflect cash flow needs. This overall plan would then form the basis for period activity planning by the product manager.

Other Information. This is a place to record information that does not logically fit into one of the other parts of the plan. For instance, the product manager might list other organizational elements that have helped prepare the plan, and include dates when they should be reminded of resource commitment and activities that must be completed to implement the total plan.

During the planning process the product manager needs to assess the uncertainty that exists relative to each element of the plan. Uncertainty is generally considered to be the absence of information. In relation to product demand in a given target market, the product manager may want to assess demand in terms of whether it is considered high, medium, or low. If uncertainty is high, he or she may want to obtain information to reduce the uncertainty to a more acceptable level. Also, planning is a continuous process and plans should be revised at periodic intervals, based on environmental feedback from control systems.

Controlling

A further examination of Figure 12-4 reveals that control is the next most significant responsibility of the product manager. This control implies that actual product performance will be compared with plans and the necessary corrective action taken. Control involves the periodic evaluation of feedback information from sources both internal to the firm and existing in the marketing environment. There are several practical ways to do this.

Salespeople.   Periodic meetings can be conducted with salespeople to probe their experiences, since they will know how customers have reacted to the product and other field conditions.

Customers.   Visits to selected sites where the product is doing well and to those where problems exist will often provide useful information. Often customer complaints will flow in of their own accord. Each complaint should be thoroughly investigated.

Observation.   The product manager should be an activist. As he or she travels from location to location, impressions should be noted and examined to disclose any patterns of behavior.

Records.   The pattern of sales in the market area can be gleaned from sales records. These patterns may reveal where some adjustment is needed as related to the market mix or other variable.

Absence of Activity.   Required reports that are not received, or the absence of telephone calls from salespeople, may disclose problem areas. Any change in the pace of activity should be examined by the product manager.

The Grapevine.   Useful information can often be obtained from the grapevine. The reliability of the information may not be high, but often this can provide a starting point for investigation.

Formal Audit or Studies.   The product manager may want to conduct a formal marketing audit for a given product that seems to be in trouble. Also, where appropriate, marketing research studies can produce useful information relative to the marketing function.

Standards.   Check points, deadlines, or standards can be incorporated to provide for periodic review to determine variances that need to be considered.

Organizing

As reflected by interactions X_6 and X_7 in Figure 12-4, the product manager has a major responsibility in identifying market segments and servicing them. This activity will involve other functional elements and will therefore require coordination. Closely related is the product manager's responsibility to serve as the leader of the marketing channel. Products, information, and funds flow through the channel; the product manager must periodically review the efficiency of the distribution system for each product, so as to ascertain needed changes.

Directing

The product manager does not directly supervise a large number of personnel, but is still responsible for the overall marketing program. This requirement includes a periodic study of the market and its attributes, such as changing tastes, demographic patterns, and the age and composition of the population. Also, the direction of the promotion plan is a critical function. Firms occasionally reduce promotion expenses when sales fall, which may or may not be a correct response. There are factors that must be examined prior to such a decision, such as where the product is in its life cycle, whether the proper promotion strategy is being used, and whether or not the focus is still on the proper market segment.

Staffing

Finally, there is the function of staffing the product organization. As mentioned earlier, the staff will be part-time in nature, drawn from other elements in the matrix. But the product manager still needs an organization chart and plan, indicating primary tasks that must be accomplished to support the product, and identifying people in the matrix who perform in each capacity.

Special Skills

The job of product manager requires special skills that the individual must possess and enhance. These skills are mandated by the product manager's position in the matrix. Each skill will be briefly considered.

Communication. The product management job requires up-to-date and timely information relative to the market and its environment, the organizational culture, and top-management commitment and support. In addition to establishing and maintaining these communication channels, the product manager must be articulate, persuasive, and a good listener.

Coordination. The success of the product manager depends on the actions of many people over which he or she has no direct authority. As a consequence the individual must continually inform managers at each level about his or her activities and needs for support. The manager must involve the functional managers in the planning process, so as to ensure resources required by the plan.

Conflict Management.   Conflict is natural to the matrix environment. From a conceptual standpoint functional personnel must serve two managers. the product manager and the functional manager. The conflict needs to be recognized and managed. From this perspective, the individual needs to be aware of the conflict resolution modes and the sources of power available.

Conflict Resolution Modes.   The basic conflict resolution modes are listed in Figure 12-7. *Confrontation* involves meeting force with force, often in the guise of "clearing the air." The product manager may decide to take on a specific functional manager. While this approach may have some appeal, usually it is not very effective. *Compromise* involves a give-and-take situation. Each participant reduces demands, reflecting that half a loaf is better than no loaf at all. This technique can be effective in many cases and is often used. *Smoothing* relates to a situation where the product manager perceives discontent and tries to pour oil on troubled waters; however, it may only drive the conflict underground. *Forcing* is the application of power, requiring an individual to adhere to the demands of the product manager without recourse. For example, one might appeal to top management to instruct a functional manager to provide a specific resource at a given time without a consideration of priorities. While effective, this technique can leave a residue of bad feelings. *Problem solving* is a working out of the conflict by the parties in a reasoned, logical manner through the use of an acceptable problem-solving routine. The *superordinate goal* involves appealing to the parties to the conflict by invoking their support for the benefit of the total organization. The appeal suggests that organizational survival is at stake, therefore the parties need to work together to ensure success. There may be situations where any one or a combination of these techniques may be appropriate. However, research has shown that the most effective modes are problem solving, compromise, and the appeal to the superordinate goal.

Sources of Power.   In a study of power sources available to product managers, Gemmill and Wilemon identified reward, coercive, expert, and referent power.[6] Generally speaking, the product manager has little direct reward power over

- Confrontation
- Compromise
- Smoothing
- Forcing
- Withdrawal
- Problem Solving
- Superordinate Goal

Fig. 12-7. Basic conflict resolution modes.

the individuals to be influenced. However, some indirect reward power exists when the product manager position is perceived as having high status and access to top management. In like measure coercive power may be germane when individuals in the matrix feel that the product manager has influence with upper management. These findings reinforce the need for top management to support the product manager in visible ways. Expert power was found to be useful for gaining support. The product manager is the focal point for planning and control information relative to the product strategy, and this knowledge and control permit the exercise of informal power. Referent power is another informal support base for the product manager, whose activities involve contact with the field sales force and other persons at all levels in the organizational hierarchy. By being fair, consistent, and positive with these individuals, the manager can become a respected figure whom people relate to and want to please. As sources of power, the study found that the expert and referent sources, if used judiciously, were more effective than the reward and coercive sources.

Negotiation.   The product manager must negotiate with parties both internal and external to the organization. Negotiation involves opposing objectives for the parties and is a conflict situation. Negotiation strategies and tactics thus require a knowledge of the conflict resolution modes previously discussed. Most often negotiation focuses on compromise by the parties, hopefully using a problem-solving approach. This negotiation skill is involved in the complete range of responsibilities for the product manager and thus is essential to the job. Negotiation is a learned skill and requires training to enhance its application.

Time Management.   The demands on the product manager's time are multiple and intense in nature. It behooves the individual to develop priorities and daily schedules permitting the careful use of time. The job by its nature is ambiguous and demanding and places a premium on the wise use of time. Appointment calendars, plan books, and other devices can assist in the planning and scheduling of meetings, appointments, visits, and other related activities.

Management of Change.   The marketplace and its environment, competitive behavior, and organizational and managerial mobility all require that the product manager be able to detect and manage change. The responsibility to cope with unexpected and expected changes is inherent in the product manager's job.[7] This situation requires a formal system to keep track of changes as they occur. Each significant change must be evaluated in the context of the internal and external organizational environments. Product changes to meet these conditions must be approached with a detailed analysis of the related costs; otherwise, product changes can be costly. As an illustration, Oxford Laboratories over a five-year period made design changes in their Oxford Automatic Dispenser to meet market demands. These changes almost doubled the cost of the product and created profitability problems.[8] Furthermore, the position of the product in the life cycle may make product extension, deletion, and new product decisions imperative, if the demands of the market and profit expectations are to be met. New product development is one way to deal with these requirements.

No consensus exists as to whether the product manager should be responsible for new product development. Different firms approach the subject in different ways. In some companies the product manager is considered the person most aware of the need for new products and so is assigned this responsibility. Various other arrangements are also used. Regardless of whether the product manager has this responsibility in total or in part, he or she must be aware of this need and at least be responsible for detecting situations where new products are required.

NEW PRODUCT DEVELOPMENT

By monitoring the control system, the product manager obtains valuable feedback relative to the market, and so receives information concerning possible new product opportunities. One disadvantage of the product manager's being responsible for new product development is a lack of time to spend in this area. Some firms assign this responsibility to new product development managers, or to teams whose main responsibility is the development and test marketing of new products. As an example, the Microwave Cooking Division of Litton Industries uses a team approach. Each team in the engineering department includes members from engineering, marketing, purchasing, manufacturing, accounting, and related functions.[9] If a given market test is successfully completed, the full-scale introduction responsibility is transferred to the product manager. Even if the product manager is not responsible for the development of new products, he is usually in a position to pass this information on to whoever is responsible for such development.

With the life cycle of products getting shorter and shorter, increased importance is being placed on new product development. There are three sources of new products: innovation, product modification, and acquisition of firms with desirable products. The focus here will be on the first two, with the product manager assumed to be responsible for the activity.

The product manager is in the best position to obtain feedback about new products, or possible modifications to present products. Ideas may be developed internally or externally to the firm. For example, many firms obtain feedback from consumers concerning ideas for new products and modifications. Some of this may take the form of complaints, but this should not be disregarded as it is a source of information. Also, research may provide information. By performing market research the company gives consumers an opportunity to provide input to new product development or product modifications. In addition to customers, other external sources of ideas include competitors, middlemen, trade associations, and government agencies. Another source of ideas can be found within the firm. Performing activities that bridge functional lines, the product manager may receive suggestions for improvement directly from employees in other departments, as for example sales personnel, service personnel, production personnel, and other executives. The company may use a company suggestion box for ideas for new product development or modification. The product manager works also with the research and development department, which may have valuable input for product development. Clearly, then, the product manager receives a steady flow of product ideas from internal sources of the firm. It is not possible for the firm to develop each idea, therefore each idea must be screened for feasibility.

Screening. The screening of new ideas is a critical activity. First, the company objectives must be considered. Is the product consistent with them? Management may develop checklists of criteria to be used when evaluating new product ideas. At each stage of the product development process, a decision is made to accept or reject the product idea. Care must be taken when making this decision, for rejection of good ideas is a lost opportunity, while accepting poor ideas leads to higher cost.

Next, the resources available must be considered. Here the technical and economic requirements of the product are examined. Does the company have the resources to meet the technical and economic requirements of the product? Aspects to be reviewed include plant facilities, equipment, sales force, and personnel. Also considered at this point is marketing feasibility: will the product complement the existing product line or mix? Other factors include marketing-mix aspects such as distribution and promotion. For example, can the product be distributed through existing channels?

Business Analysis. Following screening comes business analysis. The product is examined in terms of demand, cost, and profit. The key question is whether demand will be high enough to return a satisfactory profit. A sales forecast must be developed. When developing this forecast, three types of sales must be considered: first-time sales, replacement sales, and repeat sales. The product life cycle has an impact upon these types of sales. For example, when the product is first introduced there will be a high number of one-time sales. From the sales forecast, expected costs and profits can be determined. Data concerning costs can be collected with the help of such departments as research and development, production, and marketing. By bringing the demand and cost factors together, the profitability analysis is developed. Financial aspects such as return on investment, time to profitability, and risk ratio versus return on investment must be considered.

Development. A product idea that passes through the business analysis is turned over to the research and development and engineering departments for development. A prototype of the product is developed and tested, passing functional and consumer tests. The marketing department becomes more involved at this stage, where brand name and packaging are considered. A check should now be made as to the reliability and accuracy of anticipated costs.

Test Market. Once product performance is deemed satisfactory, the product is ready for further market testing. This is an important step. The cost of test marketing is small when compared to the costs of full-scale production and marketing. The company has the opportunity to observe actual customer reaction while varying marketing tactics. Through test marketing the company is provided with information about the opportunities and problems connected with the new product. Other data are also obtained, such as sales potential and costs. Not only are financial and market aspects considered, but other matters such as competition, legal questions, production, and research and development. Again, management should develop a checklist to evaluate important areas, so all aspects will be covered. At the end of this stage it is necessary to re-evaluate the sales forecast, as well as cost data and profit potential.

Launch. At this point there is ample information to determine if the product should be introduced into the commercial market. Launching a product into the introductory stage of the product life cycle involves a large cost outlay. A number of decisions such as when, where, and how to introduce the product are made at this time. The entire strategy of the product is developed for the product's life cycle. Again the product manager, considering the total product, must exercise a high level of coordination across departmental lines.

New product development is necessary to meet the expectations of consumers and is therefore risky. New product development is becoming harder to achieve, owing to a shortage of major new product ideas. There are a number of minor products emerging on the market, but to avoid economic stagnation, major innovations are needed. Competition is another problem area, as evidenced by Chesebrough's Vaseline Intensive Care lotion. Procter and Gamble introduced its Wondra lotion and Nabisco its Rose Milk to take market share from Chesebrough. Although these new products were developed to counter competition, neither was able to capture more than 10% of the market.[10] As a result of keen competition, new products must be aimed at smaller segmented markets, thus decreasing potential sales and profits. Competition also shortens the life cycle of a new product. There has been an increase in social and government constraints placed upon products. For example, a firm must consider consumer safety when designing a product. Government has also become involved in new product development in such areas as product design and advertising decisions. New product development is expensive. A number of ideas must be developed just to find one successful idea. Many companies cannot afford the product development process and so are beginning to focus on product modification.

Management faces a dilemma. While there is a high rate of failure, new products must be developed. Therefore new product development must be carried out in such a manner that the risk of failure is decreased. By using the information the product manager can provide, a firm will decrease not only the costs of new product development but will also the risks.

SUMMARY

The product manager approach is generally considered to have originated with the Procter and Gamble Company in 1927. In an evolutionary sense it has gained support and is now being practiced by firms such as Colgate-Palmolive, Kimberly-Clark, and American Home Products. In its general form an individual is assigned responsibility to manage a product or group of products. This position, depending on the company, can be delegated relative degrees of authority and responsibility, and exists in an organizational format termed the matrix.

In this matrix structure the marketing function is basically responsible for developing a marketing strategy for the full range of company products. A specific marketing strategy encompasses the selection of the target market for the product and deals with the related product design, price, place, and promotion decisions. These decisions are typically the ones that a given product manager must make in the context of the product management system.

The focus for the product manager is the product or products. He or she must consider the nature of the product relative to the consumer or industrial goods markets, or both. This focus involves a knowledge of the product life cycle, the organizational structure and culture, as well as the competitive environment.

In each company the product manager job should be defined and a job description prepared in written format. This ensures that the proper individual will be selected for the position. Candidates can be recruited from within the organization or from external sources. Rewards should be structured to motivate individuals to seek product manager positions; successful candidates must be trained and oriented to the requirements of the job.

The product management job places a premium on the planning and control aspects of management. The product manager is in large measure a generalist whose key responsibility is to manage a product or products by integrating the product strategy needs and available resources. The individual must possess special skills, such as the facility to communicate, coordinate, negotiate, cope with conflict, and effectively manage time.

Environmental and organizational demands require that the product manager be sensitive to the need for change. The manager's responsibility for new product development will vary from firm to firm. In some cases it will encompass only a recommendation for product addition, extension, or deletion. In other circumstances the responsibility will involve all the aspects of new product development, including idea formulation, screening, analysis, testing, and product introduction.

The increasing complexity and uncertainty of the environment will cause more and more firms to seek ways to better serve their respective markets and enhance profitability. Mechanistic organization structures are basically effective in stable environments, but organic structures are more effective in an environment characterized by instability. For survival, firms in the future will probably find it advantageous to adopt organic strucrures such as matrix and, where appropriate, the product management approach. The need for survival will dictate no less.

REFERENCES

Books

Adams, John R., Barndt, Stephen E., and Martin, Martin D. *Managing by Project Management.* Dayton, Ohio: Universal Technology Corporation, 1979.

Ames, B. Charles. "Keys to Successful Product Planning." *The Environment of Industrial Marketing.* Ed. Donald E. Vinson and Donald Sciglimpaglia. Columbus, Ohio: Grid, 1975.

Boone, Louis E., and Kurtz, David L. *Foundations of Marketing.* Hinsdale, Ill.: Dryden Press, 1977.

Constantin, James A., Evans, Rodney E., and Morris, Malcolm L. *Marketing Strategy and Management.* Dallas: Business Publications, 1976.

Cravens, David W., Hills, Gerald E., and Woodruff, Robert B. *Marketing Decision Making: Concepts and Strategy.* Homewood, Ill.: Richard D. Irwin, 1976.

Enis, Ben M. *Marketing Principles: The Management Process,* 3rd ed. Santa Monica, Calif.: Goodyear Publishing Co., 1980.

Hughes, G. David. *Marketing Management: A Planning Approach.* Reading, Mass.: Addison-Wesley Publishing Co., 1978.

Kerzner, Harold. *Project Management: A Systems Approach to Planning, Scheduling and Controlling.* New York: Van Nostrand Reinhold Co., 1979.

——— "Formal Education for Project Management." *A Decade of Project Management.* Ed. John R. Adams and Nicki S. Kirchof. Drexel Hill, Pa.: Project Management Institute, 1981.

Kotler, Philip. *Marketing Management: Analysis Planning and Control.* Englewood Cliffs, N.J.: Prentice-Hall, 1980.

McCarthy, E. Jerome. Basic Marketing: A Managerial Approach, 5th ed. Homewood, Ill.: Richard D. Irwin, 1975.

Rosenberg, Larry J. *Marketing.* Englewood Cliffs, N.J.: Prentice-Hall, 1977.

Stanton, William J. *Fundamentals of Marketing,* 5th ed. New York: McGraw-Hill, 1978.

Struckenbruck, Linn C., ed. *The Implementation of Project Management: The Professional's Handbook.* Reading, Mass.: Addison-Wesley Publishing Co., 1981.

Journals

Altier, William J. "Change: The Name of the Game." *Business Horizons* 22 (June 1979): 25–27.

"American Home Products." *Business Week* 2659 (Oct. 20, 1980): 80–95.

Ames, B. Charles. "Payoff from Product Management." *Harvard Business Review* 41 (Nov.–Dec. 1963): 141–52.

"Chesebrough: Finding Strong Brands to Revitalize Mature Markets." *Business Week* 2662 (Nov. 10, 1980): 73–76.

Clewett, Richard M., and Stasch, Stanley F. "Shifting Role of the Product Manager." *Harvard Business Review* 53 (Jan.–Feb. 1975): 65–73.

Dhalla, Nariman K., and Yuspeh, Sonia. "Forget the Product Life Cycle Concept." *Harvard Business Review* 54 (Jan.–Feb. 1976): 102–12.

Dietz, Stephens. "Get More Out of Your Brand Management." *Harvard Business Review* 51 (July–Aug. 1973): 127–36.

Gemmill, Gary R., and Wilemon, David L. "The Product Manager as an Influence Agent." *Journal of Marketing* 36 (Jan. 1972): 26–30.

Hayes, Robert H., and Wheelwright, Steven G. "The Dynamics of Process—Product Life Cycles." *Harvard Business Review* 57 (March–April 1979): 127–36.

Jackson, Barbara B., and Shapiro, Benson P. "New Way to Make Product Line Decisions." *Harvard Business Review* 57 (May–June 1979): 139–49.

Morein, Joseph A. "Shift from Brand to Product Line Marketing." *Harvard Business Review* 53 (Sept.–Oct. 1975): 56–64.

Morrison, Ann M. "The General Mills Brand of Managers." *Fortune* 103 (Jan. 12, 1981): 98–107.

"Putting Products Through an Audit Wringer." *Business Week* 2316 (Feb. 2, 1974): 60–61.

"Redrawing the Crayola Image." *Business Week* 2703 (Aug. 31, 1981): 83.

ADDITIONAL REFERENCES

1. R. H. Hayes and S. C. Wheelwright, "The Dynamics of Process-Product Life Cycles," *Harvard Business Review,* March–April 1979, pp. 129–30.

2. N. K. Dhalla and S. Yuspeh, "Forget the Product Life Cycle Concept," *Harvard Business Review,* Jan.–Feb. 1976, p. 107.

3. David I. Cleland, "Defining a Project Management System," in *A Decade of Project Management,* ed. J. R. Adams and N. S. Kirchof (Drexel Hill, Pa.: The Project Management Institute, 1981), pp. 139–40.

4. A. M. Morrison, "The General Mills Brand of Managers," *Fortune,* Jan. 12, 1981, p. 99.

5. "American Home Products—Can William LaPorte Keep Those Profits Growing?," *Business Week,* Oct. 20, 1980, p. 86.

6. Gary R. Gemmill and David L. Wilemon, "The Product Manager as an Influence Agent," *Journal of Marketing,* Jan. 1972. p. 27.

7. W. J. Altier, "Change: The Name of the Game," *Business Horizons,* June 1979, p. 26.

8. "Putting Products Through an Audit Wringer." *Business Week,* Feb. 2, 1974, p. 60.

9. David I. Cleland, "Project Management: Springboard to Matrix Management," in *A Decade of Project Management,* ed. J. R. Adams and N. S. Kirchof (Drexel Hill, Pa.: The Project Management Institute, 1981), p. 171.

10. "Chesebrough: Finding Strong Brands to Revitalize Mature Markets," *Business Week,* Nov. 10, 1980, pp. 75–76.

13. The Convergence Factor for Successful Large-Scale Programs: The American Synfuels Experience as a Case in Point

Mel Horwitch*

The recent demise of the grandiose scheme by the late Carter administration to establish a massive synthetic fuels program presents us with a unique opportunity to reassess our whole set of views on the creation and management of large-scale government-sponsored programs. These enterprises usually involve multiple and diverse actors, multisector participation, and a broad or narrow set of missions (Horwitch and Prahalad, 1981). To provide such a reassessment, the postwar evolution of large-scale programs will be briefly traced, a general framework for analyzing and managing such enterprises will be presented, and finally this framework will be applied to the American synfuels experience.

THE CHANGING NATURE OF LARGE-SCALE PROGRAMS

During the post-World War II period the character and scope of large-scale programs experienced three significant evolutionary stages. In the first stage, lasting from the end of World War II until the mid-1960s, such endeavors were primarily limited to the defense and aerospace fields. This period produced a number of outstanding successes, including the Polaris and Apollo programs. Moreover, it is from the rather optimistic experiences of this stage that many of the standard managerial lessons relating to large-scale programs originate. Indeed, as an outgrowth of this bullish mood, many practitioners and management theorists close to the aerospace industry believed that the project man-

---

*Dr. Mel Horwitch is a member of the corporate planning, policy, and strategy group at the Alfred P. Sloan School of Management at M.I.T. He has taught courses on technological innovation, technological strategy, business policy, corporate strategy, management of large-scale enterprises, production operations, and energy economics. He has worked in both government and private industry and has also been a consultant to numerous private firms and public institutions. Prof. Horwitch has written extensively on management, technology, and energy affairs. His most recent book, *Clipped Wings: The American SST Conflict,* was published in 1982 by the MIT Press.

agement techniques developed would soon transform management practices generally (Gaddis, 1959). This emphasis on applying project management methods that were basically developed in the 1950s and early 1960s for a wide array of large-scale programs persisted throughout the 1960s and 1970s (Chapman, 1973; Archibald, 1976).

The management challenges of the defense and aerospace programs during this first period were, of course, enormous. The development of management practices to plan and control such activities represented no mean accomplishment. Much of the success of the Apollo program, for example, was due to the sophisticated project management approaches that NASA developed (Seamans and Ordway, 1977).

But it is now quite clear that even the successes of the large defense and aerospace programs of this early era were due to a much more complex set of factors than mere good project management. For example, although the impressive on-time and technologically advanced Polaris program has often been attributed to its heavy use of the well-known project management tool PERT, research has shown that the success of Polaris was probably due more to a winning style in bureaucratic politics. Indeed, the Polaris managers seemed to use PERT more as a political, bureaucratic, and public relations weapon than as a direct managerial tool (Sapolsky, 1972).

Another essential factor increasing the likelihood of success was the existence of an effective champion or group of champions. The powerful blend of iron-willed commitment and skillful constituency-building practiced by leaders like Hyman Rickover and William Raborn of the U.S. Navy and James Webb of NASA was an essential ingredient for the success of their programs (Horwitch and Prahalad, 1981).

The critical objective of all of these managerial approaches during this initial stage was to build and maintain a shield that would protect a program from possible protests and opposition emanating from various sectors of society. Clearly, the defense and aerospace programs of the 1950s and early 1960s possessed an inherent advantage in achieving such a cloak, as compared with later commercialization projects. The use of national defense arguments alone during this period helped buffer many weapon systems programs from interference and damaging criticism. Invoking national prestige or providing an absolute Presidential commitment can at times play a similar role. The Apollo program benefited from this form of shield (Horwitch, 1982B).

The second stage in the development of large-scale programs took place in the 1960s. It was a period when the domain and mission of such endeavors began to change. Whereas before big programs involved mostly the defense and aerospace fields where the government was essentially the primary buyer, during the 1960s large-scale programs started to be seriously proposed for new sectors and new kinds of missions where the primary buyer ultimately became a customer in the private sector. In this new context, serious problems with the managerial methods of the earlier era began to emerge.

The major example of this period of transition was the ultimately unsuccessful American SST program, which formally began in June 1963 and was finally defeated by both the House and Senate in March 1971. Although the program started as a more or less conventional aerospace project in a relatively benign and contained environment, by its end it had become fatally exposed to the public at large and the target of an extremely effective public protest. The program had a rich array of problems, including technical difficulties, uncertain or questionable economics, bureaucratic conflict, fragmentation of control, growing environmental concerns, and general skepticism about a variety of matters at the highest levels of government and industry. But perhaps the SST's greatest weakness was the absence of a cloak to buffer the program from hostile moves in the environment (Horwitch, 1982A). For maybe the first time ever, arguments based on national defense, national prestige, or international rivalry were not sufficiently powerful to save a large-scale program, and the SST's top managers in the mid- and late 1960s—all successful and experienced project managers—were ultimately unable to direct the enterprise.

The transition stage, and the SST in particular, were the prelude for the third period of development for large-scale programs. This period began in the late 1960s and has lasted until the beginning of the 1980s. The most important characteristic of this period has been the emergence of commercialization goals for programs. No longer is the end user solely the government. No longer are such activities limited to defense or aerospace. Instead, large-scale programs with ultimate commercialization goals have been proposed and in some cases implemented in such areas as mass transit, nuclear power, and synthetic fuels.

Related to the rise of commercialization goals, the all-important protective shields for large programs practically vanished. The novel and usually fatal challenge to advocates and managers of large commercialization programs was how to guide their effort to a successful conclusion without the valuable assistance of this protective shield that their more fortunate predecessors often possessed. Under the more open and diverse conditions of the third stage, a continued emphasis solely on proper project management methods was inappropriate. (The same observation, by the way, can be made regarding the more recent commentaries that rely heavily on economic trade-offs.)

A new and more complex approach for conceptualizing and managing large-scale programs in the environment of this third stage is required. A framework for dealing with modern large-scale programs will be described and then applied to the American experience with synthetic fuels.

THE CONVERGENCE OF FOUR ELEMENTS

The dissolution of the protective shield for large-scale programs by the third stage has made managerial success extremely difficult to achieve. Success in this period depends on the convergence of at least four elements, all of which must be actively working to support the enterprise. Alone, neither effective advocacy, nor good bureaucratic and political skills, nor sophisticated project management methods are sufficient. The four key elements required for success are (1) positive *societal cost-benefit* trade-offs, (2) effective *bureaucratic-political* behavior, (3) appropriate *managerial* practices, and (4) a favorable *corporate strategic* environment. Success depends, to a greater or lesser degree, on the existence of all four elements.

The rise of the extremely influential *societal cost-benefit* perspective has paralleled the dissolution of the shield that formerly protected large programs. Researchers using this perspective differ at times (see Eads and Nelson, 1971; Baer et al., 1976; Merrow, 1978; Merrow et al., 1979; and Schmalensee, 1980); however, they exhibit a remarkable similarity on crucial dimensions. First, they generally evaluate a program at a macro or societal level and attempt to determine whether society as a whole should invest its resources in a given program or project. Their orientation is truly a societal cost-benefit one. A second characteristic is the frequent reliance on economic theory or methodology, and a third, a strong though not exclusive concern for efficiency in the allocation of society's resources. Overall efficiency is the starting point under this perspective, following which other considerations such as equity, uncertainty, or general risk to society are factored into a program's assessment. As one of the most thoughtful societal cost-benefit discussions concludes, "There are no general efficiency arguments for alternative forms of aid or for government attempts to manage the commercialization process by singling out individual technologies for special treatment; it is very hard to imagine persuasive equity arguments for such policies. None of this changes when dynamic complications or important sources of risk are considered" (Schmalensee, 1980, p. 37). A fourth usual feature of this orientation is the near unanimous conclusion to rely as much as possible on market forces and to minimize direct government intervention, so that crucial signals from the marketplace are not distorted.

The Rand Corporation, in a series of very influential studies in the middle and late 1970s, was perhaps the foremost practitioner of the societal cost-benefit approach (Baer et al., 1976; Hederman, 1978; Merrow, 1978; and Merrow et al., 1979). The most general study was by Baer, Johnson, and Merrow (1976), a large three-volume report that analyzed and evaluated federally funded demonstration programs. This study was based on an in-depth examination of twenty-four case studies (one of which was the federal coal-based synthetic fuels effort through 1975). The general conclusions of this study exemplify the societal cost-benefit perspective. Successful programs require the following: commercial feasibility, which depends on a technology "well in hand"; cost- and risk-sharing with local participants; program initiative from nonfederal sources; the existence of a strong industrial system for commercialization; the inclusion in the program of elements needed for commercialization; and the absence of tight time constraints. The Rand experts suggested starting such programs on a small scale with little initial visibility, and urged allowing sufficient slack in the schedule for slippage. They also explicitly advocated resisting "political pressure" for demonstrating a premature technology.

The contribution of societal cost-benefit thinking should not be underestimated. With the removal of the protective shield for large-scale programs, the enterprise must possess an intrinsic societal worth in order to withstand in the open the barrage of criticism that will inevitably be leveled at it. The societal cost-benefit perspective has provided a language and a rich array of concepts and methodologies to judge such worthiness. At the same time, the recent dominance of the societal cost-benefit perspective has also narrowed the range of thinking about such programs; other perspectives that are at least equally as important have been ignored.

One such perspective is the *bureaucratic-political* one. This perspective is a blend of two earlier "models": organizational processes and governmental politics (Allison, 1971), and the merging of these two approaches has been well developed (Allison and Halperin, 1972). The basis of this perspective is both organizational and political. Actions are governed on the one hand by such organization-based factors as structure, coordination, and control, standard operating procedures, standard programs and repertoires, satisficing and uncertainty avoidance in problem solving, and fractionated decision making; and on the other hand by such political factors as power, the position and ambition of individuals, bargaining effectiveness, and constituency building (Allison, 1971). It has been suggested that one way to acquire quickly an overview of the bureaucratic-political element of an issue is to ask the following questions: Who plays? What determines each player's stand? How are the players' aggregated to yield institutional or societal decisions and actions? (Allison and Halperin, 1972, pp. 46–47).

During the 1970s, in the face of the rising societal cost-benefit perspective, the importance of mastering the bureaucractic-political element was usually ignored. But this was not entirely the case. One study of the immensely successful Polaris program considered a "winning skill in bureaucratic politics" essential: "The success of the Polaris program depended upon the ability of its proponents to promote and protect the Polaris. Competitors had to be eliminated; reviewing agencies had to be outmaneuvered; congressman, admirals, newspaperman, and academicians had to be co-opted. Politics is a systemic requirement. What distinguishes programs in government is not that some play politics and others do not, but, rather, that some are better at it than others" (Sapolsky, 1972, p. 244).

A third element is also required for successful large-scale programs—effective management. The *managerial* element focuses on the ability of high-level program administrators to "get things done" (Caro, 1975). Like the bureaucratic-political element, this perspective was also neglected or, more precisely, overly simplified during the 1970s.

Usually, effective program management dealt primarily with proper application of project management tools and methods. This approach was actually most relevant for managing specific projects and not really applicable for guiding large, multifaceted programs, which could comprise various projects. Typically, according to most project management principles, a project was conceived as a "specific, time-constrained task, the performance of which cuts across the traditional lines of structure and authority within a given organization." The basic building blocks of project management were the project manager, who occupies the single point of overall project responsibility; the central planning and control systems that are exerted by the project manager and that person's staff; and the diverse and fragmented array of activities that can make up the project, which are scattered among many agencies, nonprofit institutions, and private contractors and subcontractors (Chapman, 1973, p. 3). Project management theory also gives considerable attention to the problem of managing a project at different stages of a project's life cycle, which has been characterized recently by an authority on project management as progressing from an initial stage of concept to definition to design to development to application and ending with postcompletion (Archibald, 1976).

Because of the inherent fragmentation of projects, project management thinking also dwells at great length on the problems of integration and conflict resolution under conditions of uncertainty and diversity. Matrix organization theory and methods are really offered as a way of coping with such an environment.

Finally, project management theorists often pay considerable attention to the kind of individual who would be likely to make an effective project manager. Personality traits frequently recommended include flexibility, a capacity to handle ambiguity and uncertainty, technical competence, good oral and written communication skills, and a general ability to get along with a diverse range of people (Archibald, 1976; Chapman, 1973).

Although project management methods have certainly proved useful for certain aspects of planning in certain kinds of projects, they are best suited to the earlier shielded environment of the aerospace and defense sectors. The highly normative style of most project management writing is also most applicable when a project's variety is somehow limited and contained, which is usually not the case with large modern programs. The project management approach appears more rigid and inappropriate in a highly politicized and open environment with several potentially hostile stakeholders ready to attack. Furthermore, even the discussions in project management about the personality traits of effective project managers seem stilted, antiseptic, and undifferentiated in terms of designating real personality needs. Basically, as seen in the American SST program, experienced project managers cannot by themselves save commercialization programs.

Even more critical for the successful management of large-scale programs than mastery of project management techniques is the need to have a skilled champion or set of champions who can operate in a complex, pluralistic, and potentially wide-open environment. Effective champions are rare, and are certainly much more difficult to characterize and define than the prototypical competent project manager with the checklist of personality traits that is often presented by project management theorists.

The champion must be totally committed to the enterprise and at the same time be able to identify current and potential positive and negative stakeholders. Intense commitment is necessary because of the difficult and inherent obstacles to success that practically all large-scale projects of the third period face. Yet steadfast commitment should not cloud the champion's perception of the environment. Serious areas of vulnerability should be recognized. Accurate stakeholder assessment is needed. The champion often improves the chances of success by trying to build a network of favorable stakeholders (and co-opting potentially opposing stakeholders) early in a program's life, when opposition usually has not yet reached a high level of intensity. Also, the effective champion generally possesses a broad, rich vision of relevant stakeholders (Horwitch and Prahalad, 1981). One reason that the SST program failed was that the champions had a narrow view of the essential players and generally dismissed the key and novel role of the environmentalists until it was too late (Horwitch, 1982A).

On the other hand, James Webb and his colleagues at NASA were quite adept at stakeholder management during the Apollo program. Not only did NASA gain the obvious support of the aerospace industry and related constituencies; the agency also reached out to win the backing of the educational community, the basic sciences, and the weather forecaster profession (Ginzberg et al., 1976). Indeed, the Apollo program was perhaps the best example of effectively using project management methods under the skillful direction of competent champions. One reason that the Apollo program endured with relative ease the review of the flash fire in 1967 that killed three astronauts, was the strong network of favorable stakeholders that had wisely been built during the early years of the project (Seamans and Ordway, 1977).

The final essential element for successful large-scale programs of the third period is the *corporate strategic* one. This element also has been largely ignored. Yet considerations of corporate strategy are now crucial for understanding the evolution of large-scale programs in the third stage and for managing such enterprises successfully.

Although the involvement and support of the private sector for large-scale programs since the end of World War II has always been important, until the arrival of the third period, with the appearence of commercialization goals and the disappearance of the protective shield, the backing of the corporate world in defense and aerospace could usually be taken for granted. But commercialization goals and the greater exposure to more uncertain and potentially hostile forces created significant strategic considerations for likely corporate participants.

According to one standard definition, corporate strategy is "the determination of the basic long-term goals and objectives of an enterprise, and the adoption of courses of action and the allocation of resources for carrying out these goals" (Chandler, 1962, p. 13). Because at some point, which itself could be the subject of intense debate, the private sector would assume the risk of commercializing a technology developed in a large-scale program, the program must in some fashion fit into the long-term corporate strategies of the private sector players. Therefore, corporations scrutinized large-scale programs more closely, more cautiously, and very differently in the third period than in earlier eras. Such typical strategic considerations as market opportunity, diversification, product, organizational or operations synergies, and various forms of horizontal or vertical integration were usually examined in great depth. Needless to say, this new context made real private-sector support for large-scale programs much more difficult to attain than previously, and created complex and novel challenges for program champions.

In summary, the large-scale program that emerged in the third period was, in fact, a different kind of institution from that of its defense and aerospace predecessors of the 1950s and 1960s. The likelihood of achieving success became significantly reduced, because of the dissolution of the earlier protective shield and the rise of commercialization goals. Specifically, success usually required the convergence of at least four crucial elements: meeting a societal cost-benefit trade-off; possessing effective bureaucratic-political skills; having appropriate managerial approaches; and fulfilling relevant corporate strategic goals. As will be seen in the case of synthetic fuels in the United States, attaining such a convergence proved to be an extremely difficult task.

THE AMERICAN SYNFUELS EXPERIENCE: THE ABSENCE OF CONVERGENCE

The American synthetic fuels experience is generally an excellent illustration of a large-scale program of the third period. (For purposes of this discussion, synthetic fuels are defined as those processes that convert oil shale or coal into liquid fuels or, in the case of coal, into gaseous fuels; in particular, technologies involving biomass and tar sands are excluded.) The advocates of synfuels never possessed the benefits of an effective shield based on national security or an absolute Presidential commitment. Moreover, commercialization by private industry was always the ultimate objective.

The United States has flirted with synthetic fuels since the end of World War II. At least twice, and perhaps three times, the country was on the verge of initiating a synfuels industry: during the second Truman administration, at the end of the Carter administration, and possibly during the post–oil embargo period in the mid-1970s. But each attempt ultimately failed. This lack of success was due to the absence of a convergence of the four essential elements required for large-scale programs of the third stage.

From the beginning, synfuels could not pass a convincing societal cost-benefit assessment. In 1948, due to long-term fears of excessive dependency on growing oil imports and to an immediate shortage of winter heating oil, Secretary of Defense James Forrestal and Secretary of Interior Julius Krug called for an $8–$9 billion synthetic fuels program. Four years earlier, Congress had passed the Synthetic Liquid Fuels Act, which authorized $85 million for laboratory research and the construction and operation of demonstration plants. However the oil industry, which had tolerated the 1944 law as a wartime measure, reacted with intense opposition to the much more massive 1948 proposal. The resulting conflict between Interior and the oil industry triggered a battle of numbers and cost estimates that the pro-synfuels faction ultimately lost.

The Bureau of Mines in the Department of Interior in 1949 estimated the cost of gasoline synthesized from coal at 10 cents per gallon (in 1951 this figure was increased to 11 cents by the Bureau) and shale at at 11.5 cents per gallon. These estimates, if accurate, would have made synfuels virtually competitive with domestic crude. But the oil industry, as represented by the National Petroleum Council—a group of about 80 CEOs from the industry established in 1946 by President Truman as an advisory body on energy matters—disagreed vigorously with the Bureau's figures. After a massive and detailed review of the Bureau's analysis, the Council in 1951 came out with much higher cost estimates: 41.5 cents per gallon for coal-based synfuels and 16.2 cents per gallon for shale oil.

The following year Interior asked an independent consulting firm, Ebasco Services, to review the synfuels cost estimates. Ebasco's figures were released in 1952 and were between those of the Bureau and the Council, estimating the cost of oil from coal at 28.1 cents. Ebasco also commented that it did not believe it was feasible for the private sector to finance synfuels projects as envisioned by the Bureau of Mines under prevailing conditions. Although both sides claimed vindication, the anti-synfuels faction, the oil industry, really won the first true societal cost-benefit assessment battle. This early synfuels program was terminated in the first Eisenhower administration (Vietor, 1980; Goodwin, 1981).

This pattern of synfuels faring poorly in societal cost-benefit assessments continued into the 1970s. In terms of evalating synfuels according to strict societal cost-benefit crteria, synfuels practically never appeared justified. Typical of the more modern reviews of synfuels in this genre were the series of studies by the Rand Corporation in the middle and late 1970s already mentioned. Rand analyzed coal-based synfuels as part of its overall report on 24 case studies of federally funded demonstration projects (Baer et al., 1976).

The specific discussion of coal-based synfuels placed great emphasis on the wide array of uncertainties associated with synfuels because of persisting technical problems and "exogenous" factors such as interest rates, equipment prices, inflation, changing public policy, the price of competing fuel (especially world oil prices), and the improvement of competing synfuel technologies. Rand emphasized the need to reduce the uncertainties related to synfuels and concluded that basically large-scale demonstration programs would not achieve such a reduction. An exception to this conclusion might be made if total energy independence became a national policy; but even then Rand appeared to believe that the benefit of a massive though risky synfuels program would not be worth the cost.

Two companion Rand analyses, one on the prospects for the commercialization of high-BTU coal gasification (Hederman, 1978), and the other on the constraints on the commercialization of shale oil (Merrow, 1978), echoed both the methodologies and the conclusions of the earlier Rand study. In the report on high-BTU gasification there was implicit emphasis on not being locked into a first- or second-generation technology. Reflecting a prototypical societal cost-benefit viewpoint, the report stated: "Technological innovations can also open up ways to use entirely new processes. . . . Nevertheless, there is no reason to expect the commercialization of currently developed processes to accelerate the development of third generation coal gasification technology. In fact, if resources available for coal are limited, relatively expensive commercialization activities could be expected to decrease the funds available for R & D looking to new processes with potentially superior cost characteristics" (Hederman, 1978, p. viii).

An important characteristic of this report and two subsequent Rand studies on synfuels (Merrow, 1978; Merrow et al., 1979) was the great attention given to escalating cost estimates for synfuels technologies. The report on high-BTU coal gasification, for example, mentioned that 1976 cost estimates for commercial-scale plants were at least five times higher than estimates made between 1969 and 1972. It presented four causal factors contributing to this trend: general inflation, construction cost changes above general inflation, mandatory process changes (due especially to environmental requirements), and most importantly, improved cost-estimating capability. The latter factor, which was said to contribute about 80% of the cost escalation observed, was attributed to the "enhanced engineering knowledge developed through successively more sophisticated plant designs and hands-on experience with processes," as serious commercial interest in coal gasification leads to detailed engineering designs of facilities (Hederman, 1978, pp. v–viii).

The Rand report on shale oil was very similar to its companion gasification study, though perhaps even more negative toward the prospects of synfuels commercialization. In addition to the general problem of serious cost escalations, this study also emphasized "institutional" constraints such as water availability and the limits of the learning curve in such projects. Unlike in other industrial contexts such as aircraft or semiconductor production, Rand argued that the potential for gaining the benefits of learning in shale were constrained, owing to the proliferation of process technologies, the many site-specific requirements, the constraining influence of environmental regulation, the inability of any single player to amass sufficient scale or market penetration, the capital-intensive features of synfuels technologies, and the mature nature of the technologies associated with synfuels (such as mining, upgrading, and by-product recovery) (Merrow, 1978, pp. 43–55). The overall conclusion of the study, from the viewpoint of synfuels advocates, was quite bleak: "At the present time (1978), a government commercialization effort for oil shale surface retorting would not be likely to result in a viable industry in this century" (Merrow, 1978, pp. viii–ix).

The negative assessments continued into the 1980s. A report by the Congressional Research Service for the House Subcommittee on Energy Development and Applications, released in March 1981, represented perhaps the ultimate in such views. This study examined the costs of synthetic fuels in relation to the price of oil. Seven representative synfuel processes were analyzed. In contrast to the arguments of synfuels advocates, the report concluded that as the price of oil rose, the projected cost of producing synfuels from a new plant using foreseeable technology would increase proportionately. A major assumption of this report was that oil, as the dominant source of energy, "drives" the cost of all modern industry, including the construction of huge synfuel plants. "No matter how high the price of oil rises," the report declared, "—even to $100 per barrel—a new plant built subsequent to arrival of oil at that price will not be economic as an investment prospect." (Actually, one synfuel technology, the TOSCO II shale oil process, became lower than the price of oil at around $50 per barrel; but the cost escalation of the output from this process was extremely sensitive to the cost of the raw material and capital cost investment.) The study presented a clear societal cost-benefit perspective (U.S. Congress, House, March 1981, pp. 1–3, 45–49, 57–62, and 104–11):

[E]ven with continuously increasing oil prices, a synfuels plant would not necessarily appear to be a sound investment from the private sector investor point of view at a given point in time. Without subsidies and assuming any reasonable rate of oil price increase, the length of time between expenditure of the capital investment and recovery of the capital from future higher prices may be unattractive, given the current and expected interest rates and the associated "time value of money". If oil prices should stop rising permanently, then no new synthetic fuels venture could be justified economically now or in the foreseeable future.

Synfuels advocates never could marshall credible societal cost-benifit arguments to support their positions. When the Reagan administration quickly set about dismantling the grandiose synfuels scheme established in the late Carter administration, its task was relatively easy. There was an enormous reservoir of powerful societal cost-benefit theory and analyses available for Reagan

198 II/MATRIX MANAGEMENT APPLICATIONS

energy officials to tap. Moreover, de-emphasizing direct public sector involvement and relying more heavily on the market and private sector decisions was, of course, more ideologically compatible with the new administration.

In the fall of 1981 a high-level energy official in the Reagan administration articulated rather precisely a societal cost-benefit viewpoint (LeGassie, 1981, p. 4):

Inherent in the restructuring [(of the U.S. synfuels program in the Reagan administration)] is a greater reliance on the private sector to set the pace for commercial introduction of synthetic fuel technologies. We want private firms—not the government—to choose among the slate of technologies, both liquid and gas producing, and to show their convictions in the traditional manner—through investment of private funds in anticipation of a fair profit.

But it would be an error to attribute the successive failures of synfuels in the United States only to an inability to meet some kind of societal cost-benefit test. Synfuels advocates during almost all the postwar period lacked the critical bureaucratic-political element as well. Unlike their energy rival, particularly in nuclear power, the advocates of synfuels were never able to build an effective bureaucratic-political apparatus. The fatal absence of what has been referred to as a "winning bureaucratic style" (Sapolsky, 1972) was perhaps best illustrated by the rather dismal saga of synthetic fuels in the United States in the 1960s and the early 1970s.

The coal industry by the late 1950s was a prototypical "sick" sector. Production had sunk to a very low level and coal had lost its hold on the transportation, residential, commercial and, to a lesser extent, industrial markets. The growth of the electric utility market, which practically singlehandedly would fuel the expansion of coal consumption, was just beginning (U.S. Energy Information Administration, 1981, p. 128). Furthermore, coal boosters in the industry and on the Hill watched nuclear power capture huge amounts of federal R & D funding, taking advantage of the considerable bureaucratic and political clout of the Joint Atomic Energy Committee, the Atomic Energy Commission, and the electric utility industry and its suppliers. Of course, all this coincided with, and stood in marked contrast to, the phasing out of the first synfuels program.

Not surprisingly, an attempt was soon made to bring the benefits of "big science" to coal as had apparently been done for nuclear power. But instead of the equivalent of the powerful Atomic Energy Commission, President Eisenhower in 1960 was willing to establish only the much more modest Office of Coal Research (OCR) in the Department of Interior. OCR was overseen by a General Technical Advisory Committee composed of coal industry producers and consumers and other key stakeholders, including labor. The major policy issue debated by this committee and OCR generally was whether to focus on seemingly mundane and relatively short-term matters like coal production, transportation, and air pollution, or on the apparently more glamorous and sophisticated area of synthetic fuels. OCR opted primarily for synfuels, but it never had the budgetary, technological, bureaucratic, or political resources to be successful in this field.

OCR funded a number of synfuels efforts in the 1960s. Among its contractors were the FMC Corporation, the Institute for Gas Technology, Westinghouse, and the Spencer Chemical Company (which along with its subsidiary, the Pittsburgh and Midway Coal Company, became subsidiaries of the Gulf Oil Corporation in 1963). The major program of OCR was Consolidation Coal Company's "Project Gasoline" coal liquefaction project. A pilot plant was constructed in Cresap, West Virginia, and in the mid-1960s both Consolidation and OCR released a barrage of optimistic comments and cost estimates. But "Project Gasoline" was increasingly plagued by cost increases and overruns and technical difficulties. The plant was eventually shut down in the early 1970s and later in the mid-1970s re-used as an experimental facility. OCR's major effort had failed, and the agency had very little of equal stature left in its inventory of activities.

Funding for OCR was also quite limited. In 1963, out of a total federal enrgy R & D budget of about $330 million, only about $11 million was allocated for coal R & D. (In contrast, over $210 million was spent on nuclear fission alone.) OCR's fiscal 1969 budget was only about $13.9 million (Cochrane, 1981A, pp. 367–71; Barber, 1981B, pp. 316–20; Barber, 1981A, pp. 282–85; Horwitch, 1980).

Shale oil policy and activity similarly floundered. In the Department of Interior during the 1960s several studies were commissioned and an Oil Shale Advisory Board was established. With the exception of some activity at the government experimental plant at Rifle, Colorado, and a private program by the Oil Shale Corporation (TOSCO), little was taking place in the shale oil field. In May 1965 President Johnson was advised by the advisory board to defer leasing federally owned shale oil lands, because "the art of extracting oil from shale is not yet developed to a point where it is commercially competitive," and therefore there was no need for a large-scale leasing effort (Cochrane, 1981A, pp. 371–73).

Lack of focus and resources of all kinds led to a long hiatus for synfuels in the United States (Horwitch and Prahalad, 1981). Shale possessed no focal organization. The bureaucratic hub for coal-based synfuels, OCR, was understaffed and underfunded and largely ignored. Unlike nuclear power, synfuels also lacked a powerful political ally on the Hill. Little wonder, therefore, that the synfuels advocates were totally unprepared for the 1973–74 oil embargo, which presented them with their greatest opportunity yet for creating a synfuels industry in the United States.

The absence of bureaucratic-political power in synfuels was, of course, closely related to another key failing in the American synfuels experience: the lack of effective managerial behavior, particularly in not producing appropriate champions. It is perhaps understandable that synfuels advocates could not achieve their goals in the obviously negative environment of the 1950s and 1960s. But their failure in the 1970s—a time of alleged energy crises and elaborate proposals for energy independence, including suggestions for massive synfuels efforts—deserves some comment.

One of the first major energy plans that gave synfuels a key role was a $100-billion proposal by Vice-President Nelson Rockefeller in 1975. Rockefeller urged the creation of what was variously called an "Energy Independence Authority" or an "Energy Resources Finance Corporation." Under this scheme a public corporation would have been established to invest up to $25 billion in public funds and guarantee another $75 billion over a ten-year life span.

But the Rockefeller initiative failed to take hold. One problem, in spite of President Ford's positive reaction, was the absence of what might be called an operational and effective champion in the mainstream of energy decision making. The plan was attacked on a broad front and its advocates were not in a sufficiently powerful position to defend the proposal successfully. Much of the plan was drafted by a prestigious outside consultant who had little direct experience with energy and who was really personally responsible only to Rockefeller. Most energy analysts and economists attacked the idea on societal cost-benefit grounds.

Key parts of the administration such as Treasury, the Office of Management and Budget, the Council of Economic Advisers, and the Federal Energy Administration opposed the proposal to a greater or lesser degree. On the Hill, both Democrats and conservative Republicans came out against the idea. With no operational and committed champion in place working daily for victory, the plan had little chance of passage. The initiative was subsequently killed (Corrigan, 1975; de Marchi, 1981).

Synfuels' best opportunity, of course, came in the wake of the panic created by the Iranian crisis in the spring of 1979. For perhaps the first time, it seemed that an effective set of synfuels champions was actually emerging. One group that quickly had significant impact were three old Washington hands, Lloyd Cutler, Paul Ignatius, and Eugene Zuckert, who in a memo that was published partially in the *Washington Post* and partially in the *Wall Street Journal* urged the formation of a synthetic fuels industry that in five or ten years would be capable of producing the equivalent of 5 million barrels of oil per day. They estimated the cost of their proposal at $100-$200 billion and called for a manager with an impressive track record, "a Robert McNamara-type," to take charge. They considered synfuels technologies more or less "proven" and used as historical precedents the American development of synthetic rubber during World War II, and the government's market-guarantee contracts during the Korean War for expanding the nation's capacity for aluminum, copper, and nickel production.

Just as important, synfuels—really for the first time in its history—was portrayed as having key political and psychological benefits: "It [a massive synfuels effort] would give us all the psychological lift of doing something instead of just doing without. It would employ our managerial, technological, engineering and organizing talents to achieve a productive rather than a restrictive result." (*Washington Post,* June 10, 1979, p. D3)

Support for synfuels was also building on the Hill. The most significant initiative was the resolution introduced in January 1979 by Congressman William Moorhead (D-Pennsylvania) to amend the Defense Production Act of 1950, so as to permit the Defense Department to buy the equivalent of 500,000 barrels of oil in synthetic fuels within five years. The bill enjoyed strong support from the Democratic leadership in the House.

The White House also eventually jumped on the synfuels bandwagon. President Carter's popularity was at a low level by the early summer of 1979. The Carter administration's energy policies seemed confused and ineffective in the face an apparent mounting crisis. The President was advised to seize the initiative in developing alternative energy sources both for political reasons and to avoid ill-considered, hasty, and expensive actions by Congress. Rejecting cautionary recommendations about synfuels from many in the administration, Carter opted for a bold, massive synfuels plan (Brown and Yergin, 1981; U.S. Congress, House, March, April, and May, 1979).

On July 15, 1979, Carter delivered what was undoubtedly the most important address in the history of synfuels in the United States. He unveiled a new energy policy that had as its core an $88-billion crash synfuels program. He called for the equivalent of 2.5 million barrels of oil per day from synfuels by 1990, and proposed establishing an Energy Security Corporation to provide price guarantees, federal purchase agreements, direct loans, loan guarantees, and energy bonds. (He also proposed an Energy Mobilization Board to expedite the construction of energy-producing facilities, but ultimately this idea was set aside.) (See Yager, 1981, pp. 624–33. For a comparison of the major synfuels proposals in 1979, see U.S. Congress, House, Feb. 1980, pp. 66–76.)

After considerable debate, a Synthetic Fuels Corporation (SFC) with powers similar to the original Energy Security Corporation was signed into law on June 30, 1980. The SFC was to last for twelve years. Two phases of funding and production goals were set: $20 billion and 500,000 barrels of oil per day equivalent for the first four years, and potentially $68 billion and not less than 2 million barrels of oil per day for the next eight years (U.S. Synthetic Fuels Corporation, *Summary: Energy Security Act; New York Times,* July 1, 1980, pp. D1, D4).

Grandiose visions of a massive synfuels industry by the end of the decade, resulting from an enormous program under the guidance of the SFC, turned out, of course, to be a complete fantasy. As of mid-1982, only two major synfuels projects, both backed by forms of government subsidies, were under construction; the Great Plains coal gasification project in North Dakota, and the

Union shale oil venture in Colorado. At least four huge projects have been canceled or postponed indefinitely: the Exxon-TOSCO Colony and the Gulf-Amoco Rio Blanco shale oil projects in Colorado, the Gulf SRC-II coal liquefaction project, and the American Natural Resources' WyCoal gasification project in Wyoming. As of May 1982, the SFC had yet to sign a project support agreement; perhaps three of the original 63 entries would eventually receive SFC sponsorship (*New York Times,* May 5, 1982, pp. D1, D2; U.S. Synthetic Fuels Corporation, news release of April 1981 and public project summaries).

There are those who would credit the downfall of synfuels solely to the triumph of sound societal cost-benefit thinking. But just as important was the absence of a committed and skilled synfuels champion in an admittedly hostile Reagan administration. At the very end of the Carter administration there was potentially such an individual—SFC Chairman John Sawhill. Sawhill, an experienced Washington administrator, probably came as close as anyone ever had in synfuels to the James Webb mold for directing large-scale programs. Sawhill was clearly enthusiastic about synfuels. In November 1980 he said: "The creation of a synthetic fuels industry is one of the most exciting challenges this country faces over the next two decades. . . . There is no doubt that by the latter part of this decade synthetic fuels will be powering automobiles, heating homes, and running businesses and industries as replacement fuels for foreign oil" (Sawhill, Nov. 1980).

In contrast to Sawhill the Reagan administration, with its emphasis on market forces, its de-emphasis on direct government involvement in large-scale demonstration programs, and its stress on fiscal mechanisms, made the SFC strictly an investment bank. Basically, synfuels officials under Reagan viewed proposals with primarily bankers' eyes in very narrow societal cost-benefit terms. The very notion of a committed champion mobilizing a network of favorable stakeholders—absolutely critical for the success of large-scale programs—would be alien to the Reagan government.

The final element that is crucial for large-scale programs, the corporate strategic one, has also been largely absent for synfuels. However, over the years and until perhaps very recently, an increasingly favorable corporate strategic context for synfuels has been building.

As already mentioned, it was really the opposition of the oil industry in the Truman and early Eisenhower administrations that killed the first attempt to create a massive synfuels effort. The basis of the industry's opposition was strategic: it feared the emergence of a major energy rival just as it was making a key strategic shift from mostly domestic to both domestic and foreign oil production (Vietor, 1980). But the strategic situation began to change in the 1960s. Oil companies began to diversify into other energy sources, and their views toward synfuels similarly changed. The program of the Department of the Interior's Office of Coal Research (OCR), in spite of the low level of funding, ignited some interest in synfuels. Moreover, in 1963 Gulf acquired Spencer

Chemical and the SRC liquefaction process. In 1966 CONOCO acquired Consolidation Coal and the projects in Consolidation that OCR was sponsoring. During the latter half of the decade other oil companies, including Exxon, ARCO, and Sun, acquired large numbers of coal leases, particularly in the West. In addition, Sohio acquired Old Ben Coal and Occidental acquired Island Creek. Meanwhile, Exxon established a Synthetic Fuels Research Department and clearly viewed at least its Western coal reserves as potential synfuels feedstocks. In 1968, Sun was reported to be marshalling resources for a large-scale coal synthetic fuels plant (Horwitch, 1980; Chakravarthy, 1981; U.S. Congress, Senate, June, July, and Oct. 1975, pp. 40–41).

Natural gas companies also became more involved with synfuels: Columbia Gas, Texas Eastern Transmission, El Paso Natural Gas, American Natural Resources, and Peoples Gas all planned gasification projects by the early 1970s (Horwitch, 1980). And energy industry involvement continued to intensify. In 1976, the Bureau of Mines identified 36 proposed synfuels projects in the Western United States, of which gas companies managed fifteen and oil companies seven; of the five proposed liquefaction projects in this group, three were directed by oil companies (U.S., Bureau of Mines, 1976).

Sparked by the flurry of synfuels activity in 1979–80, industry-proposed synfuels plans multiplied. In mid-1980 it was possible to identify at least six proposed commercial-scale shale oil projects (Occidental's; the Colony project of Exxon and TOSCO; the Rio Blanco project of Gulf and Amoco; Superior Oil's project; TOSCO's Sand Wash project; Union's project; and the White River project of Phillips, Sohio, and Sunedco) and one other major shale project (the Prahaho project). There were also three proposed commercial-scale coal gasification projects (the Conpaso, Wesco, and Great Plains projects), and eight coal-based demonstration-scale projects (SRC-I , SRC-II, Coalcon, ICGG, Slagging Lurgi, Cool Water Gasification, Ken-Tex, and Memphis Light, Gas and Water) (U.S. Congress, Senate, Sept. 1979; U.S. Department of Energy, June 1980; Congressional Research Service, June, 1979; U.S. Department of Energy, May 1980). Finally, the Synthetic Fuels Corporation by April 1981 received at least 63 proposals for synfuels projects (U.S. Synthetic Fuels Corporation, April 1981, and listing of public summaries of projects).

There is a certain irony in the evolution of the corporate strategic element for synfuels in the United States. Until practically the very end, corporate strategy was a true bright spot for synfuels as the strategic context grew increasingly favorable. Synfuels became part of the energy companies' long-term strategic portfolio. Oil in a sense was a gigantic "cash cow" and was used to support a number of diversification moves, including decisions to back synfuels projects. The problem for synfuels by the early 1980s was that the general strategy of the energy companies was proving faulty or ill-timed. The cash cow, oil, was in trouble in the face of declining real prices, lowering consumption rates in the industrialized countries, and a large inventory.

Synfuels was proving *too* long-term a proposition for even the largest energy company: Exxon dropped its Colony shale oil project in 1982. The corporate strategic element had come full circle. In the early 1950s, the oil industry opposed synfuels as a potentially powerful competitor; in the early 1980s, the energy industry rejected synfuels as an unnecessary burden.

CONCLUSION

The American experience with synthetic fuels provides a number of cautionary and instructive lessons. First, it should be clear that there is no automatic transfer of appropriate managerial methods from the earlier, more protected large-scale programs in defense and aerospace to the more modern and vulnerable large-scale programs for which, explicitly or implicitly, commercialization of a technology is an ultimate goal.

One issue that synfuels advocates did not really face, therefore, was the proper application of project management approaches. Their problems were of a broader and more contextual nature, dealing with such issues as uncertain technologies and economics, a hostile bureaucratic and political environment, an absence of effective operational champions, and a changing corporate strategic situation.

Another lesson that emerges is the need to take a new look at how top managers of large-scale programs are trained and selected. Clearly, the training of such people should involve much more than just techniques. A grounding in economic analysis and, even more, a *sensitivity* to the bureaucratic and political situation, to the current and potential stakeholder environment, and to the corporate strategic context are essential. Synfuels advocates generally did not possess such an awareness.

Finally, large-scale projects now require what can be termed a convergence factor. These projects are so complex that at least four very different elements—societal cost-benefit, bureaucratic-political, managerial, and corporate strategic—have to be favorable and functioning at the same time. Throughout the almost 40-year history of synfuels such a convergence never existed; consequently, even the beginnings of a synfuels industry never appeared. Achieving the convergence factor is now probably essential for the implementation of successful large-scale programs.

Bibliography

Allison, Graham T. *Essence of Decision.* Boston; Little, Brown, and Company, 1971.
Allison, Graham T., and Halperin, Morton H. "Bureaucratic Politics: A Paradigm and Some Policy Implications." *World Politics* 24 (Spring 1972) (Supplement); 40–79.
Archibald, R. D. *Managing High Technology Programs and Projects.* New York; John Wiley & Sons, 1976.
Baer, W. S., Johnson, L. L., and Merrow, E. W. *Analysis of Federally Funded Projects: Exec-*

utive Summary (#R-1927-DOC): *Final Report* (#R-1926-DOC; and *Supporting Case Studies* (#R-1927-DOC). Santa Monica, Calif.: The Rand Corporation, April 1976.

Barber, William J. "The Eisenhower Energy Policy: Reluctant Intervention" (1981A), and "Studied Inaction in the Kennedy Years" (1981B). In Craufurd D. Goodwin, ed., *Energy Policy in Perspective*. Washington: The Brookings Institution, 1981.

Brown, Benjamin A., and Yergin, Daniel. *Synfuels 1979*. Cambridge, Mass.: J. F. Kennedy School of Government case, 1981.

Caro, R. A. *The Power Broker*. New York: Vintage Paperback, 1975.

Chakravarthy, Balaji. *Managing Coal: A Challenge in Adaptation*. Albany: State University of New York Press, 1981.

Chandler, Alfred D., Jr. *Strategy and Structure*. Cambridge, Mass.: MIT Press, 1962.

Chapman, Richard L. *Project Management in NASA*. Washington: NASA, 1973.

Cochrane, James L. "Energy Policy in the Johnson Administration: Logical Order Versus Economic Pluralism" (1981A), and "Carter Energy Policy and the Ninety-Fifth Congress" (1981B). In Craufurd D. Goodwin, ed. *Energy Policy in Perspective*. Washington: The Brookings Institution, 1981.

Congressional Research Service. *Synthetic Fuels from Coal: Status and Outlook of Coal Gasification and Liquefaction*. Washington: GPO, June 1979.

Corrigan, Richard. "Energy Report: Rockefeller Presses his Plan for $100 Billion Bank for Fuels." *National Journal*, Oct. 25, 1975, pp. 1469–76.

de Marchi, Neil. "The Ford Administration: Energy as a Political Good." In Craufurd D. Goodwin, ed., *Energy Policy in Persepctive*. Washington: The Brookings Institution, 1981.

Eads, George, and Nelson, Richard R. "Governmental Support of Advanced Civilian Technology: Power Reactors and the Supersonic Transport." *Public Policy* 19 (1971): 405–27.

Gaddis, P. O. "The Project Manager." *Harvard Business Review*, May–June 1959, pp. 89–97.

Ginzberg, E., Kuhn, J. W., and Schnee, J. *Economic Impact of Large Public Programs: The NASA Experience*. Salt Lake City: Olympus Publishing Company, 1976.

Goodwin, Craufurd D. "The Truman Administration: Toward a National Energy Policy," and "Truman Administration Policies toward Particular Energy Sources." In Craufurd D. Goodwin, ed., *Energy Policy in Perspective*. Washington: The Brookings Institution, 1981.

Hederman, William F., Jr. *Prospects for the Commercialization of High-BTU Coal Gasification* (#R-2294-DOE). Santa Monica, Calif.: The Rand Corporation, April 1978.

Horwitch, Mel. *Clipped Wings: The American SST Conflict:* Cambridge, Mass.: The MIT Press, 1982. (1981A)

———. "Designing and Managing Large-Scale, Public-Private Technological Enterprises: A State of the Art Review." *Technology in Society* 1 (1979): 179–92.

———. "The Role of the Concorde Threat in the U.S. SST Program." Sloan School Working Paper #WP1306-82, May 1982. (1982B)

———. "Uncontrolled and Unfocused Growth: The U.S. Supersonic Transport Program and the Attempt to Synthesize Fuels from Coal." *Interdisciplinary Science Reviews*, Vol. 5, No. 3 (1980): 231–44.

Horwitch, Mel, and Prahalad, C. K. "Managing Multi-Organization Enterprises: The Emerging Strategic Frontier." *Sloan Management Review*, Winter 1981, pp. 3–16.

LeGassie, Roger W. A. "The U.S. Government Program and Policies for Coal Liquefaction," remarks to the Coal Liquefaction Symposium. Oct. 21, 1981, Tokyo.

Merrow, Edward W. *Constraints on the Commercialization of Oil Shale* (#R-2293-DOE). Santa Monica, Calif.: The Rand Corporation, Sept. 1978.

Merrow, Edward W., Chapel, Stephen W., and Worthing, Christopher. *A Review of Cost Estimation in New Technologies: Implications for Energy Process Plants* ("#R-2481-DOE). Santa Monica, Calif.: The Rand Corporation, July 1979.

New York Times, July 1, 1980, pp. D1, D4; May 5, 1982, pp. D1, D2.

Sapolsky, Harvey. *The Polaris System Development.* Cambridge, Mass.: Harvard University Press, 1972.

Sawhill, John C. "The Why and How of the Synthetic Fuels Industry." Speech, Nov. 1980, distributed by the U.S. Synthetic Fuels Corporation.

Schmalensee, Richard. "Appropriate Government Policy toward Commercialization of New Energy Supply Technologies." *The Energy Journal,* Vol. 1, No. 2 (1980): 1–40.

Seamans, Robert C., Jr., and Ordway, Frederick I. "The Apollo Tradition." *Interdisciplinary Science Reviews,* Vol. 2, No. 4 (1977): 270–304.

U.S. Bureau of Mines. *Projects to Expand Fuel Sources in Western States.* IC No. 8719, 1976.

U.S. Congress, House. *Costs of Synthetic Fuels in Relation to Oil Prices.* Report prepared for the Subcommittee on Energy Development and Applications of the Committee on Science and Technology. Washington: GPO, March 1981.

———. *The Pros and Cons of a Crash Program to Commercialize Synfuels.* Report prepared for the Subcommittee on Energy Development and Applications of the Committee on Science and Technology. Washington: GPO, 1980.

———. *To Extend and Amend the Defense Production Act of 1950.* Hearings before the Subcommittee on Economic Stabilization of the Committee on Banking, Finance, and Urban Affairs, March 13, 14; April 4, 25; and May 3, 1979, Serial No. 96-9. Washington: GPO, 1979.

U.S. Congress, Senate. *Interfuel Competition.* Hearings before the Subcommittee on Antitrust and Monopoly of the Committee on the Judiciary, June 17, 18, 19; July 14; Oct. 21 and 22, 1975. Washington: GPO, 1976.

——— *Synthetic Fuels.* Report by the Subcommittee on Synthetic Fuels of the Committee on the Budget. Washington: GPO, Sept. 27, 1979.

U.S. Department of Energy. *DOENEWS: A Summary of Recent Fossil Fuel Energy Events.* Issue #80-1, June 16, 1980.

———. *Solvent Refined Coal-II Demonstration Project.* Draft Environmental Impact Statement, May 1980.

U.S. Energy Information Administration. *Annual Report to Congress.* Vol. 2, 1980 (Washington: GPO, 1981), p. 128.

U.S. Synthetic Fuels Corporation. Public summaries of proposals. April 1981.

———. "61 Proposals Submitted to the Synthetic Fuels Corporation." SFC news release, April 1, 1981.

———. *Summary: Energy Security Act (P.L. 96-294).*

Vietor, Richard E. "The Synthetic Liquid Fuels Program: Energy Politics in the Truman Era." *Business History Review,* Vol. 54, No. 1 (Spring 1980): 1–34.

Washington Post, June 10, 1979, pp. D1, D3.

Yager, Joseph A. "The Energy Battles of 1979." In Craufurd D. Goodwin, ed. *Energy Policy in Perspective.* Washington: The Brookings Institution, 1981.

Section III
The Matrix Culture

In the long term, the introduction of matrix management sets in motion a series of changes that will ultimately affect the prevailing culture of the organization. Effective matrix management may be more dependent on the attitude changes of key managers and professionals in the organization than on anything else. Section III will deal with some of the cultural changes precipitated and how they can be dealt with effectively.

In Chapter 14 John W. Stuntz talks about matrix management from the perspective of a general manager. His experience in progressing from a professional to an executive vice-president in the matrix culture provides an excellent experience base for describing real-world intricacies of matrix management from the viewpoint of both the individual and the organization engaged in pursuing profitability and growth in a high technology firm. He describes the benefits and the costs of matrix and concludes that "it's working" by helping people create useful high technology products and generating more profits for the investor.

It is impossible to talk about matrix management without considering the human factor. In Chapter 15 John V. Murray and Frank A. Stickney use the need for matrix management as a springboard to talk about preparing people for the change such management requires. The authors contrast management behavior in the traditional functional and matrix organizations, using authority and responsibility as focal points to discuss managerial behavior. Murray and Stickney conclude by reminding us of the reciprocal interdependencies in the matrix design. People, they say, will be the vital factor in determining the success of organizations venturing into matrix.

In Chapter 16 James Mashburn and Bobby C. Vaught present a provocative and insightful look into the meaning of dual leadership. Initially they examine the strong bias that exists in contemporary organizations toward perpetuating unity of command. Then, recognizing the duality of group and individual behavior in organizations, they ask: Why not have two people in leadership positions? One would lead the way toward some mutually agreed-upon objective, while the second maintained the social needs of the work group. According to them, the accomplishment of both is important, if the group is to prosper and survive over the long run and reach the objective. Dual leadership provides two managers the opportunity to mutually support and criticize each other's ideas and philosophy. By having each manager assume equal authority and responsibility, all sides of problems can be examined, with exclusive emphasis on neither the task nor the people, but on both as an opportunity for meaningful, constructive conflict resolution in accomplishing organizational purposes.

In Chapter 17 Robert I. Desatnick talks about managing human resources in the matrix environment. In the context of shared accountability he states that the most single important contribution the human resource executive can make is to favorably affect the quality of management by influencing the management process in matrix-oriented organizations. He presents and explains several models that help the reader to better understand the management of human resources in the matrix context.

In Chapter 18 William E. Souder describes the pragmatic application of guidelines for motivating people within matrix organizations where highly motivated personnel are essential for success. He describes the motivational factors that should concern each manager, then examines the three important types of motivation theories: cultural, content, and process theories. After the theory of motivation is established, Souder provides some solutions to the several characteristic motivational problems in the matrix organization.

14. A General Manager Talks About Matrix Management

John W. Stuntz*

BACKGROUND

The author has worked for 30 years at the sprawling Westinghouse Defense and Electronic Systems Center, in and near Baltimore, Maryland. This industrial complex has created major high technology electronic systems such as the radars for AWACS and the F4 and F16 fighter aircraft, electronic countermeasure systems, electro-optical laser guidance systems, and space systems including weather satellite sensors and the TV camera that sent pictures of the first Apollo moon landings back to Earth. International programs have included nationwide air defense command and control systems for countries such as Morocco.

It was while Mr. Stuntz was a middle-level manager in the (then) 1500-employee-engineering department of the 4000-employee Aerospace Division that the matrix started its evolution locally, beginning in that engineering department. Now, 20 years later, a sophisticated multitiered, multidivision matrix (see Figure 14-1) encompasses not only the 10,000 engineering, manufacturing, and administration people clustered near BWI Airport, but also several thousand others in divisions and support activities spread within a one-hour drive of the central facility. It has an annual volume well over a billion dollars.

INTRODUCTION

General managers *like* matrix management! They *should* like it! It was created more or less 20 years ago to meet the needs of the general manager in the complex, dynamic defense systems business. But only the bold General Man-

*John W. Stuntz has held general management positions for 14 years in the Westinghouse Defense and Electronic Systems Center, Baltimore. He holds bachelor and masters degrees in Electrical Engineering and attended the Program for Senior Executives at MIT's Sloan School of Management. He has lectured in electronics design and currently is teaching Technological Innovation and Entrepreneurship at the University of Maryland.

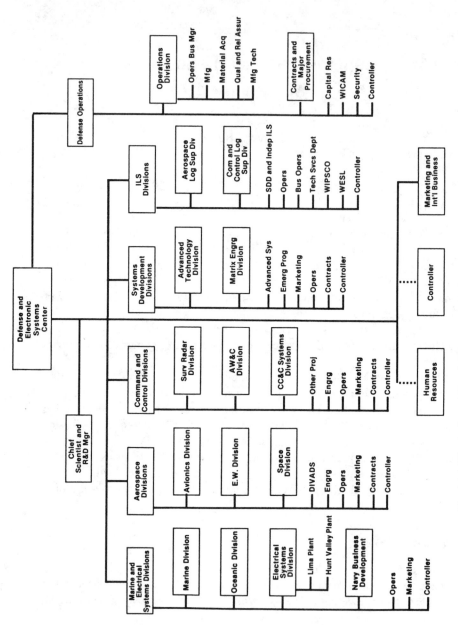

Fig. 14-1. Westinghouse Electric Corporation Defense and Electronic Systems Center.

ager (GM) would choose to initiate a matrix in his or her division. The management system changes are so pervasive—and so personal—that the effort to effect the changeover from a conventional functional organization is massive, with a cultural shock time-constant longer than almost anyone will admit.

Yet, given the particular array of management pressures that triggered the "program management matrix" in the first place, general managers *do* opt for this no-longer-new structural and cultural approach, because it best addresses precisely the class of issues that consume their attention:

- In the short term, the matrix provides for flexible use of key technical resources, both people and facilities. (Functional "fiefdoms" don't have to be reorganized to move the talent from program to program to meet fluctuating demands.)
- For the long term, the matrix expands the avenues for business benefit from broadly applicable strategic investment. (A pattern of shared resources and shared responsibilities obviates the traditional "technology transfer" issue altogether.)
- Basically, through the program management dimension, the matrix establishes mini–general managers who are extensions of the GM for a subset of the business, but without imposing the inflexibility and communication isolation of the functional resources that are characteristic of self-contained business segment departments.

A general manager must consider organization as a system, i.e., all facets for all time. In this article both near-term and long-term impacts of matrix management are discussed, relative to the three P's of the GM's job:

1. The *product,* or technical dimension: how the matrix relates to the effectiveness with which rapidly changing technology is exploited to meet the fluctuating demands of a complex array of products and customers. We will look at some issues and examples in various phases from entrepreneurial development to production and support.
2. The *profit* dimension, and some of the trade-offs between efficiency and effectiveness. We will look particularly at program visibility and control through the eyes of the GM, i.e., how the matrix affects generation of a satisfactory return on investment.
3. The *people* dimension: who benefits? How can the matrix be used as a general management development tool? What about job satisfaction and team building? And what are the special demands for openness and trust?

THE PRODUCT DIMENSION

Formally at least, the matrix started in the product dimension. In one location around 1960 we saw pressures from three directions, all product-focused:

- the need to keep up with the explosive growth of technology
- the need for availability of talent to staff large, fluctuating programs
- the need for market-focused program management

High Technology Through "Centers of Excellence"

First, the technology was eruptive. In the early 1960s transistors were revolutionizing electronics, but the *real* revolution was just starting with the birth of integrated circuits that would lead to the proliferation of digital systems, software integration, high-density packaging, factory automation, and computer-automated design, manufacture, and test. The great space age was born, spawning and demanding technical innovation from energy to structures, from sensors to navigation, from thermal design to human factors, with an emphasis on the "-ilities"—especially reliability—orders of magnitude of change beyond all prior experience. How was management to organize its technical resources to meet the new demands?

Management was faced with a dilemma. There was a need to concentrate talent and increasingly expensive facilities in each of the many cells or disciplines where technical excellence and development leadership were required. On the other hand, these emerging skills and new facilities had to be accessible to the widening array of challenging new products and programs. With the old "product line" approach, each element tried to grow digital engineers and generate its own computer architecture. There were never enough experts to go around in many of the specialties like thermal design, high-density packaging, and reliability, and those engineering managers who "had" them hoarded them!

The solution to the dilemma was an engineering matrix organization, initiated in 1961 and incrementally put in place over the following two years, that created *centers of excellence* (cells) of four types:

- In design engineering, cells were created devoted to electrical, mechanical, reliability/maintainability, and systems engineering.
- In advanced development, architecture and application-oriented cells were focused on radar, computers, electro-optical and countermeasures, and advanced systems.
- In advanced technology, there were cells in integrated circuits, lasers, solid-state microwave, and advanced sensors.
- In technical support, cells grouped laboratory technicians, drafting, engineering data, and administration.

A simplified chart of the resultant engineering organization is shown in Figure 14-2, which also indicates a Project Engineering organization. The latter is motivated somewhat differently from the four technology cells and will be discussed later.

As everyone knows, good people rather than neat organizations produce notable results. Nevertheless, most observers judge that it is more than coincidence that the annual dollar volume of business generated by this engineering unit has increased nearly 20-fold in the ensuing 20 years, while the organization unit that emerged has firmly established itself among the world leaders in many of the technologies and product types that it has pursued. The *center of excellence* is the key. Interestingly, that term appears increasingly in current public discussion of the national industrial "malaise." The Industrial Innovation Act of 1980 recommends that the Department of Commerce establish cooperative centers of excellence in generic technologies to help spark a "revitalization" of American industry. One problem has yet to be solved: how to establish a national "matrix" to finance and exploit the results!

There are four secrets of success for any such center of excellence:

1. Above all, the center provides the ability to attract, grow, and retain top talent. Over the long haul this can be achieved only if there is a critical mass in each such discipline, which implies that a talented beginner can see a career future for himself or herself while growing to the professional stature of the leading expert who personifies the standard of excellence in that center.
2. Next, the "C of E" permits the concentration of financial and facilities resources to enable this talent pool to produce. There is more to this than meets the eye. Not only is there the benefit of nonduplication of resources

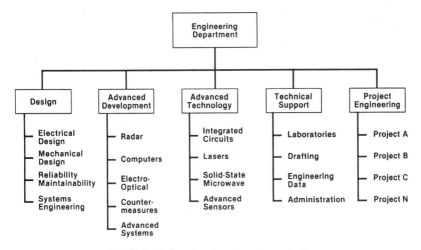

Fig. 14-2. Engineering department organization.

and effort, but also a bonus from standardization and from automatic transfer of "lessons learned."

3. Each center requires premier leadership not only effective parochially in managing that cell, but at the same time able to work well with other matrix leaders. (Not everyone can do this—least of all some of the older hands who matured in the older ways; more about this later.)

4. There must be management/organizational mechanisms to exploit the center's abilities to meet the needs of the larger enterprise. Almost by definition, no single cell can be self-contained, autonomous, or really worth much of anything by itself. Each is a block in the block diagram, a member of a team. This message is not always easy to get across!

A Matrix to Provide Assignment Flexibility

This leads to the second of the pressures for change, circa 1960. The defense and space programs individually were larger and technically more complex than had been prevalent. Each program needed a special mix of talent. When several such programs were overlaid on any division organization—each with its own gyrations, waxing and waning as it went through its start-up development, test and evaluation, production, modification, and support phases—the frequent and massive redistribution of technical talent among these competing, fluctuating demands mandated a new order as an alternative to constant civil war!

Figure 14-3 shows the wild gyration in total numbers of people employed on just one such program, the AWACS radar, in the late 1960s and early 1970s, as it went through technology demonstration, concept studies, fly-off competition, and full-scale development phases. Recognize that each of the transients shown is really the sum of many transients representing a variety of unique skills required for each phase. No management approach accommodates these fluctations easily. However, if we had not moved to a matrix earlier as a result of similar problems—if local engineer barons had had to be persuaded to transfer their top talent (historically they had *never* done so!)—well, we perceived that a competitor slower at moving his top talent lost a key job for that very reason. Those who complain about the matrix because of the frequent need for dealing with conflict must not have been exposed to the old superemotional talent wars! *Requiescat in pace.*

While the centers-of-excellence organization did provide maximum flexibility in assignment of talent, experience revealed the continued need for a remnant of the old "project engineering" group specifically assembled for each major project. In the case of the fluctuating program illustrated in Figure 14-3, the radar advanced development organization was originally "home" for the small core group who led the technical effort during early development phases. But when the project moved into major program status, it was necessary to set

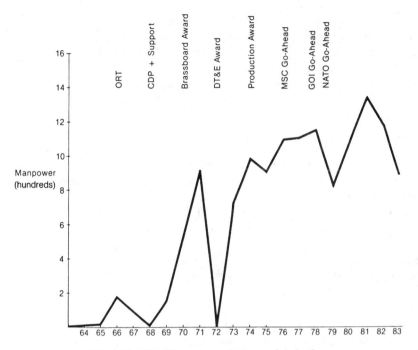

Fig. 14-3. AWACS manpower history and projection.

up an ad hoc project engineering group (see Figure 14-2), an organizational treatment that has been typical of our practice since adoption of the matrix. This is a hybrid approach—a compromise, perhaps. However, we hold the project-engineering ad hoc staffing to more or less 10% of the total engineering staffing for the program, and include therein the individuals who will be needed on that assignment for years, not months. We've found it to be a good compromise, the best of both worlds.

(The author was once advised by an expensive management consultant that precisely the compromise described "won't work," because the lack of organizational purity would lead to ambiguity of responsibility and accountability, and thus to the deadly game of "buck-passing." The consultant was wrong. The program was the AWACS radar, the time about 1971. It has become one of the most renowned successes among modern military systems!)

Another point: a classic management problem in both government and industry is the frustrating difficulty of transitioning a program from the R & D laboratory to major-project design-for-production status. During our 20 years' experience with the engineering matrix structure described here, such transition has been accomplished time after time with outstanding results. Some of the reasons for success go beyond this engineering organization, involving factors such as co-location, close organizational ties at higher level, and continuity of leadership. However, key talent from the matrix specialty groups are used in common by the early developers and by the final designers, and some project leaders invariably are moved with the project when it transitions from its advanced development home to independent major-project status. The matrix is the key: it provides the "people bridges"—always!

Market-Focused Program Management

Returning again to the scene around 1960, a third pressure on top management contributed to introduction of matrix management. While still heavily skewed toward the "product dimension" of the general manager's concerns, this third pressure went beyond product/technical management and therefore in this narrative will serve as a transition to the succeeding sections.

The third pressure was market-induced. During the creative phases of product development and introduction into the marketplace, there is no substitute for intimate interaction between the user/buyer and the developer/builder. The scope and intensity of that interaction increases as the complexity and rate of technical innovation increase. Some one person has to lead each project during these creative phases. During serial production, particularly of high-quantity nonchanging products, other alternatives are possible, such as mechanized management systems that can accept orders, prepare schedules, ship from inventory, determine charges, invoice, and collect. But clearly, especially in the defense industry, the technical dynamics of the past two decades have mandated project-focused personal leadership for the high technology system (or major subsystem) business.

The center of gravity of the effort requiring personal leadership lies in engineering, which is why the organizational innovations started there. But problems of successful transition into initial production swept the factory also into the organizational dilemma. And all the "business" functions, such as marketing, contracts, and finance, also found that they needed to focus their specialized expertise on each specific project during these turbulent innovative phases. The dilemma was that the only individual who was in a position to lead this array of functions, and the only one acceptable to the customer as "Mr. XYZ Company," was the general manager himself. But obviously the GM can't fill the role when there are many projects, and unless there are many projects, there is no way of satisfying the previously described demands for the broad range of technical excellence and the flexibility of resource assignment associated with this high-technology era.

The solution, of course, was the creation of program managers, one for each individual project, each being an extension of the general manager for directing all the functions and resources of the division to meet the needs (both internally and externally) of the project assigned.

Our first program manager appointees, back in the early 1960s, were leaders of the two biggest projects in the division. The individuals were very senior managers and they reported directly to the division general manager. As the years passed, each project, of whatever size, came to have its own program manager. We evolved networks of program and programs managers, lumping related programs for benefits of management aggregation, delegation, and review—a procedure akin to similar structuring (cells) within the functional organizational networks.

Then questions arose about organizational placement of other project-dedicated functional people. To whom should a specific project engineering group report? Should they report to the program manager or the division engineering manager? The same question arose for manufacturing, contracts, finances, and marketing. A whole new professional classification then emerged (or has nearly emerged), generally referred to as "systems management": people trained in the new science of program planning, scheduling, visibility, configuration management, documentation, and reporting. Their new classification almost has to report to a program management hierarchy—they have no other home!

The functions other than the "product" functions (engineering and manufacturing) typically involve far fewer people, hence massive dislocations are not required. These nontechnical functions are not experiencing any comparable rate of change in knowledge or techniques. Therefore choice of an "optimum" structure for them is less sensitive an issue than for the two "product" giants. The most common solution, apparently, is to tie the business management functions (such as contracts and marketing) to the program management dimension of the matrix, with perhaps a functional "VP of Contracts" (reporting to the GM) directly controlling some common parts and responsible for policies and resource adequacy of that particular function. Project engineering and project manufacturing, on the other hand, most commonly line-report to the division engineering or manufacturing manager, albeit with complete commitment to the program.

It is important to note what these program managers are *not* responsible for. Basically they do not share the general manager's responsibility for the adequacy of the resources, either physical or personal, nor for resolving the sometimes conflicting demands for their allocation. Functional managers retain those responsibilities and must be held accountable by top management. That is why divisional functional management must continue to report directly to the top, and not be subordinated to program management chains. Program managers also are not extensions of the GM for long-term strategic leadership of the enterprise, although their inputs (as well as the inputs of the functional resource managers) are integral to the strategic management process.

As a consequence of these two major limitations on the program manager's role, it is essential the GM have direct control of comprehensive management systems for integrating, directing, and monitoring all the complex elements of strategic management, including determination of the resources required to implement those strategies. Matrix management amplifies the demands on these integrating systems, and on the GM himself to direct them. (This appears to be a subtle point: it sneaked up on us many years after our transformation to the matrix.)

One more point: customers, if given a choice, rarely opt for a contractor's matrix organization for "their" project, usually preferring a nice neat, dedicated project team with "all" the resources under the direct control of "their" program manager. In part, customers are wrong to feel this way, rarely appreciating the advantages to the project of the higher quality of talent on call from those centers of excellence that couldn't be assigned long term to any single job. Nor do customers usually appreciate the advantages to their projects of the greater flexibility for fast build-up and rapid dissipation of those fluctuating teams. But in part they are right, because the greatest benefit of the matrix is to the long-term success of the *aggregate* business, enabling the building of the *best* product/technical capability, and providing the flexibility to exploit that capability for the broadest array of business opportunities. Sometimes the exploitation takes priority (and people) away from the original sponsor. No wonder it's the GM who likes it best!

THE PROFIT DIMENSION

The general manager of a profit center is responsible for generating a satisfactory continuing profit in return for the capital needed in his division. Since the division performance is the sum of the individual project performances, presuming that they are fairly burdened with shared assets and overhead expenses, the following questions must be asked:

- Is there a favorable balance between the inefficiencies of the matrix, on the one hand, and the efficiencies and enhanced competitive effectiveness on the other hand?
- Can the financial performance of individual projects be controlled adequately in a matrix organization?
- How does the GM use financial management policies to influence the performance of a matrix-organized division?

The simple answer to all three questions is, "It all depends on the specific situation"; but the following sections will discuss some of the trade-offs and nuances from a general manager's perspective.

Matrix Efficiency and Effectiveness

The first question, regarding the balance between financial efficiency and effectiveness, has been called the Achilles' heel of the matrix. Clearly there are more managers, hence a degraded manager/employee ratio. That makes "managed costs" go up, efficiency and productivity go down. Bad! Unless there are compensating competitive advantages or operating efficiencies latent in the particular business situation in question, the general manager would be making a terrible mistake to choose this complicated approach.

However, given the situation described in the prior section on the "product dimension," there is little doubt that the compensating financial benefits do exist. Many aspects of improved *effectiveness* were cited in that section:

● enhanced technical excellence
● expanded opportunities for exploitation
● accessibility of key talent
● focused program leadership

These are not minor advantages in the competitive world of high-technology systems business. Hughes, Raytheon, and Westinghouse; Boeing, Lockheed, and McDonnell-Douglas—the most successful competitors use some version of this management approach to gain market share and consequent profit potential. Many feel that there is no effective alternative. That is hard to prove, but examples abound that the matrix approach does work!

Surprisingly, there are also more significant efficiencies than are usually credited to the matrix—efficiencies that save strategic expenses, capital investment, and project costs. Strong claims? Let us see.

Strategic efficiency. Consider the modern world of digital electronics. One of the characteristics of digital architecture responsible for its competitive success is its potential for high volume use of standard module building blocks such as micro-processor chips, micro-computer boards, and standard mini-computers that can be programmed to perform a tremendous diversity of functions. In one of our locations, about 15 years ago our advanced development group created a versatile "milli" computer ("milli" denotes militarized mini). The R & D investment was substantial. Since then the "milli" has been used as the systems control center for a dozen different avionic systems. And the R & D savings continue, as iterated new technology insertion phases also derive redundant benefits from single developments. We are now repeating this success with a programmable signal processor. There are similar experiences with specialized integrated circuits, microwave transceiver modules, and low side-lobe array antennas. While there may be other ways to achieve such economy of scale from the R & D input, most other attempts to couple the lab to a diversity of market applications have been frustratingly ineffective.

Capital investment efficiency. Everything is becoming automated! That means facilities are becoming more and more expensive, but also more versatile, because that automation is programmable. Whether the capital investment is in a solid-state laboratory, a CAD/CAM design and manufacturing system, or a computer-controlled multiple-environment product qualification laboratory, the key to affordability is the matrix flexibility to apply these major investments over a very large business base.

Project cost efficiency. This efficiency is more controversial than the other two. But nothing is more efficient than the right talent at the right time, and nothing more wasteful than idle hands. Look back at Figure 14-3 and "eyeball" the waste cost of carrying that labor during slack periods. Of course skeptics raise the point that, if the people are still on the payroll, the money is still being spent. However, if this demand is only one of many flexibly served by a large functional organization, thoughtful functional management can blend the varied demands to smooth the aggregate load, so the hands won't be idle and wasteful idleness won't have to be "carried" by the project. This is not to say that there is no problem with fluctuating personnel demands. In fact, one of the GM's key concerns is management of the functional resources so that they are sufficient to meet needs, while simultaneously managing the schedule commitments (typically made by the program managers) so that everything doesn't peak at once. When you add the complication of uncertainty (most of the schedules are established during a competitive phase, so who knows what the load will really be?), perhaps the degree of difficulty is apparent. Perhaps it also becomes apparent why resource planning is such a key part of the matrix GM's job.

Financial Control of Projects

The second basic "profit" question posed earlier was, "Can the financial performance of individual projects be controlled adequately in a matrix organization?" Volumes of literature are devoted to the answer, replete with recipes of methods and practices for accomplishing that control. The world is full of cookbooks on the subject. Some years ago the U.S. Navy management "chefs" came up with PERT (Program Evaluation and Reporting Technique), a network of interacting events, elapsed times, and critical paths. PERT grew into PERT/COST, which added planned expenditures to each of the events. Fundamentally sound, PERT mushroomed, became more of a task than a tool, and pretty much collapsed from its own weight. Other variants have evolved, each with its own cookbook and legion of chefs and scullery maids. Currently in vogue is a triservice (but mostly Air Force) scheme called "CS-Square", more properly C/SCS (Cost/Schedule Control System). There is a somewhat simplified version called CSSR (Cost/Schedule Status Report)—a good compromise for smaller programs.

All these systems are effective, provided the cultists don't push them too far—provided they avoid introducing too many tiers in the work breakdown structure, too much precision and rigidity in the numbers, too much reporting detail. None is better than the quality of the planning that goes into it. (Watch out! Skilled people can "rig" these plans so that a project will look healthy when it isn't!) Actually, planning is the foundation of any system of management control, and the greatest strength of these formal systems is the discipline they impose on the project-planning process.

The systems mentioned here are all "earned value" systems, which means that for each task (or "work package") they specify what is to be done, a schedule for the work, a dollar value, and the responsible individual. Those four parameters are precisely the elements of a contract, and each work package is, in principle, an internal contract mutually agreed to by the program manager and the appropriate functional manager. Some of these internal negotiations are fierce, but they are absolutely essential, because the program manager has very little control leverage to augment the personal commitment by the functional manager to live up to the terms of that work package "contract."

If you want to find out how the program management matrix really works in the XYZ Company, investigate first the real inner workings of those little internal contracts. Some of the issues include how and when the work package "contract" was defined and priced. Did the two parties perceive that it was "captive" work, or was there some alternative source? Think how the degree of commitment and cost credibility depend on those answers.

Also, was there reasonable congruence of the functional organization to the organization of the work breakdown structure? In particular, was each work package staffed primarily by people who line-report to the manager who signed up for that "contract"? Note the consequences of negative answers: the task manager could just be a broker, and the managers who loaned people to him just body-shop entrepreneurs accountable neither for adequacy of the human resource nor for fulfillment of the terms of the "contract." Such an organization would be a mess, not matrix.

As detailed elsewhere, feedback of performance against the network of quantified work packages is routinely accomplished with the aid of modern data entry and computerized reporting systems. Visibility just isn't a problem. And using those same flexible computers, the performance data can be segregated or summed up to meet the needs of any level of management along either axis of the matrix. For the program manager, there is quantitative cost and schedule information for the total program, for all major line items of that program, and for each function's performance on the program. For the functional manager at the budget-center level, there is available not only individual task performance, but also aggregate cost and schedule performance on all work packages from all programs that the manager's organization unit has contracted for. (Please note: this is *not* the same thing as a budget statement—not at all! And it is only meaningful if the previously noted organizational congruence prevails!)

Most important, the general manager can get timely, meaningful feedback of the work performance-versus-plan of both dimensions of the matrix. Figure 14-4 is just one example of summary data charting on a single program,

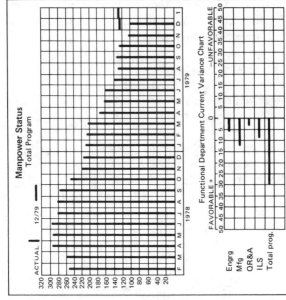

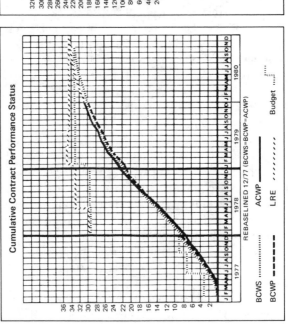

Fig. 14-4. Program summary data chart

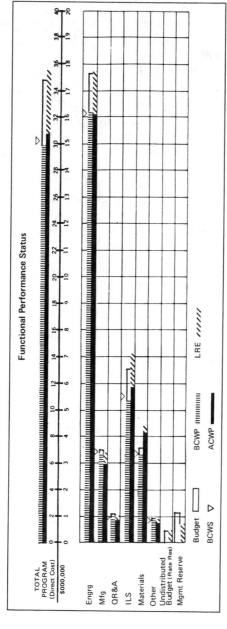

Fig. 14-4. Program summary data chart (continued)

derived as described above and specifically designed for the GM's use. From this single page (once he learns the code), the GM can see the historic curve of cumulative contract expenditures and schedule versus plan (upper left), direct charge manpower status and variance from plan by function (upper right), and finally, both current and projected final cost and schedule performance versus plan of each major function (across the bottom). This data comes right out of the C/SCS or CSSR "earned value" control systems.

Now here is where control comes from. Until this point we have described the planning, and the feedback of performance against plan. But even with these two, the program manager doesn't have the *power* to actually control the functional side of the matrix. Influence, yes; control, no! The manager derives power from the GM, who in turn can only identify the proper thrust for that power if he or she has credible visibility. Given today's management tools, a properly designed functional organization and work breakdown structure, and good matrix managers, project costs can be controlled in a matrix division. But that's a lot of prior conditions!

Influencing the Matrix with Financial Policies

How does the GM use financial management policies to influence the performance of a matrix-organized division? I could write a book, but a few paragraphs should suffice!

Probably the biggest matrix management problem is clear assignment of *accountability*. (That sentence probably should be printed in red!) This is the Siamese twin of the shared responsibility that is a primary characteristic of the matrix. But if performance can be measured, accountability can be assigned! As just described, it is easy to measure the program manager and hold him or her accountable. In fact, the manager's financial statement is a more explicit scorecard than the one the GM lives and dies by, because the business the PM manages is more precisely defined and bounded—it has a beginning and an end! The more challenging task is assignment of accountability to the functional managers, especially the middle tiers. If they are measured only by budget statements, their accountability for project performance will be poor or nonexistent. Summary statements of performance on project work packages, as described previously, are a big step forward.

A corollary situation has been observed. Suppose there is an organizational element that (1) runs programs on which it is measured, (2) manages budgets that are also measured, and (3) works on an array of assigned work package tasks on which, for reasons of organizational or financial statement inadequacy, it is not measured. What would be your estimate of that organization unit's performance in each of those categories? (One observer has referred to that situation as being a wholesaler and a retailer simultaneously.) There are no walls thicker than paper walls. Financial statements that reward parochial behavior and do not similarly credit service rendered to associates will isolate the elements and destroy the teamwork in a matrix, despite the most eloquent exhortations of the general manager—who is the ultimate victim. The best solution is accountability, measured at each functional cell, for all programs served by that cell.

A special organizational issue illustrates this point dramatically, while also raising additional issues. In the late 1960s, to accomodate growth and give more management attention to a historically neglected business segment, the product support business was segregated, with its own management, facilities, and business statements. The products it was supporting (with spares, test equipment, technical manuals, O & M training, and field tech reps) generally were the products we were manufacturing. So the support business really derived from the product division's strategic investment and start-up risks. Additionally, many of the spares that the support organization sold were built concurrently on the same production line as the prime equipment. From the beginning, the shared strategic benefit was recognized by spreading the burden of strategic expenses equally on both statements. But with independent business statements, new parochial interest showed up, and it became an increasing struggle to get spares out of a shop that was measured on its performance in shipping prime equipment! And heated debates arose on allocation of contract funds to compensate for the "unfair" skewing of risks!

The solution to the matrix management problems raised here—and they *were* matrix problems, since they arose out of an arrangement involving shared responsibilities—was introduction of financial statements sometimes referred to as "matrix bookkeeping." Very simply, where there is shared responsibility, both get credit. This is easy to do. If the financial data system already collects costs by program and includes normal work breakdown structure detail, identification and summing of common parts is straightforward. (When the statements are aggregated for reporting up the corporate chain, it is equally routine to subtract the common parts so they won't be counted twice.)

Note the benefits of matrix bookkeeping. First, it eliminated that wholesaler-retailer inequity on the spares manufacturing (and spares shipments *did* pick up, dramatically, when we introduced the approach!). Second, it blunted the "zero-sum game" associated with the assignement of contract dollars.

Zero-sum games constitute a cancer endemic to matrix organizations. The resources are shared, they are finite, and "the more I get, the less you have." Success or failure usually depends on quality and quantity of resources applied. When financial statements clearly segregate and measure the degree of success and failure on a segment of the business, but prorate the cost of resources on some basis that does not truly reflect the degree of their utilization on the business segment so measured (and there is no practical way to accomplish that totally), then you have a zero-sum game. And the general manager referees all zero-sum games. To eliminate zero-sum games in the interest of teamwork instead of internal competition for the resources, the financial policies and procedures of the division should be designed to measure specifically what each manager is to be held accountable for.

THE PEOPLE DIMENSION

Organizations are arrangements of people. It has been said that a chief executive sets goals, selects people, and motivates the people to reach the goals—that's all—and two out of those three tasks relate to the "people" dimension.

A matrix is hard on people; most seem to find it so. In a recent employee-attitude survey at a large industrial complex that was matrixed (or whatever the verb is!) 15 to 20 years ago, the matrix organization drew the least favorable reaction in the whole survey. Analysis of the responses showed that the reaction was pretty much shared by all levels. The concerns were mostly personal, rather than objective: what happens to me and my career, what team do I belong to, how do I identify with the goals and successes of the enterprise, how can I serve multiple masters, does anybody really know whether I'm doing a good job? Partly these concerns are geared to "bigness," but the people mostly ascribe them to the matrix and its inherent ambiguities (as do I). So it becomes top priority for the general manager, *especially* the GM of a matrix organization, to understand and address at least the following three issues in the people dimension:

- Selection and training of management
- Conflict, trust, and motivation
- Creating a culture

Selection and Training of Management

Some managers will never adapt. If their personal style is authoritarian, the frustration of seldom having complete authority, of having to persuade a peer (worse, *yield* to him!) as a normal day-in, day-out operating mode, can make such individuals almost totally ineffective.

Such frustration at high level tends to be cancerous, infecting peers and subordinates more pervasively and more destructively than in simpler, more segregated organizational styles. A decision to convert to a matrix, therefore, must be based on an assessment of the available management team, and must allow for a transition period sufficient to gradually replace those who can't adapt to the traumatic change. (A good early retirement policy is a great asset for conversion to a matrix!) How about the GM himself? (Or herself?) *He* can be authoritarian; in fact, in a matrix organization at times it would actually help.

Some lower management levels have the mirror image of the lifelong autocrat's problem. At the intersections of the matrix there are two-boss managers, such as the engineering manager on program X, who reports to the program manager and also to the division engineering manager. Some can't live with that. (I often relate the story of such a man who came into my office in despair, complaining, "How can I work for two bosses?" I asked him if he was married. He suddenly realized he had *three!* I reminded him that we are born with two: that old one-boss catechism is a myth.) Actually some of us, when we were very young, learned how to use the *two* bosses to good advantage—a technique that extrapolates very well to the two-boss matrix situation.

How about retraining? That always helps as a tool to induce change. There are two kinds of training: one is substantive, to impart knowledge or skills; the other is behavioral, to change attitudes and style. The second, in my opinion, is nearly worthless. The first is invaluable, maybe essential. One of the best features of this "how to" training is that it demands that substantive "how to's" be generated, documented, standardized, communicated, and disciplined. The matrix is loose, flexible, dynamic; it requires much more interaction, dialogue, and decision making across organization unit boundaries—shared responsibilities for people as well as for technology and finance. As a consequence, there is a much greater need for formal, disciplined management systems, and that, of course, is the substance of those substantive training programs.

Creation of clear, comprehensive management systems for the control of finance, tasks, and people will have far-reaching benefits, including illumination of architectural soft spots in the or ganization itself, where a little "fine tuning" of the matrix would facilitate a better procedure. Take the case of those high-level program managers and functional managers who were described earlier as frustrated by their loss of authority. A carefully tuned organization, with sensitively balanced management systems and procedures, can clarify the mutual needs of the two matrix dimensions, thereby equitably balancing the power and authority that the two hierarchies must share. The natural balance will eliminate more frustration than will the training!

But the biggest beneficiaries of disciplined management systems are the people—all the people. Most of the concerns that showed up in that attitude survey were traceable either to the lack of clear, consistent management systems and rules, or to structural ambiguities that would have been revealed in the process of the management system generation.

The giant IBM may be the world's largest matrix (although they don't seem to use that term). In fact, theirs (like most) is a multidimensional matrix. At IBM many of the pieces come together only at the very top of the corporation; others tie together at the level of a site manager; while still others are in between, with several sites sharing responsibilities for meeting business, program, or functional goals. Significantly, IBM has a very comprehensive management system for dealing with all aspects of the management task—documented, thoroughly taught, followed with consistency and discipline. It's a case in point!

Let us return to the subject of training, because there's another side to that coin. The matrix provides a new opportunity for training and development of potential GMs. Historically, the step from functional management to general management has been very difficult. One reason is the need for understanding of multiple functions. A second may be the unforgiving nature of the GM's job, the performance of which is measured regularly by a numerical scorecard. As noted earlier, the program manager's job contains both of those same ingredients. But since programs come in all sizes, it is possible to both screen and train future general managers via the program management hierarchy just as, traditionally, we have been able to develop functional managers up through their organizational tiers. Eight of the last 12 general manager appointees in our long-time matrix organization have had substantial experience (successful, obviously!) as program managers. As a by-product, think how that trend affects the motivation and influence of program managers! Positive motivation is indeed required: their job may be even more hazardous than the GM's, since the program manager is on stage playing a one-string fiddle. Without some understanding from top management, there won't be many volunteers!

Conflict, Trust, and Motivation

In any organization the actions of top management largely determine the motivation of the entire assemblage. Body language transmits unmistakable messages, just as that pattern of GM appointments did. Take the case of conflict management. As noted before, a system of shared responsibilities (the matrix) not only is a conflict generator but, given the two communication paths, it is highly likely that an abundance of these conflicts will land in the GM's lap. At that point the GM can follow one of three courses, two of which would be demotivating (perhaps demoralizing) to the troops. Hopefully the conflicts will be resolved in a timely fashion and in a manner that is interpreted to be based on merit, not personalities. The two bad alternatives would be (1) to leave them unresolved, which would lead to frustrating delays and probably a lot of ante-room in-fighting, or (2) to destroy the bearer of these bad tidings of conflict in the ranks (a technique that goes back at least to the ancient Greeks), which would isolate the front office for all time.

How do the program managers motivate the people they depend on for project success? They don't *own* them: let them try to order them around and see how far they get! (Actually, that limitation really is not unique to matrix organizations in modern high technology industry. Good engineers are as free as a bird! Too heavy a hand will send them to a competitor, probably for a 10% raise.) Unequipped to *drive* the project team, the program manager must *lead* them, which can only be done if they want to follow. And why would they want to follow? Well, this isn't a psychology text, but for most of us who have more than enough of worldly goods, the biggest motivators are our own personal pride in our accomplishments, a feeling of being a part of a team whose accomplishments we share, and—perhaps most rewarding—the respect and approval of associates whom we respect. We follow a person's leadership because we respect and trust that person, feel good about our mutual involvement with him or her and our associates in this team effort. The key words are respect, trust, mutual, team. Respect and trust almost have to be mutual: that's the way teams are built.

How do we give individuals in a very large matrix organization this sense of mutual respect and trust, of recognition for their individual contributions, of team identity and the relation of that team's effort to the larger organizational goals? Today's buzz word is "participative management." Sometimes called "Theory Z," it's not so much a fad as a necessity in the industrial ambiance just described. There are many techniques of participative management:

- A management open-door policy (even for the bearers of conflicts, and maybe especially for them)
- Management round tables
- Quality Circles
- Management's frequent presence "on the floor," listening more than talking
- A variety of formal "dialogue" programs

But the difference between a participative management program and true participative management has to do with the body language, the actions, the building of mutual respect and trust, emanating from the top. Trust starts with understanding. Given the inherently confusing nature of a big matrix, there is a need to publish explicit, meticulous organization charts, mission statements, and unit goals. And these must be explained and updated frequently. Groups within these organizations need to be recognized publicly for excellent performance in meeting their goals. And the individuals—particularly the individuals—should understand how they fit into one of these groups, and be recognized for superior contribution. In this regard, there may be no more important management institution than a credible individual "performance management system." Such a system can be the basis for a mutual understanding, between the individual and his or her boss *or bosses,* of the individual's personal performance objectives and accomplishments. It can be the basis for more credible—hence more positively motivating—recognition programs, merit increases, and promotion.

Creating a Culture

Some management scholars have asserted that matrix management is more a culture than an organizational structure. Cultures have to do with behavioral norms. A matrix can be created by the stroke of a pen, as it were, by laying out an organization (though there are layers of nuances there), by promulgating a manual of procedures (most necessary, most challenging), and by appointing managers who have the adaptability to work in this trying milieu. Real lasting success, however, depends on teamwork in a framework that makes teams hard to identify. Success depends on the *absence* of parochialism—a state that awaits nirvana! These are revolutionary cultural norms. They depend on the general manager who is at once the integrator of the loose structure and the behavioral style-setter. Meeting these demands, establishing the culture, may be his biggest challenge.

SUMMARY

At the beginning of this chapter the claim was made that general managers like the matrix. In the intervening pages we have noted that there is a lot for them to like:

- Efficient use of resources
- Technology transfusion that works
- Project performance measurement that is explicit and precise
- Focused customer attention
- Great general management development

However, nothing is free, and certainly not the matrix. Some of the costs include:

- A higher manager/employee ratio
- Diluted management control
- Motivational problems
- Complex management challenges

Also, the matrix places special demands on the general manager, who must be integrator, referee, and culture leader.

But the bottom line is that matrix management—structure, discipline, or culture—is being used, in one or another of its almost limitless forms, by most successful companies in the high-technology systems business. It's working, helping talented people to create more useful high-technology products, generating more profits for the investor. GMs, please note!

15. The Human Factor in Matrix Management

John V. Murray and Frank A. Stickney*

THE NEED FOR MATRIX MANAGEMENT SYSTEMS

The evolution of our organizations over time has been toward greater and greater task complexity, and with this greater complexity the problems in organizational design have been compounded. An organization is both a technological system and a social system, and the complexity has grown in each. However, this complexity has also affected the integration of these forces from the perspective of a contemporary organization being a sociotechnological system.[1] One of the major factors contributing to this increased complexity has been a rapidly changing technological environment having impact on all phases of an organization's operations. These forces were first felt in the aerospace and electronics industries, but have been and are still rapidly spreading to all organizations in formerly more stable technological environments. Even organizations in service industries have been experiencing growing complexities in their operations: the management and operations of banking institutions, city and county governments, social service agencies, etc., are much more complex today than ever before.

The challenge to managers over time has been how to determine and maintain an organizational structure that establishes and reflects the roles and relationships among people who are essential to the accomplishment of organizational purpose. The importance of other factors in organization design is recognized by the authors, but the focus of this analysis and discussion is on the human factors which, the authors propose, are most critical to the management and operations of any organization. The statement has often been made in the management literature that "people are our most important

---

*John V. Murray is a Professor of Management at Wright State University. He received his DBA from the University of Colorado in 1967. In addition to research activity and consulting, he has extensive experience in academic administration and systems management. His primary areas of research and consulting interest are organizational development, organizational behavior, and long-range Planning.

Frank A. Stickney is a Professor of Management at Wright State University. He received his Ph.D. from Ohio State University in 1969. In addition to his academic research activity and consulting, he has extensive work experience in project management and related activities. His primary areas of research and consulting interest are organization design, job design, strategic management, and the supervisory process.

resource"; the authors propose a revision to this idea in the following hypothesis: "People are the organization, the organization is people." They feel that it is the human factor which ultimately determines the success or failure of organizations. The management functions of planning, organizing, and controlling can be performed with the utmost degree of precision, but factors will inevitably arise that must be accommodated in a real-time sense, and it is the people and their attitudes toward organizations and work that will determine the effectiveness of this accommodation. Meaningful roles and relationships among people are therefore a critical factor in the success of any organization.

Advancing and rapidly changing technologies in all fields of human endeavor have created unique organizational relationships that must be reflected in the organizational design. Reciprocal interdependencies have often replaced the more familiar sequential interdependencies, but the traditional organizational structure with its strong emphasis on functional departmentation experienced difficulties in facilitating these reciprocal interdependencies.[2] When technology and organizational operations were less complex, the flow of work was usually sequential: when engineering completed its task, it was transferred to manufacturing, which in turn gave inputs at a particular time to marketing, etc. There was very limited need for reverse interaction. But with advancing and rapidly changing technology, the interactions and interdependencies among people and their activities became reciprocal: when engineering gave an input to manufacturing, the action by manufacturing might require subsequent actions by engineering, marketing, finance, etc., and the work flows become highly reciprocal in nature; thus there was greater and greater mutual interdependence among the parts of the organization, which increased the complexity of the integration process.[3]

This organizational design problem or challenge required an approach that would facilitate the integration of highly specialized functions operating in an environment of reciprocal interdependencies. The matrix organizational design, as it has evolved for more than 18 years since it was originally conceptualized by John F. Mee in 1964, offers the flexibility and adaptiveness to cope with these reciprocal interdependencies.[4] Most people and their managers, however, have received education, training, and experience in the traditional functional organization, with its vertical flow of authority along a clearly specified chain of command. The concepts inherent in this view have been referred to by Cleland and King as "the mainstream of traditional thought," with strong emphasis on the chain of command and its superior/subordinate relationships, clear departmentation, scalar chain, and the like.[5]

The philosophies and value systems of many contemporary managers and their people are strongly influenced by these traditional views, which often cause problems in the human reaction to the matrix organizational system. It had been assumed that, if the principles inherent in the traditional view were followed with precision, the result would be an organizational design where integration would naturally occur. This was true for many years when organizations operated in environments where the primary interdependencies were sequential. As stated before, with rapidly advancing and changing technologies and the desire for timely application, the organizational interdependencies became more reciprocal in nature and the traditional organization structure found it difficult, if not impossible, to cope with them. Examples of tasks with a high degree of reciprocal interdependency are project and product management, new product design, production and market development, deeper market penetration by existing products, acquisition of or merger with an outside company, significant reorganization of the entire company and, within recent years, strategic planning. The volume and types of information, levels of decision-making competency, and time frame for decisions in this environment pose requirements that the traditional functional organization cannot fulfill. Although there is no one best organizational design, as often implied by the traditional theorists, not just any type of design is appropriate either. However, the matrix design has inherent in it the flexibility to adapt to these organizational demands and so, during the past decade, has grown in application.

The value systems of managers and people in organizations have a strong influence on the way in which they make decisions and perform their roles in the structured set of relationships established by organization design, which for this analysis is the matrix organizational system. The authors propose that many of the human problems in matrix design organizations arise from people with values anchored in the philosophies of the traditional functional design. Since most contemporary organizations still possess a set of strong traditional values, they should undergo a process of organizational change prior to the adoption and implementation of the matrix design. This would not necessarily eliminate the human problems encountered in the past, but would substantially reduce them. This is more fully discussed in a subsequent section of this analysis.

When organizations encountered difficulties in task accomplishment, the "informal organization" was often relied on to accomplish the task through its spontaneous processes despite the formally designed structure. The authors propose that this should not be the relied-upon approach to organizational design difficulties relating to the integration of its increasingly diverse or differentiated activities. The formal organizational design should explicitly consider, reflect, and make provisions for high differentiation among parts of the organization, different and sometimes antagonistic attitudes among its participants, and the geographical separation of interdependent parts.

In the above situations the matrix organization seems preferable when tasks are greatly differentiated with a high degree of task uncertainty and the need for timely and rapid decision making. Managers using the traditional model rely upon rules, plans, direct contact, task forces, etc., to achieve the required integration—methods that often prove inadequate.

The matrix organization should not be considered a panacea, and is not warranted unless the benefits achieved through better integration processes exceed the costs incurred, especially the nonquantifiable human costs. A very thorough analysis should preceed the matrix decision. The matrix organization is probably the most costly, but extensive research by the authors, referred to throughout this discussion, indicates that it has proven the most suitable for organizations functioning in rapidly changing and uncertain technological environments with complex reciprocal interdependencies.

A guide for this analysis is proposed by Davis and Lawrence, who state that three basic conditions must exist simultaneously before adopting the matrix organization.

1. Outside pressure for dual focus
2. Pressures for high information-processing capacity
3. Pressures for shared resources

If only two of the three necessary conditions are present, there is insufficient reason to adopt the matrix structure. Davis and Lawrence further state that if these conditions are not overwhelmingly present, the matrix will almost certainly be an unnecessary complexity.[6]

The natural evolution of organizations is toward ever greater differentiation of tasks and operations, and the uncertain and rapidly advancing technological environment has accelerated this process. The successful matrix organization must be designed to support the environmental demands for both differentiation and integration, despite the fact that these two states are opposed to each other. The more differentiated the organizational units, the more difficult it is to achieve the required degree of integration. The act of achieving this required differentiation and integration is a primary organizational design consideration, and the result should be synergistic in nature if the proper design is selected. The matrix organization design should reflect conditions in not only the external but also the internal environment, where the predispositions of its members are a critical factor.[7]

Gannon believes that the highest degree of differentiation and integration can be attained with the matrix organization design. Since there must be a balance between differentiation and integration, it is essential, within the matrix structure, that techniques be established to retain differentiation but also achieve integration.[8]

Effective communication processes with organizational provisions for timely resolution of the cross-functional conflict inevitable in highly interdependent and differentiated organizations must be established and maintained, if the required integration is to be achieved.[9] The human factor is a major element in this process; unless it is considered in the matrix decision, the required trade-off between differentiation and integration will not be achieved.

Prior to adopting the matrix organization, it is absolutely necessary to make a comparative study of the organization to determine how much need there is for flexibility and what sort of cultural environment the organization has. Does the organization operate in a highly uncertain environment? Are the tasks highly complex and reciprocally interdependent? Are the rates of change rapid? Are the tasks highly differentiated? If the answers to these questions are affirmative, the matrix organization probably should be considered. Following this initial analysis, the study should proceed to examine organization attributes such as management style, the forces of the manager, the forces of the subordinates, their capability for adapting to change, etc.

Preparing an organization to operate with a matrix structure often requires a change in management philosophy. The organization design specifies the formal roles and relationships among its participants, supplemented by the naturally existing, although not explicit, informal organization. But the forces facilitating people-to-people interactions reflect the fundamental management philosophy of a given organization. The authors propose that many of the tenets in the "Theory Z" organization advocated by William Ouchi are essential to the adoption, development, and operation of a matrix organization.[10] The time for this process may extend from a few months to many years, depending upon the degree of task structure and the culture of management. Ouchi states: "A theory Z organization is egalitarian, engages fully the participation of employees in running the company and emphasizes subtle concern in interpersonal relations. It is characterized by employee cooperation and commitment to the objectives of the company."[11]

According to Ouchi, the type Z organization is successful because it has an easily understood management philosophy. This philosophy involves the sharing of decisions. The more the decisions are shared, the greater the benefits of cooperation produced in the management team. Thus the top manager has an incentive to develop over time a trusting relationship that permits a sharing of the decision-making authority with subordinates.

Ouchi believes that successful Z companies have taken the time to first establish an understanding of the true commitment at higher levels of the organization. Ouchi advocates managerial rotation or, as he calls it, career circulation as a practice for preparing personnel for the type Z organization. "In order to have the process of career circulation succeed, top level managers need to set the example. If the top managers rotate every three to five years between jobs so that the vice president of personnel takes over international sales, and the vice president of international sales takes over domestic manufacturing then each of them will, over time, bring along managers at the next level down whose skills they need. Those managers will, in turn, want to bring along some managers they know and the process will naturally trickle down through the organization."[12] Many of these concepts are relevant to the implementation, development, and operation of a matrix organization.

PREPARING THE ORGANIZATION FOR THE MATRIX

All managers throughout the organization, starting with the Chief Executive Officer, must support the implementation and operation of the matrix organization. Although not usually directly involved in the structured roles of the matrix organization, top-level managers must set the example and openly support the matrix or it will be doomed to failure. The authors propose that the success of the matrix is contingent upon two major factors: top management commitment to organizational preparation for change, and the selection of the proper type individual to be a matrix manager.

Top management must make a commitment in time and resources, and this must be clearly communicated throughout the entire organization. Time and effort must be planned for and allocated to the training and orientation of all personnel involved either directly or indirectly with the matrix organization. It will often take time, possibly two to three years or more, before the matrix organization can be considered fully implemented. "One of the basics to be learned by any organization attempting the matrix is that a successful matrix must be grown instead of installed."[13]

The matrix organization has not always succeeded, even in organizations that rightly adopted it in preference to the traditional functional organization. Often this lack of success can be attributed to the lack of patience. It takes time to make the system work. As Cleland says, "Patience is absolutely necessary."[14] To change the organization requires considerable time, perhaps years. As stated by Cleland, "the matrix organizational design is clearly the most complex form of organizational alignment that can be used and to change an existing design to a fully functioning matrix form takes time—perhaps several years."[15]

Top managers, even though not in the matrix structure, have an essential role in realizing the matrix concept. A directly involved top manager must believe in the matrix and openly sell it to those below him. He must also be aware of the necessity for balancing the power within the matrix.[16] Top managers must maintain a power balance and resolve the unhealthy power struggles that will naturally develop in a matrix organization. Top managers must at times remove weak managers (discussed below) or those who do not maintain a systems point of view, and replace them with a dominant systems-oriented manager.[17]

In the matrix organization a new role and relationship are created: the role of the matrix manager and that manager's relationships when interfacing with functional managers. The relationship between the matrix manager and the functional manager involves a sharing of responsibility and authority by the latter with the former. No longer is there a clear vertical flow of authority and responsibility along the chain of command from superior to subordinate; many times the exact nature of the shared authority and responsibility between the matrix and functional managers must be negotiated, depending upon the nature of the task. Top management must create an environment conducive to such negotiation; one factor in the performance appraisal of both the matrix and the functional managers should reflect their effectiveness in establishing viable working relationships through shared authority and responsibility.

Insecure functional managers will often oppose the matrix concept. They are probably status-oriented, highly prizing status symbols such as titles and type of office. Usually they make a concerted effort to impress others with their knowledge of management theories. To maintain their status, they will have established highly sophisticated controls that can be described as role-defense mechanisms. They will always resort to these controls to let others know that they are in command. They can be described as autocratic leaders who rely on rules and regulations, or at least on tightly controlled policies, to make sure that their role and status are preserved. Because of these characteristics, they could easily resort to a management approach that might be ethically questionable, even though they personally would probably never consider such action unethical. They are doubtless recognized by their peers and subordinates as hard workers who devote long hours to the job and are seldom on vacation or away from it. If this is the case, they may be doing this to make sure their security in the position is maintained.

The functional manager who could possibly be a barrier to success of the matrix organization must be identified. The following are some questions upper management can use in identifying such a manager. How mobile and adaptable is he or she? What would be the reactions of others in the company if the manager were fired? How is the individual perceived by others in the company? How valuable is he or she to the company? Can the person be of value to the company in another position? Is the manager perceived as a threat by subordinates, associates, etc.? To be sure, some functional managers would not see the matrix structure as threatening them; they might even see it as supportive, especially if they had an expanded role as special adviser or consultant to the team and, if so, were given a major role in implementing the matrix. But in the case of those functional managers who see the matrix as a real threat to their autonomy and security and who would therefore oppose it, upper management must obtain at least their passive support. Upper management and all key matrix and functional managers should participate in the deliberation that results in the establishment of the matrix structure.

Some of the functional managers who will not support the concept should resign or be reassigned because, if they remained, they would be a barrier to successful implementation. Contemporary managers who do not possess adaptability and a willingness to change can be highly dysfunctional to their organizations and should not continue to occupy key positions; a need for a high degree of structure and certainty are countereffective characteristics in most contemporary organizational environments.

Reeser explored some potential human problems associated with the project form of organization that also relate to matrix organizations. He contrasts these problems in a project form of organization with those in the traditional functional organization.[18] His findings indicate that insecurity regarding possible unemployment, career retardation, and personal development are problems felt by subordinates in project organizations to a greater degree than in functional organizations. He also suggests that subordinates in the project/matrix organizations appear to be more frustrated by make-work assignments, ambiguity and conflict in the work environment, and the multiple levels of management than functional subordinates. Additionally, project subordinates seem to feel less loyal to their organization than functional subordinates, and individuals temporarily assigned to project organizations appear to experience some frustration because of having more than one direct superior, a common characteristic of the matrix organization.[19] An individual working in a matrix organization receives direction from a specialty functional manager as well as from a matrix task manager.

Reeser also found that specialization is a cause for insecurity and frustration in both the traditional functional and project matrix organizations. This supports Ouchi, who also recognizes that high specialization can be dysfunctional or a hindrance to organizational collaboration and integration.[20]

The major responsibility of top management in implementing and operating the matrix organizational structure is to create an environment supportive of the working roles and relationships created by this design. Emphasis on team building, a supportive and participative internal organizational environment, deep delegation to the level of decision-making competency and information availability, and an atmosphere of open communication characterized by mutual trust and respect among all participants in the matrix structure, are prerequisites to the successful implementation and operation of a matrix organization. If the organization cannot be developed to reflect these perspectives and philosophies, it will inevitably encounter human problems significantly more complex and detrimental to its operations in a matrix design than the problems would be in the traditional functional design. These costs, although not directly quantifiable, could well exceed the benefits of operating with a matrix design, and in many instances have done so.

The second major determinant of matrix organization success is the specification of the matrix manager's role, including the critical points of interface with functional managers and the selection of the individual with the required skills and abilities, actual or potential, to fill this position. The authors challenge the concept of universality of managers—the notion that a manager effective in one situation will be effective in all others. Many individuals effective in functional management positions have met their "Waterloo" in matrix manager positions. Not all individuals are psychologically equipped to function effectively in matrix positions, therefore selection of the proper individual is absolutely essential, even if the organization must search externally. The matrix structure is becoming more and more common in contemporary organizations, and will continue to be so as organizations adopt mechanisms enabling them to cope with changing environments. As a result, organizations need to make provisions for internal development of individuals for matrix positions, since thorough knowledge of the internal operations and relationships within an organization is a critical factor in a matrix manager's success. If the manager is selected externally, the person's total effectiveness might be constrained while he or she is achieving this knowledge and understanding. Many of the concepts like managerial rotation proposed by Ouchi in his Z Organization Model will facilitate the internal development of individuals with matrix manager potential.[20]

The research done by Lawrence and Lorsch provides some insights from which criteria can be obtained for selecting an individual for a matrix manager position.[21] These individuals must possess a high degree of interpersonal competence, and leadership skills conducive to the collaborative and participative style. An autocratic style reflecting strong emphasis on rank and position emenating from the cherished chain of command and superior/subordinate relationship is highly ineffective in a matrix organization structure. The matrix structure blurs these formerly clear and explicit roles and relationships. The methods used to achieve the required integration in matrix organizational structures differ from those used in the traditional functional structure. Experience in matrix roles and relationships, supplemented by organizational development and team-building processes, can often provide this skill.

The matrix manager must possess a set of special attributes, innate talents, experience, and training. The degree to which these skills and attributes are required will vary with the scope and complexity of the matrix task. Large and complex matrix tasks should be assigned to individuals who have had previous experience on other such tasks, or to those who have demonstrated this capability and potential for further growth on smaller or less complex matrix tasks.

Human resource management and strategies resulting therefrom have become an essential component of effective strategic management. Writers, researchers, and practitioners have long specified the vital importance of matching organizational structure with strategy. But selecting the appropriate structure is futile, if the organization does not have the qualified human resources to staff this structure. An excellent example of the integration of human resource development and planning with organizational strategy and structure is contained in a conversation between Professor Charles Fombrun of the Wharton School, University of Pennsylvania, and Reginald H. Jones, retired chairman and CEO of the General Electric Company, and his Senior Vice-President of Corporate Relations, Frank Doyle.[22]

The matrix manager can be viewed as having three basic roles: integrator, facilitator, and professional manager. As an integrator, the matrix manager is responsible for a balanced orientation between the various functional areas and the specialists from those areas temporarily assigned to him or her. The manager must recognize and appreciate the need for differentiation, but at the same time work toward integrating of these diverse activities for achievement of overall objectives. In integrating the diverse work activities of the matrix team members, he or she should strive for synergism, blending the goals of the supporting functional departments in reference to the matrix task and organizational goals.

As a facilitator, the matrix manager supports team members by providing the environment where they can successfully interact with all others supporting the matrix task. The matrix manager is also responsible for creating and maintaining an atmosphere for the professional growth of the assigned functional specialists. With an atmosphere for professional growth, the team members should see the opportunities for their future growth and opportunity through effective performance and support of the matrix task. To facilitate the work of the team members, the matrix manager is responsible for balancing the various demands for each team member's support.

As a professional manager, the matrix manager combines internal and external roles, representing the team to functional managers and other administrators in the organization, as well as to personnel external to the organization such as customers. A major responsibility is involvement in the management process to ensure that continuous progress is made toward the achievement of matrix task objectives. The competition for resources presents a formidable challenge requiring analytical ability, persuasion, effective communication, and diplomacy—a high degree of interpersonal skill. When competing for organizational resources, the manager must have the strength and capability to relate the needs of the matrix task organization to the needs of the functional elements and in turn to the needs of the entire organization—a role requiring a systems perspective with a high degree of conceptual skill.

The role of the matrix manager is a highly complex one requiring three general types of skills: conceptual, human relations, and technical. The matrix manager must be able to conceptualize the matrix organization first as a system, but then as one of several subsystems within the total organizational system. With the collaboration of top management, matrix organization objectives must be derived that support organizational objectives and are also supportive of or congruent with the objectives of the other functional subsystems. The manager must be able to visualize an organizational system where the traditional line functions such as finance, production, engineering, etc., are vertical flows, while the matrix structure is a horizontal flow with concontinuous, constant crossing among all organizational elements required to achieve a matrix task with complex reciprocal interdependencies. The ability to do this requires a high degree of conceptual skill.

Another major requirement for the matrix manager is a high degree of human relations or interpersonal skill: the ability to work effectively with anyone with whom he or she needs to work, internal or external to the organization, to achieve the results of the matrix task. The manager must have a deep appreciation for the other individual's point of view, even though he or she may be opposed to it, and should constantly strive for objectivity when dealing with opposing and controversial issues. With this attitude of mutual respect and understanding, the manager can encourage associates to be fully committed in working toward matrix objectives.

The matrix manager, working with the functional managers and other matrix managers, must be responsible for balancing the various demands for specialist time based on a statement of resource needs. He or she should also have joint responsibility with the functional manager for the evaluation of the functional team members who are assigned to the matrix manager's task.

The matrix manager will often be faced with the problem of resolving conflict among staff members, between staff members and other functional departments, etc. In conflict resolution, the manager must be a leader, a diplomat, an arbitrator, and above all a humanist. This rather philosophical idealism can hardly be stated as criteria; possibly it could be better phrased by opposites. That is, the matrix manager must never be negative, but positive—and a good listener and problem solver, too.

In summary, the matrix manager must be a professional manager with a high degree of human relations and conceptual skills. The manager also must possess sufficient technical skill to be able to communicate with the technical specialists who support the matrix task. Experience and understanding of the various processes used in planning and control like task networks, budgets, information flows, decisional analysis, etc., are also essential. The effective matrix manager is indeed an exceptional individual and not the bureaucratic hierarch who often succeeds by avoiding trouble and maintaining an organizational state of equilibrium. Individuals with the above rerequisite characteristics can't just happen; they are the result of long-run professional development that must be planned by the organization, not left to chance.

ORGANIZATIONAL ORIENTATION AND DEVELOPMENT

Organizations that have operated under the traditional functional design must be prepared in advance for operations under the matrix design. The vital and critical contributions of top management and the matrix managers were discussed in the previous section. These are necessary but not sufficient conditions for effective matrix organization operations. The organization, as a system, must be prepared for this new operational orientation. The approach of this process is first from the perspective of the total organization, then from the perspective of its parts on the basis of their interdependencies and interactions, and then from the perspective of the whole consisting of a set of integrated parts. The systems approach provides the model for this process.[23] The effective operation of a matrix organization requires sufficient time to prepare its implementation, its new perspectives, and the behaviors and roles of its participants, especially its managers.[24] The design of any organization may be rational, but if the people don't understand and accept it, many problems may be encountered in its implementation and operation. Since most managers and their people have been trained and experienced in the traditional functional organization with its higher degree of certainty and structure, they find it difficult and often psychologically distressing to function in the less structured and more uncertain matrix organization.

An organizational development and change process is not a simple effort. It may require an extended period of time, often in excess of one year, depending upon the situational factors and personal attitudes of the people. The description and analysis of an organizational development and change model is beyond the scope of this chapter; however, the major factors to be considered are identified and analyzed. Team building, problem solving, participative management, and conflict management (both interpersonal and intergroup) are some of the essentials in an organizational development process. However, this process is much broader than these techniques which prepare participants to engage in organizational development and change. For a complete understanding of this process, see the work by Beer.[25]

The matrix structure significantly increases the required interactions of people and their communication networks because of the reciprocal interdependencies inherent in most matrix tasks. This increased communication often creates problems between the matrix and functional personnel because of their differing orientations: the matrix personnel define the issues from the perspective of the task, whereas the functional personnel define them from the perspective of the functional area of specialization such as engineering, marketing, etc.; but both perspectives are required for organizational performance. To deal with these inevitable problems, the confrontation problem-solving approach possesses the greatest potential for success. People air their differences without hostility or destructiveness. This process necessitates, or at least implies, a need for changes in interpersonal competence and group process, and requires substantial training and orientation aimed at changing the individual behavior as well as organizational culture.[26]

Ambiguity of roles is one direct result of the matrix structure, which creates a new set of roles for the matrix personnel, and a new set of interfaces between the matrix personnel and the functional personnel. Role definition and clarification is required prior to implementing the matrix, and provision for continuous clarification must be made, since these roles and relationships are dynamic in nature. Organizations that have not achieved this required clarification have encountered highly emotional conflicts among their people, resulting in a decline in organizational performance.[27] This illustrates the absolute necessity of extensive personnel orientation to prepare for the matrix.

Unfortunately, because of the attitudes held by some personnel ("We want to be left alone to do our own thing"), one cannot assume that achieving collaboration and integration in a matrix organization will be mechanical and rational. To attain effective collaboration, a climate of trust must be established and maintained, and this can be best achieved through confrontation, where parties as a team learn to deal with one another through open and frank interaction in a positive problem-solving environment. As a first step, the climate needs to be developed. Additionally, it should not be expected that conflict will disappear, nor is that desirable; it needs to be *managed* so as to facilitate integration without sacrificing the needed differentiation.[28]

Many other training and orientation processes can be an integral part of an organizational development and change program; the specific ones required are a function of the organizational situation determined by organizational diagnosis.[29] The point to be stressed is that these techniques are necessary but alone are not sufficient. Implementation and operation of a matrix organization require a total or systems approach to the redesign of the organization, with primary attention to the most critical part of the organization: its people. In addition to the behavioral preparation of the organization for the matrix design, there should be training and orientation in several of the sophisticated planning techniques. Some of these techniques are network planning like Program Evaluation and Review Techniques (PERT), Critical Path Method, market research and analysis, financial analysis, etc.[30] These technical skills in external and internal analysis aid in clarifying the exact nature of the matrix tasks and establish a basis on which people can relate their roles and responsibilities in the process of clarification and understanding. This must be a continuous collaborative process, because the environmental conditions creating the need for the matrix structure are ever changing and require continuous adaptation of the roles and relationships, if the organization is to remain adaptive and responsive to environmental changes.

TRADITIONAL VS MATRIX MANAGEMENT

Objective vs. Activity Orientations

One of the many characteristics of the matrix element in an organization is that it is primarily objective- or results-oriented, in contrast to the traditional functional elements. To fully explore the differences between these two organizational orientations, the extreme "activity-oriented" view will be contrasted with the extreme "results-oriented" view. In reality and in application, these concepts must be integrated. Obviously, the degree of orientation toward activity or results will often to some extent depend upon the nature of the organization and the level of management within the organization. For example, stressing results or objectives would not be as critical in an industry where uncertainty is minimal and the technology is relatively stable, in contrast to an industry with rapid technological advances like electronics. Additionally, the concepts would be less important at the operative management level where the work is highly structured like an assembly line operation, in contrast to a less-structured work environment like new product development.

One of the main problems in the traditional functional organization occurs when objectives are established for the whole organization. They often, according to Terry, "tend to get lost in the shuffle of managerial activity, their identity becomes obscured, activity is mistaken for accomplishment, and precedent or habit emphasizing 'how to do it' completely overshadows what is to be accomplished" and why.[31] The matrix or ganization overcomes this weakness because it stresses the need for setting clearly defined, predetermined objectives that prescribe definite task scope and direction. They state what results the task must aim at and what is needed to work effectively toward these targets with a system of accountability.

As previously discussed, the approach to the analysis and understanding of any organization should be the systems approach. The organization is a system, a set of interdependent, interacting, and hopefully integrated parts.[32] If the matrix design is adopted, the authors propose that the organization should be designed with two primary subsystems, one for matrix tasks and one for functional activities, with objectives providing the general framework for integration of these two subsystems. Objectives are first determined for the organization itself, then from these are derived the objectives for each of its major subsystems and in turn their subsystems. Figure 15-1 illustrates this process.

Planning

The importance of planning cannot be overemphasized. Peter Drucker has underscored the need for managers to examine tasks and proposals to determine if the right things are being done, not just things being done right. Silber notes that there is nothing so useless as doing things with great efficiency that should not have been done at all.[33]

Long-range planning and forecasting is receiving increased attention from results-oriented management, with emphasis on developing techniques and methodologies for predicting and forecasting. In this evolving management environment, it becomes increasingly important to develop techniques to reduce uncertainty; this provides for developing a complete pattern of judg-

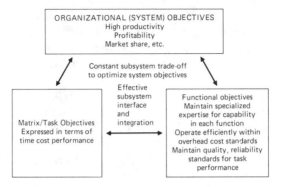

Fig. 15-1. Systems and subsystem objectives.

Source: Frank A. Stickney, "Organization Design: A Critical Factor in the Management of Resources," *Proceedings of the Project Management Institute*, Drexel Hill, Pa., 1975, p. 110.

mental forecasts to be presented to the decision maker in a form that can be readily received, assimilated, and manipulated. Such a pattern, if properly developed, would permit the top management decision maker to select areas of high potential payoff. Ideally, fewer things would be left to chance and the top decision maker would be better able to allocate today's resources to future opportunities with high potential. More specifically, the long-range forecast would aid in determining the areas of technology most likely to yield large results for a given organization. If they possess any applications to the organization, the matrix manager could act as the catalyst in introducing the technology into new products. The matrix manager's task is a specified new product manager. In the traditional functional organization, there is no one manager, except the general or top manager, responsible for introducing new or changing technology into the organization and its products.

Management based on objectives stresses causative thinking. Causative thinking emphasizes the shaping of future events and achievements instead of waiting for destiny to decide them. The future reality is conceived and made the cause of each action event. The necessary events are planned and carried out. In effect, this is thinking in reverse: results are derived by converting nonproductive present actions into related events leading to the desired future situations; the imagined future effect becomes a causative factor, with the matrix structure enhancing this process.

In the matrix organization the task is often highly technical and the market highly complex, therefore task planning is essential. The planning process should provide for specific procedures to facilitate the analysis, evaluation, and approval of the task. However, the planning process for a matrix task is inherent in both the task's structural as well as behavioral implications. This process can provide the required integration between production, marketing, engineering, etc., and allow the matrix manager to function as a referee in cross-functional conflicts. The planning process should be carried out by the people with special skills and specific responsibilities, both matrix and functional, who will implement the plan. The matrix structure facilities this type of planning on a cross-functional basis.

Planning in the activity-oriented environment is often very short-range. Activity-oriented managers respond primarily to external stimuli, therefore their thinking is for the most part "reactive." The alternatives are limited and determined by external conditions that are not controllable. Many functional managers have developed a deep concern for the security of their position, which is a natural result of functional departmentation. Therefore forecasting and planning are difficult for them, because they force them to deal with uncertainties that may be in conflict with their philosophy. They often attempt to maintain the status quo and only react in extreme emergencies, rather than being proactive and seizing opportunity. The unstructured role of the matrix manager, with its often inherent ambiguity and adaptability, is more conducive to forecasting, long-range planning, and proaction. By achieving the proper interface with the functional managers, the matrix manager can aid in reducing the tendency toward status quo and complacency. Of course the matrix manager needs the total support of top management in this process.

Organizing

Organizations usually accept the fact that strategies and strategic planning are essential to meet environmental changes; however, this is not true with organizational planning and the structuring of the organization. Organizational planning must be accomplished simultaneously with strategic planning, to assure a structure appropriate for the objectives being sought with the selected strategies.

The matrix organization is often selected because of its inherent adaptability, compared to a functional organization. Top management in the matrix organization should support this organizational flexibility and adaptation to change, and should make organization dynamics work to the benefit of the organization. Top management needs to provide the proper trade-off between stability and organizational change and adaptability. Primary emphasis should be placed on the integrated achievement of organizational objectives from both the matrix and functional perspectives; functional activities alone should be de-emphasized. If this philosophy is not communicated from top management to lower management, the matrix interface will encounter difficulties and conflict with the functional activities, which will continue to stress status quo and resist change. Organizations need a relative degree of stability as well as adaptability and change in today's fast-changing environments; the authors propose that the matrix provides a means to achieve a trade-off balance between these two forces.

Executing

Executing—sometimes referred to as directing—is a fundamental aspect of all organizations and an essential part of all the other functions of management. For the matrix manager the concepts inherent in executing have greater significance than those of strategic planning and organizing, which are more the primary responsibility of top management. However, the matrix manager can provide valuable inputs to top management from an integrated cross-functional perspective that is required in strategic decision making. The decisions of the matrix manager are more task- and operationally oriented.

Decision making is an integral part of all management activities, but is included in executing because of the highly personal nature of the process. The concepts of power and authority, delegation, leadership, and such common concepts are also included in the term "executing." Decision making is the catalyst upon which the management functions are based. If events are planned to shape the future the way we want it—i.e., to control the destiny of our organization—decisions must be made today to take advantage of future opportunities and to cope with future problems. The decision maker in the matrix organization should follow the concepts of results-oriented management, because of the complexity and uncertainty of these types of decisions. Through sequential decision making the events, milestones, and decision points in the decision network are planned and programmed for future time periods, together with alternative courses of action if outcomes are different than planned. The traditional activity-oriented manager differs because his decisions are usually not made until problems are brought to his attention; the approach is a passive one.

The matrix manager should function under the concepts of potential problem analysis. According to Kepner and Tregoe, "The aim of potential problem analysis is to find feasible, economic actions that can be taken against the possible causes of problems that have not yet happened. This gives the manager a chance to get ahead of the game in either of two ways: he can either act to prevent possible problems from coming to pass, or he can act to minimize the effects that these problems will produce if they do occur.[34]

A conscious effort, together with a methodology to analyze potential problems, is for the most part quite rare for several reasons: *First,* managers are, as noted as Kepner and Tregoe, "more concerned with correcting present-day problems than with minimizing or preventing tomorrow's."[35] *Second,* according to Terry, "Praise and recognition are rarely, if ever, bestowed on managers for things that do not happen."[36] *Third,* there is a tendency to ignore the possibility of failure, perhaps because "such thinking is sometimes referred to as negative."[37] And *lastly,* as also noted by Terry, "there is a persistent conviction by management members that plans they suggest are nearly infallible, otherwise they would not recommend them. They reason that there is little use in looking for trouble. But experience demonstrates that actions do go wrong, troubles do arise."[38] And if—as is often the case in an activity-oriented environment—management does not rationally analyze potential problem areas, anticipate possible causes, assess the probabilities, take preventive actions, and establish contingency plans, they often must react with "crash" programs or "crisis" management.

Power and authority relationships create difficult and complex problems in most contemporary organizations. Many problems today in all types of organizations result from certain of our concepts of authority. Many believe that the traditional concept of authority requiring literal obedience, as applied to certain types of organizations, are obsolete. Today, because of technological change, managers at all levels in the large organizational hierarchy are precluded from "knowing it all." They must share authority with those who have specialized ability. The concept of shared authority is a major factor in the philosophy of the matrix organization, but in activity-oriented organizations often it is not. Rather than sharing authority, the manager resorts to defensive roles to preserve his status or position. At the least, there results status quo or limited innovation, and at the worse, conflict leading to dysfunctional behavior.

In the matrix organization the power of knowledge takes on greater importance than the power of position. The power of position is more prevalent and relied upon in the traditional functional organization. As stated by Cleland, "The lines of authority and responsibility tend to be flexible in the matrix organization. There is much give and take across these lines and project staff managers act according to the situational needs rather than in accordance with position descriptions."[39] Flexibility and adaptability of authority and responsibility relationships are inherent and should be a basic philosophy in the matrix organization. But in the traditional functional organization, authority is often perceived by the manager as being invested in the position within the hierarchy. According to Scott, "Authority based on tradition is static—it supports the status quo."[40]

In contrast, results-oriented management recognizes, because of organizational complexity coupled with rapid technological advances, that there is a wide and rapidly growing disparity between hierarchical authority and specialized ability relative to decisions concerning task goal achievement. Most successful managers in highly technological and complex organizations are those with professional competence, in contrast to those whose influence is based on position within the organization. The most successful integration will be achieved if the persons who have the knowledge to make decisions also have sufficient authority to do so. The matrix manager must be delegated the necessary authority to enable matrix team members to be seen by others in the organization as having an important role in decisions. They must also be perceived as competent and knowledgeable in matrix task matters, with the authority to make the required decisions for task goal achievement.

Where matrix management lacks substance and clarity of authority and responsibility, the matrix managers will be frustrated and overwhelmed. As a result, greater prestige will be associated with, and greater competence will flow to, numerous other positions concerned with inner task influences, reviews, approvals, and the like, diluting the effectiveness of the matrix.

Authority and responsibility are often diffused in a matrix organization and could result in a lack of task control. It becomes very difficult to measure the performance of individuals, if authority and responsibility are not clearly specified. If the task is to be controlled according to plan, deviations from the plan must be communicated in sufficient depth and speed to those who know what must be done. Task matrix managers will be judged clearly by what they do in reference to problem identification and resolution. Therefore matrix managers must possess budgetary as well as scheduling and specification control. They must be committed to systems performance, to completion dates and costs, and must be able to take the actions required to achieve their task commitments. In this way it is possible to understand the exact extent of authority and responsibility of the matrix manager as compared with that delegated to others. But the ultimate relationship in a matrix organization is the interface of the matrix and functional managers with a sharing of authority and responsibility.

Most of the concepts or processes of management already contrasted will to a large degree determine leadership styles and patterns. Managers, however, can be categorized to some extent with respect to how they assess and assume the power they possess within the organization. If they assume power as that which is given them by their position, they will, for the most part, act in a passive role and be more inclined to follow an activity-oriented approach to management. On the other hand, if they visualize their power as a flexible variable, they will assume the dynamic role of being a risk taker, an innovator, and one who has the ability and power to overcome resistance to change and win compliance. Such dynamic leaders, as briefly described, will be heavily results-oriented, sometimes even to the extent of being dangerous unless they use their power wisely. *First,* matrix managers must make sure that their use of power is always task-goal-oriented; they must maintain the respect and support of their followers and must develop and maintain a moral shield resulting in mutual trust. *Second,* they must maintain a proper balance in the use of their power. They must know when to move forward, when to maintain the status quo, and lastly when to retrench. And *third,* they must be strategists rather than manipulators of people; they must be people-oriented with respect to decisions made by others. They must be able to look all ways with somewhat of an "animal sense," and be able to perceive and predict the results of decisions as they impact on individuals, groups, intergroups, and the accomplishment of their task.

Matrix managers must recognize and consider all the evolutionary change and growth around them and realize that they must be "finders" of answers in contrast to activity-oriented managers who visualize themselves as "givers" of answers. The primary purpose of the matrix structure is to have this described type of matrix manager executing cross-functional tasks. Matrix or results-oriented managers see themselves as *learning* with the group; they stress participative management. The major need for all managers is to learn from their jobs and their associates, and this is an inherent part of the matrix culture.[41] This concept, of the "learning" manager rather than the "learned" manager, is most important when we consider, to cite just one example, that engineers starting to work now will find that half of what they have learned will probably be obsolete in ten years, so they will have to learn as much as they already know to keep up with the expansion of required knowledge.[42]

Management in the matrix organization must stress the need for a free flow of communication based on fact and mutual trust. The communication network is multidimensional with communication flowing vertically, upward and downward, as well as horizontally and diagonally within the organization.[43] The communication process could also be contrasted by using the persuasion approach in contrast to the problem-solving approach. The persuasion approach associated with activities-oriented management concentrates on the regulation of communication flow through the organization as a managerial tool, so that management spends a great deal of time building communication programs that focus on persuasion. On the other hand, the problem-solving approach associated with results-oriented management assumes communication as an intrinsic component of work and problem solving, so that the emphasis is on creating a managerial climate in which trust and openness are the norm—the climate required in a matrix organization. For a comprehensive discussion of these two approaches, see the article by McConkey.[43]

The matrix organization has communication patterns that are multidimensional in nature, in contrast to the traditional functional organization where the communication network is clearly defined and limited to the vertical downward flow from superior to subordinate. In the traditional organization there is only a limited flow upward, and that is primarily information that is positive or at least neutral in nature. Accurate and timely information flow is difficult to obtain, as it is filtered through each level of the organizational hierarchy. The matrix design should provide and facilitate the timely and accurate flow of information direct to the decision point on the respective tasks—the matrix manager.

Controlling the Human Resource

In the matrix organization there is emphasis on the measurement of definite results. Objective-oriented, it is designed to guide people to do what must be done to accomplish the objective. On the other hand the traditional organization, being subjective-oriented, emphasizes evaluation and appraisal of the manager by the criteria of personality, appearance, leadership ability, and the like, in contrast to specific task achievements. All too often controlling is an end, rather than a means to the end. Considerable emphasis is often placed on arbitrary means of control such as conformity, time spent in the office on Saturdays, Sundays, etc., without regard to what the manager actually accomplishes. For additional information concerning the importance of measuring managers with respect to accomplishment of objectives in contrast to personality characteristics, see the article by McConkey.[44]

One could argue that a manager's ideas and beliefs about control of subordinates are deeply embedded in his or her value system and are therefore extremely difficult to change. However, rather than assume this pessimistic view, let us hypothesize that the manager could change provided that the reward structure emphasized and reinforced the achievement of results rather than activities. The problem is that management really doesn't know how to reward people for performing under objectives and results: the preponderance of appraisal systems emphasize rating the individual on traits or characteristics, not on performance results.

All organizations should strive to create and maintain an organizational environment conducive to high motivation, performance, and job satisfaction. The matrix organization has the potential for this, but the roles must be clearly defined and understood. The matrix organization emphasizes a philosophy of performance evaluation based on results, with everyone knowing what effectiveness means and how to measure it. There must be an immediate feedback system so that all know how they are doing and how the reward system functions. There must be an insistence on high performance visibility. The objectives must be linked both vertically and horizontally and the matrix team must be closely knit. In this type of system there is an environment very conducive to high motivation, performance, and job satisfaction.

The matrix team members must know that they are being evaluated and rewarded in accordance with the overall performance of the task. The team members have to be evaluated on their individual performance in their functional role, and also on their contributions to accomplishment of the matrix task. This is essential for providing the motivation to work toward achieving integration. A reward system needs to be designed that increases the spread of bonus and penalty for attaining and not attaining performance targets. In typical complex tasks, favorable performance bonuses can be extremely valuable to the matrix manager, because they provide the means to reinforce performance with rewards.

The use of incentives for a matrix team is essential. For example, a matrix manager may be confronted with a situation where the effort by all team members, except one, has been outstanding. When a matrix team is created, the results required should be explicitly determined. At this time, the rewards for good task performance should also be specified as an incentive for the matrix team members. The size of the rewards would be related to the team's overall success. The distribution of the reward pool would be dependent upon the matrix manager's judgment of the relative contributions made by each individual member. If the team effort as a whole is outstanding, all team members would have a chance to gain, the better ones gaining more than the others. This gives each member of the team a tremendous desire to encourage the other members toward outstanding performance. It crystallizes and pinpoints the fact that team member A's profit is dependent on team member C's performance, and gives member A a definite stake in the performance of member C, resulting in a cohesive and mutually supporting matrix team, so essential for the success of a matrix task. With such incentives, the matrix team should function on the basis of self-control by the individuals and the team themselves, as compared with the traditional hierarchical control method of activity-oriented organizations. The activity-oriented manager in the traditional organization usually has a low tolerance for criticism, with penalties for the thoughtful critics and rewards for yes men. Results-oriented management, on the other hand, encourages independent thinking with rewards for the creative thinker, the one who has the courage and conviction to be constructively critical of the present, desiring innovative ways for the future.

The matrix organization is used because of its flexibility and adaptability, while still giving the parent organization the required degree of stability in its functional activities. People assigned to the matrix team need to have incentives for the creative and innovative behavior required for the matrix to succeed; they must know that high performance will be reinforced with rewards—a commitment that must come from top management. The ultimate performance evaluation of a matrix group must be based on the degree to which the objectives of the task have been accomplished.

CONCLUSIONS

The authors propose that the matrix organization is "here to stay," because the environment of the 1980s will create situations where the matrix will often be the best trade-off between adaptability and stability. An organization basically establishes a structure within which the roles and relationships among people are defined. The matrix organization creates new roles and relationships and a different way of operating: a results orientation. All people in the organization must be prepared for the transition to the matrix, including top management, whose philosophy and support is absolutely required; the people assigned to the matrix organizational element; and those remaining in the functional activities that support the matrix. It is people—the human factor in the matrix organization design—who will determine whether the design succeeds or fails. That design is conceptually rational, but the focus on people in preparing for the matrix organization is absolutely critical. It cannot, and must not, be assumed that people will behave rationally without adequate preparation for the matrix.

The behavior patterns of most people in contemporary organizations have been developed on the basis of roles and relationships in the traditional functional organization, with its strong emphasis on activity rather than results. Therefore behavior modification is a vital prerequisite to employing the matrix design.

Matrix management concepts must be internalized, with a resulting change in attitudes and organizational climate. Before these concepts can be internalized, facilitating elements of management practice must be present. *First,* mutually understood and accepted goals must be established. *Second,* methods to review the alternatives available to achieve the predetermined goals must be established, with future plans validated by comparison of present results with expected future results. *Third,* some of the newer concepts of organizational and human behavior must be adopted. And *fourth,* management skill in dealing with the future must be developed.

Matrix management is not the key to the future because it makes the job of management easier. It is in fact more difficult because of its deviation from some of the longstanding traditional management concepts and practices. Matrix management is more challenging and demanding, and the tasks are highly complex with reciprocal interdependencies. But the rewards are greater. The future of many organizations and their management will revolve around some type of matrix design. As in the past, people will be the vital factor in determining the success of the organization, but they must be prepared to work in a matrix structure with its new roles and relationships.

REFERENCES

1. Richard A. Johnson, Fremont E. Kast, and James E. Rosenzweig, *The Theory and Management of Systems,* 3rd ed. (New York: McGraw-Hill Book Co., 1973), pp. 24–32.
2. James D. Thompson, *Organizations in Action* (New York: McGraw-Hill Book Co., 1967), Chapter 5.
3. *Ibid.*
4. John F. Mee, "Matrix Organization," *Business Horizons,* Summer 1964.
5. David I Cleland and William R. King, *Systems Analysis and Project Management* (New York: McGraw-Hill Book Co., 1968), pp. 140–45. These pages provide an in-depth discussion of the "mainstream" concepts.
6. Stanley M. Davis and Paul R. Lawrence, *Matrix* (Reading, Mass.: Addison-Wesley Publishing Co., 1977), pp. 11–18.
7. Paul R. Lawrence and Jay W. Lorsch, *Organization and Environment* (Homewood, Ill.: Richard D. Irwin, Inc., 1969), pp. 99–108.
8. Martin J. Gannon, *Organizational Behavior—A Managerial and Organization Perspective* (Boston: Little, Brown & Co., 1979), p. 86.
9. Lawrence and Lorsch, pp. 98–108.
10. William Ouchi, "Going from A to Z: Thirteen Steps to a Theory Z Organization," *Management Review,* May 1981, pp. 9–16.
11. *Ibid.,* p. 9.
12. *Ibid.,* p. 15.
13. Davis and Lawrence, p. 37.

14. David I. Cleland, "The Cultural Ambience of the Matrix Organization," *Management Review*, Nov. 1981, p. 39.
15. *Ibid.*, p. 27.
16. Davis and Lawrence, p. 47.
17. *Ibid.*, pp. 130–31.
18. Clayton Reeser, "Some Potential Human Problems of the Project Form of Organization," *Academy of Management Journal*, Dec. 1969, pp. 459–67.
19. *Ibid.*
20. Ouchi, pp. 9–16.
21. Lawrence and Lorsch, pp. 54–84.
22. "Conversation with Reginald H. Jones and Frank Doyle," *Organizational Dynamics*, AMACOM, A Division of the American Management Associations, Winter 1982, pp. 42–63.
23. Johnson et al., p. 115.
24. H. F. Kolodny, "Managing in a Matrix," *Business Horizons*, March–April 1981, pp. 17–24.
25. Michael Beer, *Organizational Change and Development: A Systems View* (Santa Monica, Calif.: Goodyear Publishing Co., 1980).
26. Davis and Lawrence, p. 105.
27. *Ibid.*, p. 109.
28. Lawrence and Lorsch, p. 13.
29. Beer, Chapters 6 and 8.
30. Davis and Lawrence, p. 113.
31. George R. Terry, *Principles of Management* (Homewood, Ill.: Richard D. Irwin, Inc., 1965), p. 24.
32. Johnson et al., pp. 115–17.
33. Mark B. Silber, "Synergy—Behavioral Sciences, Organization Effectiveness and the Training Professional," *Personnel Journal*, Feb. 1971, p. 151.
34. Charles H. Kepner and Benjamin B. Tregoe, *The Rational Manager* (New York: McGraw-Hill Book Co., 1965), p. 207.
35. *Ibid.*, p. 208.
36. Terry, p. 28.
37. *Ibid.*
38. *Ibid.*
39. Cleland, "Cultural Ambience," p. 37.
40. William G. Scott, *Organizational Theory* (Homewood, Ill.: Richard D. Irwin, Inc., 1967), p. 201.
41. Terry, p. 500.
42. Richard N. Farmer, *Management of the Future* (Belmont, Calif.: Wadsworth Publishing Co., 1967), p. 18.
43. Jack R. Gibb, "Communication and Productivity," *Personnel Administration*, Vol. 27, No. 1 (Jan.–Feb. 1964): 8–13.
44. Dale D. McConkey, "Taking the Buck out of Measuring Managers," *Academy of Management Journal*, Vol. 33, No. 4 (Oct. 1968): 35–40.

16. An Appraisal of Dual Leadership

James Mashburn and Bobby C. Vaught*

The comment is often heard that matrix management systems are not practical because they dishonor the traditional principle of unity of command—a one-person, one-boss relationship. As far as matrix or project management systems are concerned, this complaint is of course partially true. In this type of organization one person may have two or more superiors at the same time.

That this characteristic makes matrix management systems impracticable seems doubtful, considering the success of many companies using it on a temporary or permanent basis. It is not denied that the matrix system may generate problems of divided authority and loyalties, along with some confusion and possible conflict. To what degree these ambiguities develop as a result of human frailties in restructuring, or are inherent in the matrix system itself, remains unanswered. We know that organizational problems exist to some degree in all organizations. However, we also feel these same problems are not totally insurmountable.

The principle of unity of command runs deep within our social structure to the basic family unit itself. Only in the last few years have we seen a gradual change and a merging of roles within the family structure. However, for many years the father was the dominant authority figure within the family, and this role change has not developed without fears, doubts, and conflicts. From this perspective, then, it is understandable, and to some extent valuable, that we experience apprehension at a departure from such a deep-rooted principle.

These fears and doubts concerning matrix management arise out of our culture and, at the same time, out of a misunderstanding and application of the unity-of-command principle itself. Primarily this misunderstanding exists in its terminology. The word "command" implies a granting of authority to cause direction of the factors of production. It also assumes the connotative properties

---

*James Mashburn is President of Ozark Railway Supplies, Inc., at Nina, Missouri. Educated at Southwest Missouri State University and the University of Tulsa, he has been in sales and marketing for the past seventeen years.

Bobby C. Vaught is an Associate Professor of Management at Southwest Missouri State University. He received his Ph.D. and M.B.A. in personnel management from North Texas State University, and a B.S. in industrial management from the University of Arkansas. He was employed for six years as an industrial engineer in manufacturing.

of such words as "correct," "decision," and "power." Through the use of the word "command," with its implication of authority, it is often assumed that the authority structure is being threatened by a matrix management system. This, or course, is not the case. The authority, from a theoretical viewpoint, still flows from the top downward, and the organizational form remains a pyramid regardless of how wide its spread at the bottom.

Perhaps a more appropriate choice of wording for this concept would be "unity of purpose." The primary aspect of any group of people undertaking a project is their agreement on the expected outcome of the efforts involved. A loss of this unity is the real fear that all organizations face, the result of which, if not checked, could be total misfortune. Command, with its implications, may be divided as long as the purpose remains in agreement. Division or delegation of authority and responsibility may result in some confusion, but the confusion generally results from differences in methods and perception of others rather than from the division of command itself. Unity of purpose and goal is the cohesive element of successful group action.

As mentioned before, the past few years have been laced with disconcerting changes to the system of values and traditions within our society. What is expected from the workplace is changing rapidly and in all likelihood will continue to change. The methods employed to accomplish a given task have in many cases gone past individual comprehension. The work force has by necessity shifted toward more and more specialization. Complication and diversity are the rule of the day.

Until such time as we reach a plateau where humanity can catch up and absorb these changes into our perception of reality, one manager alone faces a crisis of style. At one moment coercive authority seems appropriate; at another a more understanding hand is needed. How does the manager determine the style that will best meet the needs of both employees and employer in a given situation? And more critically, how can he or she continue playing this Jeckyll and Hyde game?

The traditional relationship of one-person, one-boss per work group lacks validity, considering the velocity of change and the complexity of the work force. A compromise seems in order. In the opinion that two heads are better than one, it is suggested that a new theory of leadership is available to today's community of managers. We submit that two managers with equal responsibility and authority, but with different styles, can direct the work group more effectively than one manager alone.

In this chapter we intend to explore and clarify some of the potential uses and problems relative to this concept. To the extent that this analysis lends itself to matrix management systems, we hope it will assist those presently using it or considering its adoption. Our purpose here is to show the relevance and adaptability of two managers for one work group, as a viable alternative and a solution to the aforementioned crisis of styles.

THE CONCEPT OF DUAL LEADERSHIP

To fully understand and appreciate the usefulness of dual leadership, one must first understand the complexities of group and individual behavior. Although each person within a group is unique and has different values and views toward his or her work, certain generalities can be stated concerning group activities. These generalizations do not mean that there will be no exceptions to a new theory of leadership, only that we now have a new perspective from which to judge behavior in organizations.

What is first noticeable is that group activity is directed toward two goals. The first goal is the task of the group and involves movement toward some mutually agreed-upon objective. The second is maintenance and usually revolves around meeting the social needs of the work group. The accomplishment of both is important, if the group is to prosper and survive over the long run. This is not to deny that some groups will be more task-oriented and others more social in nature. Also, some members of a given group will be more concerned with goal achievement (as opposed to maintenance), while others tend to be more socially inclined. But as a general rule both activities will be present in some degree and form during the life of the group.

From an organizational standpoint, managers of industrial work groups should attempt to find some "equitable" balance of time and energy between these two activities. Theoretically the manager should be able to assess the needs of the members along with the needs of the organization, and then use the "appropriate" leadership style necessary to accomplish both. Situational factors such as personality, values, and needs of the group; time pressures, group cohesiveness, and top management policies; and demographics of the members' education, age, and sex—these are just a few of the potpourri of variables making effective balance difficult. Depending upon the situation, managers are required to be directive and coercive one moment and considerate and understanding the next. It is truly a gifted leader who can effectively assess these variables and adjust his leadership style to the occasion.

We suspect that in reality managers often become so frustrated on this tightrope of leadership styles (if in fact they were ever able to shift back and forth) that they finally opt for the style in which they are most comfortable. In other words, if supervisors can't completely satisfy everyone concerned, then they can at least feel comfortable while attempting to perform the job of manager. In effect they seem to be saying, "Now we are going to play by *my* rules."

At this point an odd phenomenon appears. Human behavior being what it is, the group members begin to lock the manager into his or her preferred role by stereotyping the manager's behavior. Because of this tendency to stereotype, define, and label, the group—through its expectations—make it virtually impossible for the manager to change his or her behavior, even if the manager wanted to. The task-oriented manager, with the inclination to direct, coerce, reward, and punish, is soon perceived and appropriately dubbed a taskmaster. On the other hand, the people-oriented manager tends to overemphasize individual and social needs at the expense of the task at hand. Either style or orientation seems to operate only to the detriment of the other.

As previously mentioned, we propose an alternative. As opposed to the prevailing practice of one manager per work group, we feel that in many instances the appointment of two managers with different leadership styles would be more appropriate, realistic, and effective in today's organization. By definition one manager must be task-oriented, the other people-oriented.

Both managers should subscribe to accepted principles of effective management. Although their methods might vary, they would be equal in regard to organization responsibility, authority, and accountability. They should never lose sight of the overall objectives of the group and must never purposely operate one at the expense of the other. In short, they should have a unity of purpose with the group and act together as a team toward shared objectives.

Implicit in much of the thinking behind dual leadership is the notion that satisfaction and productivity can both be obtained simultaneously. One manager, with a natural inclination toward the task or goals of the group, can place more emphasis upon planning, organizing, and controlling group performance. The other manager, with a natural inclination toward people, can act as the satisfier, consoler, and empathizer. One works toward increased productivity, the other toward improved morale and group cohesiveness. Both styles are needed for group effectiveness over the long run.

ORGANIZATIONAL APPLICATIONS

Because of the unity-of-command principle, dual leadership has not been formally recognized in most organizations, but we suspect it has existed for many years. Often a manager is formally appointed for a work group and is vested with the authority to select an assistant when or if the need arises. In many cases this assistant may never be formally recognized. A prime example of this is the bond of trust established between the boss and a strong, competent secretary. Regardless of the type of recognition involved, the assistant becomes in essence a comanager, often exerting influence over group members equal to and sometimes greater than the formally appointed manager's. The assistant frequently has the ability to penetrate the informal power structure, and through interpersonal influence can become very effective in achieving goals.

Because of this ability and influence on the part of assistants, care and judgment must be taken when choosing them. Along with problems of policy interpretation, communications, and misuse of authority, one of the most difficult problems to deal with is role perceptions. Each partner views his or her function from a different viewpoint. Quite often the assistant's role and power are not completely understood by either the manager, the assistant, or in particular the assistant's subordinates. Many of these difficulties can be minimized by the appointment of two equal managers for the group. In this situation the two are *valid* partners in their efforts, both equally responsible and accountable for the results of their actions.

When a manager is being chosen for a project or group, often two people are available, possessing fairly equal abilities and skills. Given this situation, the argument often arises as to what leadership style is best for the task or group. If two managers with different styles are appointed, the decision can be taken out of this arena and, in so doing, effort can be directed toward individual strengths and weaknesses. Thus balance within the work group can be accomplished as the situation dictates.

Since business orgainzations must earn a profit over the long run, they must be task-oriented by necessity. Because of this, many managers and supervisors are chosen for their record of achievement and task orientation. But many companies, realizing that the relationship needs should be maintained, have adopted a more participative management philosophy. Although in some cases this orientation functions at the expense of the task, it can be very effective in meeting the social needs of the company. It is unfortunate, however, that in spite of the discussion about social needs, such needs remain in a secondary position in most organizations and are often viewed by the group itself as manipulative. Through the appointment of comanagers it may be possible to achieve balance without the backlash of resentment and misinterpretation of intent.

As a company grows in size, product line, and specialization, it is normal to segregate it into smaller segments for control and efficiency. The larger and more complicated an organizational structure becomes, the more cumbersome, costly, and difficult it is to manage. To slow a natural tendency toward bureaucratic growth, dual leadership offers control of larger departments and may alleviate the need for more segmentation. This could be especially true with medium-sized growth companies caught in a transition stage.

Over the past half century many companies have grown to gigantic proportions but remain reluctant to decentralize their organizations. For those who have made this leap into uncertainty it has taken a great deal of courage, trust, and judgment at all management levels. Yet for those who have made the move to decentralization, authority is passed down to a certain level—usually the profit center—and remains guarded at that point. One of the most perplexing dangers that the decentralized organization will face is the misinterpretation of policy and objectives. As noted previously, when a manager vested with position, responsibility, and authority assumes a role requiring a set of characteristics and actions, how he or she perceives this role has a large influence on the decision-making process. Often the manager's understanding of policy and objectives differs from that of his or her actual intent. If two managers were appointed for the same position, each would have a different perception of the job. This in itself would provide an inherent check and balance to misinterpretation in the decision-making process. To improve growth in management depth, initiative, and expression of creativity at even lower levels in the organization, dual leadership could become the bridge to decentralization at the profit-center level.

One of the more immediate applications for dual leadership is within the matrix organization. Many companies have already established a dual line of authority cutting across the traditional functional design. In the matrix form of structure each member of the project team operates within a framework of function/project coordination and responsibility. Authority by its very design is split between the functional division with its emphasis on support activities, and the project with its emphasis on results. Without the accompanying change in interpersonal relations needed for this type of structure, the individual employee is likely to be caught between two conflicting supervisors.

In every dual leadership situation the matrix organization must recognize leadership's different expectations with regard to the group. Trust, open communication, and decision by participation are only a few of the prerequisites to a matrix theory of organizational member relationships. The newer, more democratic approaches to conflict resolution may be more practical—as opposed to management by command—in the matrix organization. In short, management by consensus will be the order of the day.

If the matrix structure is to operate efficiently on a day-by-day basis, then both a task and a relationship orientation are needed to fully integrate both the project and functional divisions. Individual employees should not and must not be constantly pulled between two task-oriented managers. Frustration and apathy will obviously be the result. On the other hand, two people-oriented managers may lose sight of overall company objectives by concentrating solely on social harmony within the group. Thus the appointment of two managers—one task-oriented, one people-oriented—would hopefully help integrate the project and functional divisions. As to the question of who belongs where, we leave that to the dictates of the situation. But we suspect that the project manager with his results orientation would be more task-oriented, while the support activities should be managed by a more considerate people-oriented leader.

IMPLEMENTATION

Many factors need consideration before a decision is made to adopt dual leadership. Of primary concern is the question why it may be practical. What is the task involved? Is it a complicated or specialized process? What is the makeup of the group regarding their task or relationship orientation? Can departments or functions be combined under two managers? These are only a few of the questions; many more will depend on the specific circumstances of each organization. Regardless of the reason or justification for choosing dual leadership, achieving overall objectives should remain the central issue.

In drafting a plan to implement dual leadership, caution should be exercised so that it does not become too rigid. Achievement of purpose is the key objective for organizing, and the organizational structure should be an evolving process to meet this requirement. Once the process begins and some of the goals are met, a certain internationalization takes place with regard to the organization. If the structure works, there is a tendency to view it as not needing change. This belief and commitment to the organization structure may, at the same time, be a device for protecting the group and its leaders. In short, organizations or systems tend to become the end rather than the means.

As a management tool, dual leadership must have support and acceptance from those using it and especially from those above them. What is expected, how it will work, and why it is needed should be made completely clear. Formal recognition should be given, so that employees are aware of the strategy's purpose and sanction. In this way a commitment to the concept will become the unifying element toward its successful use. Anything short of formal acceptance of dual leadership could result in failure for both the concept and the task.

Two managers having equal responsibility and authority should not create severe problems as long as their objectives and methods are in agreement. Since goals and standards must be the same for both managers, forming their agreement should result in clear, concise definitions. The design of how objectives are to be achieved should be made, as much as possible, by those involved in the work itself. It is management's responsibility to make sure that the work group understands and agrees on how it will be done. There will of course be guidance by both managers in the form of approval or veto, but the group should be held accountable for their performance and should have optimum input regarding the means of production.

Before discussing the choice of managers for dual leadership, it is appropriate to recognize the crisis situation. In most large companies a crisis of such magnitude as to cause economic catastrophe can occur at the top echelon only. But there will be times when a quick decision is necessary at lower levels and someone needs to act. In the case of dual leadership, neither manager should so depend on the other that he or she could not make the decision alone. Understanding and trust are vital in a case such as this, and both managers should anticipate and discuss such matters at the beginning.

One of the interesting aspects of dual leadership is its application to intergroup relations. Often managers of work groups are so caught up in the day-to-day activities of their immediate subordinates that they neglect the relationship with other functions within the organization. The supervisor is the key communication link between and among other vital functions or departments in the company. If no attempt is made to coordinate the immediate department's activities with overall company activities, organizational efficiency will suffer. Dual leadership could help alleviate this problem. One comanager (probably the people manager) could be primarily responsible for developing intragroup relations, while the other comanager could be more concerned with coordination problems throughout the organization. Thus a balance would be obtained between intergroup and intragroup activities.

Several companies recently have experimented with a new dimension of job design. Instead of concentrating solely on the structure, layout, and methods in the design of jobs, several innovative managements are beginning to also recognize the human element. Certainly it is a prerequisite to organizational efficiency to plan and organize jobs in industry, but to ignore the social and personal needs of the employee can lead to disaster. Consequently, terms such as job enlargement and job enrichment are often heard today in both the factory and the office. Obviously the use of a task-oriented manager (to concentrate on methods, standards, etc.) as well as a people-oriented manager (to concentrate on social and personal needs) would provide a catalyst to the already expanding field of job redesign. The engineering approach and the psychological approach can work hand in hand to improve worker efficiency *and* morale.

Lastly, one of the most touchy aspects of dual leadership must be addressed—performance evaluation. Every employee wants and needs guidance and constructive feedback concerning his or her performance. Yet we know how extremely difficult the entire performance evaluation process can be. Stereotyping and the halo effect are both good examples of how managers can distort the evaluation process. In addition, employees are very prone to be defensive concerning their past performance. For example, negative feedback from the supervisor usually leads to employee frustration, rationalization, and/ or apathy. It is even doubtful if a positive evaluation has any long-lasting effects on employee productivity. It appears that most managers are just not interpersonally skilled in the art and science of performance evaluation.

Dual leadership could help alleviate some of the traditional problems associated with performance evaluation. Using a modified MBO approach, both comanagers can develop goals and aspirations in conjunction with the subordinate. By utilizing both managers in the process, high expectations can be established, though these must not be unrealistic. Of course the methods will be determined by the employee with a certain amount of guidance from both managers during the review time period. Since the people-oriented manager is presumably more skilled in interpersonal relations, he will probably conduct the actual performance interview. Hopefully, problems can be resolved on a constructive psychological basis without the usual defensiveness on the part of the employee. If the relationship manager becomes too involved and sympathetic with employees' problems, the task-oriented manager is always in the background for balance.

Disciplinary action is not often needed in most organizations, but on occasions it is necessary. Because they are unpleasant, some managers avoid these situations until the last minute. Confrontation with employees on a group or individual basis requires self-control and a complete understanding of the problem and the employees. Once the problem has been defined and expectations are known, emphasis can be switched to methods of correction or adjustment. That two managers acting together can handle problem conditions is evident, since there is strength in numbers. The two managers, having different styles, will be perceived by the person or group in different modes: one as the authority figure and the other as an ally. After the initial confrontation, the ally plays a key role in buffering the situation and coaching the employee in achieving the desired results.

DEVELOPING THE TEAM

In choosing two managers it has been suggested that one be task-oriented, the other relationship-oriented. Because each style is necessary for the work group and its health, we feel the major benefits of group behavior will be derived in this manner. It must also be recognized that there is no *pure* task- or relationship-oriented manager. Many exist in between, but the tendency toward one or the other can be recognized through behavior and personality characteristics. It should not take an in-depth study of personalities to determine who has a required style, especially for those already employed. For new employees there are many standardized tests of personality and leadership available to determine behavior patterns. The personnel department or outside employment agencies can be extremely helpful in this regard.

The choice of two managers for a work group is one of team selection. This team may be permanent or temporary, or if necessary they can be moved to another location and work group. As with any choice of people, an investment is made, but even more so with the team. A commitment must be made by top management to give the comanagers the necessary time to connect and develop for maximum results.

Recently there has been an intense interest in individual success and fulfillment both on and off the job. Thus each of the chosen managers are likely to have some well established personal goals. These personal goals, as well as career goals, should be identified and taken into consideration. How they plan to meet these objectives as they relate to actual working conditions is important. This type of inquiry and clarification is beneficial in spotting possible degrees of harmony or discord between the two managers, the group as a whole, and the given task at hand.

The concept of dual leadership lends itself to possible benefits in the area of training and development. In the choice of managers the training needs of each member should be determined. The dual leadership situation will allow a less experienced manager to work closely with a more experienced one. Possibly a person who has been working alone could be placed on a team for exposure to group pressures and compromise situations. Obviously, management depth can be improved for future needs. Cross-training possibilities abound in an organization committed to trust and open communication through dual leadership.

Career stage is also to be investigated. As on a learning curve, there are points in time where rapid growth is evident or plateaus are reached. These areas of growth can be for a particular job, company, or individual and need to be identified. Are the two career stages compatible? Which one fits the task at hand? What career stage does the work group need? Are there possibly some motivational benefits to be gained by managing together? For example, one manager could be at the midcareer plateau. He might stay at that point or enter another rapid-growth stage. What would happen and how would that person react, if paired with a young, aggressive manager? It could have a broadening and maturing affect on both managers.

It should go without saying that the comanagers must be achievement-motivated. No manager can be extremely successful without having an inward desire to achieve goals, both organizational and personal. We also know that power is an important ingredient in the managerial process. But how managers use that power is open to question. Some like to be forceful and use the power of their office or position. Others are more considerate and prefer to use the charismatic power of personal influence to accomplish their objectives. Both approaches can be extremely effective and, used in proper balance, tend to reinforce each other in providing subordinate security with integrity. Perhaps, when choosing and developing the team, these characteristics should be taken into consideration. Although very little empirical research has been done on the relationship between leadership style and power, some generalizations can be made. Some theorists believe that power and authority flow from the top of the organization; others, that they come from the group itself. Probably it is a combination of the two. The appointment of a task-oriented manager who uses the power of his position, along with a relation-oriented manager who uses interpersonal influence, would hopefully satisfy both theories of authority and power. Again, balance is obtained.

Each member of the team will bring with him certain strength and weaknesses, and certain specializations. These should be identified and understood, to help strengthen the team. They should complement each other. Where one manager might falter, the other is there to assist. This aspect of helping each other is one that will have to develop over time, but could result in one of the most important strengths for the team and work group. Here again, interpersonal compatibility is important: both managers must want the same thing. It would be very difficult for one manager to help the other unless both recognized their interdependency. They succed or fail together.

INTERPERSONAL ASPECTS

Open communication between the two managers will be a key feature of their success, as it is in almost every working association. At the outset there will probably not be enough time in the day to discuss all necessary points. This time must be set aside and guarded, in order to develop and anticipate individual reaction and behavior patterns. Even after they have been working together for some time, it would be wise to continue a daily session for review and appraisal.

The ability to openly and honestly confront problem areas concerning all aspects of the project, work group, and their association should develop over time. Much of the beginning need for intense communication will fade and at this point care should be taken, as both will tend to make unwarranted assumptions. When this happens—and it will—both managers must be mature enough to accept these breakdowns as something that is bound to happen—an attitude that will greatly limit rifts in their relationship. We cannot overemphasize the point that these breakdowns are not personal win-or-lose situations; both managers should be equally responsible and accountable for the result.

Individual pride and self-esteem can be a sensitive area for the two managers, especially if both are high achievers. More than any other aspect of their association, pride can cause irreconcilable problems. It must be realized by all concerned that each member brings certain strengths—and weaknesses—to be applied in a very individualistic manner. An attitude of adjustment and a willingness to help each other achieve identity and personal goals can help solve many pride-related problems. Also, the ability to accept each other's faults—the realization that neither is perfect—in an attitude of good humor and improvement, will assist in building a competent, secure team.

The reverse to the problems arising out of pride and self-identification is the fatal characteristic of dependence. One of the members may become overly dependent on the other. This can develop as a result of one manager's expertise in a particular area and the other's desire to achieve or learn in that area, with corresponding attention being given by the expert. Or the group may recognize and accept one manager's style over the other's. If this dependency is recognized and accepted by both members of the team, it is not necessarily a problem area. But if recognized with resentment and avoidance, it does become a problem. There could be many ways to solve such a problem; it is best left to the two managers. The important and critical aspect is recognition of the problem and the willingness to face it honestly and openly.

As a natural course of events, disagreement and conflict will arise over a broad spectrum of activities. Some of these will be within the group, some between the two managers, and some between manager and group. We are not concerned here with the solution of these conflicts, but with their control. Competition is healthy as a challenge to agreement; in fact, situations exist where there is too much agreement. But competition can get out of hand and evolve to a personal level. Self-control is essential for limiting conflict, as it is generally for the functioning of the work group and management team. Excuses should be guarded against, but this does not mean denying the real aspects of a problem. A balance is desired in decision making; dual leadership can offer the possibility of controlled conflict to achieve such balances.

Organizations create a great amount of stress within the work environment.

Organizational variables such as rules, deadlines, conflicts, and politics are just part of a larger maze of pressures and constraints upon individual employee behavior. Often in the process of structuring the job, the organization, the group, and the roles played by the participants, we inadvertently inflict upon each individual a difficult and sometimes impossible situation. Undue stress is the inevitable result, bringing with it psychological and physiological disorders.

A certain amount of stress is needed if organizational objectives are to be accomplished. Indeed, a moderate amount of stress provides for a healthy and vigorous work environment where employees can be excited about what is happening in the workplace. It is hoped that two managers, with opposite but compatible leadership styles, will provide the amount of stress needed to achieve both effeciency *and* quality, productivity *and* satisfaction, control *and* individualism, structure *and* creativity, etc. The interpersonal aspects of this coupling appear obvious. The team of managers will have to approach each situation and subordinate with an understanding of stress and its functional/dysfunctional results.

One of the many things we are learning from the Japanese is decision making by consensus. As a general rule decisions made by a unanimous group are easier to implement, because of the group's commitment. Group decisions are also usually more thought out and thus better than when one individual acts alone. As mentioned earlier, managers have traditionally been promoted to a position of leadership because of their task orientation and their ability to accomplish objectives. However, this overemphasis on the task can lead to autocratic decision making and poor reception by the group.

From an interpersonal perspective the dual leadership model can provide a task orientation for structure, but must not do so at the expense of poor decision making. The comanagers must learn when to give authority and when to accept it, depending on the decision, the attitude of the group, and the ease of decision implementation. Once again, interpersonal communication fostered by trust and openness between the comanagers should provide the necessary atmosphere for group decision making by consensus. A certain amount of special training will be needed in order to provide both managers with the necessary interpersonal (human relations) skills for group decision making.

CONCLUSION

We recognize that problems will exist with the dual leadership method of organization. Disagreements will occur periodically. But problems also exist with the one-boss method. The question then becomes: which method will provide the opportunities for solving the problems that arise? Dual leadership hopefully provides two managers with opposing styles of supervision but equal authority and responsibility, and an opportunity to mutually support and criticize each other's ideas and philosophy. One complements the other. Because together they look at all sides of a problem, there should never be an overemphasis on the task to the detriment of the people involved, or on people to the detriment of the task. Dual leadership provides the opportunity for conflict resolution on a meaningful constructive basis.

Obviously, no two managers can be paired successfully without a great deal of commitment on their part. But as in a good marriage, conflicts can be resolved with open and trustful communications. If there is no commitment to working together for the betterment of the organization, dual leadership cannot and will not work.

17. Managing Human Resources in the Matrix Management Environment

Robert L. Desatnick*

As the demarcation between line and staff work becomes so blurred as to be virtually indistinguishable, the concept of shared accountability is gaining wider acceptance. When two or more individuals from different functions get together to collaborate on a joint decision, or to resolve a major human resources issue, or to position their organization to avoid a future problem, a form of matrix management is created.

Just as the product manager heads a product team to introduce and develop a new product, so the human resource executive in charge of management development gathers about him those line and staff executives who ultimately help prepare the successor to the chief executive officer.

But many questions arise. Just how does matrix management work in a domestic environment? In an international environment? In what ways does a product team compare to a project team? When and under what circumstances are committees useful? What types of issues in human resource management can best be solved by an interdisciplinary/line staff task force? If some form of matrix management is necessary to resolve a critical human resource issue, when should it be disbanded, if at all? How often should it be convened? And what is a sensible approach to shared accountability in a complex organizational setting?

QUALITY AND DEPTH OF MANAGEMENT

One mistake that organizations tend to make is to extend themselves into the international arena before they have their own domestic situation well in hand.

*Robert L. Desatnick is President of Creative Human Resource Consultants in LaGrange, Illinois. He has served such organizations as Booz, Allen and Hamilton, McDonald's, General Electric, the Chase Manhattan Bank, and Otis Elevator. During his 27-year career in human resource management, he has functioned as a consultant, a corporate vice-president of human resources, a director of business planning, and an MS and MBO coordinator. He has served as a lecturer and seminar leader at the Universities of Chicago, Michigan, and Minnesota. He is the author of five books on human resource management. The material for this article is drawn from Mr. Desatnick's new book, *The Business of Human Resource Management (A Guide for the Results-Oriented Executive)*, John Wiley & Sons, New York, Fall, 1983.

International operations do not manage themselves. In fact, they require a depth of management talent.

To this end, perhaps the single most important contribution the human resources executive can make is to favorably impact the quality of management, beginning at the top, through significantly influencing the management process. Taken literally, this means playing a major role in the establishment of business plans and objectives, and continuing to do so right on through the entire management cycle. Viewed exclusively from a domestic corporate point of view, the process may be visualized as in Figure 17-1. These considerations are necessary to affect the quality and depth of management and hence the results of the business.

Figure 17-1 emphasizes that building upon and improving the quality of management in any given organization requires a well-defined, systematic effort—systems approach. Because the entire organization is affected by the quality of its management, every discipline and every manager, beginning with the CEO, shares responsibility and accountability for developing the company's management talent. But if this is true, why isn't every organization doing it? Why has shared accountability not worked before? Who needs to take the initiative?

The author recalls vividly one of his first major experiences in a multinational environment. He was reporting to the French chairman of the international division of a major U.S. multinational company. Prior to one staff meet-

Fig. 17-1. A systems approach to impacting the quality and depth of management.

ing, it had been prearranged that a personnel inventory be made of all senior nationals in 29 different countries, to ascertain their strengths as well as the depth and quality of management. This inventory was to be centralized in the human resource department in Paris; then as vacancies arose in regional or staff headquarters, individuals would be drawn from the pool. This inventory also was to provide performance and potential ratings; strengths and developmental needs of each key individual; as well as plans for their development, performance improvement, or possibly career redirection.

The resistance to this project among the line regional vice-presidents was incredible. They considered this an erosion of their authority and a breach of trust regarding their abilities to manage their territories. Unfortunately, the inventory was not presented properly or positioned as a help to the line manager. After much debate, during which staff supported and line opposed the project, the issue was resolved. There would be an international human resource inventory. It happened and it worked.

In retrospect, the basic problem arose from the fact that many of these companies were losing money. Performance standards were nonexistent. Management development was unknown and most company managing directors had come up through the ranks with exposure to only one instead of several functional disciplines. Subsequent visitations revealed glaring weaknesses in quality and depth of management, but thanks to the enlightened international chairman, a process of MBO was initiated, accompanied by well-defined and challenging performance standards. The human resource executive was given the responsibility to initiate and implement this system worldwide.

This system worked because the chairman was behind it. It was he who, with the counsel of the human resource executive, initiated the process. He made it clear that senior managers were accountable for improving business results, and in the process were to make sure that the first three levels of management in every country were qualified to deliver the requisite improvement. This was a form of matrix management at its best. Harmonizing and blending together the particular strengths and talents of line executives at several levels and their staff executive counterparts, a matrix management system to develop general managers had been created.

WHAT IS SHARED ACCOUNTABILITY?

Refer back for a moment to Figure 17-1. Observe the logical progression in moving from circles one through seven. If all these processes occur and all are managed properly, an organization will be assured of having qualified staff to meet the challenges either of growth, expansion, diversification, and multinationalization, or conversely of divestiture, retrenchment, or reduction in the size and scope of operations.

When an organization has a quality staff of well-trained, broad-gauged, professionally competent line and staff executives, they *jointly* formulate sound business plans and objectives. Traditional roles are either blurred or forgotten as they work as a team to formulate and realize these strategies. This is shared accountability.

An important part of a sound business plan is the capacity to realize its market potential. In defining business objectives and priorities, an integral part of that plan must be the human resource component. Unless the people plan is part and parcel of the business plan, the latter may not be worth the paper it's printed on. People are the key ingredient in an organization's capacity.

Figure 17-2 sets forth a perspective that mandates some form of matrix management. If one formulates business objectives and priorities without a clear knowledge of the market's potential and the organization's capacity to realize that potential, plans will not be accomplished. The chairman who plans in a vacuum without this knowledge will achieve something less than what is desired. One reason for having an office of the president, executive committee,

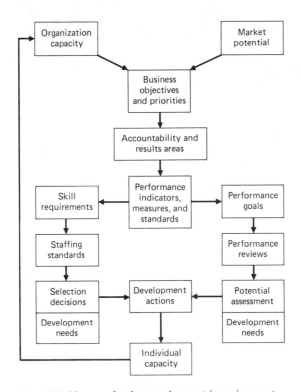

Fig. 17-2. Manager development in a matrix environment.

or planning committee is to make a better decision collectively than the CEO could do individually. The CEO's job has become too vast and too complex for one person.

If the business objectives are known to be sound and obtainable, the establishment of accountability and results areas is facilitated. This matrix type of arrangement results in realistic objectives, because staff inputs temper the enthusiasm of line managers. (In particular, marketing and sales managers traditionally tend to be somewhat optimistic.)

Once results requirements are determined, the action continues to flow downward through the various layers of management with even more precise indicators and measures and standards of performance. Note that the process envisioned by Figure 17-2 provides for significant bottom-up input as well as top-down direction. The whole process is fed with analyses and recommendations from lower and middle management, both line and staff.

To the right of the block PERFORMANCE INDICATORS is an arrow pointing PERFORMANCE GOALS. At this stage of the matirx, results areas and performance standards are translated into what the individuals are expected to contribute to overall component and/or organization results. The arrow to the left of PERFORMANCE INDICATORS points to SKILL REQUIREMENTS, which comprises the kinds of experience, education, abilities, and aptitudes required on the job. The human resource executive who manages this process on behalf of the CEO is charged with enormous responsibility as facilitator, interpreter, catalyst, agent of change, advocate, and developer of people. This executive draws out his or her counterparts and those above and below; inputs to the chairman an ongoing status of results in relation to the plan; and troubleshoots and follows up.

At the far right of Figure 17-2, beneath PERFORMANCE GOALS, is a box labeled PERFORMANCE REVIEWS. A feedback system to the individual and on up the line to senior management is necessary to avoid potential surprises and possible continuance of unrealistic goals. Perhaps some external or internal factor—fire, flood, recession—could prevent goal attainment. Individuals as well as business components need to know how well or poorly they are doing.

This chapter does not recommend substituting human resource executive or that executive's staff for the line manager; rather, it defines their mission as providing the tools, methods, and systems necessary to make things work better.

Based on the valid objective performance review, management begins to assess who in the organization have the potential to move up and become part of top management. If a valid assessment is made of the various individuals' potential, the process will yield performance-related developmental needs (or deficiencies). This will occur because performance standards and measures include quantitative as well as qualitative data.

On the far left of Figure 17-2, a complementary process is taking place. The box beneath SKILL REQUIREMENTS is labeled STAFFING STANDARDS. In some organizations the selection process, whether used for internal promotion or external hiring, is not taken seriously enough. Many managers are poor interviewers, preferring to hear themselves talk instead of listening to what the candidate has to say. Imagine the enormous impact the human resource executive has on an organization when, working with an executive search consultant, he or she manages an outside executive search. The type of candidate presented for management's final approval could well change the course and direction of the company. Only in a matrix organization are this power and influence clearly understood and accompanied by a system of proper checks and balances. As a matter of course the selection system should be designed by the human resource executive, as it needs to be related to the critical dimensions of the particular position to be filled, as well as to the culture, pace, tempo, and environment of the organization. To put it conversely, imagine how absurd it would be to hire a staff insurance executive in a senior line capacity for the fast food industry.

Staffing standards themselves should bear a direct relationship to the requirements of the position and the priorities and objectives of a particular assignment. This is usually determined by the executive officer in charge with the collaboration of the human resource executive. Thus when a selection decision is made, after proper reference investigation, the probability of success of the selected candidate is enhanced. This is so because the requirements of the position and the nature and demands of the environment and the culture have been given full consideration.

Whether a selection decision favors the internal or external candidate, the main consideration is that the best qualified person be selected. In the author's opinion, even when the specifications have been very well defined, the successful candidate will seldom exceed 60 to 70% of the position requirements. This is noted to highlight the fact that there will be developmental needs, whether for the internal or external candidate.

Near the bottom of Figure 17-2, note how arrows from SELECTION DECISIONS on the left and POTENTIAL ASSESSMENT on the right converge on DEVELOPMENTAL ACTIONS. If managed properly, the process will include a plan for developmental actions for both longer service executives and newly hired managers.

The whole process adds up to building the quality and capacity of the individual to assume more responsibility. When this occurs with an honest sense of shared accountability on the part of line and staff executives, many individuals will grow, developing their potential so as to multiply in an almost geometric progression an organization's capacity to achieve. "Synergism" is almost inadequate to describe the potential impact, when members believe in and practice matrix management.

MANAGEMENT DEVELOPMENT

Individuals defined as having high potential need to be challenged and broadened. By achieving superior results they have demonstrated that they may yet do still better, and the the best way to reward achievement is to provide opportunity for more achievement. A matrix structure presents an ideal vehicle for manager development in a large organization.

Take, for example, a cost-improvement task force. It may be structured to include several high potential middle managers from a variety of line and staff positions, and not all at the same level. Senior management approaches this group with specific cost-improvement goals. The group's charge reaches throughout the entire organization for investigative and experimental purposes. They diligently look for every cost-improvement opportunity in every function, every area, every component. They make their recommendations to top management with the full awareness of the department head who will be affected by the changes. The task force may have a life span as short as three months. The chairman is elected by his or her peers, rotating monthly.

Because each member of the task force will have an opportunity to visit every functional area of the business, an ideal development situation is created. The chairmen are chosen by their peers, who know and respect these rare wunderkind who can achieve results in difficult environments.

As individuals move away from their own functional environment, and the traditional authority they may have had over direct reports, they begin to grow and develop. The author was involved in one such situation in Caracas, Venezuela. The problem was an inability to complete elevators, resulting in missed delivery dates. The flow of production and assembly had bogged down. Charged with the responsibility to facilitate a solution by the chairman, the author created a production task force to bring together all who could possibly be part of the problem or, conversely and optimistically, part of the solution. Their authority was final and binding. There was only one caveat: the local managing director had veto power, if he wished to exercise it. He did not. Executives assigned to this task force learned team work, delegation, and priority skills. The guidelines for the task force were as follows.

THE PRODUCTION TASK FORCE

Composition and Method

The task force should include senior exectuives who contribute to the problem and/or the solution. These would include managers of sales, engineering, production, field operations, finance, human resources, and contract control. The managing director is excluded: he or she must maintain objectivity so as to analyze the problems and proposed solutions.

Analysis of the Problem

In defining the problem the task force should analyze in depth the following areas (starting with each individual for his or her own functional responsibility).

Sales

1. Availability of accurate, specific forecasts by model and unit, as well as by market segment.
2. Commitments by salespersonnel to customers, according to predetermined lead times.
3. Status of present commitments.
4. Adequacy of market coverage by qualified sales personnel.
5. Analysis of customer complaints with regard to installations, service, and quality—e.g., "trouble reports."

Engineering

1. Establishment of lead times by model and by unit for all engineering work, including drawings, specifications, parts lists.
2. Engineering cost analyses build-up for all components, in order to establish correct price. This should include price escalation for inflation.
3. Engineering project management approach for all major contracts, product scheduling systems, and controls.
4. Engineering backlog analysis vs. current and projected demands for engineering time.
5. Engineering department organization as well as right number and quality of personnel trained to do the work.
6. Engineering data retrieval system.
7. Measurement of accuracy, quality, and timeliness of data submitted in usable form to sales and factory.
8. Provision for analyses of problems caused by improper or incomplete drawings, specifications and data—part of the trouble-reporting system from the factory and field.
9. Value analysis and value engineering section within engineering department to reduce total costs and eliminate or minimize, factory bottlenecks.
10. Engineering cost and efficiency improvements program.

Production

1. Data on factory capacity, including machine loading, factory loading: analyses of factory layout and material movement.

2. Anticipated vs. actual productivity by component, by cost center, by operator, by machine.
3. Productivity per employee—direct and indirect. Existence of standard times and proper methods for all productive operations.
4. Identification of factory bottlenecks.
5. Percentage of machine utilization and analysis of reasons for underutilization.
6. Analysis of present backlog and present projected requirements.
7. Causes of parts and materials shortages, e.g., lack of qualified people, poor organization, lack of controls and follow-up, lack of raw materials, poor cash flow, lack of methods and standards, poor workplace layout, inadequate daily, weekly, monthly production schedules by contract, by machine, by cost center.
8. Existence, use, and follow-up of trouble-reporting system.
9. Analyses of total paperwork and flow as an integrated system for factory control, including purchased parts, components, supplies. This also applies to receiving and shipping departments and personnel therein.
10. Inventory control system: raw, in-process, and finished components.
11. Effectiveness of scheduling and dispatching, and of expediting functions and personnel.
12. Proper organization and sufficient trained personnel to do the work.
13. Quality standards and inspection procedures, and adequacy of these standards, procedures, controls, and properly trained personnel to do the work.
14. Actual scrap and wastage experienced and recorded, and analyzed to develop plan to eliminate or at least minimize them.
15. Proper tools, dies jigs, fixtures, and machinery in good working order.

Field Operations

1. Lead times for erection by unit and by contract.
2. Job planning and project management for major jobs.
3. Adequate number and quality of personnel for present and future work loads.
4. Problems with time lost because of delays, partial shipments, wrong parts, delivery and unpacking of parts for ease in erection.
5. Scheduling and routing of personnel.
6. Existence of standard data (time, methods) for all erection procedures by unit, by job, e.g., efficiency, productivity, and cost measurements.
7. Quality of work as measured by
   - Immediate customer acceptance
   - Immediate acceptance by service department
   - Number and types of complaints reported by day, week, month

8. Cost improvement program: timetable and follow-up for improving productivity and efficiency.
9. Major problems experienced that could be corrected through adequate training of supervisors and workers.

Finance

To assist in identifying all costs affecting, or affected by, production operations:

1. Complete cost analyses, timing and frequency of reports to sales, engineering, production, and field operations, e.g., distribution cost analysis, including expense controls, to determine how and where money is spent because of poor deliveries. Develop accurate cost data and controls and review as often as necessary with managers responsible for doing the work as it affects the factory.
2. Periodic audit of all functions to improve management and cost controls.
3. Regular follow-up of financial reports to train managers and supervisors in techniques of financial management analysis, including ratio analysis, methods for budget, cost, and expense controls in proper relation to sales and revenues, with emphasis on factory production costs.
4. Proper organization and adequate staffing with fully trained personnel.

Human resources

1. Determination of overall training needs of managers, supervisors, and employees, based on problem identification and performance in relation to planned results: understanding and acceptance of job responsibilities, priorities, performance standards, and accountability, and establishment of short-term goals by managers and supervisors for each major activity within each function.
2. Adequacy of own organization, and of manpower plans and forecasts as well as training requirements for present and projected needs.
3. Manpower inventories for each function, present and projected, including quantitative and qualitative data based on measurement of performance.
4. Assisting each task force member in applying methodology of management by objective to the solution of problems as analyzed.

Contract control

1. Function of coordinating work of all departments to meet production schedules.
2. Analysis of whole paper-work system to improve flow, timing, accuracy,

and dependability, and to eliminate confusion, duplication, overlap, and unnecessary paper work.

3. Establishment of proper lead times, agreed to by all functional managers.
4. Establishment of proper systems for control, implementation, follow-up, and immediate identification of problems that cause the system to bog down. Recommendation of immediate corrective action and follow-up, with responsible managers making sure action is taken to eliminate the cause of problems.
5. Implementation of a system of trouble reports to identify causes of problems.

General Considerations

All senior managers who affect or are affected by the output of the factory must first look to their own departments for problem identification and analysis. Each must analyze in detail the backlog in his or her own department, asking, "What can I do to help identify the causes and provide a solution to the problems?" and "Is my department properly organized to do the work?" Points to be covered include:

1. Communication of clear, written objectives, goals, and priorities for the department as a whole.
2. Job description and priorities for each individual and each activity.
3. Experience and education requirements for each job and each employee.
4. Training requirements for each worker to ensure understanding of the job, priorities, performance standards, and qualifications to do the work.
5. Manpower plan and forecast for future work load.
6. Advance recruitment and training plans.
7. Measurement of results of training and performance to identify areas for improvement.
8. System of immediate feedback to detect deviations from plans and take corrective action.
9. Acceptance by each manager of personal responsibility for the success of other managers: "How can I contribute to the solution?" as opposed to a defensive attitude.

Conclusion

After the task force, individually and collectively, has analyzed all major problems along with cost implications, it should:

1. Suggest alternative solutions agreed to by all task force members, weighing advantages and disadvantages of each to
   - Eliminate backlog of work in all departments.

- Set up an efficient system and a detailed written plan for implementation, with commitments from all department heads to one another.
2. Make a presentation to the managing director, highlighting any difference of opinion.

The managing director assures the use of the management-by-objectives approach by the following methods:

- Definition of problems and causes.
- Statements of goals to be achieved and results expected.
- Specific action steps for each goal:
  Beginning and ending dates for each step.
  Statement as to who will do the work.
  Statement as to how progress will be controlled and measured.
- Anticipation of obstacles.
- Specification of additional resources required, information, manpower, tools, etc., plus costs of each and net additions to profit if resources are committed.

Weekly group meetings should be held until the system is fully implemented and the problems solved, i.e., until delays in deliveries are eliminated and the factory is producing and shipping according to schedule with proper levels of quality.

In the case from the author's experience mentioned earlier, prior to the next follow-up management-by-objectives seminar a production task force was formed in accordance with these guidelines. They met weekly and a plan was developed by the group. It was highly successful and the production problems were solved. The detailed approach was as outlined in Table 17-1.

While it is somewhat unusual to find a human resource executive so deeply involved in multifunctional problems, the skills gleaned from conducting and organizing MBO seminars enabled the author to put his experience to good use in a pragmatic test. A matrix organization was created to resolve a critical problem. Once the problem was solved, monthly meetings were held to make sure that things kept on track.

THE MATRIX AND QUALITY OF MANAGEMENT

When a situation such as the one described above suggests an interdisciplinary solution to a complex problem, the best people with the highest potential should be chosen to staff the matrix. When the author worked at the Chase Manhattan Bank in the New York International Division, there was considerable discussion about how best to service multinational customers with a variety of banking needs.

Table 17-1. Developing Country Approach Production Task Force.

(1) PROBLEM AREAS IDENTIFIED	(2) REASONS FOR PROBLEM	(3) ACTIONS TO SLOVE PROBLEM	(4) EXPECTED RESULTS FOR PROBLEM AREAS A, B, C
A. Late deliveries of finished products. Purchase of local materials. Purchase/Import materials. Specifications. Shipping.	Inadequate knowledge of suppliers, prices, deliveries. Lack of quality control/purchased materials. No storage/purchased materials. Lacking data on material requirements/quantity, timing. Inadequate controls and organization. Unrealistic forecasts. No definition of lead time. Lack of follow-up. Errors in specifications. No real system in department.	List all local purchases. Kardex of suppliers. Price and delivery dates, from local suppliers. Minimum reorder point. Economic lot purchases. Quality control of local purchases. Control routine in receiving. Establish a storage area. Factory to control purchase orders. Procedure to follow up shipment dates. Coordinate market plan with factory's ability to produce. Make product data sheets and layouts. Review specifications before producing parts. Rent warehouse space. Private trucking firm to supplement own fleet. Reduce number of shipments for one order.	Reduce factory costs and field instruction costs by saving X + Y number of hours (because of more efficient factory costs). Increase shipments weekly from X to Y number of units. Reduce factory overtime by _____ %. Note: For each action as defined in col. 3: A beginning and ending date was estimated. The person or persons responsible for doing the work were identified and accepted the assignment. The costs and expenses of implementation were measured against the potential savings over a 5-year period.
B. Production systems. Manufacturing methods. Rationalization. Machinery, tooling.			

280

No labor "order" for each piece produced. Inadequate controls for following labor orders in process. Separate pieces not separately identified and scheduled. No "standard" pieces for production. No real production "system flow." No in-process quality control. No programming of factory production. Old, inadequate methods. Insufficient number of machines. Inadequate tooling—jigs, dies, fixtures.	Process order prepared for each piece manufactured. Control to be developed to locate an order in process. Small shop area set aside for special, unusual, and nonstandard pieces. Parts to be standardized according to marketing and engineering definition. Define series for parts production. Define machine loading. Establish QC system. Program production load according to contracts sold. Certain machines, tools, dies, and fixtures to be purchased as capital expenditure. Hire new man.	People from different functions started to work as a team for the first time and to appreciate the help-solve-problems outside own function. The task force actually tripled production and reduced total cost of production by 50%.
Stock room, kardex file: Production control Shipping.		

C. Quality and number of personnel.

Conceivably and in fact (since it did happen), one banking group would call on a customer to market merchant banking services. The next day the same customer would be visited by a commercial lending officer and shortly thereafter by a foreign exchange trader. The solution was to appoint a relationship manager who managed large accounts in a given industry. Because of the variety of banking services offered, and the differences in overseas banking practices and regulations, an exceptional person had to be selected and trained for this awesome responsibility. For instance, the individual chosen had to be a skilled negotiator as well as a most capable broad-gauge banker. This manager had to bring together heretofore independent executives who had developed their own customer relationships for their particular product lines.

This action had a favorable impact by bringing together under one roof the product, geography, function, and customer. Performance management was improved, confusion was reduced, and most of those involved ultimately benefited—the customer, the bank, and the executives in charge. There was a real growth process within a compressed time frame. The input of the human resource executive was to participate in the design of the new organization structure, helping to clarify the mission, objectives, and designation of position responsibilities. The executive also established appropriate salary grades and ranges, and assisted in designing the training needed for individuals to qualify for the new position.

If matrix management is to improve the quality of management, the broader issues of performance management within different organization frameworks must be considered up front, and with the full participation of the human resource department.

Another illustration of matrix impact on performance management comes in the exercise of decision-making authority. While much of the actual decision making in a matrix is individual in concept, it is in fact jointly performed. This ambiguity appears to be greatest in personnel matters. Specifically, the resource manager has the formal responsibility for evaluating and determining individual performance, awarding pay increases, and recommending promotions. In most instances the manager develops people, hires, fires, and makes work assignments, but in actual practice, he or she rarely does this alone.

It is the business results manager who has the authority to decide just where, and on what projects, a component's resources will be used. A product manager decides what the product and its specifications will be, when and how the work is to be done, and how much money and time will be spent in producing it. But the product manager needs the support of the resource manager to get the planning done on time, meet deadlines, and realize cost and quality objectives. Failure to obtain this concurrence from the resource manager prior to any given course of action increases the likelihood of the job not getting done.

Similarly, the human resource project manager responsible for resolving a critical personnel issue may be appointed by the chairman as the business results manager. But those who manage the needed resources can make it impossible for this manager to do the job.

Take the case of a management development task force. The company chairman becomes concerned about the lack of executive backups. He calls his human resource executive and directs him to organize a project team to start a major management development effort. He wants all high potential middle managers identified, with specific emphasis on women and minorities. Next he wants specific accelerated training programs to be designed and implemented, to help the chosen candidates move upward more quickly. This internal effort is to be supplemented with outside hirings as necessary. The chairman's last direction to the human resource executive says: "Pick whoever you want for this project; it will be in addition to their normal responsibilities."

Faced with precisely this situation, the author organized a cadre of senior executives with a multidisciplinary organization. The Executive Vice-President and Chief Operations Officer were signed up, as was the Executive Vice-President of Administrative Services. The group was joined by a zone senior vice-president, four staff vice-presidents, and two directors. The group prepared a game plan, and the human resource executive turned the project team over to the Executive Vice-President and Chief Operations Officer. He continued to actually manage behind the scenes, but did not want to give the whole company the impression that this was another personnel program.

Just as the product manager must mobilize resources and get others to give of their time, talents, material, and financial resources, so the human resource project manager had to pull together by persuasion and negotiation the best, most credible line and staff executives who could fulfill the chairman's wishes.

One of the first things this project team did was to determine those key positions for which backups should be developed. In these instances, with the company still growing rapidly, the emphasis was first directed toward the regional manager/vice-president position. This position was responsible for one of 24 geographic regions in the country with approximate sales of $200 million annually.

The team next determined which were the really essential skills, abilities, aptitudes, experiences, and education needed to perform a regional manager/vice-president's job. They asked: "How long will it take, in a carefully planned, individually tailored, accelerated effort, to produce regional managers? Can the company afford to wait?" It was established at that point that at least six candidates with regional manager potential within two to three years, should be placed in a full-time training program as soon as possible.

Candidate nominations were solicited by the project team from all officers in every function in the region and corporate areas. Each candidate submitted had to meet certain criteria: sustained outstanding performance; high potential; length of service with the company; and the ability to lead, manage, and develop people.

Approximately 25 candidates were submitted. A subgroup of the task force was formed of three to five people who on a rotating basis jointly interviewed all the candidates, using an open-ended questionnaire especially designed for this process by the human resource executive.

After a somewhat lengthy process, six candidates were chosen and assigned in carefully and highly individualized programs to provide the requisite skills that they lacked for managing a region. In two cases senior management did not want to give up their best-performing people with the highest potential. But in all cases the project team prevailed. This company was soon on its way with a highly successful embryonic succession plan. It was successful because it concentrated on the most critical slots to be filled. It worked because the chairman was behind it.

WHEN ARE COMMITTEES USEFUL?

As a general rule—one the author has followed for years—one should seek the inputs of key individuals when they are needed on vital issues that affect the entire organization. A committee brings together many years of diversified experience to analyse vital complex issues. Issues they might deal with include the company's position on affirmative action; how much to spend on total compensation and benefit packages for key executives; recommendations to the board of directors on profit-sharing and pensions; the development of a five-year plan and top management succession. Depending on the issue, its chance of recurring, and the frequency of the need to update, monitor, and audit, committees may be short-lived or semipermanent. Semipermanent examples include the compensation committee of the board of directors, the planning committee of the corporation, and the management committee. But if a committee is appointed to do a particular job such as to develop an organization's posture, strategy, and game plan for affirmative action in its broadest sense, then once that job is finished and the affirmative action department is established, the committee should be disbanded. On the other hand, administrative committees of boards of directors review legal, financial, and personnel issues as they relate to benefit plans, profit sharing, investment advisers, and results of investments. This type of committee generally meets four to six times a year, or more often as necessary. This was particularly true during the initial phase of President Reagan's Economic Recovery Tax Act, which had significant impact on savings plans, credit unions, pensions, and profit sharing; decisions had to be made as to whether to set up an in-house IRA or not.

Management committees are a unique breed. They normally meet monthly as the main communication vehicle on all matters of importance affecting the company. Topics they discuss and decide on include pricing strategy; timing and location of expansion efforts; company position on relocation and mortgage rate differentials; relationships with clients and franchises; changes in company wage and salary structure; benefits proposals; and succession planning. Their decisions are normally geared to the solution of short-term problems as opposed to three- to five-year plans, although their decisions in many instances have longer impact. This is a form of matrix management that produces balanced, best-interest decisions for the overall benefit of the company. Many of their decisions previously were the sole prerogative of one individual and hence tended to be narrowly focused and not always sound.

MATRIX MANAGEMENT OF CRITICAL HUMAN RESOURCE ISSUES

Some decisions need the involvement of several key executives, both as to whether to go ahead or not, and the specific tactics of implementation. Such issues might include:

- Whether to take a union strike during negotiations.
- Whether to launch a major career redirection effort.
- Whether to have day care facilities.
- Whether to engage third-party referral services in matters of employee assistance.
- To what degree to give employees autonomy over their work, scheduling, and participation in management decisions.
- How far and how fast to go in affirmative action.
- Whether to devise a plan to lay off and let go middle managers who have plateaued.
- Whether the company wants a sophisticated staffing strategy and a staffing plan.
- How much to spend annually on manager development.

Each of these human resource issues has major impact on the organization. The human resource executive prepares and distributes briefing materials in advance and usually chairs the discussion. A number of these issues could well fit within the monthly management committee meeting and often do.

There do not seem to be any hard and fast rules regarding when to disband a task force or committee. If the group was formed to resolve a specific short-term critical problem such as concluding a union negotiation, it is disbanded when the contract is signed. If, however, items of major impact occur on a frequent basis, such as changing trends in compensation and benefits, perhaps a standing committee is needed, even though meetings would be relatively infrequent.

A MULTI-NATIONAL ENVIRONMENT

Figure 17-3 demonstrates that there is relatively little difference in human resource issues in a multinational environment. However, because of the differences inherent in different cultures and societies, resolution of these issues is more complex. These relationships demand more flexibility and generally make communication significantly more difficult. Consequently, it takes superior managers to manage human resource and business issues in foreign environments—superior even to the strong managers needed to function in a strictly domestic matrix environment.

This is particularly the case because in developing countries the experienced

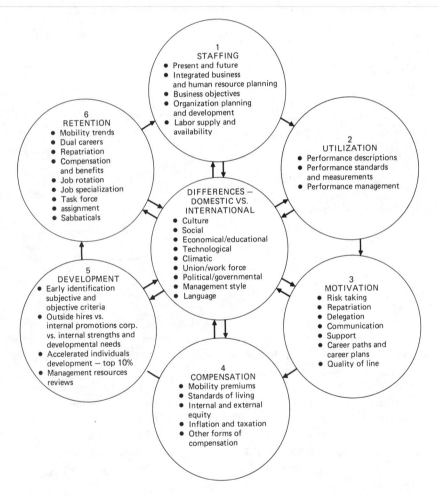

Fig. 17-3. Issue-oriented multinational human resource management.

manager will find differences in managment styles and practice. Some of these are as follows:

- Lack of business planning.
- Absence of business strategy.
- No market sophistication (competition, pricing, strategy).
- Poor cost accounting and financial management analyses and controls.
- Organization confusion in line staff relationships.
- No production/inventory controls; high receivables and borrowing.
- Lack of managerial delegation, participation, and involvement in major company decisions.
- Personnel managers functioning as record keepers, rather than as part of the management team and management process.
- Absence of people planning, forecasting, training and development.
- Too many levels of management.
- Too broad a span of controls.
- No job descriptions or performance standards.
- Inadequate and sometimes nonexistent Management Information Systems.
- MBO not really understood, or confused with the budget process.
- Inadequate budgetary controls and follow-ups.
- Poor or nonexistent communications throughout the organization.
- Lack of measurement and priority skills.

While it is not within the purview of this chapter to go into any real depth concerning differences noted in the center circle of Figure 17-3, some brief explanation is in order.

For example, in many *cultures* throughout the world, authority is inherited in both business and government. Key positions are filled from a select few families. Also, authority is inherent in the individual rather than in the position held. Most foreign citizens feel that their values, traditions, and work ethics are the best in the world, bar none. This makes it extremely difficult to impose another set of customs and social values overnight.

College and graduate education is limited to a few privileged individuals in many parts of the world. The lack of industrial development in the developing countries is generally accompanied by poverty, starvation, and high rates of illiteracy. A business education is often not prized nor valued, if available at all.

While *technology* is initially welcomed because it allegedly produces a higher standard of living for the population, many host countries fail to recognize the implications of new technology on valued traditions and the behavior patterns of whole communities. Imagine a whole village suddenly commuting many miles to work in a factory, instead of working on their own individual small plots of land. Efficient organization and ever increasing productivity may not be values accepted by the natives. After all, they didn't grow up with mechanical toys, nor were they ever taught conceptual skills, elementary budgeting, and cost control. In their culture they never learned to set and achieve goals.

In many parts of the world *union/work force attitudes* differ significantly from U.S. attitudes and practices. Much of the pay and benefits that we negotiate in this country are legislated in other countries. In many instances, unions are communistic—out to destroy capitalism and all it stands for. In other situations codetermination is the norm, not the exception. Severence pay indemnities can be enormous, even to the point where two years' pay must be given to short-term employees laid off or discharged. This mandates a great deal more care in the selection process.

Political/governmental attitudes play a major role in the future of multinationals. Governments control directly many industries. They legislate imports and exports, put forth currency and exchange regulations, and impose punitive taxes while limiting the egress of earned capital. In many cases they seek top positions for local nationals, who may not be sufficiently qualified. At times they attempt to dictate where factories will be built, and on occasion favor local suppliers. In some instances bribery and corruption become a way of life. And these are only some of the problems employers face when they go multinational. Happily, they do not all occur simultaneously everywhere. Significant, however, are the organization and control relationships between the local managing director and the corporate parent. At times it is extremely difficult for the local national to understand the imposition of a U.S. domestic policy that makes no sense in the foreign environment—for example, the mandated preemployment physical in a country where physical examinations are not given.

Human resource issues are more difficult to resolve in an environment with a different set of values. But despite those different values, such resources have somewhat similar and at times even greater expectations from their employers. Whatever happens, recognize that the world is small and shrinking almost daily. The human resource professional must reach out, extend his antenna; must do environmental scanning and recognize that anything happening anywhere in the world affects all of us in some way. As for the professional human resource manager who has become internationalized, the the world is your oyster. Learn about it. It provides a most satisfying and rewarding experience.

18. Motivating Matrix Personnel: Applying Theories of Motivation

William E. Souder*

The term *motivation* is commonly used to connote something about the directedness, strength, and persistency of a person's work behaviors. For example, we think of a highly motivated person as energetic, inner-driven, productive, eager, cheerful, enthusiastic, stimulated, inspired, animated, or goal-directed. It is common to observe that one person is "highly motivated" while another is "not very motivated."

In spite of our common use of these terms, motivation is an elusive thing, not well understood. Motivation is largely a quality within the individual: we are able to see and experience only its outward manifestations. Thus it is at times difficult to explain why some employees seem more motivated than others. Individual differences in personality, emotional levels, ability, values, wants, needs, perceptions, and aspirations are only a few of the complex variables that can influence a person's motivation. Each of these variables may be influenced by the work situation, the actions of other persons, the group setting, the nature of the organizational arrangements, and the interpersonal climate.

In spite of the intangible nature of motivation, research and experience have led to some useful guidelines for motivating personnel. This chapter deals with the application of these guidelines within matrix organizations, where maintaining highly motivated personnel is essential for success.

MOTIVATION FACTORS

Figure 18-1 summarizes the key factors that influence a person's motivation. Needs are common energizers or motivators. A needy person will often be

*Dr. William E. Souder is a well-known authority in the fields of R & D management, systems analysis, and organization behavior. He teaches systems management and organization behavior and directs the Technology Management Studies Group at the University of Pittsburgh, where he is Professor of Industrial Engineering. Dr. Souder is the author of over 60 papers and two books on R & D management, engineering analysis, and organization behavior. He is chairman of the Institute of Management Science's College of R & D Management, a director of the Product Development Management Association, associate editor of *Management Science* and *Transactions on Engineering Management,* and a member of the editorial boards of several other journals.

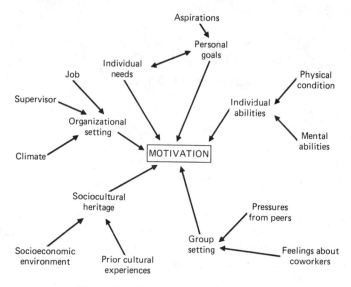

Figure 18-1. Factors influencing motivation.

motivated to seek fulfillment. Note that needs may be physiological, sociological, psychological, real, imagined, latent, or immediate. Moreover, an imagined or perceived need may be as compelling as a physical need. This fact makes it even more difficult to predict or interpret people's motives and motivations. Furthermore, a person's needs may change over time as that person matures, fulfills certain needs, acquires new tastes, or is influenced by new experiences and surroundings.

Goals and aspirations are closely interrelated with motivational processes. The goals or outcomes that a person seeks are the forces that attract or propel behaviors. Note that, in order for a goal to be a motivator, it must be personally accepted or "internalized" by that person. I can direct, encourage, or implore you to do something, but I can't *motivate* you to do it unless you truly want to. Also note that, like a satisfied need, achieved goals may not be motivators.

Intelligence, stamina, and other abilities may affect one's motives. The efforts that a person makes (or can make) to achieve his or her aspirations is often related to that person's abilities, which may thus be motivational barriers or facilitators. Also, the nature of the group setting can be a strong influence on one's motivation. Peer pressures and the felt need to "keep up with others" can induce a variety of actions and behaviors. How well one likes or dislikes coworkers and associates can also be a motivating or demotivating factor.

The culture or socioeconomic strata in which one is born and reared may strongly influence one's motivational tendencies. A person's need for achievement and sense of work ethics often can be traced to parental influence, cultural experiences, and acquired values. Similarly, the nature of the organizational setting can be either motivating or demotivating. The design of the job, the supervisor's style, the organizational climate, the opportunities for individual expression, the span of control within the organization, the challenges of the job, and the structure of the organization—all can have either motivating or demotivating influences on the individual.

MOTIVATION THEORIES

There are three important types of motivation theories: cultural theories, content theories, and process theories. Each set of theories differs in terms of its perspectives on how people are motivated and how managers can influence individual motivations.

Cultural theorists take the viewpoint that motivation is a direct consequence of one's early cultural and social experiences. For example, a cultural theorist would find connections between an individual's willingness and enthusiasm for manual labor, and such factors as his father's occupation and his parents' socioeconomic level. There seems to be some validity in the cultural theories. Differences in things that people like to do, differences in tastes and wants, and differences in value sets are often traceable to one's early socialization. However, cultural theories are less relevant today than they were a few decades ago. Today people are generally better informed and more aware of worldly events, and many traditional social barriers have been eliminated. It is not uncommon for individuals born and raised under one set of cultural imperatives to cross over into another culture, adopting and mixing the two sets of beliefs.

Content theorists believe that there are basic factors within every person that energize and direct human behaviors. Some content theorists (like Maslow) believe that these basic factors are culturally related. Other content theorists (like Herzberg) imply that the factors are universal human traits. There is an important implication for managers in these theories: to motivate personnel, a manager must appeal to a person's *basic needs.*

Process theorists believe that individuals are either motivated or demotivated by the situations that confront them. These theories focus on the processes by which behavior is energized. Some process theorists (like Vroom) view individuals as calculating thinkers who rationally assess the various alternatives available for achieving their goals, as a basis for selecting the alternative with the highest reward/cost ratio. Other process theorists (like Adams) view individuals as somewhat less economic in their orientations. These theorists see individuals as motivated by a consideration of the available rewards, the personal outlays required to achieve these rewards, and where they stand with regard to their peers in terms of various reward/outlay ratios. The implication of process theorists for managers is: to motivate people, provide environments in which they can more easily achieve their goals.

SOME CONTENT THEORIES

Maslow's Theories of Motivation

Maslow[1] has postulated a hierarchy of five needs to describe individual motivational systems, as depicted in Figure 18-2. According to Maslow, the first level of needs is physiological: the need for food, drink, shelter, and pain avoidance. The second level of needs consists of safety and security: the need for freedom from threat and outside influence. The third level comprises the needs for affiliation, friendship, interpersonal contact, social interaction, and love. The fourth level consists of the needs for self-esteem and esteem from others. The fifth and highest level of Maslow's hierarchy is self-actualization: the need to achieve self-fulfillment by maximizing the use of one's abilities and potentials.

According to this theory, individuals first attempt to satisfy the basic needs—those at the first level. When those needs are fulfilled, the individual sequentially moves on to the pursuit of the higher-level needs, and finally to the pursuit of self-actualization. The lower-level needs must be satisfied before the high-level needs begin to control one's conscious thoughts. However, a satisfied need is not a motivator. Thus people must grow, develop, and move up the hierarchy as their lower needs are satisfied, if they are to maintain their motivation. In fact, Maslow believed that individuals who were deficient in their fulfillments or blocked from moving up the hierarchy would exhibit frustration, stress, and defensive behaviors: absenteeism, withdrawal, and disagreement.

The implication of Maslow's theory for managers is: you must keep people moving ever upward toward the higher-level needs. Otherwise, managers court

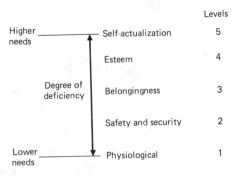

Figure 18-2. Maslow's hierarchy of needs.

the dual dangers that psychologists call "recidivism" and "substitution." In this case, individuals who feel blocked in their attempts to move up the hierarchy retreat downward, adopting lower-level need-goals and substituting them for the more appropriate higher need-goals. But since these are not their true needs, each time these lower needs are fulfilled the individuals still feel unsatisfied. The pursuit of inappropriate lower-level needs and the feelings of unfulfillment may feed on each other for long periods of time, continuing to develop such individuals into problem employees. Maslow felt that attempts to satisfy the higher-level needs would generally have a better chance of success, since they reinforce the development of higher-level skills and human development. In fact, Maslow saw self-actualization as an ever expanding need with no upper bound. Once people taste self-actualization, they will (theoretically) continue to want more of it and to grow endlessly.

Though several attempts have been made to test Maslow's concepts, most have been inconclusive, perhaps because the theory is a difficult one to test. However, these studies have reinforced the notion that most individuals do indeed have physiological needs, and that a hierarchy of human psychological needs also exist. But there is considerable individual variation: new employees may have different need structures from near-retirement employees; some individuals may have more intense psychological needs than others; etc.

Herzberg's Theory of Motivation

Based on a series of empirical research studies, Herzberg[2] identified a set of extrinsic job conditions that resulted in dissatisfied employees when these factors were not present. He called these "dissatisfiers" or "hygenics." Curiously, Herzberg found that when these factors were present they did not necessarily motivate. Rather, their presence was found to be necessary to reduce gripes and to maintain a level of "no dissatisfaction." Herzberg's "dissatisfiers" are listed in Table 18-1.

In his studies, Herzberg also identified a set of intrinsic job conditions that led to high levels of motivation and productive outputs in employees. However, he found that when these conditions were not present they did not necessarily

Table 18-1. Herzberg's Two-Factory Theory.

Dissatisfiers or Hygenics	Satisfiers or Motivators
Salary	Achievement
Job Security	Recognition
Working Conditions	Responsibility
Status	Advancement
Company Policies and Procedures	Nature of the Work Content
Quality of Technical Supervision	Potentials for Individual Growth
Quality of Interpersonal Relationships	

create dissatisfied employees, though motivation and productivity fell off precipitously. Herzberg labeled these factors the "satisfiers" or motivators, as listed in Table 18-1.

Herzberg's theory has been criticized on a number of grounds, partly because of the small samples he used in his studies and his attempts to generalize these findings to all work situations. Though his theory may seem somewhat oversimplified, it has been tested and found to be applicable in several cases. The concept of satisfiers and dissatisfiers is highly useful, though the actual satisfiers and dissatisfiers may vary with the individual, the situation, and the nature of the job.

McClelland's Theory of Motivation

McClelland[3] believes that most needs are learned or acquired from one's surroundings. He distinguishes three needs: achievement, affiliation, power. According to McClelland, the high-need-for-achievement person likes to take the responsibility for solving problems, wants to see things get done, and is very task-oriented. In contrast, the high-need-for-affiliation person is much more concerned about interacting socially with others and achieving high-quality interpersonal relationships. The high-need-for-power person focuses on exercising power and authority, and is largely concerned with influencing others and winning arguments.

The main theme of McClelland's theory is that these needs are learned through coping with one's environment. Since needs are learned, behavior that is rewarded tends to be repeated. Thus managers who are rewarded for achievement behavior learn to take risks and achieve goals. Similarly, a high need for affiliation can be traced back to a history of receiving rewards for being sociable. As a result of this learning process, according to McClelland, individuals develop unique configurations of needs that affect their behaviors and performances.

McClelland's theory has significant implications. If the needs of employees can be accurately measured, organizations can dramatically improve their selection and placement processes. For example, an employee with a high need for achievement can be placed in a position that will best enable that person to achieve. High-affiliation-need persons can be placed in jobs that best utilize their social orientations.

McClelland's theory has been subject to considerable experimentation and testing. Though achievement, affiliation, and power needs have indeed been repeatedly identified, and individuals can be characterized by such profiles, the theories seem to give a somewhat oversimplified picture of motivational processes. The concepts seem to be valid, but the theory does not go far enough in including the many other factors and variables that influence motivation. Most people can exhibit varying degrees of all three needs (achievement, affiliation, and power), depending on the circumstances and situations that confront them, on which of these needs has most recently been satisfied, or on which of these needs has most recently been threatened.

SOME PROCESS THEORIES

Vroom's Theory of Motivation

According to Vroom,[4] an individual's motivation is a function of three factors. One factor is the relative importance of the ultimate outcomes: the goals or rewards sought by the individual. Vroom calls this the value or "valency" of the outcome. A second factor has to do with whether or not the individual perceives that various events will be instrumental in leading to a high-valency outcome. Vroom calls these events "instrumentalities." A third factor is the individual's belief that exerting a particular effort will lead to the desired instrumentality. Vroom calls this factor the "expectancy." To illustrate Vroom's model, consider the example in Figure 18-3. Suppose an individual strongly desires a promotion: it has high valency. Further, suppose the individual perceives that an outstanding performance appraisal (the instrumentality) will lead to that promotion. The individual expects (the expectancy) that hard work and devotion to the job will lead to an outstanding performance appraisal. Thus, according to Vroom, people are motivated by a combination of high valencies, expectancies, and instrumentalities. According to this theory, individuals make a kind of benefit/cost calculation before exerting purposive efforts. They are motivated to the extent that the expected benefits exceed their apparent costs or the efforts required to get them.

Vroom's theory indicates that individuals will be motivated where they can see a connection between their efforts and their level of performance, and

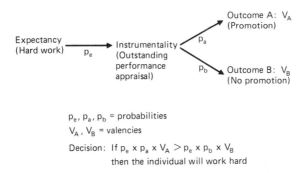

Figure 18-3. Vroom's model for a promotion situation.

where this performance leads to an outcome that has high value. Thus the implications are that motivation will be encouraged where management does three things. First of all, managers must provide instrumentalities. They must show that there are close connections between the level of performance and the outcomes. Second, management should strive to facilitate expectancies—should make it very clear to the employee that effort will indeed lead to high levels of performance and that this has high value. Third, management can facilitate motivation by showing the employee that management's own desired outcomes also have high value for the employee, i.e., that there are connections. For example, instrumentalities will be clarified when management specifies how employees can be promoted, and sets policies that provide many different alternative ways for personnel to actually be promoted. Expectancies will be facilitated when managers show by words, policies, and examples how performance leads to promotion.

Vroom's theory has been tested many times with mixed results. Though the theory seems to be valid, its greatest practical difficulty is providing valencies in organizational settings. When individual goals and organizational goals are congruent and highly valued, Vroom's theory appears to hold up rather well. For example, if personal wealth is the individual's central goal, then a corporate profit-sharing plan may be highly motivating. The theory breaks down if individual and company goals are not congruent, or where the individual's needs are satiated. In general, it is questionable whether people are as rational and calculating as Vroom's model assumes.

Adams's Theory of Motivation

Adams[5] proposes a kind of "equity" theory of motivation, in which employees make comparisons of their efforts and rewards with those of others in similar situations. According to Adams, equity exists when employees perceive that the ratio of their inputs (or efforts) to their outputs (or rewards) is equivalent to the ratios that other employees are achieving. Inequity exists when employees perceive that these ratios are not equivalent. Adams allows for the fact that there may be psychological differences, and that an individual's own ratio of outputs to inputs may be greater or less than others (in mathematical terms). What is sought is equality in the psychological interpretation of the ratios.

Adams's theory says that people will compare their ratios to others, and to various referenced persons around them. The existence of a perceived inequity will create tensions to restore equity. Individuals may then attempt to increase or decrease their outputs or their inputs. Or if neither of these actions is possible, then the individuals may stay away from the work situation, delay the work effort, or create other behavioral problems.

Adams's equity theory has been tested to some extent. The major difficulty with the theory is that individuals vary in their tolerance for inequity, and in their proclivities to take actions to restore felt inequities. When these aspects are not taken into account, the theory has low predictive powers. Of course the theory is not a complete theory of motivation: it deals only with the equity dimension.

OTHER THEORIES

Dissonance Theories

According to dissonance theories, feelings of conflict are set up within the individual whenever beliefs or assumptions are contradicted by new information.[6] This conflict produces feelings of discomfort. The individual may then attempt to soothe the discomfort in several different ways. One way is by internally reconciling the differences; another, by convincing oneself that there are no differences. A third way is to generate various defense mechanisms. for example, if a particular supervisory management style results in poor motivation, the manager might adopt a different style. On the other hand, the manager might choose to simply ignore the evidence, or to demand more evidence. In other words, dissonance explains the common human tendency to screen out unpleasant data. There is a need to make inconsistent or dissonant experiences consistent with what we already know. If workers have negative perceptions of themselves, they may actually *need* negative outcomes to achieve a consistent result. Thus work climates focusing on incompetence inspire incompetence in performances. Similarly, work climates focusing on self-competency inspire competent performances.

Self-Implementation Theories

According to self-implementation theories, the higher one's perception of one's personal competence, the more effective one's performance.[7] Supervisors who can create an environment of self-confidence will therefore increase the chances of getting competent performance from their subordinates. In short, nothing succeeds like success, and nothing fails like failure. In this context Korman identified three specific types of self-competency: the chronic, the situational, and the social. In the chronic case individuals have a generalized feeling regarding their level of competency that tends to sustain itself over time. In the situational case individuals come to feel that they are competent only in selected areas or tasks, based on their prior successes or failures. In the social case individuals develop their feelings of competency on the basis of various social contacts and what they perceive others think about them.

Theory of Personal Causation

In 1968 DeCharms[8] set forth a theory holding that the primary motivation of human beings is the need to effectively cause changes in their environment. According to this theory, most people do not want to have their lives determined for them, do not want to be manipulated, and do not want to be pawns. People value behaviors that they themselves originate, and devalue behaviors that are imposed upon them. For example, students are more likely to cherish and remember their own insights, rather than the insights a teacher spoonfeeds them in a lecture. In the same way, a supervisor who encourages employees to originate their own work and who actively listens to their ideas is likely to have higher-motivated employees.

Reinforcement Theories

It has long been known that behavior that is reinforced will tend to be repeated.[9] An application of this idea, called "behavior reinforcement," is beginning to be used for motivating employees. This idea is built around the concept of operant conditioning, which calls for the immediate positive reinforcement of desired behaviors. The closer the positive reinforcement follows on the desired behavior, the more likely it will be repeated.

In most organizations an annual pay raise is given. Thus employees generally receive their rewards long after, and unconnected with, any particular desired behaviors—just the reverse of the policy prescribed under behavior reinforcement. This is why Herzberg considers pay to be a hygiene factor.

Contingency Theories

Cultural, content, and process theories of motivation all miss one sailient aspect: the *interaction* between the individual and the situation. Experience shows that many highly motivated individuals can become very demotivated when moved to a new job, new group, or new organizational location. Similarly, a demotivated person can become a "ball of fire" when placed in a new environment. Clearly, the fit between the individual and the surroundings can be a salient factor.

Contingency theories focus on the fit between the person and the situation. Though contingency theories are presently in their formative stages of development, they promise to add enormously to our understanding of human motivation. Contingency theories believe that there are some truths in the cultural, content, and process theories, but that their validity depends on the situation and on how the particular individual interacts with that situation.

APPLYING THE THEORIES IN MATRIX ORGANIZATIONS

Matrix organizations are governed by a fundamental equation:

$$\begin{array}{c}\text{Effectiveness} \\ \text{of a Matrix} \\ \text{Organization}\end{array} = \begin{array}{c}\text{Effectiveness} \\ \text{of the Matrix} \\ \text{Structure}\end{array} + \begin{array}{c}\text{A Matrix} \\ \text{Culture}\end{array}$$

$$+ \begin{array}{c}\text{Matrix} \\ \text{Motivation}\end{array}$$

A high degree of motivation is essential in a matrix organization, where getting things done depends on the initiative, ingenuity, and team spirit of the individuals who make up that organization. Unlike classical "top-down" organizations, where individual behaviors are dictated, matrix structures require individuals to take the initiative in defining their own roles. Highly motivated individuals are therefore the key to a productive matrix organization.

Matrix organizations typically encounter seven motivational problems. Let us look at each of these problems and pose some solutions.

Motivating Professionals

Engineers and scientists desire challenging work, are often self-directing, desire some freedom to work on their own, value recognition from their peers, possess a high level of initiative and ingenuity, and like to be kept informed.[10] Thus professional personnel seem to be motivated by the work itself, by company policies related to the manner in which the work is performed, and by participation in strategic decisions about their work environment. Professionals place considerable importance on responsibility, technical achievements, and peer respect.

Two types of motivations should be considered for professional employees. First, incentives and incentive programs should be devised to promote professional growth and development. This allows for the creation of new work opportunities that enhance freedom of imagination and creativity. Second, policies should be considered that foster increased time for personally chosen research, and that increase the individual's recognition and achievement in the eyes of persons outside the organization. Motivation is enhanced when the work is challenging to the individual, when there is variety in the work assignments, and where extternal contacts are encouraged. Specifically, management should make sure every professional has at least one major opportunity every year to tell a gathering of peers, coworkers, and top management what he or she is doing, the technical mountains that need to be climbed, and how this can be donè. Management must make sure that each professional produces at least one patent, paper, or report each year in which he or she can take a personal pride of exclusive authorship or creation. And management must not only base pay and other rewards on supervisory judgments, but on peer reviews and evaluations by colleagues. Thus there are some elements of both the content and cultural theories in these prescriptions.

Managing the Project-Functional Interface

Tensions between the project and functional suborganizations are common in matrix structures. One way to avoid much tension is to use a dual performance evaluation program to make sure that employees who report to two bosses are evaluated by both. That way the employee cannot be penalized by a supervisor who says he or she has been recalcitrant in following orders, when in fact the employee was given conflicting instructions by another supervisor. Another way to defuse tensions and build confidence is to maintain a good career tracking and skills inventory for all employees. This will give the employees the comfort of knowing that their strengths are on record in a skills bank, ready to be tapped as promotions become available.[11] These types of solutions are directed at basic needs for safety and security, i.e., the lowest levels of needs in Maslow's hierarchy. As we have seen, people cannot be expected to move on to the pursuit of their higher needs, i.e., self-actualization, until these lower needs are fulfilled.

Achieving Collaboration

A matrix organization cannot be successful unless it has high levels of collaboration. This means that individuals must share a great deal of information, work together on planning, consult with each other before making decisions, and think more about the whole organizaton than they do about their own departments or functions. To achieve such collaboration, the main ingredient is trust. Individuals must learn to rely on each other and to accept each other's judgment, especially when such judgments must be based on unique competencies and knowledge. Without trust, the organization will revert to reliance on the chain of command for its authority. Under these circumstances, individuals will share less, make more unilateral decisions, and use forcing and coercion as a way of solving conflict.[12]

To develop and maintain trust, matrix personnel must be prepared to take personal risks in sharing information and in revealing views and feelings. When a few people reveal their uncertainties and feelings, others tend to open up, and eventually the entire organization becomes an open system. Reluctance to share and trust has always been one of the greatest obstacles to matrix systems.

However, unless these qualities are highly perfected in the organization, a matrix structure can never be successful.[13]

Maslow's and Herzberg's theories tell us that trust and collaboration cannot be achieved unless the personnel's basic needs have first been fulfilled. Vroom's theory tells us that we can encourage trust and openness by showing personnel how these behavior patterns can lead to benefits for them, or how their personal goals can be achieved through increased trust-based behaviors. Thus the prescription is straightforward: set an example by opening up with the employees, and tell them how they can meet their own goals through increased openness and collaboration. Fortify these statements with various tangible and intangible rewards (see below). This is in keeping with McClelland's theory and with the reinforcement theories.

Use of Incentives and Rewards

Money has often been successfully used as a motivator. However, incentive and bonus plans are invariably controversial and difficult to administer, and they may not motivate some individuals. In fact, as Herzberg, McClelland, and Adams all tell us, money may actually demotivate some individuals.

This is not to say that money should not be used. Rather, great care and skill are needed in setting up effective monetary reward systems. The key is fully understanding the varying needs and desires of the individuals within the organization. In general, it has been found that total organizational salary plans de-emphasize individual performance and generally have a leveling effect. Individual bonus plans are thus more motivating than salary plans. Bonus plans typically are related to current performance, while salary plans are usually viewed as related to past performance. However, if management is attempting to relate nonpay rewards to performance, then group and total organization plans will be much more effective. This is because personnel tend to encourage improved performance among their peers throughout the organization, if they view their rewards as dependent upon the performance of others.[14] Hence we see the relevance of the theories of Maslow, Adams, and Vroom in these results.

Experience has shown that direct feedbacks and the timely evaluation of performance motivate employees. Opportunities to learn and to grow psychologically within the job environment, and a certain autonomy that permits people to schedule their own work, have been found to motivate employees. Providing some uniqueness in each job (so that each individual can develop a certain element of uniqueness) can also be important.[15] In general, these types of nonmonetary rewards are usually more motivating than monetary reward systems.[16] This is consistent with Maslow's ideas about higher needs, Herzberg's concept of satisfiers, and the ideas behind the self-implementation theories of motivation.

Obtaining Commitment

A matrix means that more and more responsibility must be taken at lower levels. Thus greater involvement can lead to greater commitment. Therefore we would expect a matrix to have high-potential human energy. This should be felt in terms of higher individual productivity and greater organizational effectiveness. The organization should experience higher levels of control, even though individuals may be controlled less.[17] Obtaining commitment is an ideal situation for the application of Vroom's theory, dissonance theories, and self-implementation theories. If management can show employees how commitment helps them reach their goals (Vroom), then employees will be motivated to become more committed. If the project manager inspires feelings of competence (dissonance theories), then the employees will generally feel more committed. Feelings of competency tend to be contagious and feed on themselves (self-implementation theories).

Handling Paper Work Burdens

Matrix personnel, especially the engineers and technical professionals, often consider the technical aspects the most important. They often feel they spend too much time on reporting, planning, scheduling, and budgeting. These are inherent difficulties in the matrix. Over time, these feelings tend to sap the personnel's enthusiasm and motivation.

This is a clear case where the project managers and the supervisors can influence motivation through their interpersonal interactions with the employees. Paper work is an essential part of any job. By explaining the necessity and importance of paper flows, by appropriately counseling the employees, rationalizations and dissonant feelings can be reduced (dissonance theories of motivation). Paper work can be assigned in such a way that it provides new challenges and training for employees, thereby increasing their areas of competency (self-implementation theories).

Handling the Two-Boss Dilemma

Reporting to two persons, a project manager and a functional manager, need not create any special frustrations for the employee, if the theory of personal causation is applied. The essential feature is to delegate sufficient authority to the employees for them to feel they control their own destiny. This means that the project manager should be concerned only with *when* the work will get done and not *how*. How the work gets done should be left to the responsible individuals. Similarly, the functional manager must be concerned with making sure that the individuals within the matrix have sufficient supporting resources to carry out their assigned tasks.

It has been said that there may be more satisfaction for engineers in a functional organization, where engineers find solutions to detailed technical problems on their own, though the problems are assigned to them. Thus it has been claimed that their motivation for technical excellence has more opportunities in a functional or classical organization structure. However, the opportunities for self-expression need not be any less in a matrix organization, if the individuals within that organization are delegated sufficient authority to select their own ways of carrying out their assigned work.[18]

SUMMARY AND CONCLUSIONS

There are many different theories of motivation. While none is universally correct, none is necessarily incorrect. Rather, motivating employees is a matter of selecting the "right" theory, contingent upon the situation and the individual to be motivated.

In general, it may be concluded that people are motivated by anything that will help them maintain and enhance their self-esteem, i.e., by supervisory styles of interacting with them (self-implementation theories). People are generally motivated by active listening that shows understanding and respect for their feelings and ideas (theory of personal causation). People are motivated by goal-setting activities that provide them with clear, obtainable steps to improved performance (Vroom). The use of reinforcement and operant conditioning techniques can show employees the kinds of behavior that will make them successful, and can encourage such behavior until it becomes a regular practice (reinforcement theories). Employee motivations are generally enhanced when they have clear and open communications with their supervisors, and receive periodic feedbacks on their performance (dissonance theories). Some individuals may have unfilled basic needs, while others may respond to appeals to esteem, self-actualization, and other higher-level needs (Maslow and Herzberg). Some persons may respond to social opportunities; others, to opportunities to display their power (McClelland). Still other individuals may become highly demotivated when they feel they are victimized by inequities in the reward sysysttms (Adams).

Motivation is thus truly an individual matter—a function of the interpersonal relationship between the supervisor and the employee. The supervisor must know the employee well enough to interpret his or her needs and minister to them. This is all the more important in a matrix organization, where effectiveness depends on the individual initiatives and motivations of those who make up the organization.

REFERENCES

1. A. H. Maslow, "A Theory of Human Motivation," *Psychological Review,* July 1943, pp. 370–96.
2. Frederick Herzberg, B. Mausner, and B. Snyderman, *The Motivation to Work* (New York: Wiley, 1959).
3. D. C. McClelland, *The Achievement Motive* (New York: Appleton, 1953).
4. V. H. Vroom, *Work and Motivation* (New York: Wiley, 1964).
5. J. S. Adams, "Toward an Understanding of Equity," *Journal of Abnormal and Social Psychology,* Nov. 1963, pp. 422–36.
6. L. A. Festinger, *A Theory of Cognitive Dissonance* (Stanford, Calif.: Stanford University Press, 1957).
7. Alfred Korman, *Industrial and Organizational Psychology* (Englewood Cliffs, N.J.: Prentice-Hall, 1971).
8. R. DeCharms, *Personal Causation* (New York: Academic Press, 1968).
9. B. L. Rosenbaum, *How to Motivate Today's Workers* (New York: McGraw-Hill, 1982).
10. Don Pelz and F. W. Andrews, *Scientists in Organizations* (New York: Wiley, 1965); M. K. Badawy, *Developing Managerial Skills in Engineers and Scientists* (New York: Van Nostrand Reinhold, 1982).
11. "How Ebasco Makes the Matrix Work," *Business Week,* June 15, 1981, pp. 126–31.
12. S. M. Davis and P. R. Lawrence, *Matrix* (Reading, Mass.: Addison-Wesley, 1981). See also "How Ebasco Makes the Matrix Work."
13. See Davis and Lawrence.
14. William E. Souder, *Management Decision Methods for Managers of Engineering and Research* (New York: Van Nostrand Reinhold, 1980), pp. 64–78. See also Davis and Lawrence, and Badawy.
15. See Davis and Lawrence, and Souder.
16. See "How Ebasco" and Rosenbaum.
17. See Davis and Lawrence.
18. See Rosenbaum, and Davis and Lawrence.

Section IV
Implementing Matrix Management

The design of the means by which matrix management is achieved is a most challenging task. Changing the organizational structure to accommodate the matrix form is only the beginning. The common denominator of matrix success depends so much on the ability of key professionals and managers to develop the knowledge and skills necessary to provide leadership during the conversion to matrix. The attitudes these individuals reflect are important in setting an example as to how matrix is to be perceived and practiced in the organization.

This section deals with some of the major considerations involved in the design and development of strategies by which matrix management is implemented. In Chapter 19 Harold Kerzner describes some of the typical obstacles, problems, questions, and answers encountered. He draws on the experiences of many managers at different levels and organizations to present a "hands-on" approach for dealing with the organizational change set in motion by matrix management. A manager interested in preparing for the introduction of matrix management will find practical guidelines in this chapter.

Once matrix management has been introduced, the challenge of making it work on an ongoing basis concerns many executives. functional managers, matrix managers, work package managers, general managers, and so forth. The matrix works through a complex of *interfaces.* In Chapter 20 Linn C. Stuckenbruck presents an insightful look into the management of these interfaces. He examines them in terms of how to build a working communication network that functions across all the matrix organizations.

Can matrix management improve productivity in organizations? This question is examined by Stephen F. Miketic in Chapter 21. He suggests some guidelines for a matrix management approach to productivity improvement, as well as some pitfalls to avoid. Miketic concludes with some illustrations of actual use of matrix management for productivity improvement.

In Chapter 22 David Wilemon discusses the concept of matrix teams with their advantages and problems. He examines team situations where a breakdown in the team management process occurs. Then the author presents a team development model that can improve the functioning of matrix teams. He concludes by noting that technical performance is likely to be suboptimal, unless the team of specialist experts can work together effectively in the matrix ambiance.

Stephen E. Barndt's Chapter 23 describes the role of communication in the matrix organization, where decision making is less programmed than in a more traditional organization. He emphasizes the impact of heightened communication dependency on the many responsible managers of the matrix. Barndt prescribes some approaches that can lessen the upward-filtering distortion and blockage of important information in the multidimensional communication networks found in a typical matrix.

In Chapter 24 Paul C. Dinsmore looks at the pragmatic pros and cons of matrix management. He examines the common complaints about matrix and probes what lies behind them. His concluding theme that "matrix seems to work better under certain circumstances" reminds the reader that matrix is not a panacea and that it does have problems.

Chapter 25 by Panagiotis N. Fotilas deals with the introduction of semiautonomous production groups. Fotilas describes the nature of such groups and analyzes the obstacles encountered in introducing them. Then, through some actual cases, he examines strategies for dealing with these obstacles. Finally Fotilas looks at how these groups help to further humanize work and productivity.

In Chapter 26 Prakash Sethi and Nobuaki Namiki use their knowledge of Japanese management traditions and practices to suggest how consensus decision making can be applied in the matrix environment. They examine the nature and techniques of the consensus decision process and how the matrix manager can use the technique to improve team performance in a matrix.

19. Matrix Implementation: Obstacles, Problems, Questions, and Answers

Harold Kerzner*

Within the past 20 years there has been a rather well-hidden "organizational revolution" not only domestically, but even on the international scene. This revolution stemmed from the fact that commonly used organizational structures proved inadequate in responding to an ever changing environment. Simply stated, the complexities of modern business had increased to such a degree that companies were forced to search for and implement an organizational structure that could rapidly respond to any changes in the environment or marketplace.

THE EMERGENCE OF PROJECT MANAGEMENT

This organizational revolution has been characterized by the acceptance of project management as a way of life. As might be expected, the exact degree of acceptance varied with the nature of the company's business, the maturity and experience of the company's top management, and the risks that the company was willing to incur in investigating new organizational structures that are dynamic in nature.

Project management can take on many forms and shapes, such as:

- Fragmented
- Formalized matrix
- Informal matrix
- Partial project management
- Task force management[1]

---

*Dr. Harold Kerzner is Professor of Systems Management and Director of the Project/Systems Management Research Institute at Baldwin-Wallace College, Berea, Ohio. Dr. Kerzner has published over 35 engineering and business papers and six tests: *Project Management: A Systems Approach to Planning, Scheduling and Controlling; Project Management for Executives; Project Management for Bankers; Cases and Situations in Project/Systems Management; Operations Research, and Proposal Preparation and Management.*

There are many (if not infinite) versions of project management, perhaps one for each company, dating back to the late 1940s and early 1950s. It was not until the late 1950s that project management came into the limelight, and even then the growth rate was relatively slow. There were three reasons for this slow rate of acceptance:

- Executives appeared to be afraid of change and preferred the status quo.
- Many companies operated in an environment where technology changed very slowly, and therefore project management was deemed unnecessary.
- Companies that needed project management wanted to be followers rather than leaders, preferring to learn from the mistakes of others.

By 1960 only the aerospace, defense, and construction industries were willing to accept project management as a way of life. The reason for this can best be seen from Figure 19-1. Companies heavily burdened with complex tasks and operating in a highly dynamic environment were the first to accept project management as a way of life. These industries (construction, aerospace, and defense)—the pioneers in project management—were almost entirely project-driven organizations.[2]

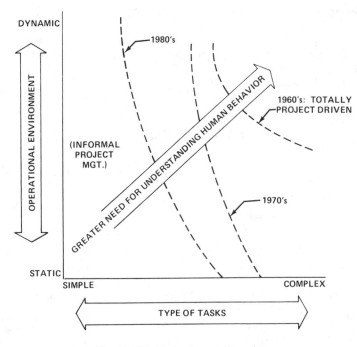

Fig. 19-1 Matrix implementation scheme.

The significant fact about these pioneering industries is that, once having realized the necessity for project management, they plunged into complete, formalized project management head first, rather than in a piecemeal manner. The majority of the companies even went so far as to implement the full-blown matrix with total cost control on the horizontal line.

As could be expected, many of the companies found it extremely difficult to implement the matrix, especially given their lack of training in interpersonal skills. Again returning to Figure 19-1, we see that the necessity for improved interpersonal skills increases as we approach the totally project-driven organizations. Unfortunately, training in interpersonal skills and communicative skills followed rather than preceded this massive organizational restructuring. Yet many companies took their lumps and survived, mainly because the employees (primarily the white-collar workers) were willing to give the new system a chance, realizing that the matrix was a necessity for staying in business.

The 1960s proved both feast and famine. In the early 1960s funding was plentiful for aerospace, construction, and defense industry contractors and subcontractors. By the late 1960s, however, much of the money for project-driven organizational subcontractors dried up, and several industries contemplating the acceptance of matrix management postponed their plans.

Yet even with the feast and famine conditions of the 1960s, more and more companies adopted the matrix approach to management. The majority of the companies following the "pioneering" industries were high technology companies performing a continuous stream of complex tasks. This can be seen in Figure 19-1. The matrix was being readily accepted as a necessity for doing business, yet few companies realized its true value.

During the 1950s and 1960s, literature on project/matrix management was scarce, and those articles that did appear were almost entirely construction, aerospace, or defense-oriented. One of the best articles to appear at this time was by C. J. Middleton.[3] The article did an excellent job of explaining the advantages and disadvantages of matrix implementation. (See Tables 19-1 and 19-2.) The article also went on to point out several common mistakes made by executives in the implementation phase. The most common mistake appeared to be the establishment of several layers of managerial supervision for both the vertical and horizontal functions. This results in an increased overhead rate, but may be cost-effective if the increased control gives profits high enough to more than compensate for the increased overhead rates.

Full implementation of project management, specifically matrix management, may take as much as two to three years. Organizational restructuring and interpersonal skills training can be accomplished in one year, but the establishment of a (computerized) project management cost and control system often requires departure from existing cost-control systems and usually brings with it the greatest resistance to change.

Table 19-1. Major Company Advantages of Project Management.

ADVANTAGES	PERCENT OF RESPONDENTS
• BETTER CONTROL OF PROJECTS	92%
• BETTER CUSTOMER RELATIONS	80%
• SHORTER PRODUCT DEVELOPMENT TIME	40%
• LOWER PROGRAM COSTS	30%
• IMPROVED QUALITY AND RELIABILITY	26%
• HIGHER PROFIT MARGINS	24%
• BETTER CONTROL OVER PROGRAM SECURITY	13%
OTHER BENEFITS	
• BETTER PROJECT VISIBILITY AND FOCUS ON RESULTS	
• IMPROVED COORDINATION AMONG COMPANY DIVISIONS DOING WORK ON THE PROJECT	
• HIGHER MORALE AND BETTER MISSION ORIENTATION FOR EMPLOYEES WORKING ON THE PROJECT	
• ACCELERATED DEVELOPMENT OF MANAGERS DUE TO BREADTH OF PROJECT RESPONSIBILITIES	

Source: C. J. Middleton, "How to set up a Project Organization," *Harvard Business Review,* 45 (March–April 1967) pp. 73–82.

By the 1970s matrix management and project management took their position center stage and the word "matrix" became common. The full advantages and disadvantages of the matrix (see Tables 19-3 and 19-4) were now becoming clear, as more and more companies began searching for ways to implement the matrix, even if fragmented.

By the mid-1970s journals were abounding with literature on matrix management and project management as identified below: "Project teams and task

Table 19-2. Major Company Disadvantages of Project Management.

DISADVANTAGES	PERCENT OF RESPONDENTS
• MORE COMPLEX INTERNAL OPERATIONS	51%
• INCONSISTENCY IN APPLICATION OF COMPANY POLICY	32%
• LOWER UTILIZATION OF PERSONNEL	13%
• HIGHER PROGRAM COSTS	13%
• MORE DIFFICULT TO MANAGE	13%
• LOWER PROFIT MARGINS	2%
OTHER DISADVANTAGES	
• TENDENCY FOR FUNCTIONAL GROUPS TO NEGLECT THEIR JOB AND LET THE PROJECT ORGANIZATION DO EVERYTHING	
• TOO MUCH SHIFTING OF PERSONNEL FROM PROJECT TO PROJECT	
• DUPLICATION OF FUNCTIONAL SKILLS IN PROJECT ORGANIZATION	

Source: C. J. Middleton, "How to Set up a Project Organization," *Harvard Business Review,* 45 (March–April 1967), pp. 73–82.

forces will become more common in tackling complexity. ... There will be more of what some people call temporary management systems as project management systems where the men who are needed to contribute to the solution meet, make their contribution, and perhaps never become a permanent member of any fixed and permanent group."[4]

Articles such as this made it clear that project management was a viable tool to achieve complex corporate objectives, by simply defining a project as a group of people thrown together temporarily to achieve a corporate goal or objective. The literature identified the project manager as the "leader" of the project team, even though the team could include employees of a higher rank or pay grade than the project managers.

One of the best articles to appear at this time was by William Goggin:[5]

Although Dow Corning was a healthy corporation in 1967, it showed difficulties that troubled many of us in top management. These symptoms were, and still are, common ones in U.S. business and have been described countless times in reports, audits, articles and speeches.

Our symptoms took such forms as:

● Executives did not have adequate financial information and control of their operations. Marketing managers, for example, did not know how much it cost to produce a product. Prices and margins were set by division managers.

● Cumbersome communications Channels existed between key functions, especially manufacturing and marketing.

● In the face of stiffening competition, the corporation remained too internalized in its thinking and organizational structure. It was insufficiently oriented to the outside world.

● Lack of communications between divisions not only created the antithesis of a corporate team effort but also was wasteful of a precious resource—people.

● Long-range corporate planning was sporadic and superficial; this was leading to over-staffing, duplicated effort and inefficiency.

Goggin was typical of the new breed of executive that emerged in the late 1960s and early 1970s, and that was characterized by sufficient flexibility to accept change and adapt to it. This adaptation required departure from the traditional business organization form, which was basically vertical and achieved successful unification of efforts through a strong superior-subordinate relationship.

Goggin then went on to describe the advantages that Dow Corning expected to gain from the multidimensional organization:[5]

● Higher profit generation even in an industry (silicones) price-squeezed by competition. (Much of our favorable profit picture seems due to a better overall understanding and practice of expense controls throughout the company.)

• Increased competitive ability based on technological innovation and product quality without a sacrifice in profitability.

• Sound, fast decision making at all levels in the organization, facilitated by stratified but open channels of communications, and by a totally participative working environment.

• A healthy and effective balance of authority among the businesses, functions, and areas.

• Progress in developing short and long-range planning with the support of all employees.

• Resource allocations that are proportional to expected results.

• More stimulating and effective on-the-job training.

• Accountability that is more closely related to responsibility and authority.

• Results that are visible and measurable.

• More top management time for long-range planning and less need to become involved in day-to-day operations.

Goggin's article was the first to identify the effects on the total corporation of implementing a matrix, and provided good indications that the matrix could be multidimensional and multinational. The matrix was here to stay.

ORGANIZATIONAL FORMS FOR A MATRIX

As mentioned previously, the 1970s brought with them a variety of matrix structures. The most common structure was the formalized matrix structure as used in project-driven organizations, shown in Figure 19-2. There are two key points. First, many projects cannot be managed by one person alone and thus the project office emerged. The project office consisted of one or more assistant project managers who had the right to communicate directly with the customers. The people in the project office could be there either full time or part time. Full time was preferred because of the dedication and commitment expected of project office personnel.

Almost all project offices had a project engineer who was responsible for integrating all engineering work on the project. The project manager was chiefly busy controlling time and cost, whereas the project engineer's responsibility appeared to be primarily technical performance. The "marriage" of the project manager with the project engineer seemed to be the best of both worlds.

The second feature seen in Figure 19-2 was the creation of the position of Director: project managers. This position was usually filled by either a division manager, a director, or a vice-president. The importance of this position was

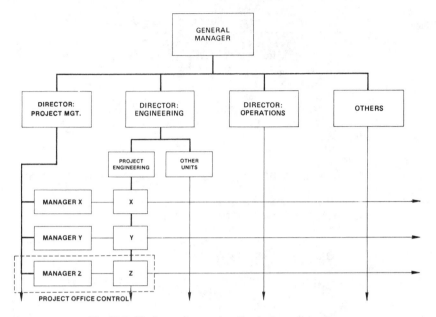

Fig. 19-2. Placing project engineering in the project office.

that all project managers now reported to one functional group. The manager of project managers now became a very powerful position. This manager was responsible for

- Selecting and assigning project managers.
- Working with executives to establish uniform company project priorities and making sure that this information was transmitted to all project managers.
- Assisting project team members in the resolution of project conflicts.

Because of these responsibilities, the manager of project managers was often viewed as being the most powerful executive in the company, since he or she appeared to be directing the operations of the entire company. This held true primarily for project-driven organizations.

Most non-project-driven (and some project-driven) organizations accepted fragmented project management in which only one department or division would implement project management. For example, it was quite common for R & D to have a matrix, whereas all other functional lines performed vertically. Another example occurred in manufacturing divisions, where fragmented matrices exist for capital equipment projects. In many companies the fragmented matrix became the "seed" for the full, formalized matrix. Thus when one Fortune 500 Corporation met with tremendous success in using frag-

Table 19-3. Advantages of a Pure Matrix Organizational Form.

- THE PROJECT MANAGER MAINTAINS MAXIMUM PROJECT CONTROL OVER ALL RESOURCES, INCLUDING COST AND PERSONNEL.
- POLICIES AND PROCEDURES CAN BE SET UP INDEPENDENTLY FOR EACH PROJECT, PROVIDED THAT THEY DO NOT CONTRADICT COMPANY POLICIES AND PROCEDURES.
- THE PROJECT MANAGER HAS THE AUTHORITY TO COMMIT COMPANY RESOURCES, PROVIDED THAT SCHEDULING DOES NOT CAUSE CONFLICTS WITH OTHER PROJECTS.
- RAPID RESPONSES ARE POSSIBLE TO CHANGE CONFLICT RESOLUTION AND PROJECT NEEDS.
- THE FUNCTIONAL ORGANIZATIONS EXIST PRIMARILY AS SUPPORT FOR THE PROJECT.
- EACH PERSON HAS A "HOME" AFTER PROJECT COMPLETION. PEOPLE ARE MORE SUSCEPTIBLE TO MOTIVATION AND END-ITEM IDENTIFICATION. EACH PERSON CAN BE SHOWN A CAREER PATH.
- BECAUSE KEY PEOPLE CAN BE SHARED, PROGRAM COST IS MINIMIZED. PEOPLE CAN WORK ON A VARIETY OF PROBLEMS (I.E., BETTER PEOPLE CONTROL).
- A STRONG TECHNICAL BASE CAN BE DEVELOPED, AND MUCH MORE TIME CAN BE DEVOTED TO COMPLEX PROBLEM-SOLVING. KNOWLEDGE IS AVAILABLE TO ALL PROJECTS ON AN EQUAL BASIS.
- CONFLICTS ARE MINIMAL, AND THOSE REQUIRING HIERARCHICAL REFERRAL ARE MORE EASILY RESOLVED.
- BETTER BALANCE BETWEEN TIME, COST, AND PERFORMANCE.

mented project management for short-term facilities projects, the matrix was expanded to include large-facility expansion programs and ultimately became effective for ongoing business activities.

There is one modification to the fragmented matrix that is now being used by several electronics companies. Each department within the company has its own departmental matrix. When the company decides either to undertake an in-house project or to bid competitively for external funding, one of the departments is selected to head the project. The project manager then has the right to treat each of the other departments as "internal subcontractors." If the project manager wishes to look for subcontractors externally, he can do that also, if the external price is lower. This technique is designed to stimulate competition among departments and lower costs. Although competition is generally good, in this type of situation detrimental effects to the total company can result, if each line organization operates independently and keeps the best information to itself. This is shown in Figure 19-3, where the total company looks like a composition of small organizational islands. This type of structure does not lend itself to the systems approach to management, as long-term goals may be difficult to achieve.

Many companies wanted project management but did not feel that all

Table 19-4. Disadvantages of a Pure Matrix Organizational Form.

- COMPANY-WIDE, THE ORGANIZATIONAL STRUCTURE IS NOT COST-EFFEC-TIVE BECAUSE MORE PEOPLE THAN NECESSARY ARE REQUIRED, PRIMAR-ILY ADMINISTRATIVE.
- EACH PROJECT ORGANIZATION OPERATES INDEPENDENTLY. CARE MUST BE TAKEN THAT DUPLICATION OF EFFORTS DOES NOT OCCUR.
- MORE EFFORT AND TIME IS NEEDED INITIALLY TO DEFINE POLICIES AND PROCEDURES.
- FUNCTIONAL MANAGERS MAY BE BIASED ACCORDING TO THEIR OWN SET OF PRIORITIES.
- ALTHOUGH RAPID RESPONSE TIME IS POSSIBLE FOR INDIVIDUAL PROB-LEM RESOLUTION, MATRIX RESPONSE TIME IS SLOW, ESPECIALLY ON FAST-MOVING PROJECTS.
- BALANCE OF POWER BETWEEN FUNCTIONAL AND PROJECT ORGANIZA-TIONS MUST BE WATCHED.
- BALANCE OF TIME, COST, AND PERFORMANCE MUST BE MONITORED.

aspects of it were necessary for their organization. Elaborate planning and control systems, the continuous need for paper work, and the delegation of authority to the project manager were the major obstacles to implementation. Companies solved this problem by implementing informal project management. The key to such management rests with the department managers, who must be willing to

- Work with the project managers, who have not been delegated a great amount of authority.
- Accept the project managers' informal planning and informal requests.
- Communicate functional decisions and true project status to the project managers, without having to write complex reports.
- Work closely and honestly with other functional managers.

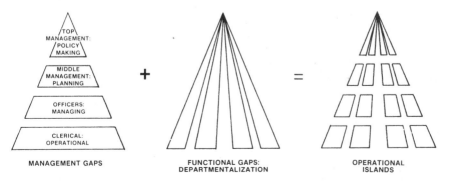

Figure 19-3. Why are systems necessary?

This last item requires that all functional managers escalate problems to the surface and be willing to ask other functional managers for assistance. To the best of the author's knowledge, only non-project-driven organizations have successfully implemented informal project management.

In the late 1970s the task force concept of matrix management came into existence. One of the major advantages of the matrix is that program cost is minimized by sharing key people on several projects. Unfortunately, customers who were giving large dollar-value contracts were insistent that key people be assigned full time. As one executive remarked: "Why the hell shouldn't I want those key people full time? What's another two million in salaries compared to six hundred million, when the integrity and quality of the project is at stake?"

Today, task force management is project management where the key people are assigned full time to the project. Task force management is easily married to matrix management, because the task force project appears as simply one more horizontal line on the matrix. Some companies have created separate vertical lines or divisions for large task force projects. This can work well if, and only if, all functional employees are assigned full time. Care must be taken, however, that the disadvantages of pure product management do not occur.[6]

SELLING EXECUTIVES ON PROJECT MANAGEMENT

With the turbulent beginning of the 1980s, more and more executives of companies functioning in a dynamic environment felt the need of better control over resources and thus the need of selected project management (see Figure 19-1). Many small companies tried to implement a matrix, but soon found that the implementation problems for the small company were much more difficult. There were several reasons for this.

- Small companies generally cannot afford the luxury of maintaining a line function of full-time project managers.
- Therefore in the small company functional managers and employees are often required to wear "two hats." This creates a severe conflict because rewards are received vertically, not horizontally.
- Small companies generally have more projects (i.e., perhaps customers) than large companies, but these are smaller in nature. Project managers are thus required to handle multiple small projects where each can be of a different priority. This again produces conflict within the organization.
- Project managers in small companies have limited resources to select from and therefore have to perform with what is available. Because of this, the project manager has a greater need for interpersonal skills.
- Executive meddling is more pronounced in the small company, because the company may not be able to withstand the failure of even one project. Thus communication channels (to the top) are generally shorter.

- Project offices are virtually nonexistent in the small company, and much greater pressure is placed upon the project manager to be a jack-of-all-trades.

Because of these conditions small companies have had much more success with informal and partial project management than with a matrix.

Executives of the 1980s have been much more inquisitive concerning matrix implementation than their predecessors, even though the advantages and disadvantages have been extensively published in the literature. The remainder of this chapter deals with questions that executives have asked the author. For the most part, the questions were asked in a closed session, because quite often the executives wished functional employees to hear neither the question nor the answer. In each case the executives were contemplating a change to a matrix organizational structure. When reading the questions, remember that executives of the 1980s are operating under greater pressure and with more risk and uncertainty than the executives of either the 1960s or 1970s, and therefore have had to be "sold" on the project management approach.[7]

Can our people be part-time project managers? The question suggests that the executives wanted to manage the projects within the existing resource base of the company. The answer to the question depends, of course, upon the size, nature, and complexity of the project. It is generally better to have a full-time project manager responsible for several small projects than to have many part-time project managers. Executives, as well as functional personnel, will never be convinced that the matrix works until they see it in action. Therefore it is strongly recommended that the first few projects be "breakthrough" projects with full-time project managers. This implies that, initially, these project managers will be staff to a top-level manager rather than within a newly developed line group for project managers.

If we go to project matrix management, must we increase resources, especially the number of project managers? The reason for this question is obvious: the executives do not wish to increase the manpower base or overhead rate. Matrix management is designed to get better control of functional resources so that more work can be completed in less time, with less money and with potentially less people. Unfortunately, these results may not become evident for a year or two.

Initially, executives prefer to select project managers from within the organization, arguing that project managers must know the people and the operation of the organization. Even today some companies require that all project managers first spend at least 18 months in the functional areas prior to becoming project managers. However, there are also good reasons for filling project management positions from outside the company. Sometimes newly transferred project managers still maintain ties to their former functional department, so that impartial project decision making is not possible.

Let's assume that we set up a separate staff function called project administration, which is staff to one of our executives. Can we then use our functional people as part-time project managers reporting vertically to a line manager and horizontally to project administration? With proper preparation and training, most employees can learn how to report effectively to multiple managers. However, the process is more complicated if the employee acts as a project manager and functional employee at the same time. When a conflict occurs over what is best for the horizontal or vertical line, the employee will usually bend in the direction that will put more pay in his pocket. In other words, if the part-time (or perhaps even full-time) project manager always makes decisions in the best interest of his line manager, the project will suffer. The most practical solution is to let the functional employee act as a part-time assistant project manager rather than as a part-time project manager, because now the functional employee has someone else to plead his case for him and is no longer caught in the middle.

The preceding question has serious impact on how employees are treated. If an employee reports to multiple managers and some managers treat him like Theory Y while others treat him like Theory X, decisions will almost always be made in favor of the Theory Y managers. People who report to multiple managers must understand that, even if they are Theory Y employees, in time of crisis they will be treated as though they were Theory X. This type of understanding and training must be given to all employees performing in a project environment.

Which vice-president should be responsible for the project administration function? Assuming that the company does not want to create a separate position for the Vice-President of Projects, we must find out whether or not a dominant percentage of people (on all projects) come from one major functional group. If, say, 60 to 70% of all project employees come from engineering, then the Vice-President for Engineering should also control the project administration function, because then there will exist a common superior for resolving the majority of project conflicts. Having to go up two or three levels of management to find a common superior for conflict resolution can create a self-defeating attitude within the matrix.

The assignment will become more difficult, if functional dominance does not exist. We must now decide who dominates the decision-making process of the company (is the company marketing-driven, engineering-driven, etc.?). The project administrative function will then fall under the control of this line function. Without either of these degrees of dominance and assuming a project-driven organization, it is not uncommon to find all project managers reporting under marketing with the Vice-President for Marketing acting as the project manager.

Is it true that most project managers consider their next step to be a vice-president? Most project managers view the organization of the company with the project managers on top and executives performing horizontally. (Rotate Figure 19-2 to the right by 90 degrees.) Therefore project managers already consider themselves executives on the project and naturally expect their next step to be executives in the total company.

However, we should mention that many project managers are so in love with their jobs that money is not an important factor, and they may wish to stay in project management. Project managers are self-motivated by work challenge and therefore many have refused top-level promotions because they did not consider the work to be as challenging.

Can we give our employees (especially engineers) a rotation period of six to eighteen months in the project office and then return them to the functional departments, where they should be more well-rounded individuals with a better appreciation and understanding of project management? On paper, this technique looks good and may have some merit. But in the real world the results may be disastrous. There are four detrimental effects. First, employees who know they will be returning to their line function will not be dedicated to project management and will maintain a strong allegiance to their line function. As a result, of course, the project will suffer. Second, when employees know the assignment is temporary and brief, they usually walk the straight and narrow path and avoid risk whenever possible. Risky decisions are left to other project office personnel or even their replacements. Third, upon the rate at which technology changes, employees may find themselves technically obsolete upon returning to the functional group. But the fourth and last point is the most serious. Employees may find themselves so attracted to project management that they want to stay. If the company forces them to return to their functional departments, there is always the risk that they will update their resumes and begin reading the job market section of the Tuesday *Wall Street Journal*. Simply stated, a company should not put people into project management unless willing to offer them a career path there.

How much control should a project manager have over costs and budgets? Executives in accounting and finance are very reluctant to delegate total cost control to the project managers. Project managers cannot be effective unless they have the right to control costs by opening and closing work orders in accordance with the established project plan. However, if the project manager redirects the project activities in a manner that causes a major deviation in the cash flow position of the project, then that manager must coordinate activities with top management in order to prevent a potential company cash flow problem.

What role should a project manager have in strategic and operational planning? First of all, project managers are concerned primarily with the immediate execution of an operational plan, therefore they are operational planners. However, because of company-wide knowledge obtained during functional operations and integration, the project manager becomes an invaluable asset to the executives during strategic planning, but primarily as a resource person. Project managers are not known for their strategic project-planning capability.

How much authority should an executive delegate to a project manager? Generally, project managers should be given as much authority as they actually need (or perhaps slightly more) to get the job accomplished. Unfortunately, this is not often the case. The key factors that management considers in the delegation of authority include:

- The maturity of the project management function
- The size, nature, and business base of the company
- The size and nature of the project
- The life cycle of the project
- The capabilities of management at all levels

The first item indicates that during matrix implementation the project manager may have less authority than anticipated, because the executive is reluctant to give up total control. This is to be expected. The last four items indicate that even after the matrix becomes mature, not all project managers are equal when it comes to authority. The following list identifies the types of authority that can be delegated to or withheld from the project manager:

- Focal position
- Conflict resolution between the project manager and functional managers
- Influence to cut across functional and organizational lines
- Participation in major management and technical decisions
- Collaborating in staffing the project
- Control over allocation and expenditure of funds
- Selection of subcontractors
- Rights in resolving conflicts
- Voice in maintaining integrity of project team
- Establishing project plans
- Providing a cost-effective information system for control
- Providing leadership in the preparation of operational requirements
- Maintaining prime customer liaison and contact
- Promoting technological and managerial improvements
- Establishing the project organization for the duration
- Cutting red tape

The exact amount and type of authority depend on the mood of the executive. On high-risk projects the project manager expects to have more authority, but usually has less because the executive (i.e., project sponsor) appears to be calling the shots.

What working relationships should exist between executives and the project manager? The answer to this question involves two parts: internal meddling and customer communications. Executives are expected to work closely with the project manager and take an active role during the conceptual and planning stages of a project. However, after the project begins the implementation phase, active participation by executives becomes executive meddling and can do more harm than good. After planning is completed, executives should step back and let the project manager run the show. There will stlll be structured feedback from the project office to the executive, and the executive will still be actively involved in priority setting and conflict resolution. The exception to this would be when the executive is required to act as the "project sponsor," as shown in Figure 19-4. In this case, the client wants to be sure that the project is receiving executive attention, and feels confident on seeing one of the contractor's executives looking over the project. The project sponsor exists primarily as the executive-client contact line, but can also serve as an invaluable staff resource.

Executives must not be blinded by the partial success they may achieve through executive meddling during the early days of matrix implementation. The overall, long-term effect to the company could be disastrous, if executives feel they can effectively contrlol both vertical and horizontal resources at the same time.

Where do we find good project managers? First of all, project management is both an art and a science. The science aspect includes the quantitative tools and techniques for planning, scheduling, and controlling. The art aspect involves dealing with a wide variety of people. The science portion can be learned in the classroom, whereas the art portion can come only from on-the-job experience. Perhaps the most important characteristics are interpersonal skills and communicative skills.

Experience is usually the best teacher in project management. Figure 19-5, for example, shows the job description of a construction project manager. This implies that construction project managers will probably be much older than

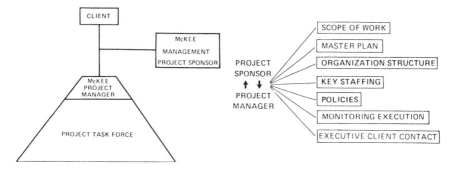

Fig. 19-4

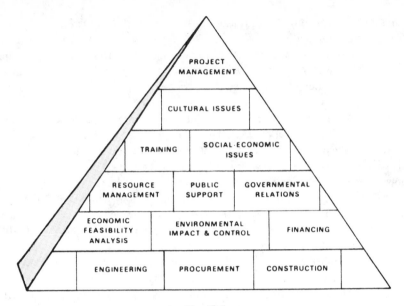

Fig. 19-5

their MIS project management counterparts. In project management, experience rather than gray hair or baldness equates to maturity.

Most companies have such qualified people internally and often produce disastrous results by "forcing" such people to unwillingly accept a project management assignment. Project management generally works best if it is a voluntary assignment. This usually brings with it loyalty and dedication. Unfortunately, many people enter project management without fully understanding the job description of the project manager. If the employees are promoted into project management and then "want out" or fail, the company may have no place for them at their new salary. Sometimes it is is better to laterally transfer employees to project management under the stipulation that rewards will follow if they produce.

What percentage of a total project budget should be available for project management and administrative support? The answer depends upon the nature of the project. Management support may run from a low of 2% to a high of 15%.

My company has 50 projects going on at once. The project managers handle multiple projects, each with a different priority, and can report to anyone in the company. Will a matrix give better control? The matrix will alleviate a lot of these problems, provided that all the project managers report to one line group. This will give uniform control of projects and will make it easier to establish priorities. If it becomes necessary to get better control over the project managers, then the projects should be grouped according to the customer or to similar technologies, not necessarily dollar value.

In a matrix, people are often assigned full time to a project. What happens if a functional manager complains that pulling a good employee out of his department will leave a large gap? In a matrix the employee is still physically and administratively attached to his or her functional group. And even with a full-time project assignment, the employee will probably still find sufficient slack time to assist another project, even if only in a consultant capacity.

On some of our projects the first step is a cost-benefit analysis, to see if the project is feasible. Who will do this in a matrix? On some projects, the job-related characteristics are more important than the project manager's personal characteristics. In this case, it may be better to have project managers who are trained in this area, rather than having the cost-benefit analysis performed by another group. Project managers should be actively involved in any planning or decision making that may be bottom line–oriented.

How do we make sure that everyone in the company knows what the priorities are? Priorities should be transmitted to both the project and functional departments through the traditional structure within the matrix. Even with the establishment of priorities, project managers will still fight for what they believe is in the best interest of their project. This is to be expected. Initially, during the implementation of the matrix, it may be necessary to have all priorities documented.

There is a risk within the matrix that the slippage of even one project could cause alteration of all other project priorities. Even though some project managers may control their project so closely that they can obtain daily status, continuously changing priorities on a daily or even weekly basis can destroy the functioning of the matrix, because the functional managers may now be forced to continuously shift resources from project to project.

We have had an explosion of Operations Support Systems (the mini-computer era). How do we manage these projects? Can we use matrix management? Matrix management works best for projects cutting across more than one functional group. Multifunctional MIS and data base packages can be very effectively managed by a matrix. Banks are a prime example of industries where matrix management may exist primarily for such projects.

One major risk should be considered. A controversy exists over whether the programmers or users should be the project managers. The usual arguments are that the programmers don't understand the users' needs, and the users don't understand why it takes so long to write a program. Many companies have established a project management group to handle such conflicts. Each project is headed by a project manager and two assistant project managers, one from programming and one to represent the users. Conflicts and problems are now resolved horizontally rather than vertically. In this situation it is possible for one project manager to handle several projects at once.

How does top management control the responsibilities that each person will have on a project? Neither top management nor project management controls the responsibilities. The functional managers still control their own people. Project managers can fill out a linear responsibility chart (LRC) as shown in Figure 19-6, to make sure that every work breakdown structure element is accounted for. However, the functional managers should still approve the amount of authority and responsibility that the project manager wishes to delegate to functional employees, since the project manager should not be able to upgrade functional employees without the consent of the functional manager. The exception would be the project office personnel that may report full time to the project manager and also be evaluated by the letter. During the implementation phase of a matrix, the executives may wish to be actively involved in the LRC establishment since it is part of the planning process, during which executives are expected to be closely associated with the project.

How do we ensure effective and timely communications to all levels? The project manager, being the focal point for all project activities, should be able

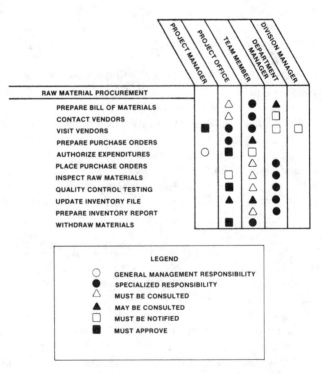

Fig. 19-6 Linear responsibility chart.

to provide timely project information to everyone, including executives, at a faster rate than the traditional structure itself. The ability to provide effective and timely communications should be part of every project manager's job description.

How do we get top management committed to project management? Regardless of how much literature exists in the area of effective project management, executives will not become committed until they see the system operating effectively and producing the expected dollar value of profit on the project's bottom line. To effectively observe and comprehend the problems, executives must understand their new role in a project management environment and should attend the same "therapy" training sessions as middle management.

We need an awful lot of front-end work (i.e., planning) on projects. We are living in a world of limited resources. We need commitments from our people, not just promises. How do we get them? When the functional managers realize that project management is designed for them, and not for executives or project managers, then the functional managers will start giving commitments that they will live up to. The functional managers must be convinced that the matrix is not simply an attempt on the part of the project manager to control their empires, but that the project manager and matrix exist to support the functional managers in getting better control over their own resources, so that future commitments can be kept.

How do we resolve problems in which project team members are not fully informed about their own roles? The responsibility here rests with both the project and the functional managers. Planning tools such as the linear responsibility charts can be used, but the "bottom line" is still effective communication. This is why one of the major prerequisites for a project manager is to be an effective communicator and integrator.

How do we convince people to escalate problems and not bury them? In a matrix organization the critical point is the project/functional interface. Both the project and the functional managers must be willing to escalate problems and ask for help, especially on the horizontal line. When the project manager gets into trouble, he goes first to the functional manager to discuss project resources. When the functional manager gets into trouble, he goes to the project manager to seek additional time, additional funding, or a change in specifications. Project personnel must realize that the project is a team effort and that everyone should pitch in when problems occur.

Many people refuse to escalate problems for fear that the identification of the problem will be reflected in their evaluation for promotion. The matrix structure is designed not only to put forth the best team for accomplishing the objectives, but also to resolve problems. Because the matrix approach encourages the sharing of key people, employees may find that the best corporate resources may now be available to assist them temporarily. Executives and functional managers must encourage people to bring forth problems, especially during matrix implementation. This encouragement should probably be done verbally with personal contact, rather than through memos.

Is it true that if we go to a matrix, many of our functional people will start communicating directly with our customers? When you have a matrix structure, customers are very reluctant to have all information flow from your project office to their project office, for fear that your project office is filtering the information. Therefore the customer may request (or even demand) that his technical people be permitted to talk to your technical people on a one-on-one basis. This should be permitted as long as the customer fully understands that

- Functional employees reflect their own personal opinion. Official company position can come only through the project office or the project sponsor.
- Functional employees cannot commit the company to additional work, which may be beyond the scope of the contract. Any changes in work must be approved by the project office.

Functionally, employees should contact the project office after each communication and relate to the project office what was discussed. The project office will then consider whether or not a memo should be written to document the results of the discussion.

The purpose of the question-answer session is to convince the executives that a change might be for the better. Unfortunately, with matrix management the executives themselves may have to change their own management styles. The following are the most common arguments that executives give for avoiding change:[8]

- *Why change?* I must be doing something right to get where I am. I may have to start working differently. Can I succeed?
- *Balance of Power.* I understand the balance of power and my role within top management. Why change it? I might lose my present power.
- *Loss of Control.* I presently generate change on projects and in policy areas. Why change? I won't be able to control recommended changes.
- *Need for contact with projects.* I'll lose my ability to perceive appropriate adjustments in organization policies when I lose detail involvement in projects. Why change?
- *Excessive delegation.* It isn't good practice to have key decisions delegated below the top men. Why change?
- *Coordination.* Coordination responsibility is a key management job. Why delegate it to project managers?

If the executives are willing to accept change, then the next step is to discuss methods for implementation. The executives must understand the following strategies and tactics, for implementation to be effective:

- Top management must delegate authority and responsibility to the project manager.
- Top management must delegate total cost control to the project manager.
- Top management must rely upon the project manager for total project planning and scheduling.
- Only the project managers must fully understand advanced scheduling techniques as PERT/CPM. This may require additional training. Functional managers may use other scheduling techniques for resource control.
- Top management must encourage functional managers to resolve problems and conflicts at the lowest organizational levels and not always run "upstairs."
- Top management must not consider functional departments as merely support groups for a project. Functional departments still control the company resources and, contrary to popular belief, the project managers actually work for the functional managers, not vice versa.
- Top management must provide sufficient training for functional employees on how to report and interact with multiple project managers.
- Top management must take an interest in how project management works.
- Top management must not fight among themselves as to who should control the project management function.

The project manager also has strategies and tactics that must be understood during implementation. The following key points should be carefully considered by the project managers:[9]

- Start with a "breakthrough project" that the administration can keep pace with in the new project management format.
- The new project manager must be sure that traditional types of functional and project information are available to top management for traditional problem solving. He should take this information forward voluntarily, ahead of top management's knowledge of the problem, preferably more quickly than the traditional line of communication.
- Allow every administrator to retain his traditional power within the hierarchy during the implementation phase.
- Project managers should carefully and thoughtfully develop only policy

recommendations that can be easily accepted by the administration as being in concert with the organization's goals and objectives, and that are easy to implement and are readily accepted by those outside the organization.

- Project managers must push for change, but not at a rate that in itself builds opposition.
- Project managers should keep schedules and other tools in the background of their involvement with top management. (The tools of project management are of far less interest to top management than the results obtained through them.)
- It is extremely important that project managers decode all their reporting documents to meet the style of the executive with whom they are trying to communicate.
- Project managers should be sure to recommend as general policy changes only those items applicable to a broad range of projects. Exceptions should be clearly indicated as meeting clearly defined project objectives.

It should be readily apparent from these key points that during implementation the project managers could easily frighten executives to such a degree that all thoughts of matrix implementation will be forgotten.

The last points to be described to executives concern "blockages" that some executives face even after the implementation phase is completed.

- Top management directly interfaces a project only during the idea development and planning phases of a project. Once the project is initiated, the executives should maintain a monitoring perspective via structures feedback from the project manager.
- Top management still establishes corporate direction and must make sure that the project managers fully understand the meaning.
- Top management must try to control those environmental factors that may be beyond the control of the project manager. These factors include such items as external communications, joint venture relationships, providing internal support, and providing ongoing environmental intelligence.
- Top management must have confidence in the project managers and must be willing to give them projects that are difficult as well as easy to perform.
- Top management must understand that in order for work to flow horizontally in a company, a "dynamic" organizational structure is necessary. Not all activities can flow parallel with the main activities of the company.
- Upper-level management must not try to take an active role in this "new" concept called project management.
- Upper-level management must be familiar with their new responsibilities and interface relationships in a project environment.

There is no sure-fire method today for the successful acceptance and implementation of matrix management. The best approach appears to be an early education process (including questions and answers) to persuade executives, project managers, and functional personnel to at least give the system a chance. This type of approach should be acceptable to perhaps all types of companies in all industries where the matrix is applicable.

REFERENCES

1. Further definitions of these structures will be provided later in the chapter.
2. The terminology should be clarified. A project-driven organization is one where each horizontal project has its own profit and loss statement and is considered a separate horizontal cost center within the company. In the project-driven organization the functional units exist solely to support the projects, and corporate profitability is measured on the horizontal rather than the vertical line. In the non-project-driven organization (such as manufacturing), profits are measured on the vertical line and the projects exist to support the functional departments.
3. C. J. Middleton, "How to Set Up a Project Organization," *Harvard Business Review* 45 (March–April 1967): 73–82. Copyright © 1967 by the President and Fellows of Harvard College. All rights reserved.
4. "Business Says It Can Handle Bigness," *Business Week* Oct. 17, 1970, p. 115.
5. William C. Goggin, "How the Multidimensional Structure Works at Dow Corning," *Harvard Business Review,* Jan.–Feb. 1974, pp. 54–65. Copyright © 1973 by the President and Fellows of Harvard College. All rights reserved.
6. Organizations with vertical task forces look like pure product management. For the disadvantages to this, see Harold Kerzner, *Project Management: A Systems Approach to Planning, Scheduling and Controlling* (New York. Van Nostrand Reinhold, 1979), p. 51.
7. The conceptual phase of project management implementation begins with the functional managers, who identify the need for project management because of their problems with resource control. The next step, therefore, is to sell top management on the concept. This is usually best accomplished through outside consultants whom executives trust to give an impartial view.
8. Adapted from John M. Tettemer, "Keeping Your Boss Happy While Implementing Project Management—A Management View," *Proceedings of the Tenth Annual Seminar/Symposium of the Project Management Institute,* Oct. 8–11, 1978, Los Angeles, pp. Ia-1 through IA-4. Reproduced by permission of the Project Management Institute.
9. Ibid.

20. Interface Management, or Making the Matrix Work

Linn C. Stuckenbruck*

INTRODUCTION

The matrix is a unique organizational form. Its most unique aspect is its complexity when compared to a simple hierarchical organization. The matrix provides great challenges in the planning and control processes because of the number of people that must be consulted. Since the matrix is a project organization, or some other type of organization, superimposed upon a conventional functional organization, the complexities of its reporting relationships are readily apparent. In addition, the maxtrix has many very complex nonreporting people interfaces, which will be one of the principal subjects discussed in this chapter.

The matrix is a complex organizational form that can become extremely complicated in very large projects. The matrix is complex because it evolved to meet the needs of our increasingly complex society with its very large problems and projects. The conventional hierarchical functional management structure usually finds itself in difficulty when dealing with large, complex projects. The pure project organization is a solution when the project is very large, but it is not always applicable to smaller projects. Therefore management, to obtain the advantages of both project and functional organizational forms, has evolved the matrix. But the matrix is not for everyone. It should be utilized only if its advantages outweigh the resulting organizational complexity.

The type of organization involved is not important; the complexity of the resulting matrix is essentially the same. As pointed out by Cleland and King, many different types of organizations may be involved in matrix situations.[1] The organizations can have operations as diverse as the construction of an office building and manufacturing toys. Also, projects vary in size from the development of a major aerospace system to a small research project. Yet in spite of this variation most projects have one thing in common: the utilization

---

*Dr. Linn C. Stuckenbruck is with the Institute of Safety and Systems Management at the University of Southern California, where he teaches project management and other management courses. Prior to this he spent 17 years with the Rocketdyne Division of Rockwell International, where he held various management and project management positions. He holds a PhD. from the State University of Iowa and has recently published a book, *Implementing Project Management—The Professionals Handbook*, with Addison-Wesley Publishing Company.

of some form of the matrix organization. It has also been pointed out that the matrix may have more than two dimensions: a three-dimensional matrix occurs when an organization such as a multinational company superimposes an additional geographically oriented organization upon a two-dimensional matrix.[2] Then the reporting relationships become even more complex.

This chapter will primarily consider the two-dimensional matrix, which is the common organizational form and exhibits most of the problems of complexity. How does one manage a matrix? What is so complex about having two bosses? Is the project manager or the functional manager the "real" boss? Does an employee in a matrix really have two bosses, or is one of them the "real" boss? It has been suggested that the functional manager is usually considered to be the "real" boss by most matrix employees, since the functional department is the employee's home to which he or she returns at the completion of the project.[3] If so, how does the project manager get the job done? In any case the real authority, responsibility, and accountability of both managers and of the employees become quite blurred. However, the project can be successfully completed in spite of the added complexity of a multiple command system.

Cleland and King suggest that the role of the two managers can be clarified and their specific responsibilities defined.[4] The project manager is responsible for the *what, when, why,* and/or *how much* of the project, and the functional manager for the *how, where,* and *who.* In practice this proves somewhat simplistic. What good project manager will fail to get deeply involved in the *how, where,* and *who* of the project? Particularly, *who* is going to work on the project? After all, the project manager wants only the best people. Likewise, what good functional manager will fail to get deeply involved in the *what, when, why,* and *for how much* of the project? Particularly regarding *for how much,* the functional manager has to make the basic cost estimates and then perform the promised cost.

It becomes evident that the matrix works in spite of unclear and ambiguous authority, responsibility, and accountability roles. In studying various matrix organizations, the author has observed that there appears to be a strong correlation between the successful operations of the matrix and the degreee of harmony within the organization. Perhaps the reporting relationships are not nearly as important to the successful functioning of the matrix as are the personal relationships among the participants or players in the game of matrix.

SYSTEMS INTEGRATION

Every manager has the job of creating within his or her organization an environment that will facilitate the accomplishment of its objectives.[5] This axiom is certainly true in the matrix organization, whether the emphasis is on a discipline, a project, a product, or all of these at the same time. In addition, every manager is responsible for the managerial functions of planning, organizing, staffing, directing, and controlling. Every manager has these functions, whether they are in a matrix organization or not. Is the project manager any different? For the project manager one can substitute program, product, area, or other type of manager involved in a matrix. How then does the job of project (or other) manager differ from that of the functional, line, or discipline manager?

The project management concept is based on vesting in a single individual the sole authority for the planning, resource allocation, direction, and control for the accomplishment of a single time- and budget-limited enterprise. But this statement does not indicate any major difference between the job of the project manager and that of the line or discipline manager. What does make a difference is the complexity of most projects, and the project manager's necessary preoccupation with the intergration of his or her project.

As already indicated, projects vary greatly in complexity. However, all but the simplest projects have a common element: they must be integrated. Integration can be defined as the project manager's most important responsibility: ensuring that a particular system or activity is assembled so that all the subsystems, components, parts, and organizational units fit together according to plan as a correctly functioning, integrated whole. Of course all levels of management have this goal, but the project manager must be preoccupied with it and has the direct responsibility to ensure that it occurs. This job of integration is most important and most difficult when the project is organized in the matrix mode. The project manager's problem is that of interface management, and what he or she does to solve the problem can be described by the more general term *systems integration.*

Why is systems integration difficult in the matrix organization? What is so different about the matrix? Since the matrix is such a complex organizational form, all decisions and actions of the project manager become very difficult, primarily because the manager must constantly communicate and interact with many functional managers. The project manager discovers that the matrix organization is inherently a conflict situation. The matrix brings out the presence of conflicting project and functional goals and objectives. In addition, the project manager finds that many established functional managers contributing to the project feel threatened; stresses result that eventually lead to conflict.

Integration doesn't just happen; it must be made to happen by carefully planning, and by designing it into the system. It is more than just fitting components together: the system has to function as a whole. Integration cannot be an afterthought, nor does it consist only of actions that can be accomplished after the subsystems are completed. The critical actions leading to integration must take place very early in the life cycle of the project, particularly during the implementation and planning phase, to ensure that integration takes place. In a "pure" project organization, there is no question as to who initiates these actions, since the project manager runs his or her own empire. In a matrix organization, however, the project manger encounters particular difficulties and problems in carrying out the function of integration.

The matrix organization has evolved to cope with the conflict inherent in any large organization—the need of specialization versus the needs of coordination.[6] These divergent needs in the hierarchical organizational structure lead to inevitable conflict between functional and top management, and often to nonoptimizing decisions, since all major decisions must be made by top management, who may have insufficient information. The matrix organization evolved naturally out of the need for someone who could work problems through the experts and specialists. The project manager has assumed the role of "decision broker" charged with the difficult job of solving problems though experts (functional mangers and their specialists), all of whom know more about their particular field than the project manager.

Recognizing that the matrix is a complex organizational form is just the first step. The next step is getting this complex organization to function. Its successful operation, like that of any organization, depends almost entirely on the activities of the various people involved. In a matrix, however, the important activities are concentrated at the interfaces between the various organizational units. There are many such interfaces, although two stand out as probably the most important. These are the interfaces between the project manager and top management, and between the project manager and the functional managers working on the project.

INTERFACE MANAGEMENT

Matrix organizations will not automatically work, and an endless number of things can go wrong. However, most matrix problems occur at the interfaces between the project manager and his functional managers, or between two other people involved on the project. The project manager must effectively work across these interfaces in order to accomplish the integrative function. Strictly speaking, the project manager cannot be said to manage these interfaces because one or more of the people involved do not work for the manager. The manager must manage the interfaces in spite of severe handicaps.

The project manager carries out systems integration primarily by carefully managing all the many diverse interfaces within the project. According to Archibald, "the basic concept of interface management is that the project manager plans and controls (manages) the points of interaction between various elements of the project, the product, and the organizations involved." He defines interface management as identifying, documenting, scheduling, communicating, and monitoring interfaces related to both the product and the project.[7]

The complexity resulting from the use of a matrix organization gives the project manager many more organizational and project interfaces to manage.

These interfaces are a problem for the project manager, since whatever the obstacles encountered, they are usually the result of two organizational units going in different directions. An old management cliché says that all the really difficult problems occur at organizational interfaces. The problem is complicated by the fact that the organizational units are usually not under the project manager's direct management, and some of the important interfaces may even occur outside the company or enterprise.

PEOPLE INTERFACES

There are many types of interfaces within any organization, and many more within a matrix. Archibald lists three interfaces that the project manager must continually monitor for potential problems: personal or people interfaces, organizational interfaces, and system or product interfaces.[7] Project management is more than just management interfaces; it involves all three types.

People interfaces involve the people within the organization who are working to carry out the project. Whenever two people are working on the same project, there is a potential for personal problems and even conflict. If the people are both within the same line or discipline organization, the project manager has limited authority, but he or she can demand that the line supervision resolve the personal problem or conflict. If the people are not in the same line or discipline organization, the project manager becomes a mediator, with the ultimate alternative of insisting that line management resolve the problem or remove one or both of the individuals from the project team. Personal interface problems become even more troublesome and difficult to solve when they involve two or more managers.

Organizational interfaces are the most troublesome, since they involve not only people but also varied organizational goals, and conflicting managerial styles and aspirations. Each organizational unit has its own objectives, its own disciplines or specialties, and its own functions. As a result, each unit has its own jargon, often difficult for other groups to understand or appreciate. Thus misunderstanding and conflict can easily occur at the interfaces. These interfaces are more than purely management interfaces, since much day-to-day contact is at the working level. Pure management interfaces exist when important management decisions, approvals, or other actions will affect the project. Organizational interfaces also involve units outside the immediate company or project organizations such as the customer, subcontractors, or other contractors on the same or related systems.

System interfaces are the product, hardware, facility, construction, or other types of nonpeople interfaces inherent in the system being developed or constructed by the project. These will be interfaces between the various subsystems in the project. The problem is intensified because the various subsystems will usually be developed by different organizational units. As pointed out by Archibald,[7] these system interfaces can be actual physical interfaces existing between interconnecting parts of the system, or performance interfaces existing between various subsystems or components of the system. System interfaces may actually be scheduled milestones involving the transmission of information developed in one task to another task by a specific time, or the completion of a subsystem on schedule.

The critical interfaces in the matrix are between people—the personal or people interfaces. People interfaces can best be thought of as communication links. Whenever there are communication links, there is usually a certain amount of resistance that blocks or limits communication. This resistance acts like an attenuator or filter in limiting communication. Communication blocks can be caused by a number of factors such as:

1. Different perceptions of the goals and objectives of the project
2. Different perceptions of the goals and objectives of the individual organizational units involved in the project
3. Competition for facilities, equipment, materials, manpower, and other project resources
4. Personal antagonisms or actual personality conflicts among personnel on the project
5. Resistance to change on the part of people working on the project

Why are people interfaces so important, particularly in the matrix? In the matrix getting the job done depends less on a authoritarian direction than upon personal persuasion and personal contacts. So the key is people—getting people to work together. In other words, getting the interface to work. People interfaces don't automatically work; they must be carefully nurtured and encouraged to function.

There are more important people interfaces in the matrix than in a simple hierarchical organization. A simple matrix is shown in Figure 20-1 as it is usually depicted, based on reporting relationships. This organization chart based on reporting relationships ignores not only the unofficial organization but the people interfaces, many of which may be more important to getting the job done than the reporting relationships. Even in a simple two-dimensional matrix the people interfaces become very complex; the complexity of the interfaces in a three- or four-dimensional matrix would be almost indescribable. Figure 20-2 attempts to graphically depict the multitude of people interfaces. The double-ended arrows represent an interface between two people. If all the people interfaces were put on the figure it would be impossibly busy, with arrows criss-crossing and running in every direction. However, the figure can be kept relatively simple by depicting primarily the various types of people interfaces found

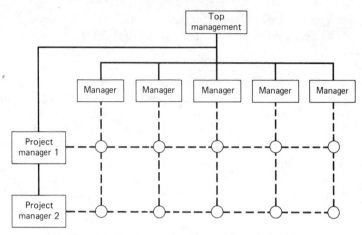

Fig. 20-1. Matrix organization reporting relationships.

in a matrix organization. As indicated in Figure 20-2, there are eight types of people interfaces:

1. Between top management and a project manager
2. Between top management and a line manager
3. Between two line managers
4. Between a line manager and a project manager
5. Between two project managers
6. Between a project manager and a worker (specialist)

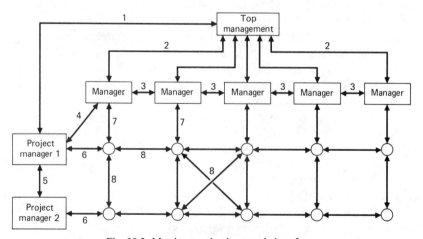

Fig. 20-2. Matrix organization people interfaces.

7. Between a line manager and a worker (specialist)
8. Between two workers (specialists)

The first type of interface, between top management and a project manager, is extremely important to the project manager. A good working relationship with ready access to top management is essential for resolving major problems, removing obstacles, and setting priorities. The project manager who doesn't work very hard to cement his or her relationships with top management will probably be a project manager only once. It is very easy for a project manager to spend all available time keeping the project on the track, thus neglecting the top management interface. The interface between top management and line management is also important but usually functions very well, since this is a normal hierarchical reporting relationship. If a conflict develops, top management merely finds someone more compatible to their views.

The interface between any two line or functional managers is a little more of a problem. If such an interface has poor communications or exhibits real or potential conflict, it may be a long time until top managements learns of it or anything about it. Such nonfunctioning interfaces—i.e., between two managers who do not communicate, but do not reach the point of actual conflict—might remain undetected for years. It is often entirely up to the project manager to detect and try to resolve such nonfunctioning interfaces for the good of the project.

Probably the most critical interface in a matrix organization is that between the project manager and the line or function managers. This is a difficult interface and a naturally adversarial situation because these two managers do not necessarily work for the same boss, and because their roles, goals, and objectives are so different. However, practically every project decision and action must be negotiated across this interface. This interface relationship can be one of smooth-working cooperation or of bitter conflict, depending on the personalities of the respective managers. Unfortunately for project managers, they can accomplish little unless this interface works very well indeed; they must depend on the cooperation and support of the line managers. The key to success in the matrix lies in the successful functioning of this interface. Arbitrary or one-sided decisions on the part of either the project manager or the line manager can only lead to poor performance and/or potential conflict. Cooperation and negotiation, as initiated by the project manager, will lead to project success.

MAKING INTERFACES WORK

Making interfaces work or managing the interface is more than an exercise in group dynamics; it is a problem in communications. Obtaining good communications involves getting people to talk to each other, and removing barriers to communication. With the help of top management, the project manager can do five things to ensure that these two important goals are accomplished:

1. Clarify organizational goals and objectives
2. Clarify individual roles and responsibilities
3. Reduce conflict
4. Provide a climate conducive to good communications
5. Provide a channel for feedback

Clarifying Organizational Goals and Objectives

Everyone working on the project must be made aware and appreciative of the overall project goals and objectives. Organizational goals and objectives should originally be stated in the project charter and other project planning documents. The *what, why, when, how, who,* and *where* of the project should be clearly stated in project documents. However, even though these goals and objectives are clearly stated, the project manager and top management must continually be on the lookout for people who do not understand or who choose to misinterpret goals and objectives.

Line and project goals and objectives are often completely contradictory, and cause much needed negotiation. In fact, it would be strange if they were not contradictory, since line and project organizations exist for different purposes. The line manager's principal goals may be to maintain his or her empire and to develop maximum quality products. The project manager's goal has to be to get the job done within performance specifications, and within budget and schedule limitations. People working in a matrix have to learn to live with contradictory goals and objectives. Both project and line management must work to achieve harmony in spite of these conflicting goals and objectives. The matrix organization can be said to deliberately utilize conflict to get a better job done. Teamwork and problem solving must be emphasized, rather than worrying about who solves the problem.

Clarifying Individual Roles and Responsibilities

As the project manager must be made aware of his or her authority, responsibility, and accountability, likewise *everyone* else on the project team must also be made aware of their roles and responsibilities. As minor as their roles may be, everyone on the project must know what is expected of them. No one should be left hanging, which will inevitably lead to insecurity, friction, and potential conflict. Such a tool as linear responsibility charting may be very useful in clarifying individual roles and responsibilities, particularly among managers.

Reducing Conflict

Whenever there are two managers, each with certain prerogatives of power, there is a potential adversary climate and potential conflict. Whether conflict actually develops depends on the circumstances and people involved. However, conflict is most likely to develop between project managers and the line managers with whom they are negotiating for the use of people, services, and other resources. Conflict between various project managers may also develop over priorities and allocation of resources. However, conflict can develop almost anywhere in an organization and for almost any reason.

What causes conflict? Conflict in a matrix can be of two types: conflict with the organization, and conflict between individuals. By definition, organization describes and limits personal behavior. If the demands of the system are not compatible with the capacity and desires of the individual, the latter is likely to experience frustration, which can lead to low job satisfaction, poor job performance, and often conflict.

One of the problems in the matrix is the presence of multiple chains of command, which can lead to inconsistant, contradictory, and sometimes very conflicting demands on the individual. For example, project personnel will recieve instructions to develop the best possible product, plus instructions to do it in two weeks with no overtime. Conflict inevitably results. The conflict may be primarily inside the person but it will certainly lead to frustration, and often to "foot-dragging" and conflict with the system.

Conflict between individuals may be due to differences in age, maturity, personality, background, education, and the ability to communicate. Poor communication, however, is certainly the basic cause of most misunderstandings and conflicts. This may be due either to the inability or reluctance of one or both persons to express themselves, or to their inability to understand the situation or see the other side of the picture. In either case a problem exists that must be resolved by some member of mamagement. Such personal conflicts are much more likely to occur in a matrix organization, because of the diversity of the people involved and the number of contacts to be made. Quite often the project manager is the one person able to see both sides of the conflict, and so must resolve it for the good of the project. Thus the project manager becomes the conflict resolver.

There are many ways to resolve conflict; however, the route usually selected is an appeal to higher authority. Someone, usually a common superior, is called upon to resolve the conflict. What happens when one of the parties is a project manager working in a matrix organization? Going to a common superior is a valid alternative, but it is a step that project managers are often reluctant to take, usually feeling that the resolution of conflicts is part of their job of systems integration. Going to top management too often with problems may not only wear out your welcome, but may be interpreted as a sign that your project is in trouble.

Conflict resolution in a matrix organization is increasingly more difficult as the matrix becomes more complex, and authorities and responsibilities become indistinct. It becomes harder to determine who exactly is the common superior to appeal to.

Project managers must be continually aware of potential and real conflict situations, and intercede whenever necessary. Most conflict can be resolved if the people involved can be persuaded to talk to each other. However, talking is not necessarily communication; it may only be arguing, which perpetuates the conflict. True communication is a two-way street that involves first listening and then understanding on the part of both parties. But before that, people must be induced to talk to each other.

Providing a Climate Conducive to Good Communications

How do you get people to talk to each other? You could take the people involved and lock them in a room and not let them out until they have talked out all their problems. Of course, this is not very practical in the real world. Basically, the challenge facing management is to provide a climate conducive to good communications at all levels in the organization. It's not just a problem in group dynamics, it involves management encouragement of a team project and the removal of barriers. What specifically can the project manager and top management do to provide a climate conducive to good communications? Two obvious answers should be considered immediately: (1) encourage discussions by keeping an open-door policy, and (2) put the people who should be communicating geographically close together.

Management can do a great deal to create an open, nonsecretive atmosphere by encouraging an open-door policy. Anyone wanting to see the boss should be able to do so without too long a wait. This includes the project manager, who is in every sense one of the bosses.

One of the most effective methods for getting people to talk to each other is to put them in close physical proximity during their working hours. This ploy is definitely applicable to managers; just put the managers who must be in constant communications as close together as possible. But even more important are project teams. If it is critical that the project team members be in constant communication, put them all together—in the same room, if possible. One company's solution to this problem is shown in Figure 20-3. Not only are the project teams together in separate rooms, but the project managers and the line managers are close together. Probably the most important aspect of this matrix arrangement is that the project managers and the line managers are close to each other and will be the most likely to communicate. As projects

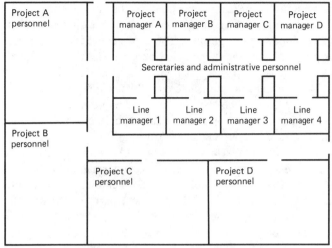

Fig. 20-3. A tight matrix office arrangements.

come and go, the individuals will be constantly changing, but the physical arrangement can be maintained and maximum communication assured.

There are many other ways by which management can encourage communication in their organization. One of the more controversial approaches is meetings. Much has been written about the ineffectiveness of meetings, but all meetings do not have to be bad. Meetings can take many forms, from unscheduled, informal meetings in the hall to formal project review meetings. All of them have their place as long as they accomplish their purpose: getting the right people together at the right time to communicate. Meetings should be scheduled whenever more communication is needed. Likewise management should recognize that informal meetings are usually more effective than formal meetings and so should be encouraged.

Providing a Channel for Feedback

Feedback is necessary to ensure that all project team members are aware of how their behavior is perceived by others. Feedback is a two-way street in that it is equally important for the project personnel to know how they stand, as it is for the project manager to know what is happening with the project. Both channels of feedback must be maintained. First, it is important that the project manager perceive what is going on in the project. The project manager must provide a channel of feedback by (1) formalizing official lines of communication through progress reports and project performance review meetings, and

(2) encouraging informal communication lines back to the project office. Only personal contacts aggressively maintained can assure that the informal communication channels are kept open. The project manager must maintain an open-door policy and arrange for regular personal status review meetings with all project personnel.

Next, it is important that project personnel know how they stand with regard to their project performance. To neglect this important feedback would be a mistake on the part of the project manager; the project personnel must be continually made aware of how they are performing.

Also important is the relationship of conflict to communications. It must be recognized that disagreement and conflict are a normal and indispensable part of an organization—particularly of a matrix. To eliminate disagreement would be to eliminate the vital feedback loop so necessary for the operation of the matrix. If no one disagrees, the boss never finds out what is going wrong in the organization. The project manager must be willing to accept disagreement and seriously consider what its consequences will be to project performance, and must be able to separate true feedback from mere static.

CONCLUSION

The matrix organizational form proliferated widely in the 1970s and will become even more important in the 1980s. The matrix has gone far beyond its original application with project management; it is now a permanent part of such organizations as multinational companies. In any case, the matrix is unique in regard to the number and complexity of the interfaces involved. It takes two people to create a problem. Whenever there are two people there is an interface, and in the matrix there are a great number of such people interfaces.

Interface management is the particular form of management necessary to ensure smooth operation of a matrix organization with its multiple command system. Interface management is the job of the project manager, or other matrix manager, to assure that there is a working communication network functioning across all the interfaces in the organization. If the communication systems work, the matrix will work.

REFERENCES

1. David I. Cleland and William R. King, *Systems Analysis and Project Management,* 2nd ed. (New York: McGraw-Hill Book Co., 1975), p. 183.
2. Stanley M. Davis and Paul R. Lawrence, *Matrix* (Reading, Mass.: Addison-Wesley Publishing Co., 1977), p. 45.
3. L. C. Stuckenbruck, ed., *The Implementation of Project Management—The Professional Handbook* (Reading, Mass.: Addison-Wesley Publishing Co., 1981), p. 83.

4. Cleland and King, p. 237.

5. Harold Koontz and Cyril O'Donnell, *Principles of Management: An Analysis of Managerial Functions* (New York: McGraw-Hill Book Co., 1972), p. 46.

6. Leonard R. Sayles, "Matrix Management: The Structure with a Future," *Organizational Dynamics,* Autumn 1976, pp. 2–17.

7. Russel D. Archibald, *Managing High Technology Programs and Projects* (New York: John Wiley & Sons, 1977), p. 66.

21. Matrix Management for Productivity

Stephen F. Miketic*

Today American industry is facing its first real challenge in over two hundred years—reaffirming leadership in productivity growth. This challenge is very critical because the "window for action" is time-limited, and international competition has momentum and is fierce. This chapter of the *Handbook* places this challenge in perspective. Then it addresses how matrix management can be used as a key approach to achieving a substantial quantum improvement in productivity—the real payoff. Some guidelines are suggested for a matrix management approach to productivity improvement, as well as some pitfalls to avoid. While some of the data repeats other chapters, here it is applied to a real-world business environment. The chapter concludes with several illustrations of actual use of matrix management for productivity improvement.

THE PRODUCTIVITY CHALLENGE IN PERSPECTIVE

Background

In the last two decades, substantial progress has been made in introducing into American industry computer-controlled machinery, processes, and controls. Unfortunately the 3–4% annual increase in productivity growth rate in the 1960s slowed in the 1970s. In the late 1970s the rate of productivity growth was minimal and even negative. Current statistics indicate a very small 1–2% growth rate for the start of the 1980s Concurrently the world leaders in productivity growth rate, Germany and Japan, are experiencing growth rates respectively of about 4% and 7% annually. It is forecast that the productivity of the Japanese worker will reach the level of the American worker before the end of the decade. As an example, it is openly discussed today that the Amer-

---

*Stephen F. Miketic's many years of in-depth management experience range from operating small plants to managing major international businesses and ultimately to serving as Vice-President of Manufacturing at Westinghouse Electric Corporation. Mr. Miketic has since founded and is now President of the MS Management Services Company, a consulting firm specializing in all phases of management operations, including strategies for managing productivity. Currently one of his key interests is promoting, planning, and hopefully implementing America's first really automated factory of the future.

ican automobile industry is facing a $1500/car disadvantage, because of the recognized productivity and lower cost of the Japanese worker over his counterpart in the United States. It is obvious that we have not taken advantage of the improved manufacturing technology, management controls, and human resources available to us.

Of great significance is the aging of America's factories. The latest addition of the *Economic Handbook of the Machine Tool Industry* indicates that one-third of the machine tools in use are over 20 years old. Two-thirds are more than ten years old, and only 3% are N/C or computer-controlled.

Even more fundamental, perhaps, are some observations of industry made in recent years. Machine tools generally cut metal less than 15% of the time and are operated for one but less than two shifts per day, five days per week on the average. Seldom does a worker operate more than a single piece of equipment. Manufacturing cycle efficiency, defined as the percentage of time a part is worked on compared with its total time in the plant to completion, is extremely low in job shops—on the order of 5% or less. Few machine tools are available for use 90% of the time or better because of breakdown, lack of material for parts production, or operator absence. Gauging generally is not built into the processing cycle but handled as a separate operation. Less and less skilled operators are entering industry. Apprentice training has all but disappeared from most industries. Factories with one or several N/C type tools find it difficult to attract and retain the programming skills so necessary to make the sophisticated equipment function efficiently. Only recently has programming become more simplified and the machine tool industry become aware of supplying loading devices to improve the cutting time of high-cost machining centers.

In addition, little focused attention has been given to reducing inventory costs in American industry. The Japanese "KANBAN" theory of producing for today only what is needed is a manufacturing philosophy that only now is being recognized. It encompasses not only material control but also process controls. An example is the normally accepted practice in Japan that no die change in stamping, punching, or forming should exceed a few minutes. Major stamping presses in the Japanese automobile industry disturb production for less than three minutes to effect a die changeover. Contrast this with 30 minutes to several hours in many American industries.

Another significant factor is machine tool and process maintenance. The Japanese in recent years have advanced greatly, achieving higher levels of performance than experienced in America. For example, machine tool availabilities in excess of 95% are now routine. It is my personal observation, resulting from several trips to review Japanese industry, that the Japanese factory employs less than one-half the machinery maintenance personnel used in comparable American plants. Yet few domestic plants experience the high reliability of equipment of Japanese factories.

Much has been written on the extensive use of robots in foreign industry as compared to domestic applications. Suffice it to say that robotics technology in America is as advanced, if not more so, than anywhere in the world, but currently is only slowly being exploited for its huge potential by American industry. Welding and painting applications, and a small number of material handling applications in hot or toxic environments, constitute the bulk of robot use in America today. The explosion of robot applications to processing cells of all types and to assembly operations offers real opportunities for major productivity increases.

Productivity Improvement: The Japanese Approach

It is interesting to note that the progress just cited for foreign industries has been achieved more by technology transfer than new technology development. The Japanese, particularly, have become experts in accumulating knowledge of world technologies and in implementing state-of-the-art technology in an efficient manner. Much has been written of the Japanese approach of thoroughly discussing ideas in depth throughout all levels of the organization. One Japanese executive referred to this process as letting information "soak" down through the organization. This seemingly slow, matrix-type approach to decision making has been so perfected that in recent years it has actually shortened the total time needed from idea conception to implementation. Many American executives are realizing that Western-world decision making may be speedier than that experienced in Japan, but that implementation generally is much slower. Initial communication of ideas in American industry usually encompasses only a limited number of people in the organization, as contrasted with the "get everyone involved" Japanese approach. This latter approach evolves into the significant asset of *commitment* so often referred to as a big element of Japanese productivity.

In putting the productivity issue into perspective, it is important to note the adversary attitude of industry, labor, government, and academia in America. This contrasts with the cooperative attitude displayed particularly in Japan, and to lesser degrees in the more productive European nations. Significant also is the commitment to a high-quality product, as contrasted to the American approach of not having enough time or money to design and build the product right the first time, but having time and money available to "fix" poorly designed and manufactured products in the long term.

Additionally, the Japanese labor force is a well-trained one. For each day of training of an employee in American industry, his or her counterpart in Japan receives at least ten days annually of formal training, and in some businesses about twenty. Intercompany movement is almost nonexistent in Japan, so training is well received and "pays off" because of a long-term stable work force.

It is significant to recognize the training and quality of the management structure in Japan as contrasted with American industry. Exposure to many functions is a way of life for the rising management employee in Japan. For about ten years the young professional expects and is content to work in a variety of functional responsibilities. When a promotion to an executive level takes place, this accumulated expertise serves well in the decision-making process. In American industry the rise to the executive level is generally less structured and training much less comprehensive. Too many executives in American industry have detailed experience in only one or at best two functions and can be classified as generalists. Thus decision making is based more on instinct and less on perceiving in depth the real issues of the situation evaluated.

Productivity: Office Automation Potential

The preceding data has focused on shop operations and a few operating philosophies. The great improvements in productivity achieved in the last decade by foreign industries were based on the above factors. A very vulnerable area worldwide is the productivity of the salaried work force associated with all industries. This area has hardly begun to be exploited anywhere in the world. While American investment in industry per shop employees averages about $30,000 per individual, the investment per office employee is about $2500. Yet the ratio of shop to office employees has changed greatly in the past two decades. This ratio of shop to office employees of 3:1 and 2:1 is rapidly moving to 1:1. In addition, the total cost of wages and benefits paid to office employees in large companies is about 1 to 1.5 times those of all shop employees. Studies indicate that shop employees in American industry physically work about 5.5 to 6.0 hours per eight-hour day, while office employees work about 4.5 to 5.0 hours. While these are harsh statistics, they do indicate that opportunities for productivity improvement in both shop and office areas are immense. Indeed, the automated and electronically oriented factory of the future so often referred to today is not a fantasy, but a real opportunity for improving productivity several magnitudes.

Productivity: Challenge to American Industry

The challenge before American industry is best realized by a brief review of what has happened in recent years to some of our basic industries. No longer is our steel industry an example for the world. Much more technologically advanced mills and associated productivity improvement are evident in Japanese and Korean mills. The latter now are evaluated as the best in the world. In the last decade we have lost the consumer electronics business to the Far East and may well lose our lead position in semiconductor technologies. The

Japanese have in the last decade licensed semiconductor technology and are now designing integrated circuits of their own.

While the machine tool industry has been a key factor in American industrial supremacy in this century, in less than 20 years the Japanese have risen to an extremely tough competitive position, threatening the very existence of this basic industy. The most advanced machine tool plants in the world today do not exist in America, but in Japan. A frightful example is the tremendous growth of the Japanese automobile business. Today that industry is challenging the large United States automobile firms and proving to be a tough competitor.

Will this competitive edge be tempered in the future? On the contrary, it is predicted that competition will become more intense. Two examples should suffice. In the past several years a consortium of Japanese industry and government has contributed manpower and money to design a small automated factory capable of parts production through a series of laser, welding, and assembly processes. This is planned as a pilot batch type factory that can machine and assemble a preselected family of geared parts. Another example is the ten-year development project aimed at designing the next generation of major computers, based on yet-to-be-invented concepts.

Productivity in Perspective

In summary the Japanese, through an organized program, have put to use the world's state-of-the-art technologies in many key industries. Aggresssiveness and commitment to detail have resulted in their building high-quality products at lower costs than world competition. They have developed a base of engineering excellence in their factories and have long recognized the need for cultivating all the human resources of a business for maximizing productivity. Today Japan is organizing to move into new technology development and already has demonstrated significant strength in this area. For example, their genetic research is outstanding.

In placing productivity in perspective this discussion has centered on the Japanese, since that nation has achieved the greatest productivity growth rate in recent years. It is significant to note that many other nations—Germany, France, and Italy, just to name a few—also are becoming significant challengers to American's past industrial superiority. Is the battle for productivity lost? In my estimation the answer is a resounding, "No, not yet!" The window for action is limited in time and demands organized actions over the remainder of the decade.

Evolution of America's Productivity Status

Let us quickly summarize the current status of America's productivity effort. As previously stated, the American worker is still the most productive in the world. However, current trends indicate that Japan and perhaps Germany may reach an equivalent level by the late 1980s. America's technology still leads the world, but other nations are now developing strong technology bases in key industries. The strong adversary relationship of government, industry, labor, and academia in America has been very detrimental to producitivity improvement. However, this is now of great concern and being openly discussed by all sectors of the economy. Suffice it to say that a change is forthcoming and the adversary attitude is slowly changing to one of cooperation. Thus recent changes in tax laws will make more capital available and so should stimulate new plant investments and research and development, two critical factors for improved productivity.

A lack of professional talent in America's factories has been evident for a decade or more. Young professional engineers have not found factory positions attractive from a recognition or compensation base. As a consequence, manufacturing engineering courses in universities have all but disappeared, replaced by computer science programs. This need is now recognized and action is being taken. However, a talent gap in manufacturing engineering will continue to exist for many years, and available talent will be taxed very heavily. Some of this gap will be made up by talent in universities being diverted from pure research and development projects to application projects for industry. Examples are the robotics and manufacturing process automation projects now being carried on at the Massachusetts Institute of Technology and Stanford and Carnegie-Mellon universities. Closer cooperation of academia and industry is coming to fruition very rapidly.

Just as the Japanese in the past decade have descended upon American industry to study its capabilities and successes, so American teams are now traveling the world. We are learning that investments in capital equipment and research and development can and must be supplemented by a fuller use of the human resources of each company. This latter approach has been a key factor in the Japaneses effort for productivity improvement.

Summary

Currently American industry is losing the battle for productivity. Plants are old and growing older; two-thirds of the equipment is over ten years old, and only 3% is computer-controlled. The quality of products and services is not equivalent to foreign manufacturers in many product lines. Salaries and wages are among the highest in the world and have exceeded productivity gains substantially in the last decade. Industry is beset with regulations from all levels of government, adding excessive cost burdens, though many regulations benefit very little, if at all, any segment of society.

On the positive side, American workers are still the most productive in the world, despite the recent trends in productivity growth rate. They are generally well educated and resourceful, and desire to use their knowledge to improve productivity. American industry still has technological leadership in key product areas, and our country is rich in natural resources. Our ability to innovate is outstanding. However, we need to recognize that the "time window" for action to improve productivity by orders of magnitudes is limited. Yet crisis management is not warranted, for it is neither creative nor productive.

To get "back on the track" now demands quick and correct short-term decisions to maintain the viability of many industries. We need to plan actions to "buy time," so as to get our house in order for the long pull. The patient is in critical condition but needs to survive in the near term to take advantage of longer-term opportunities. Recognizing our assets and deficiencies, how do we put it all together to once again achieve commanding industrial leadership?

While some base technologies in American industry are being challenged, we still enjoy substantial advantages in many areas—for example, in office automation and communications. In summary, we have a solid foundation on which to build a strong productivity effort—one that can leapfrog world competition. The time window is limited and the necessary actions must be "rapid and revolutionary," not "slow and evolutionary." Recognizing our assets and deficiencies, how do we put it all together to once again achieve commanding industrial leadership?

POTENTIAL CONTRIBUTIONS OF MATRIX MANAGEMENT

These is no simple solution to the productivity delimma. The solution is very complex and demands the optimal use of the physical and human resources by all sectors of the American population—industry, government, labor, and academia. This section is devoted specifically to only one aspect: how matrix management can help meet this most serious challenge. A few key contributions of matrix management will be discussed; these have originated from business experiences of the author.

Overcoming the Talent Gap

In the past decade there has been a tendency in American industry to emphasize growth. This has resulted in unprecedented opportunity for advancement of personnel rapidly to management ranks with little time for development of "experienced" expertise in any single function. The productivity of many companies has suffered because of inattention to the fundamentals of designing and manufacturing quality products at a competitive cost. In some instances this experience gap has been supported by the build-up of staff departments supporting the executive structure. This has resulted in attendant communications problems relating from delay of decisions and low morale—all productivity losses. In addition, too few managers have been developed in the factory area for lack of incentives, primarily lack of top management to recognize the significance to the success of the organization of a sound and well-managed manufacturing base. The Japanese for years have appointed top-flight professionals to keep the manufacturing function strong, and today are reaping the dividends from experienced manufacturing engineers whose positions are highly recognized for their contribution to their companies. Most American companies today recognize this deficiency, and universities now are reorganizing their curricula to develop this needed talent. However, the process is time-consuming. This talent gap will last through the 1980s.

In addition, competition is becoming worldwide and growing in intensity. Because of transportation costs and political and material resources demands, there is more pressure to serve the world from a number of locations. This further strains the already thin management expertise of companies and adds the requirement of starting businesses in foreign environments, usually requiring wholly different operating techniques and procedures. Matrix management can serve to improve this situation significantly. Specific projects can be organized with a matrix setup or a permanent matrix organization, the key managers being those with the greatest experience and expertise.

Such moves initially may add to productivity for several reasons. First, a good manager is stimulated by an appointment to a high-level matrix project. Second, unrecognized talent is usually discovered in the function to adequately fill the void caused by the appointee working on the matrix project. Third, the best available talent from each function becomes a part of the project, resulting in much-improved quality of decision making. Less experienced staff can be assigned to serve the needs of the matrix managers.

One example of matrix management involved building a new factory to handle an increased volume with modern state-of-the-art technology. To achieve this objective required expertise not available in the existing manufacturing function—for example, office and shop automation techniques and engineering product design, all linked with computer science procedures. As a result, the new plant was designed with a matrix organization of the manufacturing, engineering, and computer science managers, all of them having key roles. Another example involved building a factory abroad to better serve a sector of the world and to eliminate trade tariffs. This company had in place a country sales and strategic planning staff, but had no plant-operating experience available. A matrix organization of domestic experienced plant personnel and country managers was formed. From this, a plant resulted that was strategically located and sized to fit the needs of the company in the designated world-market area. In addition, the in-place foreign-based personnel guided the plant-planning people in local customs and personnel approaches with which they were unfamiliar. As a result, the plant performed better than planned in its first year of operation.

Saving Time

A very significant resource of any business today is time. The rapid pace of technology development—both product and process—can be described as "revolutionary." World competition is rapidly expanding to many products once considered basic to American industry: for example, automobiles, steel, electronics, and machine tools. The viability of many segments of American industry is threatened by the growing productivity growth rate of foreign nations. Hence the rate of key decision making in many businesses must be accelerated, if the businesses are to remain viable. A matrix organization formed of experienced people can examine world opportunities and technology status, and render suggestions best suited to the company's strengths. Most companies do not have an abundance of experienced talent available to perform this important task. Matrix management may be the only way to evolve plans for significant strategic discussions within the existing time constraints.

One company, as an example, was manufacturing office products with annual domestic growth rates in excess of 15%. Management became aware that the products had great international appeal but would require some changes in design to be acceptable abroad. A matrix project group was formed of domestic functional personnel and international marketing representatives. Within a matter of months the blending of expertise of domestic designers and in-country sales people produced a strategy formation that resulted in only slight modifications of existing products for overseas applications.

Individual functions of the business may suffer because of the demands of the matrix organization. However, the pace of technological progress and competition demands the best of information today for top-level policy decisions. Many productivity experts believe that American industry must initiate actions to offset the growing productivity advantage of Japan and Germany within this decade or risk becoming second-rate. The pace of competition and advanced technology is mind-boggling. A matrix management approach to major decision making is a prime way to accelerate the rapid pace of decision making now or soon to be demanded of most businesses. Neither time nor costs permit us the luxury of developing new talent to address this challenge. To be sure, assigning the best personnel to work on matrix-oriented projects will lessen the effectiveness of plant functional organization to a varying extent. In my estimation the response has to be one of priorities. Simply stated, the key challenges must be addressed by the best available talent.

Synergy of Knowledge

One of the most overlooked assets of American industry has been that of its human resources. Modern management principles that evolved in this century generally limited even minor decision making to levels high in the management structure. Such principles tended to separate the "doer" from the "decision maker." In fact, some management consultants openly state that the technical methods experts reduced to a low level the authority and decision making of the foreman or first line supervisor. The prime point to be made is that the decision-making process does not involve those who really perform the work or who work on the lower levels of product design or process design.

Only in the last decade has this "oversight" been acknowledged. Many of the participative management approaches originated to tap the knowledge available, the ultimate purpose being to evaluate a problem from many aspects and to expand in a positive manner the utilization of all the know-how and ideas of everyone in the organization. Involving more people slows the decision-making process but tends to improve decision making through exposing the issues to more in-depth analyses. In like manner, a matrix management approach increases the involvement of various functions and can be envisioned as a "shotgun" rather than a "rifle" approach to problems.

Let us review an example that supports productivity improvement through a synergy of knowledge. For years one manufacturer of defense products found that antenna systems used in numerous products were designed and manufactured individually for each product line. Currently, there were seven different types of antennae, all with some common design attributes. As product contracts were completed, layoffs of skilled people in some antenna-manufacturing areas resulted, while other product departments were looking for skilled workers to fulfill their antenna demands.

A reorganization resulted in a common antenna-manufacturing organization, supported by product line engineering reviewing designs and manufacturing through a matrix setup. The shop attitude changed from product production to a group technology orientation. Technology people were attracted to shop operations, forming a vintage of shop expertise and front-office technology experts. This resulted in the involvement so demanded by many of today's employees and opened up great human resource knowledge previously untapped. As a result, today all of the basic features of any antenna system are standardized and specific applications are simplified. Shop productivity has resulted in substantial gains through volume improvement and reduced fluctuation in the skilled work force.

A Better Decision Base

Two by-words for productivity improvement are *communication* and *involvement*. Matrix management by its very nature implicitly involves communication with a broad spectrum of personnel. It tends to submit many viewpoints and to bring to the decision-making process a fuller evaluation of the facts of each situation. Through the previously discussed synergy of knowledge comes involvement and a multifaceted review of the problem at hand. The cross-functional lines approach to fact gathering results in many more people being informed of critical issues in the company. The Japanese concept of Quality Circles has been highly successful and is paying off in improved productivity resulting from lower costs and improved quality.

An excellent example of involvement recently occurred in a large Fortune 500 corporation. In less than ten years over 1500 Quality Circle groups have been formed in shop and office areas of the company. These are groups of four to ten people with a common interest who meet regularly to discuss job-related problems and opportunities. In the words of the board chairman: "Here people tell their managers how they think quality, productivity, and the work environment can be made better. And managers are listening, because nobody knows a job better than the person who is doing it." From my personal observation, this company is experiencing remarkable productivity advances. One Quality Circle composed of word-processing clerks was instrumental in originating a centralized electronic filing system that in my judgment is unsurpassed in any other company.

In like manner the matrix approach to top-level strategic decisions leads to decisions based on a fuller evaluation of facts from many viewpoints, rather than a "rifle-shot" judgment of a few top people. Thus the judgment gap is reduced and higher-quality decisions should result. Because of the demands that matrix organizations place on the time of key executives, there is a natural tendency to give priority to key decisions to work on those areas of maximum importance to the company. As the Japanese have experienced, the initial decision making is slowed by a matrix approach because of broadening people involvement. However, the approach generally turns into a great asset to project implementation, once a decision is reached. Both decision making and implementation are significant today because of the rapid pace of technology development and shortened product life.

Productivity through Commitment

Matrix organization is most complicated. It operates effectively and productively only if built on a strong interpersonal relationship of people—on trust in one another and commitment to the good of the company. If the matrix operates in this manner, it stimulates the contribution of each person involved, since each can give of his or her maximum intelligence and effort. Involvement is limited only by one's personal decision of how much he or she can or wants to contribute to the process. On the other hand, top-management involvement is essential to successful operation of a matrix organization. Because of the freedom this type of organization gives to participants, more frequent involvement of top executives to keep the team "on track" is a prerequisite. The broadness of involvement and the more frequent communiciation with the executive level of the organization tend to support commitment and dedication to company objectives. One of American's most successful generals had a personal philosophy regarding commitment. He told his staff: "Any challenge might be met with a number of strategies. However, even the least desirable will work if there is commitment to the plan." Commitment, probably more than any other factor, has resulted in Japan's achieving today the world's highest growth rate in productivity.

For many years a large multinational organization enjoyed only modest success in penetrating world markets for its products produced domestically. Several years ago the top executives decided on a matrix approach to stimulate interest. An international task force was formed to develop a corporate plan for increasing international business up to 50% of domestic sales within a five-year period. Task force members were selected from all corporate functions and businesses. The task force worked diligently for four months, visiting key areas of interest worldwide. The result was a matrix-oriented international segment for the company that now receives positive support from the domestic profit centers. Former task members are now committed to "making the system work."

Accelerating Leadership Training

Productivity and leadership must be viewed as partners. It has often been stated that productivity starts at the top of any organization, and that maximizing producitivity demands the best of everyone in the organization. The complex interpersonal relationships and self-development of ideas in a matrix organization afford an excellent opportunity for leadership development. To be objective and successful, members of a matrix group must operate from the viewpoint of a general manager. Great attention must be paid to evaluating situations from many functional areas, without being too dominated by one's functional expertise. Matrix management functions on trade-offs and compromise, not line authority. Those who can survive under the complex human relationships and the close scrutiny of line management supervisors and ambitious peers, and have the ambition to present viewpoints objectively, become excellent candidates for top-management positions.

For example, in one business enterprise the manufacture of a multiplicity of product lines was centralized in one plant. Product line managers "matrixed" the operations organization. The manager of the plant functioned well under the matrix setup and quickly was advanced to a general managership, his talents to operate under complex interhuman relationships having become evident much sooner than under a vertical organization alone.

Summary

Not only can a matrix management approach aid productivity; this complex management form may *have* to be employed. Lack of expertise and management talent is a real problem in many industrial organizations. The intensity of world competition and technology growth does not leave time for filling the talent gap in the conventional manner. We need to spread available talent and direct it to priority challenges. This can be done through a matrix management approach, either by project or on a continuing basis. Through synergistic pooling of knowledge, a better decision base, and increased involvement, productivity of the organization can be materially effected. However, as mentioned throughout this text, matrix management requires a base of interpersonal relationships that generally does not exist in industry today. While the use of matrix management may be demanded by today's business environment, the advantages discussed for productivity improvement need to be balanced by a discussion of some cautions in applying matrix management approaches.

CAUTIONS REGARDING MATRIX MANAGEMENT

The pros and cons of matrix management are adequately discussed elsewhere in this handbook. In the ensuing pages only what the author considers the most critical issues will be discussed, all resulting from real operating experience.

Executive Understanding and Commitment

Of primary significance is top management's understanding of matrix organization and its commitment to it. This means involvement in the process to assure that progress is being made and priorities established. In addition, it requires that all personnel be knowledgeable of how matrix management is to be used in the organization and be given formal training in interpersonal relationships. For matrix management to be a positive contribution to productivty requires a working environment that is based on trust and knowledge of each person's responsibility in an organization, and a freedom to speak one's ideas freely, knowing that one's views will be accepted constructively. On the other hand, matrix management can actually impair productivity, if the organization structure is in conflict, with key managers striving for power.

An excellent way to show commitment is for top management to back up the matrix manager with the human, physical, and financial resources needed to effect results. Giving the matrix manager "the bucks to get the job done" is readily discernible in any organization. In simple words, a sound matrix approach can be a powerful productivity tool, but it requires as a foundation a solid, stable, trustworthy, and capable vertical structure. It all starts up at the top. The additional positive gains of synergy attributed to a matrix approach are attainable only if the existing vertical organization is sound and the matrix principle is backed up by top management. It should not be attempted in an organization in conflict, or where interpersonal relationships do not exist on a high trust basis. The inherent confusion and conflicts of the matrix approach can be challenging to a sound organization, but will result in chaos in less disciplined organizations.

Shared Responsibility

A sharing of responsibility for the project or decision at hand must be an every-day attribute of the matrix members. Successes and failures must be accepted as a group responsibility. Unlike in a vertical organization, no single matrix manager should or can be viewed as the key to success or failure of the task. This all-encompassing attitudinal development is a key condition for matrix success and must be stressed by top management. Notwithstanding this fact, the reporting status of matrix managers must be explicitly known and understood.

Of significance, also, is top management's evaluation of fitting a matrix manager to the task at hand. For example, a project encompassing high technological development might best be handled in the research and development stage by an individual who is research-oriented. However, the implementation stage of product manufacture might be handled better by a more shop-oriented manager. The need to change managers in a matrix setup to best accomplish the task at hand must be recognized. Matrix management should be group-focused, rather than individual-focused as in vertical organizations.

Recognition and Compensation

Commensurate with the development of an attitude of shared responsibility is the need for recognition and compensation of managers involved in the matrix structures. In most stiuations the standard management compensation system is based on structured systems tied to traditional evaluations of functional contributions to the business. Since in a matrix organization sharing of responsibility and a general manager's viewpoint of each challenging situation is promoted, functional compensation practices result in conflict. The area of compensation dichotomies must be recognized in a matrix organization and addressed. The solution is not simple. However, it cannot be neglected for an extended period of time, without having a negative effect on the organization. The same thought applies to nonfinancial recognition of the matrix managers for their contribution. It is important that all be recognized on an equivalent basis for success or failure. If one of the managers of a matrix organization is not pulling his or her load, it is best that the person be removed from the group and reassigned.

Need for Priorities

Matrix responsibilities are termed cross-functional, yet they must not detract from but be synergistic to the vertical organization. The time pressure on many people engaged in a matrix responsibility and also managing functional activities is commonplace. For those that hold multifaceted responsibilities, priorities must be established to reduce conflicts occasioned by time demands. To those fully assigned to matrix responsibilities, knowing the priorities of the company gives better insight on how demanding to be of the human resources of the various functional managers. This is particularly critical when the functional departments have only limited talent that can be applied to maxtrix needs.

From a second standpoint it is important for top management to establish dates for achieving principal objectives of the organization. The matrix's free-thinking group-involvement approach to problems needs to be controlled much more than the procedures of a vertical organization structure. Periodic progress reports from functional and matrix managers afford an excellent tool for top management to assess program progress and coordination of objectives. Priorities of projects as well as dates for decision making are a must, if a matrix organization is to be effective.

Summary

The above-mentioned cautions are few in number. However, they represent significant concerns that must be addressed, if matrix management is to attain a quantum productivity improvement in a company. All the answers to the potential problem areas are not known. Much research is needed on implementing matrix management as a test for great improvements in productivity. The purpose of this chapter is not to overwhelm the reader with the potential challenges of a matrix setup and its use. The intent is to point out that top management's involvement in making the matrix work is not a task of small significance to be delegated to others. Top-management involvement and commitment are demanded, to assure successful operation.

THE NECESSITY OF MATRIX MANAGEMENT

Today American industry is faced with the formidable challenge of improving productivity growth in a quantum manner. Increasing world competition compounded by a revolutionary growth of technology complicates the task. Yet the very viability of America as an industrial leader is at stake. The time-window for significant actions is limited. Matrix management is one tool that cannot be overlooked in meeting this challenge. The aforementioned time-window, the talent gap that exists in many industries today, and the need for better decision making demanded by rapid technology change—all can be addressed by a soundly developed matrix organization that supplements the vertical management structure. However, matrix organization is still in its infancy. It uses the human resources of an organization to a fuller extent than a vertical organization structure. But it also demands the development of interpersonal relationships to a degree not existent in most companies today. In the author's view, the decision is not one of whether or not matrix management is to be used for productivity improvement. The challenge to the future viability of America as a first-class industrial nation demands the use of all of our technology and ingenuity. Matrix management cannot be bypassed. The challenge before us is to make matrix management work and to recognize its pitfalls early in the game. We must maximize its positive contribution by recognizing drawbacks. Matrix management can be a major factor not only in improving decision making but in implementing those decisions. The real payoff will be substantial gains in the use of the physical and human resources of the business—a quantum jump in productivity essential for the future of many of our businesses.

MATRIX MANAGEMENT AT WORK

As examples of matrix management application to productivity improvement, four real-world situations will be cited. The first example involves an expansion to international activities by a multinational corporation. The second summarizes a domestic-based industry that has become a worldwide leader, largely through more productive use of its resources, human and material. Example three involves an academic institution and example four deals with an industrial situation.

Example One

A domestic division of a multinational corporation decided to build a plant abroad to gain some cost advantages in penetrating the European, Middle Eastern, and African markets. The division represented a business unit in the vertical structure of the corporation. In place also was a corporate-level matrix organization located in Europe and responsible for the strategic planning for the market area that the plant was to serve. Although the matrix organization in Europe was relatively new in origin, from the initiation of the project on, close cooperation was enforced on key managers of both organizations. Responsibilities for plant and market development were defined. In fact, the managing director of the new plant was mutually selected by both organizations. The domestic division was charged with plant design, construction, and implementation of production. The international segment assisted with market development and coordinated country-wide management, financial, and compensation policies. The result was a plant built on time and within budget—an objective rarely achieved in building domestic plants in industrial environments that are much better understood. Despite this excellent productivity accomplishment, some hang-ups were reported by key managers. A few resented the constant pressure for project help on the part of matrix members, but admitted that this forced team action. Matrix decisions were hindered by the difficulty of arranging meetings of both the domestic and international groups. When the meetings were arranged, time permitted a discussion of only one or two items. One manager commented: "The project was successful, but major decisions took too long to make and required three times the effort. The demands of the matrix are great, but they seem to work. Today we seem to be working together a lot better."

Comment. This project represented a very difficult one, for it involved building a major facility in a foreign environment. The international organization had not been in existence long enough to be smoothly functioning. However, the synergy of knowledge of plant operation by the domestic division personnel and the knowledge of foreign relations and financing of the international organization were exploited. Today both organizations are more mature and the matrix approach is accepted and recognized for its merits. The key to this success story was the direction of cooperative effort in the two organizations forced by the executive level. From a productivity standpoint the new plant planning, construction and start-up were recognized as one of the better ones in the corporation's history.

Example Two

This example covers a domestic plant operation that has grown in the last decade to supplying highly sophisticated products worldwide. This business designs and manufactures product lines involving advanced electronic technologies. The product lines are short-lived and involve not only sales of basic products but also sales of technology and spare parts.

Until a few years ago all these income sources were incorporated in a single business. The main product lines were the principal sources of revenue and received the most attention. A decision was made to move toward a matrix operation, and spare parts and technical information were made separate businesses. Subsequently, a matrix project organization was structured throughout the whole business.

In the last decade the business has grown five-fold with a modest build-up of total manpower. How was this accomplished? Much of this growth and quantum improvement in productivity is attributed to the use of a matrix management organization. The matrix members are credited with placing great attention on spare parts and technical information, areas previously neglected because the major income was from the basic product lines. In addition, the rapidly expanding business is a high technology one. The synergy developed by a matrix approach used available talent more fully and forced some standardization of component design. In several instances a specific product division transferred component manufacture to a new organization and level-loaded skills that previously had fluctuated with product life cycles. One information-generating function was reduced from 600 to 200 people and volume of output doubled. On-time shipment of products has materially improved and quality is the highest in the industry. The time factor for introducing new lines has been decreased substantially. International relationships within the business have been improved through shortening the communication gap to top management afforded by the matrix setup. The above factors contributed to a quantum jump in divisional value-added productivity by everyone in the business unit.

Comment. This is a real success story. The operation has grown fivefold in a short time to almost a billion dollar business, and the company is a technology leader worldwide. Matrix management has become a way of life and gets much credit for this success. However, the organization is replete with problem situations. The conventional accounting system makes it hard to recognize the contributions of the conventional and the matrix organization. The interpersonal relationship level of the entire organization is inspiring, but much new knowledge is needed for improving the effectiveness of the matrix and line organizations. Frequent structural changes occur and are confusing. Holding matrix and divisional managers responsible in an equitable manner remains difficult. In this example a matrix organization approach has been in place for many years; its success has been demonstrated, but it is not without its problems.

Example Three

This example involves an educational institute that is directing the design and implementation of a high technology, automated factory. The project involves a matrix management approach supported by a consortium of academic, industrial, and government expertise. The project's objective is to originate a commercially operable enterprise whose productivity gain will be orders of magnitude for conventional but modern plants.

The project has the commitment of the executive staff of the university and is supported by personnel who are recognized worldwide for their expertise. The team leader chosen by the university is an individual with years of industrial experience in planning, constructing, and operating technically oriented plants. Industry and government are represented by top-flight personnel. The matrix organization has been formed, has its first funding, and is proceeding with a positive attitude, with no single faction dominating the project.

Comment. This project is relatively new but has been planned as a matrix to bring together the synergy of expertise that does not exist in any segment of the consortium. All parties are committed to the project's success, objectives have been mutually determined, and an air of cooperation has developed. The project is staffed with real experts. While only time will tell, the project is off to a test start and has a high potential for success.

Example Four

This example concerns an industrial situation. A segment of a large corporation consists of four businesses, all reporting to a corporate executive. The businesses are all different and range from high to moderate technology. The executive is a very intelligent and progressive individual, dedicated to improving the productivity of all his operations in a quantum manner. He has observed that none of the four businesses is by itself capable of automating plant and office operations as effectively as would occur if all manufacturing operations were combined into a new support organization serving all the business units. Planning has proceeded for more than two years to bring about this reorganization. The third high-level executive in two years is now directing the planning effort.

The technical aspects of a new plant design have been developed and substantial management information systems have been proposed to tie control to the parent plants. Despite frequent communication and involvement of the business unit leaders and key managerial personnel, resistance to yielding manufacturing responsibility to a new organization has been intense on the part of the business-unit leaders. In summary, progress on this project is slow and team members have been prone to leave for other assignments.

Comment. Under a matrix management organization, it is important for all levels of management to be committed to and to support the project. In this example, the executive's enthusiasm and vision is outstanding. The project has been well defined. However, the too-frequent change of project leaders, plus the resistance of several business-unit managers to sharing manufacturing responsibilities, have stalled the project. It is likely that at least two of the business-unit managers will have to be replaced, for they cannot adjust to the degree of organizational change proposed by the excecutive.

22. Developing High-Performance Matrix Teams

David Wilemon*

Most management scientists agree that matrix management has been a significant management innovation. In the last two decades some of the largest and most complex projects have been managed by a matrix management approach: Apollo, the Space Shuttle, various pharmaceutical development projects, high performance military aircraft, unmanned satellites, and large civil engineering projects around the world. Since the early 1970s, matrix management has also been widely used to implement computer systems projects. A vice-president of a large high technology company recently made this observation:

> Matrix management has some benefits other managerial systems don't have. It can be highly adaptive to changing organizational environments, technologies, and market conditions. It also can help you utilize human and capital resources more efficiently. Frankly, it there were no such thing as matrix management, we would have had to invent it to successfully complete our more complex projects.[1]

As the above statement suggests, matrix management can offer some major advantages in complex program administration.

Matrix teams, however, are not without their problems. Consider the following situations encountered by three different matrix teams.

Situation A. A matrix team charged with developing a new product for the surgical field was unable to gain an agreement on which technical approach to use. As a result of the ongoing conflict and lack of progress, two key team members left the company out of frustration. The project never got off the ground.

Situation B. A matrix manager racing to get his team off to a quick start, failed to discuss adequately the overall project management plan with them.

---

*Dr. David Wilemon is a professor and Director of the Innovation Management Program in the Graduate School of Management at Syracuse University. He is widely recognized for his work on conflict management, team building, and leadership skills in project-oriented work environments. He has studied various kinds of project management systems in the United States and in several foreign countries.

As a result, confusion developed within the team over direction and project priorities. A great deal of "wheel-spinning" resulted, leading to intense frustration and apathy.

Situation C. A matrix team formed to implement a management information system in a large manufacturing organization experienced trouble from the start. In team meetings two members of the team dominated the discussions. Although the project leader had a printed agenda, it turned out to be useless since the meetings, once started, quickly got off track. The less assertive members of the team eventually found excuses for not attending the team meetings—morale and team productivity dropped.

In each of the above situations a breakdown in the team management process occurred. Problems such as those mentioned above can often be avoided or minimized by using effective team development (team-building) strategies. This chapter presents a team development model that can improve the functioning of matrix teams. The model was developed from the author's experiences in helping a number of project and matrix teams function more effectively in various organizational settings.[2]

WHAT IS TEAM DEVELOPMENT?

In recent years an increasing need to develop more effective matrix teams can be traced to four major factors:

1. Increasing emphasis on achieving more productive project performance.
2. Greater need for synergistic problem solving to achieve novel solutions, often at the state-of-the-art.
3. Turbulent and unpredictable task and operating environments.
4. Complexity of tasks, requiring a highly coordinated and disciplined team effort.

Team building is the deliberate process of creating and then maintaining effective team functioning. The process of team development takes time, energy, commitment, and hard work, but the results can be enormous in terms of higher performance and morale, increased productivity, and more effective problem solving. In a team development program we are concerned with the usual task performance measures—e.g., cost, schedule, technical achievement—but also with how well the team members work together. It is important to recognize the generally high correlation between a team's "internal health" (the interpersonal relations among members) and the team's task performance. Poor interpersonal relations among the members of a team can cause poor performance, and vice versa. Many matrix teams that perform at a sustained high level often possess a high task orientation (output focus) as well as an effective interpersonal orientation. Thus, a major objective of team building is to help teams learn to manage their task, as well as the interpersonal issues and concerns that can develop.

WHEN IS TEAM BUILDING USED?

There are a number of situations in which team building can be beneficially used. Frequently, for example, team building is employed to get a new team off to the right start. One team leader in a major pharmaceutical company made it standard practice at the start of a new project to spend three to four days with his team dealing with potentially significant team issues. His objective was to anticipate and avoid likely team problems and conflicts, and to establish approaches for handling such matters if they did develop. In addition, the team undertook several strategic planning exercises to minimize future detrimental surprises.

In other situations matrix managers may use team building to help their teams deal with major new challenges. A large company building a chemical plant ran into financial difficulties and decided to take on a partner in a joint venture. Several team-building sessions that focused on how the teams could coordinate their efforts more effectively to accomplish the goals of the venture facilitated the fusion of the two project teams.

Team building can also be used in the case of an established team that finds itself in a "downward spiral" or in a "failure mode." In such situations, team building can help determine the causes of the problem and develop new approaches to alleviate the situation. Finally, team building is often used to help a team achieve even higher performance levels. In such a case, the purpose of team building is to help the team "fine-tune" its functioning.

Unfortunately, many matrix managers have used team building as a one-time solution to a specific "team problem." While a single application certainly can be productive, team building is most useful when it is considered an ongoing process.

THE MATRIX TEAM DEVELOPMENT MODEL

There are generally a number of phases in a team development program. A descriptive model that conceptualizes the components of a team development process is presented in Figure 22-1. Although the model denotes a general linear relationship among the various steps, each phase is in fact interrelated. Moreover, there is "process behavior" at work in each of the development phases. Process behavior describes the interpersonal interactions among the team members as they pursue various tasks: communicating, interacting, interpersonal relationships, and even intergroup relationships (cooperation and competition with functional departments). While some team development specialists see a clear distinction between the task issues a team faces (goal set-

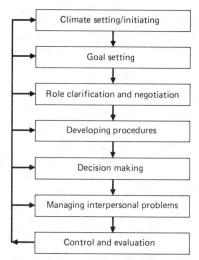

Fig. 22-1. The team development model.

ting) and process issues (communicating), task and process issues are not easily separated. It is difficult, for example, for teams to "set objectives" without a high degree of interpersonal communication.

The approach taken to illustrate the team development model will (1) describe each phase and its role in developing matrix teams, and (2) give examples of how each phase can be used.

Climate Setting/Initiating

Climate setting is establishing a positive environment for team building to occur. It is the process used to help a team reach a risk-taking state where it is willing to begin the process of team development. Climate setting is important for an effective team-building process because:

- Some team members may not understand the goals of team building, how it works, the costs, or the benefits.
- Some team members may be suspicious of team building and the motives of the matrix manager.
- Some team members may feel that the "pressure to produce" does not allow for their investment in team building.
- Some team members may prefer to be an "individual contributor" rather than a "team player."
- Other team members simply may be content with the status quo.

Take, for example, the matrix manager in a large electronics firm who believed it important to initiate a team development program for his newly formed team. Important factors in his decision were the complexity of his program; the state-of-the-art technology that would be used and its potential for producing conflict within the team over technical alternatives; and his division's spotty past record in managing unproven technology. The first step of his team development program was to get the team together for a two-hour discussion on the task and its importance, and why he thought team building could increase the team's effectiveness. In explaining team building he discussed with his team (1) the broad goals of team building, (2) how team building can contribute to project objectives, and (3) what it requires in terms of time and personal commitment. This matrix manager was fortunate to have participated in other successful team-building sessions in another division of the company. When the matrix manager has little or no experience in team building, he or she may want to enlist the assistance of an internal or external company consultant to help with the climate-setting process.

In the climate-setting/initiating phase it also is important to allow those with questions and concerns to express themselves freely. The more open the discussion, the more likely it is that any developing anxiety will be kept to a minimum. If some team members believe that team building is simply "sensitivity training," care should be taken to explain that the purpose of team development is to increase the team's performance. It accomplishes this objective by focusing on key task issues as well as interpersonal issues that can contribute to or thwart performance. Depending on the situation, team building can be a preventive as well as a cure for task and interpersonal problems.

Another important activity in the climate setting/initiating phase is for the matrix manager to help each team member feel that he or she is an integral part of the team, and that his or her contribution will be important for the team's overall success. This is a crucial step in establishing an attitude of commitment. Suggestions for helping achieve this feeling of ownership are seeking advice and ideas from team members, sharing information, giving credit, and using the team members' ideas. A session in which each team member has a chance to discuss his or her background and experience can help the team better understand each member's contribution to the team's mission.

One approach that can be used in the climate setting/initiating phase is to ask the team to discuss such questions as these:

- What concerns you about this project?
- What concerns you about your role?
- What expectations do we have of one another?
- What do I or we need to do to meet those expectations?
- What issues do we need to discuss to get this project off to a successful start?

From such questions, the team may start developing ways to prevent problems before they occur, or establish a framework for handling problematic issues if they do occur.

When a good climate has not been set, it can lead to foot-dragging, low ego involvement, fear, and half-heartedness. By contrast, an effective climate sets the stage for tackling the other important team issues.

Goal Setting

After the climate setting/initiating phase, new matrix teams normally will need to address, clarify, and agree on the goals to be accomplished. These goals include those for the overall project as well as for the various subtasks. Unless these goals are clearly established, understood, and accepted, the project will not be successful nor will the potential of the team be realized.

For the team to grasp the dimensions of the overall project, it is important to develop a broad but comprehensive "mission statement" about what the team expects to accomplish. From the mission statement, specific objectives need to be developed for cost, schedule, and performance targets. Again, these objectives are for the overall project as well as for the subtasks.

One approach for developing clear goals is to list the proposed goals at a team meeting. This can be done with the total team or with subcommittees responsible for a particular task or a technological subsystem. Once the goals are listed, the entire team looks for agreements as well as disagreements. Then the team examines the conflicts or disagreements and works through each conflict area. An action plan needs to be developed to handle those conflicts that cannot be disposed of in the team meeting. The action plan should identify who will be involved in resolving the goal conflict, the process to be used, and when a resolution is needed.

When project goals become so complex that they lose meaning, it is important to organize and restructure the objectives to reduce their complexity. This increases the probability that a common shared perception about goals and objectives can be achieved. The following criteria can be used in evaluating the team's goals:

- Are the goals clear?
- Are they measurable?
- Are they time-based?
- Can they be effectively communicated?
- Are they specific?
- Can they be related to responsibility?

Agreement on these and other criteria should be reached before the team undertakes its goal-setting sessions. The process of developing and reaching a consensus on goals is a cornerstone of effective team building. Unclear goals can lead to confusion and wasted motion. Also, unclear goals—or too many goals—often produce stress.

Role Clarification and Negotiation

Once the goals have been clearly identified and agreed upon, attention should be focused on the roles and responsibilities of the team members in accomplishing the task. Unless roles are defined and accepted, only frustration, conflict, and wasted effort will occur. Roles include the formal understanding of the responsibilities each team member has, as well as the informal agreements and understanding among team members. A matrix manager explained the importance of clear role definition this way: "In an R & D project like mine you have a lot of different groups that must be coordinated to achieve project objectives. We have five different labs we're dealing with, as well as our client. In addition we have two sub-contractors supporting us. I have to get clear on each group's responsibilities. Unless I do this, I'm going to run into some show-stopping problems real quick." As indicated, the clear articulation of roles is necessary to avoid conflict or situations where ambiguous roles leave the project vulnerable.

To identify and negotiate roles and expectations, the team can work together by focusing on these questions:

- What roles should be performed to achieve the goals established for the project?
- Who should perform these tasks (individuals or work units)?
- What skills are needed, and what skills do we have within the team? How will we deal with skill deficiencies?
- How much freedom should individual team members have in defining and carving out their own roles? How will individual roles be coordinated to achieve a unified team effort?
- How will work assignments be allocated?
- What are the "gray areas" in defining responsibilities?

Answers to the preceding questions can identify those areas where agreement can be reached, as well as areas where more information is needed or where role conflict is clearly evident. A team could also use the above process as an initial step in role clarification. Once conflicts over roles are identified, other roles can be negotiated. In essence, role negotiation is a process whereby each team member articulates his or her needs to those on whom he or she is dependent, and vice versa. The process may require a high degree of problem solving to effect a clearly agreed-upon role definition. Figure 22-2 illustrates the logic of role clarification.

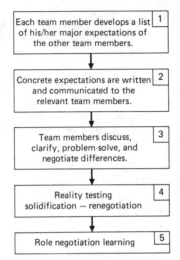

Fig. 22-2. The role negotiation process.

The final phase of the model, "role negotiation learning," represents the learning process experience by the team. The highly effective team will learn from the role negotiation process and will probably be able to handle such situations more easily and effectively if and when they develop.

In formulating clear role messages, the sender needs to communicate specifically what he or she wants from the other team members in order to perform his or her responsibilities. As the project progresses over its life cycle, other issues are likely to create new situations where role negotiation also can be helpful. Roger Harrison presents a highly useful framework for negotiating expectations from team members that can be used at most any point in a project's life. His framework suggests that each team member communicate to the other team members the following information:[3]

- If you were to do the following things more or better, it would help me to increase my own effectiveness.
- If you were to do the following things less, or were to stop doing them, it would help me to increase my own effectiveness.
- The following things which you have been doing help to increase my effectiveness, and I hope you will continue to do them.

This approach allows the members of a team to fine-tune their expectations of one another and to make changes as needed. The alternative to effective role negotiations is role ambiguity, role conflict, and even project failure.

Developing Procedures

Another step in the development of a highly effective matrix team is the formulation of procedures to guide the actions of the team. Carefully formulated procedures that routinize some areas of team functioning can save time and help achieve a smoothness in team operations. A matrix manager commented about the usefulness of well-developed project procedures this way:

It's very hard for me to imagine that a matrix team could function without clear, well-thought-out procedures. Some issues that a team faces occur over and over. For these repetitive issues, you need some guidelines and policies on how to handle them, so when they do arise you and your team know what to do. Good procedures leave you time to plan ahead, manage the day-to-day nonroutine affairs, and focus on trouble spots. Procedures also eliminate a lot of conflict because they give the team a road map about how to deal with a lot of important issues.

Examples of procedures often used in a matrix environment are:

- How and when will our status review meetings be conducted?
- How will information be distributed among the team members?
- How will cost, schedule, and performance data be tracked?
- How will we manage our interface with the client group?
- Who should attend team meetings?

Other procedures might be formulated in the technical, procurement, financial, personnel, and legal areas.

It is suggested that the team should hold a problem-solving session early in the life of the project to identify specific procedures needed. In this procedure-defining session, some team members should also be particularly prepared to discuss areas where they are knowledgeable and have had experience. In setting procedures, it is important not to reinvent the wheel. If the company or division has existing procedures and policies to handle key project areas, then these procedures should be used. When none exist, the team will have to develop them. Each procedure should pass the close scrutiny of the team in terms of its usefulness, in order to exclude those that are detrimental and cause unnecessary red tape.

Decision Making

The fifth major issue of the team development model is to determine how the matrix team should handle its decision-making processes. Several suggestions will be made to help matrix teams address those factors that can shape their decision-making process.

First, the team needs to be aware of areas where decisions are the responsibility of an individual and involve no one else. There are few things more unproductive than having an entire team try to make a decision that should be made at the individual level, or having an individual try to make a decision that should be the team's. The decision-structuring approach requires that major decisions be examined to determine who should be involved in the decision and how they will be involved. For example, decision A may require that the matrix manager and the R&D manager *jointly* made the decision. In making the decision they need to *consult with* the manager of engineering and to *inform* senior management, the client, and the other team members. Decision structuring also helps minimize dysfunctional conflicts and increases the efficiency and effectiveness of the decision-making process. A more detailed view of the decision-structuring process is illustrated in Figure 22-3.

A second suggestion for enhancing a team's decision process in a matrix environment is to have regular project review meetings where major issues can be discussed with all the team members. These review meetings are useful for tracking the progress of the project and for information exchange, as well as for handling problems in "real time." One matrix member spoke about the beneficial aspects of project review meetings as follows:

> Our status review meetings are regular and they are intense. Even though they are not particularly enjoyable for me, they are extremely beneficial. For example, we are able to deal with a lot of technical information fast, because we have all the key players there and we can make decisions on the spot. When we can't make a quick decision, everyone knows when the decision will be addressed and by whom. Our review meetings also keep me from being surprised and surprising others.

Final suggestions for increasing the productivity of the matrix team's decision-making process include the following:

1. When appropriate, involve and/or consult with the younger, less experienced team members on important decisions. Such a step helps them feel like an integral part of the team (which can be highly motivating) and lets them see the role decisions play in overall project performance, as well as hone their own decision-making skills.
2. Brainstorming and "what if" sessions can be excellent vehicles for generating needed information for decision making and contingency planning. Some highly creative approaches for handling complex situations and decisions can result from brainstorming within the team. This can be a motivating team-building experience.

Nature of the Decision	
Scope Importance Routine Nonroutine	

Who should be involved?	How involved?
1. Matrix manager	1. Decision maker
2. R & D manager	2. Decision maker
3. External consultant	3. Provide information
4. Manager of engineering	4. Consulted
5. Team members, senior management, and the client	5. Informed

Decision Process Used
1. Explore the decision and its potential impact on team goals, functional support areas, and project scope.
2. Collect or develop information for decision making.
3. Develop and examine alternatives.
4. Select the most appropriate alternative.
5. Evaluate the decision.

Fig. 22-3. The decision-structuring process.

3. Some matrix teams and/or team members have a tendency to force a decision too early by jumping to a conclusion. One approach for minimizing or avoiding this outcome is to establish criteria for an acceptable decision, and to make certain that proposed solutions meet these criteria.
4. Still another approach is to establish a norm that for important team-wide decisions everyone will have an opportunity to make a contribution to the decision.
5. Finally, subcommittees can be used to seek out and propose solutions to the larger team. The advantage of this approach is that only those with the expertise are involved in finding a solution.

Table 22-1 illustrates a number of common barriers to effective decision making within matrix management environments.

Managing Interpersonal Problems

A major component of a team development program is how a team plans and is prepared to deal with the inevitable interpersonal issues that develop within a team. Examples of interpersonal concerns are:

- Conflicts
- Poor communication

Table 22-1. Major barriers to effective matrix team decision making.

BARRIERS	SUGGESTED SOLUTIONS
Language: differences in technical terminology among team members	• Clearly define and explain technical concepts • Actively listen to the other party • Check out for communication clarity; ask questions
Myopic views, and prior experiences of team members that limit an open and creative search for alternatives	• Brainstorm alternatives • Encourage openmindedness • Examine applicability of prior experiences to the current decision
No sense of urgency in making and implementing important decisions	• Clearly communicate and follow up why the decision needs priority • Explain consequences of not meeting deadlines • Follow up quickly with responsible parties when deadlines are missed
Cultural differences among work groups, e.g., R & D and Marketing	• Promote awareness of cultural differences and discuss what promotes divergence in work group cultures and the consequences • Cross-functional training • Promote team climate and accomodate differences
Less dominant "experts" not encouraged to become fully active in the decision-making process	• Use effective "gatekeeping" in order for all team members to participate fully in decision making • Develop team awareness of the importance of a high degree of individual participation
Treating a "symptom" as the real problem	• Fact finding • Identify what the problem is not • Be sensitive to the problem's false signals (the symptoms)

- Lack of commitment
- Lack of support
- Lack of trust and openness
- Detrimental competition
- Aggressive and dominating behavior
- Lack of sensitivity to team members' feelings and needs

In preparing to deal with interpersonal concerns, the team leader and team members must be aware of potentially detrimental behavior that can occur within the team. If the team leader has had prior team experience, he or she can brief the team on the kinds of interpersonal issues likely to arise. Also, an internal or external organization development consultant can play a valuable role in exploring with the team the kinds of conflicts likely to develop. One matrix team leader managing his first project discussed this process as follows:

> One of the first actions I took after I was appointed project manager was to talk with our organization development people about what to expect in terms of interpersonal problems within the team. That conversation proved so helpful that I asked them to discuss these issues with the entire team. I'm certain it saved us a lot of grief later, because we were far more sensitive to the kinds of interpersonal problems that can develop within a team.

The most effective way to handle interpersonal problems is for the team leader and team members to develop the capability to deal with these problems as they arise. This capability gives the team the resources and confidence to handle a variety of interpersonal concerns, and gives the team leader an opportunity to influence and shape the behavior of team members. One matrix manager commented: "You have a great opportunity when you're dealing with your own team to model appropriate behavior. If you have a disagreement or a conflict with a team member, how you deal with it can influence them in their dealings with others. You also have an opportunity to coach your team members in dealing with a wide variety of interpersonal problems."

Table 22-2 illustrates one approach which this author has found highly use-

Table 22-2. Framework to help sensitize a team to potential interpersonal issues.

WHAT KINDS OF INTERPERSONAL PROBLEMS ARE WE LIKELY TO FACE WITHIN THIS TEAM?			
POTENTIAL PROBLEM AREAS	CONSEQUENCES TO THE TEAM	HOW TO *PREVENT* THIS PROBLEM FROM DEVELOPING: WHO, WHEN, & HOW	HOW TO *MANAGE* THE ISSUE IF IT DOES ARISE: WHO, WHEN, & HOW

ful in helping teams become aware of and handle various relational problems. The team leader (if experienced) or with the assistance of a consultant asks the team to discuss openly the kinds of interpersonal problems which they believe may develop within their team. If several issues are mentioned, then the team leader may want to ask the team to pick a few representative issues. These should be listed on a flipchart or blackboard to give them visibility. The next step is to take each of the identified issues and discuss the likely consequences to the team's functioning if the problem occurs. The purpose of this step is to help promote awareness of the potential consequences of the issue if it does arise. The third step is for the team to discuss ways that the identified issue can be prevented from developing or to minimize its affects if it does develop. This step should sensitize the team to actions they can take to minimize these and other interpersonal problems within the team. The final phase involves discussing, in specific concrete steps, how best to deal with the problem if it does develop. In these final two phases, the who, when, and how should be identified.

Time spent dealing with these issues in a simulated fashion can help prepare the team for managing real issues when and if they do develop, for two primary reasons. First, the simulation creates an awareness that the development of interpersonal problems is a natural occurrence in the management of complex projects. Second, it promotes thinking about alternative ways to prevent and/ or cure these problems if they do develop.

It is also recommended that the team spend time discussing not only the project but how well the team works together in accomplishing its goals. Here the norm should be established that it is "O.K." for team members to discuss personal issues and concerns as they relate to the team's goals and tasks. The team leader should encourage team members to be direct and specific when discussing behaviors that can undermine team functioning.

For example, if Jim continually misses important deadlines and is repeatedly late for meetings, the project leader might address Jim's behavior as follows: "Jim, when you missed yesterday's scheduling deadline and were late to today's status review meeting, I got awfully frustrated because we're unable to work together as a fully coordinated team. I need and want the benefits of your expertise."

Finally, the team leader needs to be able to sense when interpersonal problems are emerging. In effect, the team leader needs an early-warning system to signal potential problems. Examples of behaviors to observe include:

- Team members begin manufacturing excuses to avoid important team meetings.
- There is a notable increase in the stress/anxiety level of team members.
- Interest in the project or the functioning of the team wanes.
- Overall progress falters.

Control and Evaluation

A crucial task for the highly effective matrix team is to evaluate and control its effects, so that necessary performance adjustments can be made. Several options available to matrix managers can help them appraise and control their team's functioning. One approach is to evaluate periodically how well the team is meeting its established budget, schedule, and performance goals. This can be accomplished via the formal status review meetings or simply by having the team informally discuss its performance. The team should discuss not only its task but also how well the team works together. Identified problems should be discussed and dealt with by the team.

The use of open-ended questions also can dramatically help a team focus periodically on barriers to effective team performance. Examples such as the following can be used:

- What are five things that impede our team from functioning at a high, sustained level?
- How can we help one another to be more effective in accomplishing the goals of this team?
- What factors limit us from achieving the established performance goals, and what can we do about them?
- What are this team's strengths and weaknesses in terms of (1) task performance, and (2) the process by which we work together?
- How does this team's functioning affect my own proformance?
- How are our relationships with other areas of the organization and the client affected by the way we work together?
- What do I like and dislike about how this team functions?
- How are we increasing our capabilities to diagnose and solve our problems as the project progresses through its life cycle?

If one or two questions like those above are routinely addressed, along with more routine task-oriented concerns—cost, schedule, and performance control—the team is more likely to be able to control its desired performance levels. The author's experience with a number of teams suggests that the higher the level of trust within the group, the easier it will be to discuss these questions productively. Moreover, as a team becomes used to discussing these issues and getting desired results from the process, the discussion is likely to become more natural and productive.

SUMMARY

Matrix management is here to stay. Like other organizational forms, however, it is not without its problems. One major concern lies with the importance of developing a sense of ownership and collaborative effort regarding the various supporting functional specialists, since the matrix manager may be competing with others for these specialists' time and commitment. There is also a strong need for the matrix manager to devlop a sustained level of technical accomplishment, which comes largely from the combined efforts of specialists within the team. Technical performance is likely to be suboptimal unless these experts can work together effectively. The team-building process proposed here can help transform these functional specialists into a cohesive, highly productive team.

BIBILOGRAPHY

Davis, S., and Lawrence, P. *Matrix,* Reading, Mass.: Addison-Wesley Publishing Co. 1977.

Francis, D., and Young, D. *Improving Work Groups: A Practical Manual for Team Building.* University Associates, San Diego, Calif. 1979.

Gordon, W., and Howe, R. *Team Dynamics in Developing Organizations.* Kendall/Hunt Publishing Company, Dubuque, Iowa 1977.

Merry, V., and Allerhand, M. *Developing Teams and Organizations.* Reading, Mass.: Addison-Wesley Publishing Co., 1977.

Schein, E. *Process Consultation.* Reading, Mass.: Addison-Wesley Publishing Co. 1969.

Steele, F., and Jenks, S. *The Feel of the Work Place.* Reading, Mass.: Addison-Wesley Publishing Co., 1977.

Tubbs, S. *A Systems Approach to Small Group Interaction.* Reading, Mass.: Addison-Wesley Publishing Co., 1978.

REFERENCES

1. Correspondence with author.
2. Quotes used in this chapter are from field interviews conducted with matrix managers in a variety of technology-based organizations. The purpose of these field interviews was twofold: (1) to explore some of the problems team leaders experienced in managing teams; and (2) to examine some of the experiences project and matrix managers have had with team development programs.
3. Roger Harrison, "Role Negotiation: A Tough Minded Approach to Team Development," in W. W. Burke and H. A. Hornstein, eds., *The Social Technology of Organization Development* (La Jolla, CA., University Associates, 1972), p. 95.

23. The Matrix Manager and Effective Communication

Stephen E. Barndt*

COMMUNICATION DEPENDENCY IN THE MATRIX ORGANIZATION

The communication of information is essential to cooperative goal-directed behavior in all organizations. However, in the more organic, less structured matrix type organization where decision making is less programmed, and both message frequency and content are highly variable, it is even more important. In varying degrees, this heightened communication dependency affects the responsibile managers of projects, task forces, products, product lines, and international operations that rely on matrix relationships for access to resources and that are significant to their parent organizations in terms of profit or cost potential.

The effective manager of such a matrix organization performs multiple roles. One role is that of leader, guiding team workers, coordinating their efforts, and helping to solve problems. A second important role is that of integrator, tying together tasks within the team and coordinating with all interested individuals or groups outside the team. A third critical role, related to the leader and integrator roles, is that of communicator. It is in this role that the manager comes to grips with the team's communication dependency and the effectiveness with which he or she performs in this role determines, in part, the effectiveness of leading and integrating. That is, the extent to which the manager is able to effectively obtain and disseminate accurate information will show up in the performance of the team in terms of focus, coordination, and integration of effort. Figure 23-1 depicts the matrix manager's roles and communications relationships.

The matrix manager's role as a communicator is on the one hand particu-

*Stephen E. Barndt is an Associate Professor of Management at the School of Business Administration, Pacific Lutheran University. Dr. Barndt, who earned his Ph.D. degree from the Ohio State University, has directed research into management communications and various behavioral aspects of project management and has published articles on those subjects. In addition, he has coauthored texts on operations management and project management. Dr. Barndt's matrix management experience includes performing as an assistant program manager and as an R & D project administrator.

Portions of the material presented in this article were published in "Upward Communication Filtering in the Project Management Environment," *Project Management Quarterly,* Vol. 12, No. 1 (March 1981): 39–43.

Fig. 23-1. Communication and matrix manager roles.

larly crucial, and on the other hand particularly demanding. As implied in Figure 23-1, communication is a critical function that links the various activities that are important to the success of the project, task, or product. It is an important means of integrating activities within the matrix team and, at the parent organization level, of integrating the efforts of any one matrix group with the efforts of others and the organization as a whole. Assuring effective communication between the matrix team and external sources and users of information, as well as within the team, rests largely on the shoulders of the matrix manager.

These two types of communication—within the boundaries of the matrix team and across those boundaries—depend to a large extent on verbal and often informal communication. For example, research studies of communication channels used by managers, scientists, and engineers in research and development organizations reveal that about 55% of communication is by informal means.[1] This use of informal channels and often spontaneous communication results from the relatively fluid and frequently nonformalized nature of the matrix organization, particularly projects and task forces, and the inherent shortcomings of those formal channels that are established. Because of the prevalent use of and reliance on more informal channels, the matrix manager is in a position to communicate by virtue of his or her central location and in a position where the success of the undertaking will depend in some large degree on the effectiveness of such communcation. On the one hand, being in a position to receive reports, requests, and complaints is a great advantage. On the other hand, unless it is used to advantage to give individuals within and outside the team pertinent information at the proper time, the project's success and the individual's career may suffer.

COMMUNICATION PROBLEMS

Harold Stieglitz suggested three barriers to communication: individual differences, corporate climate, and mechanical barriers.[2] Individual differences include backgrounds—cultural, social, educational, economic, and occupational, as well as physiological—that show up in our different value systems;

languages; our ways of interpreting events, actions, and observations; and the inferences we make. The corporate or organizational climate—the general feelings of mutual interest and respect among the organization and its members—is reflected in perceptions of trust and fairness, equitability of rewards, recognition, quality of supervision, and support from the top down. Mechanical barriers have to do with clarity of reporting relationships and responsibilities, work load, physical layout and dispersion, use of specialization, and work interdependencies.

The communication barriers, singly or in interaction, create communication problems for organizations. Three major classes of problems that can be expected to hamper the timely and effective flow of information in matrix organizations are depicted in Figure 23-2. These are encoding and decoding inaccuracies, filtering and distortion, and structural blockages. As implied in the figure, filtering occurs when not all the available information is transmitted. Distortion involves a slanting of the content of messages so that only selective information, fairly well directed toward one particular meaning, is passed. Encoding and decoding inaccuracies occur when a sender, who has in mind one meaning, encodes a message using visual or verbal symbols and the receiver, in decoding the message, affixes another meaning. Errors can occur when the sender incorrectly encodes the message or when the receiver incorrectly decodes it. Structural blockages result when an individual is isolated with no communication channels, or is uncertain what information is to be transmitted to each of several possible destinations, or possesses information but either does not know that it is needed or does not know when it is needed.

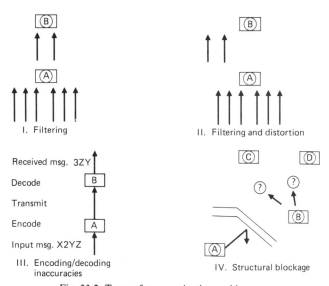

Fig. 23-2. Types of communication problems.

Filtering and Distortion

Formal organizations rely on a free flow of information vertically, up and down the scalar chain; horizontally; and laterally. However, free flow does not mean an open pipeline that transfers all information. Instead, organization members are expected to act as filters, passing on to supervisors and subordinates only information needed for efficient performance of duties and satisfaction of personal needs. The extent to which supervisors and subordinates alike filter out unnecessary information and transmit important information has to affect organizational performance.

Studies have confirmed that less than the total of all information is passed along and have provided some quantitative measures of the extent of this filtering. For instance Boyles and Wicker, using a behavioral laboratory experiment involving managers acting out roles in research and development projects, found that only approximately 54 to 65% of availahle information was passed upward from subordinates to supervisors.[3] In another study, O'Reilly and Roberts found that subjects who were role-playing managers in a staff specialist department (personnel) communicated, on the average, approximately 73% of available information.[4] Thus, allowing for differences in settings, tasks, and information available, one should be prepared to see only about 55 to 75% of the problem-related information potentially transmittable to immediate supervisors actually transmitted.

The filtering of from 25 to 45% of information in upward communication is not, in itself, a matter of concern. In fact, if the information filtered out is unneeded and no important information is excluded, the communication process is working just as it should. Unfortunately, some important information is likely to be withheld for one or more conscious or nonconscious reasons. Both the Boyles and Wicker and the O'Reilly and Roberts studies indicated the possible magnitude of such filtering of important information (see Table 23-1). In

Table 23-1. Information passed upward.

	PERCENT INFORMATION SENT TO SUPERVISOR		
	BOYLES AND WICKER STUDY		O'REILLY AND ROBERTS STUDY
COMPARISON	1ST EXPERIMENT	2ND EXPERIMENT	
Total information passed as percent of total information possible	54	65	73
Important information passed as percent of important information possible	63	77	72
Important information passed as percent of total information passed	83	85	—

both studies, independent judges preidentified important items of information as such. Using the percent of the total important information available that was actually transmitted upward as a measure, the two studies yielded similar results. In the O'Reilly and Roberts research, only 72% of availabe important information was transmitted, while in the Boyles and Wicker research only 63% was transmitted in one experimental setting and only 77% in another. Thus, to the extent that the role-playing experiments reflect the filtering (and distortion) in the work setting, supervisors may be denied one-fourth or more of the information they should have to perform their jobs.

An additional indication of the failure to accurately and usefully communicate to superiors is provided by Boyles and Wicker. Their data show that only 83 to 85% of the total information passed upward was classified as important. This means that, in addition to denying supervisors some of the information they need, subordinates also send along unimportant information.

A considerable, although incomplete, research-based body of knowledge exists concerning the factors associated with blocking, withholding, or otherwise passing on to supervisors less than the total amount of information possessed by a subordinate. This research has explored a number of variables posited as being related to the content and accuracy of upward communication. These variables include mobility, status, ascendency drive, level of insecurity, trust, influence, autonomy, and organizational climate.

With respect to mobility opportunity and status, Kelley found that the task relevance of upward communication was related to both the communicator's status position and possibility for upward mobility.[5] That is, the more unpleasant the position—e.g., low status and nonmobile—the greater the tendency to transmit irrelevant information. Similarly, Cohen concluded that content of communications, in terms of length and amount of irrelevant information, was related to the rank (based on power and status) and upward mobility of the communicator.[6] In particular, low-ranking nonmobile individuals tended to communicate more irrelevant information.

When upward mobility aspiration was tested as an explanatory variable, the conclusions were mixed. For example, both Read[7] and Komsky and Fulk[8] found support for the existence of an inverse relationship between upward mobility and accuracy of upward communication. In addition, Athanassaides found, with respect to two occupation groups, that distortion of upward communication was positively related to ascendency drive (similar to mobility aspirations) and level of insecurity.[9] Further, O'Reilly found indications that less intentional distortion was associated with low mobility aspirations. Thus, it would appear at this point that the more upwardly mobile the subordinates, the greater the extent of inaccurate upward communication. However, O'Reilly's actual conclusion, based on weak statistical support for this relationship, was that mobility aspirations are of little importance in explaining intentional distortion.[10] This conclusion was also made by Burchette, Porter,

Blalack, and Davis as a result of their study of communication in four organizations.[11]

Support for the existence of relationships between both accuracy and openness of upward communication and trust in one's superior and the influence of the superior, has been provided by Read[7]; O'Reilly and Roberts[4]; Burchette, Porter, Blalack, and Davis[11]; and O'Reilly.[10] The degree of trust was found to be strongly and directly associated with the degree of open, accurate, complete, or undistorted communication in all studies. Influence of the superior, although directly correlated with open or undistorted communication, was considered to be of much lesser importance by O'Reilly[10] and Burchette, et al.[11]

Finally, autonomy and organizational climate have been investigated as factors affecting upward communication. Athanassaides found autonomy of authority in performing work to be inversely related to the distortion of upward communication.[9] That is, where authority was more widely distributed among organization members, less distortion occurred.

More recently, Muchinsky presented weak support for the belief that accuracy of communication is positively related to several dimensions of organizational climate.[12] Significantly, Muchinsky found a direct correlation between organizational climate and trust. This adds support to the belief advanced by communication theorists and many practicing managers that a supportive climate builds interpersonal trust. Based on the research results presented here, we can carry the causal inference further to suggest that trust leads to more open, more complete, less distorted communication.

Encoding and Decoding Differences

The encoding-decoding problem is the familiar but difficult one of semantics, or meanings that we attach to symbols, where one person's or group's meanings are not the same as another's. The result is the all-too-frequent feeling that the "other side" are not listening again, or only hear what they want to hear, or are deliberately acting counter to one's wishes and hence to the organization's best interests.

Despite the paucity of empirical research aimed at measuring the extent of and reasons for encoding and decoding errors, personal experience with this interpersonal barrier to accurate communication is so widespread that it is generally recognized. Hayakawa,[13] Haney,[14] Thayer,[15] and Rogers and Roethlisberger[16]—to name just a few among a great many—have written on the subject. For example, Haney describes the process of interpersonal communication, the factors that influence differing perceptions of situations, and the role of defensiveness, and prescribes several practices to lessen miscommunication. These practices include internalizing questions concerning whether a statement is one of fact or merely an inference, which particular subject a statement pertains to, and what other meanings could be attached to the message. In general Hayakawa, Haney, Rogers, Roethlisberger, Thayer, and other authors clearly and often dramatically reveal how different interpretations of messages can come about because of honest differences in meanings assigned to symbols. The actual extent and degree to which the affixing of different meanings to messages occurs within and affects business organizations is probably highly variable, depending a great deal on similarity of backgrounds, values, objectives, and experiences of the communicating individuals. Differences in meanings can be fairly great even when individuals have much in common. For example, in a study involving communication between executives, questions concerning discussion during interaction revealed differing perceptions of what had been the subject of previous discussion in one-third of the cases.[17]

Two conditions likely to exist in the matrix organization heighten the possibility of encoding-decoding problems. First, the matrix team normally consists of a number of individuals from identifiably distinct occupational specialties, The specialized language or jargon that goes with each should be expected to cause problems, when messages from one specialist are interpreted by another. In addition, the loyalties to professions and the acquired values of various specialist group members are likely to result in group frames of reference that view such subjects as cost, time deadlines, and technical perfection in different ways.

A second condition increasing the chance of miscommunication is the frequent reliance on verbal communication. Such communication is often in response to an urgent need and hence is nearly spontaneous and probably unrehearsed. These conditions preclude the careful selection of words and a setting for the conversation where both speaker and listener can concentrate on communicating.

Structural Blockages

Confused communication-reporting relationships can result when channels and requirements are not clearly established and understood, when spatial arrangements make communication difficult, when work flow does not dictate a mutual interest in coordinating, and when individuals are under severe time pressure. All these conditions can exist in matrix organizations.

The temporary nature of many matrix-type arrangements and, in some cases, lack of precedent, lessens the feasibility of establishing formal channels for the issuing of instructions, reporting of results, and coordinating work efforts. Further, to the extent that departments, sections, groups, or individuals are isolated from one another, the opportunities for informal and impromptu communication are greatly lessened. This isolation can be a result of geographical or physical separation; social isolation; lack of work association (being outside the mainstream of activity); or being on different organizational levels, (the lower the level, the greater the potential for isolation).[18] The combination of functional group separation and isolation, coupled with the lack of formally prescribed channels in many matrix management applications, means an inadequacy of reliable communication pathways.

Based on research, Simpson developed a tentative typology of communications frequency and degree of mechanization.[19] In essence, he suggested that a low degree of mechanization is associated with a high level of vertical communication, as instructions and progress reports are frequently needed to control and track the effort. A high level of vertical communication is also to be expected when level of mechanization is very high, as in automation, because of the need for prompt corrective action when machines break down. On the other hand, medium mechanization, as in the typical assembly-line type of operation, would be associated with a low level of vertical communication, because the machines pace the operation, thereby minimizing supervision. Matrix management situations are generally expected to involve low mechanization and a high need for vertical communication, yet unlike many industrial situations the various work tasks frequently stand alone and lack procedural interdependency. As a result, although communication is needed for control, there is little inherent in the tasks to encourage the individuals accomplishing them to communicate.

The pressures of time that frequently accompany projects, task force efforts, new product development, new ventures, and other types of matrix management applications present two related kinds of difficulties. One is the lack of time to communicate or prepare messages. When coupled with considerable freedom concerning what is communicated and how, this lack of time encourages individuals to postpone or forget about communications whenever other duties become particularly demanding. Another difficulty is the tendency for some individuals to at times become overloaded with work as a result of unforeseen events, scheduled compression, and compounding of problems under conditions of substantial uncertainty. The result is the same: lack of explicit communcations requirements often means that team members simply never get around to communicating some of the information that is needed.

COMMUNICATION ROLES

Given the nature of matrix management functions and structure; the importance of communication to the project product, or task; and the communication problems inherent in organizations in general and in matrix organizations in particular, two distinct communication roles emerge for the matrix manager. Each of these roles has associated needs with respect to developing a desired quality and quantity of communication.

One role is that of internal communication facilitator. Here the manager makes sure that necessary information is routed among team members in the form needed and at appropriate times. This role requires the full, free disclosure of information by team members. Therefore the principal needs for the matrix manager are to reduce the filtering of important information, reduce the distortion of information, establish an open and favorable climate, and establish appropriate channels for passing messages.

The second role is that the boundary spanner, wherever the matrix manager constitutes the interface between the team and important external contacts. As such, the matrix manager is a focal point both for information provided by the team and needed by outsiders, and for information needed by members of the team but deriving from outside sources. In carrying out this role, the manager must elicit accurate information and attain a high level of accuracy in the decoding and re-encoding processes, so that unintended meanings are not attached.

IMPROVING COMMUNICATION

As the needs discussed above suggest, if the matrix manager is to effectively carry out the communication roles, he or she must receive full disclosure of accurate and useful information from team members. Yet a restricted flow of information in upward communication is highly probably in the matrix management environment. There are three conditions that point to this high probability. First, team personnel do not always know what is important or unimportant. The lack of clearcut channels of communication and the lack of reporting formats can leave them uncertain as to just what is wanted and when. In addition, team members typically lack knowledge of the total effort and thus of what may be newly important at any given time. Second, in the absence of formal channels and reporting formats, the channels that are characteristically used may inadvertently cause filtering. In an unrehearsed face-to-face or telephone conversation, it is easy to forget, or to be led to a different topic, before a given topic has been concluded. Also, signs of impatience or tiring on the part of supervisors may cause subordinates to reduce message content in order to "wind it up." Third and last, team personnel may fear that they will be penalized for passing on unpleasant news or information that may reflect unfavorably upon themselves. The changing composition of the matrix team, and the specialized and sometimes isolated nature of their work, means that many team members may not get to know their supervisor well enough to know if they can trust him or her with such information. The extent to which supervisors can cope with these three conditions that favor filtering, distortion, or blocking information will be reflected in an improvement in full disclosure of important information.

As early as 1957 Simon provided insights concerning several conditions, each of which will result in the upward communication of information.[20] First, he indicated that its transmission must not result in consequences unpleasant for the sender. Second, if the superior would obtain the information anyway, the subordinate is likely to go ahead and pass it on first. Third, if the superior needs the information to deal with his or her own superiors, would be embarrassed without it, and makes it known that he or she would be unhappy if it weren't provided, the subordinate is likely to pass the information along.

Key Considerations

However, as sound as Simon's precepts may be, there are more basic considerations in developing a useful flow of communication. They key aspects of any effort to effectively improve the upward communication of important information are recognition, creation of a supportive climate, and creation of channels of communication.

The most crucial step toward reducing filtering and distortion must be a recognition that important information is filtered out or withheld in the upward communication process, and a desire to improve on the situation. The figures given earlier, indicating that 23 to 37% of important information is filtered, provide a useful baseline for deciding whether or not to even attempt a solution. If these figures are acceptable, then the matrix manager can probably dismiss filtering as a problem area. If they are not acceptable, then the manager needs to (1) internalize the question "Am I getting it all?" when receiving a communication, and (2) effect changes to improve the chances of getting it all.

Variables for Manipulation

Research on communication provides a basis for classifying behaviorally and structurally derived variables in terms of their probable impact on the transmission of information. The results of such a classification are presented in Table 23-2.

The variables involved in encoding and decoding differences are many and very complex. Generally, manipulation is not practical. For example, changing peoples' value systems would be a major and probably futile undertaking. Similarly, developing a common language with single meanings for words is beyond the power of the matrix manager. The only variable the manager can feasibly work with is his or her own degree of understanding of others on the work team, and the meanings they are likely to attach to symbols (and what symbols they are likely to use for particular meanings). To the extent that the manager can empathize with and develop an understanding of team members' frames of reference, he or she will more accurately anticipate what senders meant by the symbols used in message transmission, and will more accurately anticipate the meanings that receivers will place on messages. Since it is very difficult to deeply understand others, this variable in shown in Table 23-2 as having at

Table 23-2. Potential for manipulation to improve upward
communication.

VARIABLE	POTENTIAL		
	HIGH	MODERATE	LOW
Individual			
Understanding of Others' Frames of Reference		X	
Understanding of Languages		X	
Behavioral			
Ascendency Drive (notably aspiration)			X
Influence			X
Security		X	
Trust	X		
Structural			
Mobility Opportunity			X
Status			X
Organizational Climate	X		
Autonomy		X	
Security			X

best moderate potential for manipulation. Developing an understanding of the languages used by others on the job has moderate potential also. This is not because the matrix manager could not, given the time, learn to be a cost estimator, controller, accountant, and chemical engineer, for instance, but rather because of the infeasibility of doing so. At best, the manager can increase his or her *degree* of knowledge of the specialist fields employed in the matrix effort.

Several other variables—those that relate to the setting within which messages are communicated—are potentially manipulatable. Trust and organizational climate are considered to have a high potential for manipulation. This is because of a combination of strong relationships found through research, and the opportunity of managers to do something on their own about them. The inability of the individual matrix manager to do much about (1) the subordinate's desire for advancement, (2) the manager' own influence as perceived by others, (3) the status differences among jobs that relate solely to the relative position of the job in the hierarchy, (4) the career mobility associated with a job or specialty, or (5) the security attached to a job, constitutes the principle reason for categorizing these variables as having low potential. Although the feeling of security and the autonomy of work performance were supported as upward communication influences in only one study cited here, the creating of a nondefensive, secure environment and the providing of limited opportunities for self-direction are within the capability and authority of many managers. For this reason, a climate of autonomy and security are classified as variables of moderate potential for lessening filtering.

Prescriptions for Action

Efforts should focus on those variables classified as having high and moderate potential to improve upward communication. Actually, an understanding of languages, understanding frames of references, a supportive organizational climate, and trust between supervisors and subordinates throughout the organization are the only variables that need be directly addressed. In matrix organizations team members typically already have a reasonably high degree of autonomy of action and trust. This in itself will lessen perceived threats to an individual's security.

The effectiveness with which information obtained from various sources is translated into a language understandable by intended receivers depends in large part on the matrix manager's familarity with technical languages. Developing a *working knowledge of the various technical specialties* involved is very important from the standpoint of encoding and decoding. Understanding frames of reference and reducing perceptual differences is attainable, in part at least, through several means. Special outside training sessions using behavioral training models such as transactional analysis (TA) or the Johari window are sometimes available.[21] A do-it-yourself everyday practice that can help the patient manager is *active listening in a nonevaluative manner,* where a deeper appreciation of what the other person is thinking and feeling is sought. Haney has also suggested that paraphrasing or asking questions is a good way to help clarify meanings whenever they might be in doubt.[14]

Since a supportive organizational climate and trust seem to be closely related, we will cover actions that can be taken by the matrix manager, in the normal course of duties, to improve both at the same time. First, the matrix manager (and other team supervisors) needs to *develop an effective presence.* The supervisor needs to be seen in a friendly, businesslike manner and to project an image of professional competence. The findings by Burchette, Porter, Blalack, and Davis, indicating that trust in the superior is positively correlated with the percent of information received from the superior, the percent of time in contact with the superior, and the perceived accuracy of information received from him or her,[12] also indicate that supervisors should devote time and attention to downward communication. To the extent that supervisors are willing to communicate their feelings, concerns, problems, and hopes to subordinates, such downward disclosure can be expected to serve as a model for the subordinates' upward disclosure. Of course the supervisor must be careful not to assume an undue role of confessor and counselor.

Second, the supervisor should *visibly demonstrate commitment to the job.* Again, this means assuming a role as a model to team members. These team members may become committed to the project or effort, even though they may be there only for a relatively short time and have no personal commitment to the matrix manager. A personal commitment to the effort, and hence a personal desire to see it not get into trouble, can provide an incentive for more complete upward disclosure.

Third, the supervisor must *take time to listen* and not appear reluctant to do so, or seem to lack the time to listen to everything the subordinate has to say. The supervisor should accept what the subordinate has to say for its importance not only to the supervisor but also to the subordinate. Finally, the supervisor needs to *consistently reward open upward communications* through acknowledgement, praise, and public credit.

As pointed out by O'Reilly, even with trust there is a bias against passing unfavorable (though important) information upward.[10] Therefore the manager needs to work further to provide a structure more conducive to upward communication. This structuring does not necessarily mean setting up rigid rules, procedures, offices, and otherwise attempting to prescribe exactly what will be communicated, and when and how. Rather, those less formal kinds of structural adaptations are suggested that matrix managers and supervisors can easily make on their own. The first such adaptation is for the matrix manager to *assume a major role as a communicator.* The manager needs to take actions to *become established as a communication focal point,* seeking information, asking questions in a nonthreatening, nonevaluative manner, and passing on information to subordinates. If the matrix effort is large, he or she may not be able to act as the total communications focal point. In this case, effort should be expended to *identify those persons or positions that seem to be communication nodes* and establish more frequent and open communication with them. If no such communication liaison positions exist and if communications problems so warrant, capable communicators may be shifted so as to place them where they are most likely to receive and pass on messages. As much as possible of this personal communication by managers and liaisons should take place in the team members' workplace or on neutral ground, rather than in the manager's office. Such on-location communication activity tends to reduce the structural barriers that frequently attach to status and power differences.

A last structural change would be to formally and informally *establish the general and specific kinds of information that are considered important.* This does not mean creating a catalogue of new standard reports. Standard reports are often not very applicable even in standard situations, let alone in project or task-type nonstandard situations. Therefore it is probably better not to require such reports, which waste subordinates' time and cause them to question the supervisor's competence. A more appropriate technique is to simply *disseminate updated ideas about what is important* in meetings, conversation, and memoranda or bulletins on a recurring basis as needed.

Isolated individuals or groups in the matrix, who for one reason or another cannot feasibly be moved, can be included in the communication network without setting up elaborate or formal reports or mechanisms. All that is needed is either to establish liaisons where they can visit the isolated members, or to take the extra effort to personally talk with them more frequently. The essential is not to formalize the contacts or their regularity, but rather to include communication visits as an important item on the agenda and see to it that they are carried out.

CONCLUSION

Lessening the extent of upward filtering, distortion, and blockage of important information can be accomplished by matrix managers without significant added cost and without special behavioral training. With the more complete and accurate information thus obtained, matrix managers can more effectively perform their vital roles of internal communication facilitator and boundary-spanning agent. All that is needed is a commitment to bring a better communicator within the context of the manager's job as presently defined.

On the matrix manager's part, this commitment involves:

1. Recognizing that upward filtering and distortion do occur and that they are extensive, and wanting to lessen them;
2. Thinking of communication as a major supervisory role and spending a lot of time on it;
3. Developing a working familiarity with the jargon used in the various specialties represented on the matrix team;
4. Being a model of commitment to the job and to team members;
5. Letting matrix team know the kinds of information that are important;
6. Becoming a communication focal point, identifying other focal point persons, and communicating with them;
7. Actively interacting with and communicating information to subordinates;
8. Actively seeking information, listening, and asking questions and;
9. Rewarding upward communication.

REFERENCES

1. Jerry C. Wofford, Edwin A. Gerloff, and Robert C. Cummins, *Organizational Communication* (New York: McGraw-Hill Book Co., 1977), p. 390.
2. Harold Stieglitz, "Barriers to Communication," *Management Record* 20, (Jan. 1958): 2–5.
3. David J. Boyles and Harvey L. Wicker, "Deliberate Filtering of Upward Communication and Awareness of Self and Others in Air Force Subordinate-Supervisor Dyads," unpublished master's thesis, no. LS54 26-77B, Air Force Institute of Technology, 1977.
4. Charles A. O'Reilly III and Karlene H. Roberts, "Information Filtration in Organizations: Three Experiments," *Organizational Behavior and Human Performance* 2 (1974): 253–65.
5. Harold H. Kelley, "Communication in Experimentally Created Hierarchies," *Human Relations* 4 (1951): 39–56.
6. Arthur R. Cohen, "Upward Communication in Experimentally Created Hierarchies," *Human Relations* 11 (1958): 41–53.

7. William H. Read, "Upward Communication in Industrial Hierarchies," *Human Relations*, 15 (1962): 3–15.
8. Susan H. Komsky and Janet L. Fulk, "A Multi-role Approach to Perceptions of Message Distortion," paper presented at the 41st Annual Meeting of the Academy of Management, San Diego, 1981.
9. John C. Athanassaides, "The Distortion of Upward Communication in Hierarchical Organizations," *Academy of Management Journal* 16 (1973): 207–26.
10. Charles A. O'Reilly III, "The Intentional Distortion of Information in Organizational Communication: A Laboratory and Field Investigation," *Human Relations* 31 (1978): 173–93.
11. Hugh T. Burchette, Robert O. Porter, Richard O. Blalack, and Herbert J. Davis, "Gatekeeping and Upward Communication: Another Test of Three Contributing Factors," paper presented at the 37th Annual Meeting of the Academy of Management, Orland, Fla., Aug. 1977.
12. Paul M. Muchinsky, "The Interrelationships of Organizational Communication Climate," proceedings of the 37th Annual Meeting of the Academy of Management, Orlando, Fla., Aug. 1977. pp. 371–74.
13. For example see S. I. Hayakawa, "How Words Change Our Lives," *Saturday Evening Post*, Dec. 17, 1958, pp. 22ff., reprinted in Richard C. Huseman, Cal M. Logue, and Dwight L. Freshley, *Readings in Interpersonal & Organizational Communications*, 2nd ed. (Boston: Holbrook Press, Inc., 1973), pp. 13–24; or S. I. Hayakawa, "Conditions of Success in Communication," in William V. Haney, *Communication and Interpersonal Relations*, 4th ed. (Homewood, Ill.: Richard D. Irwin, Inc., 1979), pp. 234–45.
14. William V. Haney, *Communication and Interpersonal Relations*, 4th ed. (Homewood, Ill.: Richard D. Irwin, Inc., 1979).
15. Lee O. Thayer, *Administrative Communication (Homewood, Ill.: Richard D. Irwin, Inc., 1961)*.
16. Carl R. Rogers and F. J. Roethlisberger, "Barriers and Gateways to Communication," *Harvard Business Review* 30 (July–Aug. 1952): 46–52.
17. Tom Burns, "The Directions of Activity and Communication in a Departmental Executive Group," *Human Relations* 7 (1954): 73–79.
18. Keith Davis, "Management Communication and the Grapevine," *Harvard Business Review* 31 (Jan.–Feb. 1953): 43–49.
19. Richard L. Simpson, "Vertical and Horizontal Communication in Formal Organizations," *Administrative Science Quarterly* 4 (Sept. 1959): 188–96.
20. Herbert A. Simon, *Administrative Behavior*, 2nd ed. (New York: The Free Press, 1957).
21. These techniques generally involve getting to better understand others through getting to know yourself and through the development of open, rational, and nonevaluative behavior. A good general description of transactional analysis and the Johari window as tools for self-enlightenment is contained in Chapter 5 of William V. Haney, *Communication and Interpersonal Relations*, 4th ed. (Homewood, Ill.: Richard D. Irwin, Inc., 1979).

24. Whys and Why-Nots of Matrix Management

Paul C. Dinsmore*

INTRODUCTION

The idea of the camel is said to have been conceived by a committee whose initial task was to design a horse. The give-and-take compromise of the committee members resulted in something that was awkward-looking and cumbersome, but highly effective in the inhospitable desert environment for which it was created.

The matrix may be the organizational cousin to the camel. Surely the original organizational premise was not based on the ambiguous conflict-raising structure that finally evolved. Yet the matrix model emerged, and despite its ungainliness it has proven effective in given complex organizational habitats—often as inimical as the camel's desert locale! Another point common to the matrix and the camel is that outside their natural habitats they are equally strange creatures, both clumsy and very much out of place.

Matrix management is chock full of pros and cons. There are justifications for heralding the matrix as the salvation for curing organizational blights. On the other hand, an accusing finger can often be pointed straight at it as the culprit for some critical forms of corporate strife. There are reasons for brickbatting the matrix, as it can be incredibly complex, highly frustrating, and sometimes just plain unworkable. Yet there is justification for jumping on the organizational-change bandwagon, stemming sometimes from the inoperability of other traditional organizational forms. Full-fledged horn-tooters for the matrix are relatively rare creatures, even though many support this management form as a solution that somehow meets a need, despite the difficulty of putting it into practice. The matrix can be equated with the institution of marriage, which some question and criticize, but which, in spite of difficulties and frequent confrontations, has proven a highly satisfactory form when successful "consensus management" is practiced by the husband and wife (who play the matrix manager roles in the relationship).

---

*Paul C. Dinsmore has occupied executive-level managerial positions for various multinational companies in the engineering design and construction industries. He has also served in consulting capacities in the fields of project management and executive time management. He has presented professional papers, written numerous articles on management applications, and has completed his second book on project management.

The swing to matrix is seen in an ever increasing number of organizations and groups. Varying approaches reflect individual company peculiarities, cultural differences, and specific phases of organizational development. Some matrix organizations evolve naturally owing to the type of work, or the nature of the people performing the work. Many project-oriented tasks take on characteristics of this hybrid organization, even though the participants may not be aware of it and the term "matrix" may never be used. Other organizations swing toward matrix because the approach being used does not meet the need. This reflects a philosophy that says, "What we're doing now doesn't seem to be working, so let's try something else"—like the tennis ground-stoker who, taking a beating in the back court, decides to change to forceful volleying at the net as a change in game plan. Others may go with the matrix in an attempt to stay on the foreward fringe of managerial advances in organizational development.

The move to the matrix is often made for the right reasons, yet sometimes the change is not really justifiable. Even when made for the proper reasons, no guarantee for success exists; the matrix is "fiendishly difficult," as Drucker reminds us, to put effectively into operation.[1]

In this chapter, the *whys* and *why-nots* of matrix management are explored. What justifies the risk and potential pain of making such an organizational change? Can the need for the shift be seen ahead of time, or should circumstances come to a head to help spear on the organizational evolution? What solid reasons indicate that the new form should be more effective than the present one? What circumstances point down the matrix path? On the other hand, what are good reasons for not going with the matrix? Which situations would better be met by another structure? What opposing forces push the pendulum back toward more traditional approaches? The basics behind these questions will be discussed.

THE MAKE-UP OF MATRIX PROS AND CONS

In greater or lesser degree, most organizations contain the ingredients necessary for matrix management. Some are loosely woven, with extensive informal communication channels and participative decision-making processes. Others are "tight ships" that move information, people, and things only as determined by all-encompassing systems, standards, and procedures. Still other organizations are hybrid forms or mutations containing both the raw material for a smooth matrix operation and an array of stumbling blocks that can render it inoperable.

Strong justification is required before putting any organization into the matrix mode and its subsequent push-pull tussle. Solid reasons must be mustered to offset the risk and natural human resistance to an organizational approach so complex that (as Figure 24-1 suggests) it can't be represented in

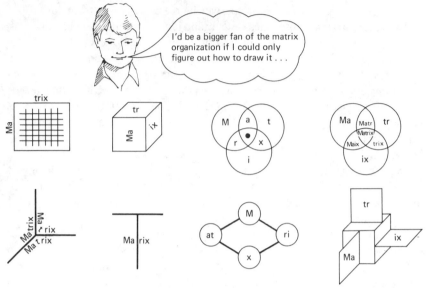

Fig. 24-1. The Matrix is hard to draw.

a reasonable graphical form, much less by the neatly positioned boxes and lines of the conventional organization chart. There are reasons that appear to justify the matrix, but not all these, alone or in certain combinations, are sufficient to inspire the move.

Davis and Lawrence, in viewing the situations calling for the matrix, boil down their conclusions to three basic conditions: (1) outside pressure for dual focus, (2) pressure for high information-processing capacity, and (3) pressure for shared resources.[2] The authors conclude that where at least two of these conditions exist, the matrix structure should be seriously considered as an organizational form; where only one or none of those conditions appears, a shift toward the matrix should be sharply questioned. While the conditions, standing alone, might not be enough to justify a swing toward the matrix structure, each one carries a strong incentive to move in that direction. For instance, outside pressure for a dual focus may represent a need to meet customer needs through a central spokesman, even though the work is developed on a discipline basis within the company; such pressure could well result in a project coordinator who then would operate in a matrix style within the company. The matrix can also relieve the pressure for high information-processing capacity, since its parallel-channel, less formal communications patterns give it a sizable advantage over other forms. The problem of pressure for shared resources can also be well met by the matrix, which makes for more efficient utilization of resources than, say, multiple-project structures that duplicate certain functions. The cumulative affect of these conditions, however, is considered more than sufficient to offset the difficulties of dealing with the conflict-prone, fuzzy organizational compromise called the matrix. Another approach to analyzing matrix pros and cons is to discuss some justifications heard along "managers' row" in support of, or in opposition to, that organizational form. Some common views have been paraphrased in the following text and grouped under *why* and *why-not* sub-headings. Each view is then described with regard to how it supports or opposes the matrix organization.

Whys

Reasons for adopting the matrix are born largely out of the incapacity of other organizational forms to "fill the bill" under certain circumstances. Some of the classic day-to-day complaints result from sticking to traditional organizational forms when the times and circumstances call for a matrix stance. Here are some justifications for giving serious consideration to the matrix.

What we're doing isn't working. A change in organizational strategy is one of several points to be analyzed when things aren't going right. The overall "game plan" and the quality of personnel and motivational programs are other aspects that should be thoroughly explored. Prior to any quick organizational change, the move should be checked against the basics. Matrix-testing questions such as the Davis and Lawrence conditions need to be posed: "Is there pressure for a dual focus?" "Is there a high information-processing requirement?" "Is there a need for shared resources?" If things aren't working because of affirmative answers to these queries, then the matrix may indeed offer the path to better performance.

We've pyramided into a giant bottleneck; things don't move. The hierarchy, or classic bureaucracy, isn't known for record-breaking speed at problem solving. While some functional organizations work better than others, all suffer from "vertical-itis," or forcing the communications flow up and down the hierarchy. The manager who strives to keep tabs on everything happening in his or her area becomes the organizational bottleneck, inadvertently getting mired down in the middle of a busy communications intersection. A solution to excessive verticalization is to strike a balance with horizontal communication practices. Such a slant toward opening more agile horizontal channels naturally breeds a shift in power, bleeding authority from the functional manager's previously sovereign domain. This change—needed to break up the managerial bottleneck—results in a matrix approach, redistributing the balance of power and opening "horizontalized" communications lines. Thus the matrix is a valid approach for loosening up the organization to "make things move."

We're not meeting our customers' needs. Many a customer or client who has tried to accompany the evolution of a design, systems, or equipment-manufacturing package in a functional organization, knows the real meaning of a "wild-goose chase." Functional groups set up along discipline lines all know why their part of the project isn't further along—because some other functional group hasn't come through with its part! The customer trying to follow up on the project by identifying problem areas at the supplier's plant, becomes involved in an in-house pointing contest but gets no satisfactory answers to the questions asked. Getting no overall picture of what is happening, he or she feels the urge to cry out: "Who's in charge around this place, anyhow?" or "Can't anybody give me a straight answer?" A classic matrix application will fill the void by establishing a project *coordinator* to centralize customer contacts and be responsible for giving a single coherent response to customers' questions. The coordinator, needing basic information, will also be acting in matrix fashion within his or her own organization, helping to hurry the project in and out of interdisciplinary interfaces. For bettering response to customer needs, the matrix offers a suitable solution.

Our technical people feel they're not getting their say. A strongly project-oriented organization often leaves some of the technical personnel out of the decision-making mainstream. If further technical input is in the interest of the project, a shift to matrix would help balance the functional-versus-project power ratio, thus allowing the technical people to manifest themselves as they feel inspired. Without such a power base, a functional manager in an aircraft industry might lament: "I feel confident that the plane will be finished on time and within budget, but I don't think it will fly!" The matrix gives everybody, including the technical people, "elbow room" to maneuver about the project structure and add their contribution—a simple final touch, or in graver circumstances the raising of a red flag and a call of "Back to the drawing boards!" The matrix, then, is an appropriate solution for bolstering technical personnel who are being given the "steamroller" approach by a management-heavy project group.

There's disgruntlement in the ranks. Job satisfaction can be enriched by going to the matrix, although the route is mined with treacherous pitfalls. Take the case of the functional organization that is frozen up so tight by bureaucracy that virtually nothing moves: project participants tied into such a situation are bound to feel frustrated, so a slant toward the matrix should jolt loose some positive project action. Another example is the technical specialist being trodden under by the go-go project-type organization; a change to the matrix would give him or her a more active voice in the project picture. Therefore the matrix, when structured for the right situation, can attenuate certain types of disgruntlement such as "Nothing moves around here" or "There's too much bureaucracy," but it may also create some matrix interaction complaints like "What's-his-face keeps butting in all the time" or "There are too many people running this show." Dissatisfaction in the organization needs to be weighed; if the basic cause is an inadequate structure for meeting project requirements, then the matrix should definitely be considered.

We need to reconcile exacting technical requirements with the need to get the job done on schedule and on time. The matrix is the epitome of the compromise: a solution that doesn't leave any of the parties turning backflips of glee, but that is fair and represents a decent "halfway point." Any fast-tracking activity involving reasonably complex technology requires technical watchdogs with enough "clout" to keep the project sound. On the other hand, it also needs a group of goal-oriented "chargers" who are bent on meeting project milestones and budgets. The sometimes contradictory goals of (1) *developing the best technical solution* and (2) *getting on with the job* can be put into a state of "unstable equilibrium" when harnessed into the matrix organization. The matrix thus sets the stage for an organized push-pull that hopefully will best meet the needs of the target activity.

Things are too formal—everyone keeps writing letters and memos. Functional organizations are great breeders of written correspondence, and project organizations can spew forth reams of correspondence at a client. The matrix structure, however, tends to tear down formal communications channels and in their place offer simpler approaches such as *the telephone chat, the problem-solving meeting,* or *face-to-face negotiating.* Since many of the communications routes in the matrix are "dotted-line" or "diagonal," formal communications are sometimes impractical. Also, since the matrix musters up consensus-type approaches, a verbal tack at getting a decision is often more effective than trying to "go through channels."

We're caught in a quagmire of red tape. The matrix tends to blow red tape off into the sidelines. Concern for protocols, duplicated filing systems, obsessive sequencing, and extensive rubber-stamping tend to drift into the background as the "loose" matrix network moves people across the structure looking for solutions. This doesn't mean, however, that the matrix is necessarily swift and efficient; on the contrary, at times it can be exceedingly slow and cumbersome. Yet the eventual slowness stems much more from the complexity of interactions in the network than from excessive bureaucracy. A move to matrix can therefore offer benefits when it comes to cutting red tape.

Why-Nots

Reasons *not* to use the matrix organization also abound. Some of the negative stances result from natural resistance to change. Others are on more solid ground, reflecting justifiable reservations. Here are some common examples.

"What we're doing now is working okay." There is a strong justification for the *status quo* when it is getting the job done. If straightforward organizational approaches are meeting the need, then there is no reason to take on more complex forms; organization, after all, is simply a *means* to channel human and material resources so that the company goals are met in the most effective manner. The matrix is rarely adopted when another approach is working rea-

sonably well; it is often quite appropriately considered a "last-ditch" effort to be used when all else fails. Even if the characteristics of the company seem to call for a matrix approach, an organization operating satisfactorily should not be "matrixed"without weighing the risks of actually *reducing,* instead of bettering, organizational effectiveness.

"We've always done it this way." This may sound like an antiquated excuse for resisting organizational change, but, depending on how well-ingrained the existing procedure is, it may be advisable to give the organizational shake-up some second thoughts. A consequence/benefit analysis helps give perspective on the advisability of such a move. For instance, what are the consequences of making the move? Will the employees quit, productivity plummet? Will a morale problem arise? Will the cost of training be prohibitive? On the other hand, what are the potential benefits favoring the organizational change? Will profits increase? Service improve? Is the change one part of an overall philosophical shift that can be done gradually over time? "We've always done it this way" is surely not sufficient justification for resisting needed change, but it's a red flag that needs to be taken under advisement so as to stave off an unwarranted change for change's sake.

"The matrix is too confusing." The matrix can indeed be difficult to grasp and operationally confusing. Matrix structures are almost impossible to draw realistically, although some diamond-shaped structures and criss-cross shapes are sometimes shown. The fact is, the matrix breeds ambiguities and conflicts while operating in a state of managerial limbo. The consensus philosophy and balanced-power setting set team members to feeling their way along the foggy fringes of interface, forcing shaky formal authority to give way to competence and group synergy. If the personnel being targeted for a matrix shift are not prepared for working in undefined, nebulous situations, then the "too-confusing" syndrome may be big enough to make a matrix move impractical. Unless a program is developed to train project people to work in the fuzzy matrix world, the organizational structure might well be left as is.

"It's too difficult to determine who's in charge." From outside the organization, the matrix can make communications simpler—for instance, in the case of the client who is attended by a specific coordinator or matrix manager. But within the matrix the power relationships are often murky at best and, to further complicate matters, constantly shifting. For technically oriented subjects, the center of attention naturally shifts toward the technical groups. Informal "power surges" are common in the matrix, as the person most competent on a specific topic moves more freely within the matrix than, say, in a functional group. Just as such power grows, however, it also wanes, when other areas of expertise become more pertinent within the context of project priorities. For "Indians who are used to one chief," the matrix presents a confusing array of figures who at times may seem to pass the chief's headdress from one to another, and on other occasions seem to huddle under it all at the same time.

For such personnel, a switch to the matrix can represent a definite drawback that may be reflected by lower morale and a subseqeunt dip in productivity.

"The matrix appears to duplicate certain functions." The matrix has a built-in redundancy that in the right situations can generate dividends, and when applied out of context only breeds head-butting and duplicated costs. For instance, if a systems manager in a remote international office is tied to both the home-office systems director and the country manager, the matrix relationship can bring benefits if (1) complex technology is used, (2) tie-ins with other company systems are required, and (3) the volume of systems operations is large. On the other hand, if the systems are relatively simple and are restricted to isolated local use, participation of the home-office systems director might be unnecessary; the organization might work better "un-matrixed" and therefore nonredundant.

I like neatly defined responsibilities." Matrix structures play havoc with the definition of responsibilities. Individual responsibilities tend to become joint or group causes because of the integrated effort going into them. For personnel accustomed to neatly drawn action lines, the matrix must look like a toss-up between an "organizational mess" and a "messed-up organization." The properly structured matrix offers the virtue of creating a "we're-all-responsible-for-final-results" atmosphere. By the same token, however, the matrix structure can involve so many people that it dilutes responsibility to the point where *no one* feels responsible.

"Conflicts are a constant in the matrix." High on the list of matrix why-nots is the tendency to breed conflict. Not that all conflict is necessarily bad—on the contrary, the matrix proportions a controlled-conflict ambience that can prod projects along their schedule, avoiding periods of apparent calm followed by major crises. The conflict generated by the matrix might be likened to intensive diplomatic skirmishing between major international powers (controlled conflict), as opposed to war and peace situations (crisis management). Conflict, then, is part of the matrix game and must be plugged into the overall formula. Since the matrix is a compromise organization, conflict is one of the negative offshoots that must be constantly reconciled among the team members.

"Our functional department heads will never stand for sharing their power base." The switch from a functional organization to a balanced matrix can be traumatic for old-time department heads—and even more traumatic for the company, if those old-timers are well entrenched. Managers accustomed to very strong power bases are often reluctant to give up the "ego trip" of centralized command. A move toward the matrix is only feasible after an introductory phase, some artfully-aimed training for the functional managers, and a very skillful performance on the part of the new matrix managers. Unless steps are taken to set such a stage, strong resistance on the part of functional managers (who probably control most of present operations) is sufficient to put in question the decision to set up a matrix organization.

The *why-nots* are in some cases fully justifiable reasons, and in others only excuses or conscious road-blocking provoked by inner power struggles. A con that is applicable in one circumstance might not hold water in another that appears identical. Subtle differences in the organizational structure, company culture and traditions, and power tilts can be enough to scuttle a well-touted *why-not*, just as matrix *whys* are susceptible to rapid about-faces caused by almost imperceptible shifts in organizational personality.

COMPARING PROJECT, FUNCTIONAL, AND MATRIX PROS AND CONS

If ups have corresponding downs, ins their outs, and the Arctic its Antarctic, then do the matrix *whys* discussed in this chapter correspond to specific *why-nots*? For every matrix advantage is there a counteracting disadvantage? Is there a see-saw relationship between the pros and cons as in the case of the project-versus-functional structures?

Hemsley and Vasconcellos in *The Design of the Matrix Structure for R&D Organizations* describe the advantages and disadvantages of the matrix organization in terms of reconciling the *whys* and *why-nots* of two basic organizational forms: the functional and the project structures.[3] The authors point out that the matrix results in a compromise between the two types, balancing the weaknesses of one with the strengths of the other, achieving satisfactory results by almost all criteria of effectiveness, whereas the two basic forms tend to yield polarized results, either very positive or very negative. The reasons behind this polarization are outlined by the authors as follows:

(1) *Due date achievements* tend to be better in project organizations because of the centralized control over output-directed activities.

(2) *Technical results* tend to be better in functional organizations because of the greater concentration on the technical aspects.

(3) *Resource utilization* tends to be better in functional organizations because facilities can be utilized in various projects; this helps to reduce cost levels.

(4) *Cost performance* tends to be better in the project form because of the presence of management information systems concentrating more on cost control against budget than is generally possible in a functional organization.

(5) *Job satisfaction* of staff preferring to specialize tends to be better in functional organizations because they can concentrate on their speciality more effectively. For staff who prefer to develop broader capabilities the project form tends to be preferable.

(6) *Client feedback and control* tend to be better in project organizations because the project manager is well placed to provide the client with appropriate information.

(7) *Development of technological expertise* tends to be better in the functional form for single disciplines because of the possibility of focusing better

on the single area, but better in the project form for multi-disciplinary technology because of the greater interactions between the corresponding areas.

The matrix organization positions itself in an intermediate position between the extremes given in the project and functional structures, offering the advantages to a somewhat lesser degree, while attenuating the corresponding disadvantages. Shown another way, the *pros* and *cons* of project and functional structures can be lined up as shown on Table 1.

SOME FUNCTIONAL ORGANIZATION PROS	SOME PROJECT ORGANIZATION CONS
• Superior technical quality	• Tendency to relegate technical quality
• Better human resource utilization	• Periods of poor resource usage
• Job satisfaction high for technical personnel	• Job satisfaction reduced for technical personnel
• Greater technical motivation for single-discipline or traditional technologies	• Lesser motivation for single-discipline or traditional technologies

By inverting the position of the project and functional organizations, a similar chart results as indicated on Table 2.

SOME PROJECT ORGANIZATION PROS	SOME FUNCTIONAL ORGANIZATION CONS
• Ease in tracking schedules	• Difficulty in tracking schedules
• Cost control facilitated	• Cost control hampered by diffused control
• Strong client liaison	• Weak client liaison
• Stronger stimulus for multidisciplinary technologies	• Weaker stimulus for multidisciplimary technologies
• Job satisfaction strong for management-oriented personnel	• Job satisfaction weak for management-oriented personnel

By comparison of the advantages and disadvantages of each type of organization, it becomes obvious that each *pro* of the functional organization is paired with an opposing *con* in the project structure; and conversely, each con of the functional organization is related to a countering pro in the project structure.

While "cross-breeding" the functional organization with the project structure yields, to an appreciable degree, all the previously listed advantages of the functional and project structures on a toned-down scale, it also, according to Hemsley and Vasconcellos, spawns some disadvantages peculiar to the matrix itself. Those matrix *why-nots* are: (1) conflicts result from the dual authority structure and power ambiguities; (2) the organization is more difficult to put into practice; (3) management of the matrix is much more complex; (4) the indefinitions of organization exert a greater pressure on project team members.

Some strong points evolving from the matrix are: (1) communications become easier and more flexible; (2) once consensus agreements are reached

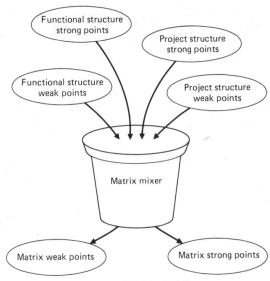

Fig. 24-2. The Matrix Mixer.

by matrix managers, putting ideas into practice is facilitated; (3) the organization can shift and adapt itself to new conditions.

The matrix structure thus strikes a compromise between the project and functional organizations. This cross-breeding results in trimming down some of the negative points while simultaneously reducing some of the positive points. As in all hybrid forms, new strong and weak points are born.

Matrix advantages and disadvantages are distinct. Each can be broken down into related *pros* and *cons* in project or functional organizations. The matrix *whys* and *why-nots* therefore are *compromise* positive and negative points and do not lend themselves to the same type of correlation found in the project-versus-functional comparison. (See Figure 24-2).

A CLOSER LOOK AT THE MATRIX ORGANIZATION

A close look at the subtleties of the matrix is required to complement the analysis of matrix management pros and cons. The matrix seen from afar may appear to have a relatively clear format, but a view from point-blank range can present a more nebulous view. The eyeball-to-eyeball perspective of the matrix can be likened to a cloudbank seen from the inside. The overlap and interfaces between the clouds are hard to see, but they are there! A close view of project phases, disciplines, and organizational structures puts into perspective the blurry interfaces that are so much a part of the matrix.

By nature, the matrix is not a fully stable organizational form; even though it is predicated on a "balance of power," two-boss concept, it tends to shift and adjust itself to the prevailing organizational winds—which are not always as "balanced" as theory would have it. The sway in the matrix results from new conditions or characteristics that set off reshuffling and subsequent accommodation. Other adjustments may be imposed by internal power struggles, company tradition, and overriding personalities. These are subjective factors that exercise strong influences on the matrix and can make the organizational pendulum swing far out of balance. Inadequate strategic planning and lack of training can also bend a theoretically symmetrical matrix into strange forms. Therefore, even though the matrix has the capacity to adjust itself to better meet the goals for which it is designed, it can also be pushed out of shape for less-than-objective reasons.

Either by design or as a result of the push-pull forces inherent in the organization, the matrix, in its search for equilibrium, tends to settle or pause for substantial periods in one of the following modes: project matrix, balanced matrix, or functional matrix.

Project Matrix

The project matrix adopts the basic matrix format, yet is slanted strongly toward a project philosophy. The project-oriented matrix coordinators and/or managers thus assume more decision-making power than the functional managers. Goals such as meeting due dates and performance within budget take on strong importance; the matrix channels exist, but the structure has strong project overtones.

The project matrix is particularly applicable for activities requiring limited technical resources that can be drawn periodically from an existing "pool." It is also an appropriate shift for intermediate phases of some projects, when the big project "push" is needed after the technical parameters and basic concepts have been firmly set.

Balanced Matrix

The balanced matrix distributes the influence and decision-making power equally between functional discipline managers and project (or task) coordinators and managers. Decisions are negotiated on an even footing, and consensus-type solutions are widespread. This results in an ongoing trade-off between overall quality (of the product, service, or project) and task-oriented goals (budget and milestones dates).

The balanced matrix is the model used in the discussion of matrix management *pros* and *cons* given previously in this chapter. Interestingly enough, the degree of conflict in a balanced matrix is often greater than in the two variations (project matrix or functional matrix). The predominating power is more clearly defined in those cases, whereas the balanced matrix corresponds to an ongoing fencing match in which tension-causing parrying for position is a constant.

Functional Matrix

The functional matrix is at the opposite end of the spectrum from the project matrix, in that the functional discipline manager exerts a stronger influence on overall activities. In the functional matrix, budget and achievement dates are subordinated to the greater concern for overall quality. This form sometimes results from a matrix born out of a well-entrenched functional organization, where tradition and resistance to change prevent the new structure from assuming a balanced posture.

The functional matrix can be appropriate where cost and schedule performance must indeed be subjugated to overall quality. However, on fast-tracking tasks requiring stricter budgetary controls, the technical matrix does not stack up well against the other matrix alternatives.

MATRIX VARIATIONS

The matrix can be built into various types of organizations to meet widely ranging requirements. Industrial applications abound for the matrix reflecting those situations in which an entire company is shifted into the matrix mode. Such industrial companies may have operations that are concentrated geographically, making for a compact matrix, or operations that operate all about the globe, stretching the matrix over both organizational and international boundaries. Projects are also good targets for the matrix. For instance, a large project organization may structure itself on a matrix basis, or the project may be plugged into an overriding home-office matrix. Double matrix approaches are also used on some big projects. Functional departments may be structured as a matrix internally, but relate to other departments functionally; or upper management may operate on the matrix while lower levels stick to functional or project organizations. The variations are many, and the twists peculiar to given organizations are surely more commonplace than the theoretical "balanced matrix" that academically spreads its branches evenly throughout the organization. Here are some examples of different matrix organization applications and the logic behind the type of usage.

Company-Wide Concentrated Matrix

Companies concentrated in one plant or geographic area frequently share resources effectively in a matrix structure; several projects can be performed by personnel from "pools" of specialists. In such a structure, information can be processed at a speedy rate. The matrix relationship for concentrated groups often occurs between discipline-oriented groups and task-oriented coordinators.

Company-Wide Scattered Matrix

The matrix spread over a wide geographic area also offers resource-sharing advantages, as home-office specialists can be matrixed into regional activities as needed. The matrix also meets the dual needs of regional focus and backup of technical expertise and direction. For the spread-out organization, overall company goals can be well met by effectively balancing regional management and main office expertise.

Functional Departments Structured on the Matrix

Functional departments can well be structured in matrix form, while maintaining a departmental posture in relation to the rest of the company. For instance, a department that requires rapid information flow and needs to strike a balance between project quality and schedule, yet is not allowed matrix relationships with other departments because of the in-company hierarchy, might find itself in this position. Although sometimes strange bedfellows, the project matrix and functional organizations coexist in most major firms.

Matrix Relationships at Middle-to-Upper Management Level

Another matrix criteria is for middle-to-upper management to intermingle on a horizontal basis, while departmental groups maintain their functional posture. The higher managers mix freely and come up with consensus-type decisions. Lower-level personnel, however, remain relatively isolated in their functional departments, with little liberty for crossing boundaries and discussing common problems with colleagues from other departments.

Project Operating under Overall Matrix Structure

This form adopts a project manager or coordinator, who with a limited administrative staff conducts the project by utilizing departmental personnel "loaned out" on a matrix basis, or by performing services from within the departmental structure. The project manager (or coordinator) has matrix privileges of working horizontally across the departments and negotiates with the department head necessary additional resources.

Independent Project Using Internal Matrix Structure

A project set up along functional lines ("engineering design," "procurement," and "construction," for instance) and beefed up with area coordinators working across the organization, is operating on a matrix basis, but within a project structure. Large projects are sometimes structured internally along matrix lines because of the rapid processing of information and the need to balance technical quality with schedule and cost goals. Project structures using the matrix often start with a strong functional slant, gradually shifting authority toward area coordinators or other task-oriented matrix power centers. Some projects go all the way to the double matrix, which means that the project is broken down into distinct packages with each package having area coordinators. The "double" part of the matrix comes in with general area coordinators who tie together the efforts of the "package" area coordinators.

AREAS OF APPLICATION FOR THE MATRIX

Matrix structures are applicable in various fields and cut across both service and manufacturing lines. Although the organization is not limited to particular industries, applications have become more widespread in some fields than others. The following uses of the matrix, classified by area of specialization, are given as examples.

Research and Development (R & D)

R & D work is performed by project, on a functional basis and in matrix form. The R & D efforts that are drawn into matrix form are those pressured by the need for optimizing resources and calling for a balance in the functional (quality-oriented) and project (result-oriented) approaches. Because of the high degree of specialization required in R & D, the matrix can be particularly useful in helping to give due importance to completion dates and budget requirements.

Architecture and Engineering Design

A classic application of the matrix structure is used by many architecture and engineering design firms. Such companies have a strong functional need (a permanent staff of specialists in given engineering disciplines such as soils mechanics, piping design, structural steel, electrical installations, etc.), along with a requirement for meeting project-oriented demands (client interface, milestone dates, and cost requirements). The matrix is particularly applicable for medium to small projects that can be handled "under one roof." Large engineering contracts with broader logistical needs tend to be organized more independently (as projects), although available technical manpower is often drawn from existing functional areas.

Made-to-Order Fabrication

Certain made-to-order fabrication operations lend themselves well to the matrix, provided the right combination of project-versus-functional pressure exists. If pressure for resource sharing is minimal, the organizational structure may naturally tend toward project organization, if, on the other hand, the operation leans toward an assembly-line activity, then a strongly functional approach may be more appropriate. A shipyard, for instance, might well use any one of the standard organizational forms: project, matrix, or functional. For the manufacturing of one giant supertanker using sophisticated technology, a project structure could be appropriate; for fabrication of hundreds of almost identical vessels, the functional appraoch might best meet the need. For a mixed combination of vessel types, however—all of them competing for common internal resources in terms of technical personnel, equipment, and materials, yet having distinct client requirements—the matrix offers a way to mediate the conflicting pressures.

Project Management of Superprojects

A sizable number of projects fit into the once exclusive club of capital projects with cost estimated in excess of one billion dollars. The spillover into the billion-dollar bracket reflects the growing trend toward bigness in an attempt to take advantage of economies of scale. Technological advances in materials handling, and improvements in logistics management, have paved the way for superprojects as well. Inflation has also effectively lowered the "entrance fee" into the superproject fraternity: a big-time investor might reflect, "A billion dollars today doesn't go as far as it used to!" In spite of the inflationary erosion, however, the billion-plus project still requires special treatment.

Because of their very bigness, superprojects are often organized on a project basis. In relation to the overall project cost, skimping on management costs is rarely justifiable, and a centralized project structure is the logical approach. Within those centralized project structures, however, the diversity and complexity of project activities, disciplines, and logistics require that the organi-

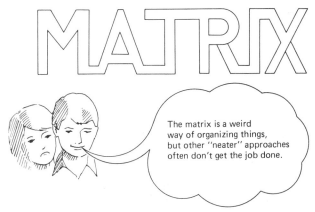

The matrix is a weird way of organizing things, but other "neater" approaches often don't get the job done.

Fig. 24-3. The Complex Matrix.

zational structure be adapted to the specific needs. The matrix within a project structure, discussed previously, lends itself well to meeting such internal super-project requirements. Such a matrix gives organizational flexibility to the gigantic project that, if not broken down into smaller operational groups, runs the risk of becoming a giant project bureaucracy. In the "matrix within a project," area or discipline coordinators are created to work horizontally across the project organization, creating greater flexibility within the overall project structure.

Other Matrix Applications

Davis and Lawrence point out that industry is the area where the matrix has been most widely developed.[2] However, in the service sector the matrix is being widely utilized where two or more of what the authors call "environment forces" exert pressures on the organization. These environment forces are (1) geography, (2) specialized functions or services, and/or (3) distinguishable clients. Since these three forces can exist in sundry fields, the matrix has also found widespread acceptance in such service enterprises as insurance, law, health care, education, governmental agencies, and others.

CONCLUSIONS

The matrix is an awkward compromise solution at best. It is rarely fully satisfactory to the parties involved. Discipline specialists are required to tone down their technical fervor in view of pressing achievement dates and budgetary requirements. Task-oriented personnel are forced to adjust schedule and cost goals to the realities of the technology being managed. The result for the project or activity sandwiched into the matrix, however, may well be a highly desirable compromise characterized by a series of trade-offs between discipline technology and time/cost variables.

Whys and *why-nots* for using the matrix abound; with the advantages come disadvantages as well. One of the best justifications for the matrix is, "It seems to work better under certain circumstances—often where other forms have failed." On the other hand, when applied out of context, it can be disastrous. The matrix accumulates the *pros* and *cons* of the functional and project organizations (albeit in attenuated from) and generates some of its own strong and weak points.

The matrix is pliable and can be used in varied situations. The basic forms are the project matrix, balanced matrix, and functional matrix. These variations in turn may be found in sundry applications such as the company-wide concentrated matrix, the company-wide scattered matrix, functional departments structured on the matrix, matrix relationships at middle-to-upper management level, projects operating under overall matrix structure, and independent projects using the internal matrix structure.

Matrix structures are found throughout industry and service-oriented organizations. Research and development, architecture and engineering design, made-to-order equipment, and superprojects are examples of applications that are becoming increasingly widespread.

The matrix can be complex and confusing, and sometimes seems an organizational monster; in given situations, though, it has proven an effective organizational form. For all its awkwardness and unorthodox qualities, the matrix—like the rarely admired camel—is indeed the best beast to ride in certain settings.

REFERENCES

Notes

1. P. F. Drucker, *Management: Tasks, Responsibilities, Practices* (New York: Harper & Row, 1974), p. 598
2. S. M. Davis, and P. R. Lawrence, eds. *Matrix* (Reading, Mass.: Addison-Wesley Publishing Co., 1977), pp. 11–24
3. J. Hemsley and E. Vasconcellos, eds. *The Design of the Matrix Structure for R & D Organizations*, PMI Proceedings, Project Management Institute, 1981, pp. 1–12

Bibliography

Cleland, D. I., and King, W. R. *Systems Analysis and Project Management*. New York: McGraw-Hill Book Co., 1968.
Knight, K. *Matrix Management*. Westmead, Farnborough, Hants, England: Gower Press, Teakfield Ltd., 1977.
Miles, R. E., and Snow, C. C. eds. *Organizational Strategy, Structure, and Process*. New York: McGraw-Hill Book Co., 1978.
Ouchi, W. *Theory Z*. Reading, Mass.: Addison-Wesley Publishing Co., 1981.

25. Managing the Introduction of Semiautonomous Production Groups

Panagiotis N. Fotilas*

THE NEED FOR MEANINGFUL WORK ORGANIZATION PATTERNS

The significance of the worker's job content, and of his or her having a reasonable degree of autonomy in carrying out the job and a direct influence on the work environment, have long been underestimated. The relevance of these factors in the worker's attitude toward the job and in the proper functioning of the industrial enterprise has yet to be fully understood. The employee has usually been regarded as an individual only interested in wages and shorter working hours. Work has often been organized on the basis of maximum specialization and repetitiveness, with clear attempts to reduce the variety of jobs and training time to a minimum. Little room was left for personal initiative, with consultation and communication among blue-collar workers being particularly limited.

This traditional work organization philosophy yielded remarkable results in the increase of material prosperity, but this growth in prosperity was accompanied by growing tensions, since the general level of education has improved enormously. Factory workers began looking for a more meaningful job rather than just putting their manual ability in the company's service. Workers became more conscious and critical of their work situation, and in so doing disclosed some basic weaknesses in the traditional systems. These inefficiencies can be characterized as *social-psychological* and *technical-economic-organizational*.

Social-psychological inefficiencies include:

- High absenteeism
- High turnover of personnel
- Monotony

*Panagiotis N. Fotilas is Assistant Professor of Industrial Economics and Management at the Graduate School of Business of the Freie University of West Berlin. Holding a diploma engineering (M.S.) degrees in mechanical engineering and economical engineering and a Ph.D. in economics, Dr. Fotilas has served as consultant to various industrial corporations. In addition he has lectured for the Greek Productivity Center and the Institute of Technology in Athens, Greece. Dr. Fotilas has written mainly in German and Greek. His current research interests include the impact of microelectronics on West German industry.

- Stress
- Sabotage
- Strikes
- Protests, etc.

Technical-economic-organizational inefficiencies include:

- Much rejection and waste
- Many breakdowns
- Planning and stock problems
- Much maintenance required
- Many accidents
- Low production flexibility
- Ergonomic problems
- Bad process control
- Low work efficiency
- Labor market problems (no qualified applicants when needed), etc.

These inefficiencies can be regarded at the same time as the driving forces for "humanizing" the work through "work structuring." One of the methods applied by some industrial corporations to make work more suitable for human beings is the introduction of semiautonomous production groups (SPGs).

An SPG is a small group of 2 to 20 workers with a specific set of technological equipment laid out in a demarcated area of the shop floor. The members of the groups share a common target in terms of completing a wide range of components or a complete product (job enlargement, job enrichment, and/or job rotation). The group has no direct supervisor and enjoys a certian degree of freedom with regard to decision making. SPGs are laid out and organized in such a way that they can vary their production rhythm within the time limits imposed by management without affecting the rest of the groups and the prescribed production output.[1]

The main aim of introducing SPGs in industry is to increase job satisfaction and productivity. This chapter presents a down-to-earth approach to business problems, and by no means suggests that managers should adopt a laissez-faire policy or sentimental attitudes toward running their companies.

HUMAN OBSTACLES IN IMPLEMENTING THE SPG CONCEPT

In most cases investigated, the initiative for introducing SPGs in production organization was claimed at top management level. However, a closer inquiry indicated that, up to a certain point, the nature of the innovation ran counter to the values to which some managers had been accustomed. The main efforts to introduce the new form of work organization had in fact come from industrial engineers, ergonomists, sociologists, psychologists, and external consultants.

Management Attitudes

Undoubtedly, a fundamental question to be raised by even the most benevolent management is that concerning the costs of the change and the profitability of the proposed system. It is not easy to persuade all executives by simply presenting the results of some investment calculations, particularly when these calculations indicate that in the short term the costs of the innovation could be higher than the conventional solution. Every innovation is subject to a certain risk, and it is completely understandable that those running a company will try to avoid or minimize such risks.

Another reason for top management scepticism toward change is their familiarity with the existing technical-organizational system. Particularly in the case of a paced assembly line, the supporters of the innovation will have to display considerable ingenuity in persuading management that even this production system has disadvantages that can put the profitability of the system into question.

A prerequisite for the implementation of SPGs is the democratization of the existing organizational system: i.e., the delegation of authority, the decentralization of decision making, and the participation of workers in the process of innovation.

Managers who are used to running their companies in an authoritarian style tend to regard *delegation of authority* as a loss of their own power and competence. Especially in the case of SPGs, delegation includes a transfer of some important tasks to the lowest level of the existing hierarchy—the blue-collar workers. The negative attitude of some leading members of the management can be reinforced by the fact that this category of employees is often regarded as incapable of acting autonomously. Granting a certain degree of autonomy is considered extremely risky, since the management still retains overall responsibility. Delegation of authority also entails an element of personal risk, as it inspires a fresh comparison of abilities among personnel irrespective of position in the existing structure.

Problems arise not only with managers who are too egoistic to delegate authority, but also with those who, although willing to share, do not know how to go about it. The latter often choose the wrong people for a task and expect responsible results without providing the necessary means.

Delegation of authority alone—i.e., an enlargement of the operation field of the organizational subsystems—is not enough to permit a successful introduction of the alteration. A certain degree of *decentralization*—a delegation of decision making (not departmentalization)—seems absolutely essential for establishing SPGs within the production organization.

The latitude of decision making in SPGs is actually limited in matters concerning the workplace itself and its closest environment. Despite these restrictions, some managers cannot cope with the idea of letting workers decide, for example, over their time schedule, working methods, workplace design, etc. Workers are often accused of lack of discipline, immaturity, low educational level, etc.—all factors seen as prohibiting the transfer of decision making at the lowest organizational level.

In everyday discussions of *participation of workers,* the term "participation" is often interpreted as codetermination. Codetermination—*Mitbestimmung,* or joint management—was first introduced in West German steel and coal mining industries in 1951. At that time, employees were given half the places on the board of directors, which gave them genuine decision-making power at the top level of the corporation. This degree of power sharing may appear somewhat drastic in America, but participation here means only codetermination of the workers in solving day-to-day problems directly affecting their work station and environment.

This kind of participation is considered absolutely necessary in introducing SPGs. In attacking a business problem, it always makes sense to mobilize all the knowledge and experience available in the organization. This, of course, is not a new principle, and many companies use the skills of all the managers and staff specialists who can contribute to the solution of emerging problems. But what is new nowadays is the value placed on the knowledge, experience, and skills of the lowest organizational level. The individual who actually works every day with certain mechanical equipment in a certain production process has considerable experience and is aware of the everyday problems. His or her contribution in redesigning the workplace, planning the work, and solving problems may be regarded as useful to say the least.

Since most people are eager to express an opinion concerning their work situation, participation will provide them with the opportunity to exert influence on matters affecting them personally. In addition, participation establishes stronger links between top management and workers, thereby reducing the possibility of industrial disputes.

Despite these considerable advantages of participation, some managers see in it an erosion of the hierarchical structure, a diminition of their authority, and an inadmissible delay in the decision-making process.[2] But even if participation is accepted as a means of helping to implement change, there are still factors impeding the application of this method. If participation is to function properly, it must be genuine and not merely an artificial veneer of good will. It is sometimes believed that workers "participate" by organizing meetings in which decisions are already made by managers or other experts. The problem is that managers and experts "already know the solution of the problem." To participate in a conference with no preconceived notion can be a difficult and challenging task, requiring some alterations in traditional management thinking patterns.

On the other hand, it is not easy for workers to participate in a conference discussing problems primarily regarded as management matters. Therefore effective participation requires some training. Weekly morning meetings and the formation of participative groups of representatives from different hierarchical levels can all contribute to the feeling of real participation.

The introduction of SPGs in industry is based on assumptions about the individual that differ from those of the traditional theory of organization. This fact has by no means always been recognized by management, and in these cases most attempts at implementing the innovation have failed. Investigations reveal that successful applications occurred in companies where top management was completely committed to the new approach and deeply involved in applying it.

Deverticalization and Supervisory Attitudes

If certain tasks or job elements are transferred from superior levels of the hierarchical line to production workers, a revision of the existing organizational structure seems necessary. One of the most common consequences when introducing SPGs in industry is the so-called deverticalization, which can be interpreted as a shortening of the line. In most cases the tasks of the superior levels now disappearing are transferred to the subordinate levels.

However, the SPG concept affects only the bottom of the hierarchical pyramid. Production departments—particularly in West Germany—consist frequently of four levels: departmental head, foreman, charge hand, and production workers.

Figure 25-1 presents two possible steps in deverticalizing the existing hierarchical structure. The first step includes a reduction of the number of foremen. For example one foreman is now responsible for three SPGs while the charge hands are absorbed from the production groups. The job content at the level of charge hands is clearly changing. Their function of chasing materials

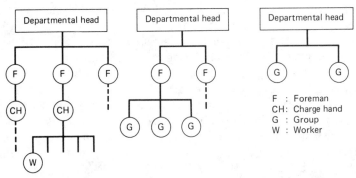

Fig. 25-1. The impact of SPGs on an existing organizational structure.

has been eroded to such an extent that this job has been abolished as a separate task.

The groups come under the control of the foreman, who now has to control considerably more workers than in a traditional system. However, this kind of supervision is facilitated, in that many of the foreman's previous responsibilities have been transferred to the SPGs.

The second step of deverticalization may be achieved by abolishing the jobs of both foreman and charge hand. Normally the leader elected by the group represents that group and establishes direct links with the head of the department. However, it has been found that this kind of deverticalization is mostly theoretical. In the few cases where this method was applied, it has been impossible for the group to cope with all the problems that have arisen, even when the group members possessed ample experience and knowledge.

The main driving force for a reduction in the supervisory personnel is undoubtedly based on simple economic considerations. Simultaneously, foremen and charge hands have come under particular fire for their lack of skill in handling people. Furthermore they have been frequently attacked for their tendency to issue orders without taking into consideration the opinions of the workers. In Norway, for example, these negative attitudes become so strong that the SPGs came to be interpreted as simply the abolition of supervisors.

The supervisors, facing all these attacks and fearing loss of their jobs, in many cases tried to impede the implementation of the SPG concept. A prerequisite for successful application of the innovation is the maintenance of an unhindered and clear information flow from the top level to the bottom and vice versa. In many cases supervisors tried to prejudice or even block the communication channels between the upper management and the shop floor. Especially during the experimental phase of introducing SPGs, supervisors used to manipulate information in both directions, creating a feeling of insecurity particularly among the workers. The result of this behavior was a clear negative attitude of the blue-collar employees toward the planned or already introduced innovation.

Deverticalization and in particular the reduction of supervisory personnel sometimes constitutes a fictitious advantage for the company. Saving money by firing supervisory personnel can be regarded as a simple but at the same time oversimplified economic policy. General business policies and also moral considerations prohibit the dismissal of old, loyal, and often very skilled personnel. Even if one forgets the moral commitment, the trade unions can cause enough trouble to counterbalance the benefits of a dismissal.

A reasonable solution to the problem consists of waiting for such workers' normal retirement from the company and not replacing them. Meanwhile supervisors must be placed in the very center of the organization's change. The introduction of SPGs should therefore occur with the active participation of supervisors, not without them or against them. Inquiries prove that where foremen were included in developments, the innovation succeeded, and where they have been kept out, the projects have failed. At the same time new roles have to be created for them, such as:[3]

- Control of the qualitative and quantitative SPG input and output
- Evaluation of group performance
- Production planning and control
- Coordination of SPGs
- Training, both initial and remedial
- Counseling on jobs
- Counseling on personal problems
- Innovative planning
- Intervention in case of conflicts, etc.

During the experimental phase of such innovations and despite the new tasks defined for supervisors, a certain job overlap was registered, because of the unwillingness of some foremen to relinquish their previous roles. In most cases, serious efforts by all personnel involved soon led to the solution of these and similar problems. However, since autonomy in production groups is limited, supervisors will always be needed, while the realization of their new tasks will depend greatly on the cooperation of both the higher and lower levels.

Blue-Collar Resistance

Attempts to restructure existing organization patterns usually face a certain degree of resistance from the employees affected. This almost natural reaction to any planned change can be reinforced if the workers realize that the innovation was initiated by the top management, and in particular when endeavors to rationalize production organization become evident. Rationalization is often interpreted as losing one's job, while the managerial initiative is regarded as at least suspicious. In the case of SPGs, the dual aim of increasing job satisfaction and productivity is clear. However, it was found that in companies where efforts to improve productivity failed to increase job satisfaction, the whole job-restructuring program turned out to be a failure.

One of the main reasons for the workers' caution toward the SPG concept was a fear of slipping below their normal income level. Working in an SPG means the individual will have to abandon his or her "own" workplace (job enlargement, job enrichment, job rotation) and become more of a general than special worker. A decrease in specialization, and the dependence of the individual upon the group activities, were frequently interpreted as losing the privilege of influencing one's own working rhythm. The usual employee arguemnt was that working individually under the traditional piece-rate wage system made possible a degree of control over one's own monthly income.

At the same time, piece rates have often created uncertain income conditions, and for some individuals they have caused considerable stress. The traditional piece-rate system harmonizes with short-cycle work. However, as soon as productivity is determined more by machines than by humans, and quality requirements assume great importance, the advantages of the traditional piece-rate systems tend to diminish.

Since the form of remuneration must be consistent with the type of work organization chosen, several companies introduced new wage systems quite successfully. These systems are mainly based on the performance of the SPG, because group wages constitute an excellent method to increase cooperation among group members. In order to guarantee a stabilization of income and productivity by reducing the stress caused by the piece-rate systems, the new wages are based on some form of group bonuses. However, the fixed portion of the wage accounts for 70 to 90% of the total and in contrast with the fluctuating bonuses can be regarded as a monthly salary.

The SPG concept is not founded on the assumption that every group member has the same qualification and does the same job. The fixed part of the wage is based on job evaluation data and on the skills and competence of the individuals, whereas the group bonus is often divided equally among the group members. Furthermore, job enrichment provides individuals with a number of opportunities to improve their monthly wage by learning and doing some additional tasks.

Last but not least, there is another problem that has not been solved satisfactorily up to now: namely, what are the promotion opportunities for good group members after the deverticalization process. Only a few workers can be promoted to foremen, so specialists will have to make considerable efforts to solve this emerging problem.

Gaining commitment for the implementing the SPG concept among blue-collar workers requires their active participation in all initial discussions and debates concerning the innovation.

TECHNICAL OBSTACLES IN IMPLEMENTING THE SPG CONCEPT

The prerequisites for forming SPGs in industry depend heavily on the design of the production system already in existence. Different kinds of systems call for different methods and practices for organizing work. The following systems are grouped according to German classification.

1. Assembly site
2. Functional layout
3. Traditional group layout
4. Manual flow line
5. Paced assembly line
6. Process-oriented layout

Assembly Site

This type of organization is used when the product in process is too large or too heavy to be moved frequently; e.g., turbine construction, big Diesel engines, etc. The same design is found even when the product in process is produced in relatively short series, as in aircraft construction and shipbuilding. The product itself is at the center of the entire process, and small specialized or all-round groups work on the project.

Functional Layout

All machines of the same type (all lathes, all milling machines, all drilling machines, etc.) are laid out together in the same section, which means that there is a clear differentiation between the drilling department, turning department, etc. The products of this kind of production design are components, usually produced in batches.

Each worker is specialized in one process and operates independently. The individual is tied to the apparatus and his or her contribution to the completion of the final product may be regarded as modest. The large number of components in process, the complicated material flow, the large intermediate stockpiles, and the extensive planning and intensive control efforts needed to cope with the system, indicate that the introduction of modern work group organization might alleviate some of these inefficiencies.

Sober analysis shows that the transition to group production is quite difficult. The main obstacle lies in the nature of the work itself, which might be described as inherently individual. The possibilities of changing this functional layout to the SPG system are limited, because of:

- High costs in removing and reinstalling machinery. In order to create SPGs where job enlargement and job enrichment may be applied successfully, different types of machinery have to be transferred to the same department.
- Tooling and machinery surpluses. All groups need nearly the same sets of equipment to produce a larger range of components or a complete product.
- More space needed for buffer stocks. Group independence in work rhythm is guaranteed by the buffer stocks placed between the different groups.
- Certain types of equipment like cyanide baths are too expensive and dangerous to be installed in every group. The limitation of group independence is obvious especially for products that have to undergo a similar kind of treatment.
- High training costs.

Some possibilities of modifying work organization in the direction of SPGs, without changing the design of the existing production system, are found in tasks with a common feature, like transport services and maintenance and machine adjustment.

Despite the complications involved in introducing SPGs in already existing functional layouts, 22% of the cases investigated proved to be of this category.[4]

Traditional Group Layout

In this layout, different types of machines are installed in demarcated areas of the work shop: for instance, a lathe, a mill, a grinder, and a drill form a machine group in a certain area. In most cases, all the components made are divided into families, and worker-machine groups are organized in such a way that most parts of the family can be processed in one group only. The main savings from this type of design come from the simplification of the material flow system.

Here the transition from traditional to SPG layout is mainly a matter of training the workers involved. Some additional problems may arise if the number of group members exceeds the size of the usual "small group" of 2 to 20 members. In such a case a restructuring of the group size seems necessary.

Manual Flow Line

Here the products are made on flow lines in accordance with standard designs. According to the official definition, these lines are considered "nonpaced" lines. The product flow from station to station is effected by means of flexible transport equipment (such as fork-lift trucks) or even conveyor belts. However, if belts are used, they do not prescribe the work rhythm of the staff involved.

Since some kind of pacing always exists in production operations, we will differentiate between the mechanical pacing and operator or self-pacing. In the case of the manual flow line the mechanical control of flow rate is removed, and small intermediate stockpiles placed between the stations facilitate the variation of the output rhythm.

This kind of production system is vulnerable to changing market demand. Faced with fluctuating and often highly diversified demand, managers have to provide for a variation or for an efficient adjustment of the system's capacity. The introduction of SPGs seems to facilitate such activities.

Since workers depend to a certain degree on one another and are not very closely tied to their stations, the creation of SPGs along the existing line should not cause particular problems. Buffer stocks between the groups provide for a certain degree of independence. Communication between group members might possibly be hindered by the physical arrangement of the equipment. Nevertheless, job enrichment and job rotation can lead to higher job satisfaction and productivity.

The division of the assembly line into smaller subunits (SPGs) by removing and reinstalling machinery should be favored only if the existing production system cannot cope with the diversified market demand for parts, components, and complete items. The high costs of a total change of system are obvious. However, different groups may simultaneously produce different items. The increase in the system's flexibility leads to a remarkable improvement in capacity management and makes possible the acceptance of small and at the same time highly diversified orders.

Paced Assembly Line

This type of line is usually found in assembly work. In contrast to manual flow lines, a mechanical pacing effect is provided by continuously or intermittently moving belts, conveyors, and similar automatic transport equipment. The length of the cycle time is determined by the velocity of the conveyor belt. An excessive operation time may mean that a product will pass a working station without being completed. A prerequisite for such a production system is that the batches or series of the item to be produced are large enough to guarantee an acceptable capacity utilization. The entire process is usually adapted to a small number of different types of a single item.

The advantages of the paced line—short throughput times, clear material flow, efficient utilization of factory equipment, short training times, etc.—are well known. However, a brief reminder of the main disadvantages of the system might help to understand the efforts undertaken by several firms in transforming lines into SPGs. Some of these disadvantages are:

- Vulnerability to disruptions. Stoppages and breakdowns may be avoided only by means of a permanent high-level readiness.
- Low degree of system flexibility. The adaptation of the system to changing market demand, both in quantity and quality, is complicated and costly.
- Monotony.
- Stress.
- Waiting time due to balancing losses.
- Waiting time due to system losses.

Without any further discussion of the first two points, it is worth mentioning that in recent years the much-praised advantage of specialization—which can lead to high manual dexterity and consequently to productivity increases—has come to be regarded as the main cause of monotony and stress. A discussion of balancing and system losses follows, since the introducttion of SPGs may contribute remarkably to their reduction.

Balancing losses. In general, individual workers seldom operate all at the same speed, so that time losses result from differences in average speeds among the staff involved. These losses are called balancing losses. The products in process can never travel quicker than the worker with the longest average oper-

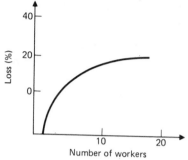

Fig. 25-2. Balancing losses and line length. Source: H. G. van Beek, "The Influence of Assembly Line Organization on Output, Quality and Morale," *Occupational Psychology* 38 (1964): 164.

ation time. Accordingly, the entire line has to wait for this worker. Figure 25-2 shows that shorter lines are more efficient than longer lines.

System losses. Additional losses are caused by the variation in the operating speed of the individual worker during any given period. Even if there is no difference among the average speeds of different workers, the fact that individuals vary their work rhythms results in so-called system losses.

SPGs and Light Lines. Typical examples of light-paced assembly lines are found in industries producing television sets, typewriters, small electrical motors, etc. Workers on such lines are normally unable to stop operating for more than a very short period of time, unless replaced by a colleague.

In an effort to alleviate or even abolish the disadvantages mentioned previously, SPGs may be formed along the line without removing and reinstalling machinery. Buffer stocks placed between the different groups guarantee a certain degree of group independence. Reduced pacing and increased cycle time allow greater autonomy for the workers. Moving belts are actually used as means of transport. Pacing may even be completely abandoned, which means that the line has the same characterics as a manual flow line organized in accordance with SPGs principles.

Figure 25-3 visualizes the relationship between line length, buffer capacity, and line output efficiency. Larger buffer stocks improve line efficiency up to a certain point, while for a given buffer capacity shorter lines permit higher output.

One basic requirement for the proper operation of SPGs is to arrange the groups in such a way that interpersonal contact is facilitated. The conveyor should not impede this desired proximity. At the same time, a clear and unambiguous common target has to be given to every group. The integration of jobs like inspection, maintenance, and even paper work may lead to job enrichment. The relative freedom of the group members to leave their workplaces permits the internal allocation and sharing of responsibilities.

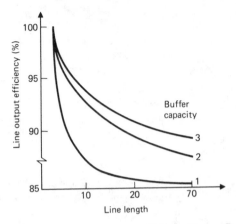

Fig. 25-3. Line length, buffer capacity, and output efficiency. Source: R. Wild, *Work Organization: A Study of Manual Work and Mass Production* (London: 1975), p. 109.

When the desired results cannot be obtained by a simple modification of the production system, a complete restructuring of work organization can be taken into consideration. If, for technical reasons, a division of the line is out of the question, then shorter conveyor belts can replace the existing ones. There are instances where conveyor belts were completely abandoned, transportation between and within the SPGs being achieved by other means. This last method of work structuring provides ideal preconditions for a successful application of the SPG system.

SPGs and Heavy Lines. The assembly of automobiles is one of the most common examples of heavy-paced lines. The modification of light lines toward SPGs, as described above, may be theoretically applicable for heavy lines as well. Simple investment calculations point out that, for example, the use of an expensive conveyor system simply as a means of transportation is irrational. Yet the slightest effort to remove even a small part of the machinery to facilitate group formation may cause enormous complications. Furthermore, any thought of a complete transition from heavy line to SPGs should be avoided a priori, because of the prohibitive expense involved.

Although one wishes to avoid generalizations, in this case it has to be stressed that heavy lines are not suitable for the formation of real SPGs. However, attempts such as prolonging the time cycle, introducing short breaks, etc., can be made to increase individual job satisfaction, while at the same time keeping or even increasing the efficiency of the line.

Limitations on Changing Paced Assembly Lines to SPGs. The main limitations imposed on a modification or transition from paced lines to SPGs are similar to the limitations mentioned with regard to functional layout. They can be summarized as follows:

- More space needed. Assembly lines are usually installed in long and relatively narrow buildings. Buffer stocks and also tooling and machinery surpluses require additional space. Depending on the way in which SPGs are laid out, an extension of the building may be necessary.
- Large amount of material in process. The introduction of buffer stocks between SPGs may neutralize a classical advantage of the line; the small quantities of material in work. The capital tied up increases accordingly.
- Problems with the transportation system. The low capital utilization becomes evident, if moving belts or other conveyors are used only as transportation means. The costs resulting from an abolition of moving belts and the emerging capital outlay for purchasing new transportation facilities may be regarded as high.
- Stoppage of the production process. In contrast to other production systems, the transition from paced lines to SPGs requires a complete stoppage of the production process until the alteration is completed. The high costs emerging are obvious.
- Instruction and training problems. Staff instruction and training constitute a serious problem, since the majority of line workers are unskilled. Consequently, long and costly instruction and training periods must be taken into account.

Process-Oriented Layout

This kind of production system is to be found in bulk material production. The items made may be measured in units of volume, weight, area, or length. Typical products are mineral oil, steel, textiles, wires, and food. The production apparatus is almost obligatorily designed according to the "natural flow" of the process leading to the production of the desired item (there is no gasoline before refining crude oil). These items are normally made on mechanized or automated lines.

Despite high mechanization and/or automation, the prerequisites for introducing SPGs are generally satisfactory. The first experiments toward this kind of work organization date back as far as 1953 and took place at Calico Mills in Ahmedabad, India.[5] Clear and common production targets and the obvious interdependence of the workers favor the formation of SPGs. The main tasks of the groups include control, inspection, and service and repair activities. The introduction of SPGs may be accomplished without high expenditure, since most of the limitations imposed on other production systems do not apply to this type of organization.

However, difficulties may arise in combining tasks with fundamental differences in their job content, like monitoring with maintenance duties. Another—perhaps the most important—limitation may derive from the long geographic distances between workplaces inherent in the system. Yet despite these difficulties, more than 27% of the SPGs formed to date belong to this category.

Finally, it should be noted that the combination and coexistence of several production systems within the same factory is not uncommon. However, the arguments for and against the formation of SPGs here still apply.

New Plants Designed for SPGs

The main opposition to the traditional production organization system is based particularly on the negative effects of the paced line. More than 41% of the cases investigated concern a transition (modification) from paced assembly lines to SPGs.

Hoping to increase productivity and to improve the quality of working life, several firms that normally produce on paced lines have erected new plants laid out especially for SPGs. The following typical examples highlight the experience gained in this sector.

The Saab-Scania Case.   The Saab-Scania factory at Södertälje, Sweden, was one of the very first examples of a new plant in which new concepts of assembly were introduced. The plant was to build the Saab 99 petrol engine.[6]

The main premises for building this plant were:

- 110,000 engines to be manufactured annually
- The engine to have relatively few and light components
- Only a few versions of the engine to be manufactured

At the same time the work organization was intended to:

- Reduce the extent to which an individual was tied to his or her task
- Increase the individual's possibilities of developing with the job
- Increase the influence of the individual on the distribution of work
- Improve the flexibility of the system
- Increase the ability to handle disruptions
- Simplify rebalancing
- Facilitate training of the workers involved

Previous positive experiences with group work favored the introduction of SPGs. Figure 25-4 presents the layout eventually selected. This layout permits the formation of SPGs in which group members can assemble a complete engine. A large conveyor is located to the side of the SPGs. The incoming

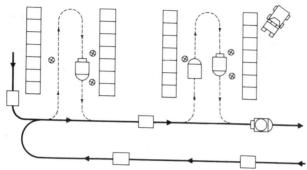

Fig. 25-4. SPG layout at Saab-Scania. Source: J. P. Norstedt and S. Aguren, *The Saab-Scania Report* (Stockholm: SAF 1973).

preassembled components and the outgoing completed engines are carried to and from the SPGs by the same transport system. U-shaped floor guide tracks make possible the disconnection and diversion of the engines into the group area. The trucks bringing the materials to the racks approach from the opposite side of the conveyor.

Each group is responsible for the internal distribution of work inside the group; there is no mechanical pacing during the actual assembly. The workers' maximum cycle time is 30 minutes, instead of the 1.8 minutes prescribed by the conventional system.

The company reports that the required investment was higher owing to the expensive transportation system and the greater space needed. But overall, the minuses were balanced out by the pluses. Productivity targets and objectives inducing higher job satisfaction can be regarded as met.

The Volvo Kalmar Case.   The Volvo Kalmar plant can be considered as unique to the automotive industry. Right from the beginning it has been possible to incorporate the atmosphere of a small workshop in a large plant. P. G. Gyllenhammar, Managing Director of Volvo, set up the following targets: "This is a factory that, without any sacrifice of efficiency or financial results, will give employees the opportunity to work in groups, to communicate freely, to shift among work assignments, to vary their pace, to identify themselves with the product, to be conscious of responsibility for quality, and to influence their own work environment."[7]

The plant was laid out to produce about 30,000 cars per year, working on only one shift. The design, however, permits a transfer to two-shift operation and a production of up to 60,000 cars. The factory was especially constructed for SPGs and consists of four six-sided units located next to each other.

Prepainted bodies are freighted from Torslanda to Kalmar. Production is divided into body assembly, chassis assembly, and final assembly. While the body is assembled on the second floor, the engine, gearbox, axles, and exhaust system are assembled on the ground floor, which is also where the body is united with the chassis. The control function is carried out parallel with the assembly process, and workers are informed immediately of the results of these quality checks. The components inventory is located in the center of the building.

During the entire assembly procedure, the body is carried on a battery-powered carrier that functions both as a transporter of bodies and as a working platform. When assembly work starts, each carrier is given a designation registered in the central computer. The computer constantly tracks the carrier. Furthermore, the platform may be steered manually. Platforms are electrically powered and guided by magnetic tracks embedded in the floor. Assembly work on the underbody of the car is facilitated by the ability to tilt the carrier as much as 90 degrees.

There are approximately 30 SPGs in the plant, each consisting of 15 to 20 persons. There is a supervisor for every two or three groups. Work can be organized in two different ways. In *straight-line assembly* there are four or five stations placed in series (see Figure 25-5). The mechanics, in groups of two, can carry out part of the work at a set station or follow on the carrier in the direction of production flow, so that the entire function is fully assembled by the same group.

In *dock assembly* the carrier glides automatically into a dock, where the team carries out its task on stationary vehicles (see Figure 25-6). The main difference from straight-line assembly is that the entire work cycle is done at a single station. Volvo reports that most workers have a positive attitude toward group organization and have a strong feeling of belonging to their team. The system permits a higher degree of influence by employees on their own work than a conventional plant, while almost everybody wants to share responsibility for the quality of their own work. Volvo quality standards are attained to the same degree as in other plants. Assembly times remain unchanged in comparison to conventional plants, and lower absenteeism and personnel turnover have occurred. Higher investment is being offset by advantages such as fewer super-

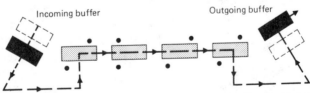

Fig. 25-5. SPG straight-line assembly at the Volvo plant in Kalmar, Sweden. Source: S. Aguren, R. Hansson, and K. G. Karlsson, *The Volvo Kalmar Plant: The Impact of New Design on Work Organization* (Stockholm: 1976).

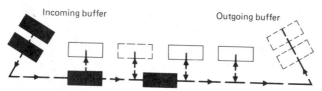

Fig. 25-6. SPG dock assembly at the Volvo plant in Kalmar, Sweden. Source: S. Aguren, R. Hansson, and K. G. Karlsson, *The Volvo Kalmar Plant: The Impact of New Design on Work Organization* (Stockholm: 1976).

visors, the ease of adapting the production system to changing demand, and low absenteeism and turnover.

SPGs, HUMANIZATION OF WORK, AND PRODUCTIVITY

Experiences with SPGs demonstrate that, given a reasonable combination of economic, social, and technical factors, this method of organizing work can increase job satisfaction while maintaining, if not increasing, productivity.[8] Here higher productivity means more output per unit of input, with only the variables of input and output connected with the change taken into account. Investigations have revealed that the SPG concept includes both positive and negative economic aspects. An increased efficiency of the new production system is based on:

- Lower absenteeism
- Lower turnover
- Same or even higher quality levels
- Fewer supervisors
- Reduction of overtime
- Increased flexibility of the production system toward changing market demand

But at the same time these benefits are reduced by:

- Machinery and tooling surpluses
- Need for more space
- Higher wages because of the enlarged tasks
- Higher training costs

On the other hand job satisfaction—often tightly interwoven with medium and long-term economic targets—has improved remarkably. The SPG concept has

- Increased the autonomy of the workers
- Promoted cooperation among them
- Stimulated their interest in their jobs
- Enlarged their range of responsibility
- Facilitated the development of additional individual skills
- Reduced monotony and stress

Top executives are primarily interested in the costs and financial benefits of the innovation, often regarding human aspects as relevant but still secondary.

The investment costs for implementing the SPG concept range between 10 and 300% above those for a conventional system. For example, Volvo estimated that the investment costs for the Kalmar palnt were only 10% higher than those of a normal car assembly plant, while FIAT reported that the investment required for its Rivalta plant was four time as large as the amount needed for a paced assembly line.[9] However, there are marked indications that in a total calculation the minus items are being balanced out by the plus items.

REFERENCES

1 P. N. Fotilas, *Arbeitshumanisierung und teilautonome Produktionsgruppen: Wirtschaftliche, soziale und technische Aspekte* (Berlin: Erich Schmidt, Berlin, Bielfeld, München 1980), p. 22–23.

2 P. N. Fotilas, "Semiautonomous Work Groups: An Alternative in Organizing Production Work?," *Management Review*, Vol. 70, No. 7 (July 1981): AMA, p. 53.

3 *Ibid.*

4 P. N. Fotilas, *Arbeitshumanisierung*, p. 266. All percentage figures concerning SPG applications given in this chapter, if not otherwise attributed, are taken from p. 266 of this book.

5 A. K. Rice, *Productivity and Social Organizations: The Ahmedabad Experiment* (London: Tavistock Publications; 1958).

6 J. P. Norstedt, and S. Aguren, *The Saab-Scania Report* (Stockholm: SAF (Swedish Employers' Confederation), 1973).

7 S. Aguren, R. Hansson, and K. G. Karlsson, *The Volvo Kalmar Plant: The Impact of New Design on Work Organization* (Stockholm: SAF and LO (Swedish Trade Union Confederation), 1976), p. 5.

8 E. Gaugler, M. Kolb, and B. Ling, Humanisierung der Arbeitswelt und *Produktivität* (Ludwigshaften: 1977); W. Rohmert and F. J. Weg *Organisation teilautonomer Gruppenarbeit* (Munich: Friedrich Kiehl Verlag GMBH Ludivigshafen (Rhein), 1976).

9 H. Strebel, and P. N. Fotilas, "Teilautonome Gruppen in der Fertigung," in *Handbuch für Rechnungswesen* (Stuttgart: Taylorix Fachverlag, 1980), p. 539.

26. Japanese-Style Consensus Decision Making in Matrix Management: Problems, and Prospects of Adaptation

S. Prakash Sethi and Nobuaki Namiki*

The environmental context—the economic, social, and political forces that impact business—has been undergoing rapid change in the last two decades. These changes have caused the external environment to become more complex, have increased the degree of uncertainty and risk, and have created the need for assimilating and evaluating vast amounts of new information. Managements of large organizations, especially corporations, have had to develop new technology to process this information, and new forms of organizational structure to cope with and manage in the changing environmental context. Matrix form of management has been one such response.

It is often said that the more things change, the more they remain the same, and that each generation views its own era as that of rapid change surpassing all previous ones. Thus we find that some of the elements of matrix management are indeed quite similar to those of the highly traditional and eminently successful Japanese style of management, which in turn may have some similarity—although scholars strongly disagree on this point—with the older Western style of management based on paternalism, which we seemed to have discarded in an effort to develop more rational structures to cope with the growing complexity of large organizations.

The objective of this chapter is threefold:

1. To provide a brief description of Japanese-style consensus management.
2. To compare and contrast various elements of Japanese-style consensus management with those of matrix management.
3. To evaluate the relative merits of Japanese-style consensus management and the matrix management in terms of their suitability for various organizations under different external environmental contexts.

---

*S. Prakash Sethi is a professor and Nobuaki Namiki is a doctoral student at the School of Management of the University of Texas at Dallas.

The essence of any management style and organizational design has to do with its suitability in "managing" people so as to motivate them to perform tasks, undertake responsibilities, and carry out activities that are in the best interests of the organization. No system of organization will perform well unless it is congruent with the economic interests, values, and beliefs of the employees and their sociocultural mileau. Therefore any comparative analysis of different management systems must be cognizant of the different sociocultural contexts that gave rise to those systems and be cognizant as well of where they are to be applied.

MATRIX MANAGEMENT

We shall provide a brief description of matrix management in terms of its various elements, so that it can be compared to Japanese-style consensus decision making. The origins of matrix management can be traced to the aerospace industry, where a flexible and adaptable system had to be developed to manage resources and develop procedures that could satisfy unique project requirements, and at the same time keep up with rapid technological advancement and innovation. The matrix form of management was developed to maximize the strengths and minimize the weaknesses of both the then existing organizational structures, the functional and the project type.

The matrix form of structure helps coordinate and direct two conflicting interests—e.g., marketing and technology—that a conventional organizational structure, whether of functional or product type, may not be able to cope with efficiently. It also helps facilitate a speedy and timely trade-off and balance of conflicting interests in decision making.

The matrix organization superimposes a lateral or horizontal structure of a project coordinator on the standard vertical hierarchical structure. A major feature of the matrix form of organization is that some managers report to two bosses, as opposed to a single boss in the traditional structure, and have a dual rather than a single chain of command.

There are three conditions for the successful establishment of a matrix organization.[1]

1. *It is absolutely essential that managers be highly responsive to two sectors, such as markets and technology.* An organization in such a situation has to balance the power of managers who are project-oriented, and those who are oriented to such specialties as engineering and science. Both orientations must be simultaneously involved in a myriad of trade-off decisions involving schedules, costs, and product quality. Thus a high degree of both differentiation and coordination is required. The dual command structure of a matrix serves to induce this kind of simultaneous decision-making behavior.

2. *There are pressures for very high information-processing capacity among members.* This situation may come about for three reasons. First, there may be a high degree of uncertainty about external demands placed on the orga-

nization, which are continuously changing and are relatively unpredictable. Second, the complexity of organization tasks may have increased, usually through simultaneous diversification of products, services, and/or markets. Third, interdependence among people and tasks may have increased. When all three generators of information load—uncertainty, complexity, and reciprocal interdependence—have increased, the matrix form of management may be appropriate.

3. *Managers must deal with strong constraints on financial and/or human resources.* An organization may be under considerable pressure to achieve economies of scale in human terms and high performance in terms of both costs and benefits, by fully utilizing scarce human resources and by meeting high-quality standards. Most organizations have limited human resources, particularly when expensive and highly specialized talents are required, and limited financial resources, when expensive physical facilities and capital equipment are required. Conventional functional or product organizations tend to resist the sharing of equipment and the rapid redeployment of specialists across organizational lines. An organization under such a pressure of sharing and redeploying limited resources may utilize a matrix system of management, because its dual command structure helps induce efficient resource allocation decision-making in a simultaneous fashion.

Matrix organization, however, is not simply a design of structure. It is a combination of four factors: (1) matrix structure—dual chains of command; (2) matrix systems, operating along two dimensions simultaneously—i.e., planning, controlling, appraising, and rewarding along both functional and product lines at the same time; (3) matrix culture; (4) and matrix behavior.[2] The matrix culture includes a tradition of change, the open and frequent exchange of ideas and positions on issues, and a sense of challenge and experimentation. A matrix organization differs from the traditional structure in that it places less emphasis on authority and more emphasis on people's behavior to get the jobs done.

The process of decision making in matrix management is that of conflict resolution, which is managed through group consensus. The first condition for the establishment of matrix management, a dual environmental pressure, and the second condition, the complexity of the project, make conflict inevitable. As long as a conflict does not surface, a matrix manager makes decisions autonomously. When conflicts do arise, a matrix manager follows a three-stage decision process: (1) the conflict is brought into the open; (2) the conflicting positions are debated in a spirited and reasoned manner at a meeting of all those involved in the conflict; (3) the issue is resolved and a commitment made in a timely fashion.[3]

In all these processes a high order of interpersonal skills and a willingness to take risks are needed among involved persons. Moreover, these processes call for a minimum of status differential from the top to the bottom ranks.

As shown in Figure 26-1, there are three kinds of managers in a single-cell

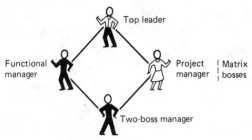

Fig. 26-1. A simplified matrix structure.

matrix:[4] (1) the matrix leader; (2) matrix bosses, i.e., functional and project (or area and business) managers who must share subordinates with other matrix bosses; and (3) a subordinate manager with two bosses. In the conflict-resolution processes, those in the matrix depend mostly on certain necessary behaviors.

Role Model for the Top Leader

In the conflict-resolution process of decision making, the role of the executive is not to make a decision, but to set the stage for decision making by others.[5] The existence of a matrix is an acknowledgment that the top executive cannot make all the key decisions in orderly and timely fashion: There is too much relevant information to be digested, and too many different points of view to be taken into account. But the executive must set the stage for the group decision making, and must also set the overall strategy. The strategy should be strictly followed by the lower echelons and there should be no room for the other matrix managers to challenge the executive's strategy. The executive must also assure the openness of the group discussion, being open to dissent and displaying a willingness to listen and debate.

In most cases the manager with two bosses is the one who obtains relevant information about markets, technological changes, or customer's requirements. This manager must be encouraged to reveal all his or her information and opinions, which may often conflict with those of the two matrix managers. There is a possibility that one of the two may view this as disloyalty and want to punish the manager. Unless this is prevented by the top manager, certain information and opinions may never be revealed. The top manager must act to "balance the power" of the two matrix managers, so as to protect the integrity of the matrix system. One method is to establish dual budgeting and personal evaluation systems.[6]

There are five other means to balance the power among the matrix bosses; adjustment to pay levels, adjustment of job title, access to the top manager at meetings and on informal occasions, location of office, and reporting level, or open communication channels to the top manager.[7]

Role Models for Matrix Managers with Two Bosses

In a matrix system, the top manager acts as an impartial chairman in the group conflict-solving discussion. Most of the decision making is entrusted to the matrix managers and the manager with two bosses (hereinafter referred to as the two-boss manager).

When a problem or conflict arises and is brought into the open, all members involved voice their opinions as to the desired strategy and their resons for favoring it. There is an assumption that discussion and debate are "healthy and that higher quality solutions will develop, if people with different expertise and orientations relating to a given task get together to thrash out their differences."[8] Management differences are valued; people should feel free to express their views, even when others may disagree. A manager should be willing to abandon his or her position, upon recognizing that another's opinion is better for the group. This process demands a high level of interpersonal skills for all involved, and requires substantial training and development efforts to change both organization culture and individual behavior.[9]

The individual attitudes and behaviors required by the matrix organization are quite unlike those required in line-and-staff organizations. People are required to be "more open, take more risks, work at developing trust and trusting others."[10] Some individuals do not fit in a matrix but may work comfortably in a conventional organization, owing to their personality, career background, and experience. They should be encouraged to leave the matrix organization. One of the development methods is a team-banding meeting in which common objectives and expectations for the group and the individuals are developed in the discussion. This meeting also helps individuals understand how the matrix works and what kinds of behaviors are needed, and helps them judge whether or not to stay in the matrix.

A candidate for the matrix boss should:

- be a "generalist" with a fair amount of knowledge in all functional specialities
- be able to assess the judgment of the functional specialists
- not be biased toward other functions
- be motivated and skilled to work collaboratively
- be willing and able to use personality and expertise as a source of influence, even where formal power is available
- be willing to involve others in decision making
- be receptive to a participative problem-solving process[11]

Both the matrix bosses and the two-boss managers must share certain characteristics:

- inclination toward collaboration and interpersonal skills
- competence in handling interpersonal relationships
- capacity to work with dual pressures
- capacity and skills to engage in problem solving in groups
- capacity to develop and maintain a broad organizational perspective[12]

JAPANESE MANAGEMENT AND CONSENSUS DECISION MAKING

The next stage in our analysis presents the essential elements of what in the United States is generally considered as the Japanese style of management. These Japanese management practices include decision making by consensus, a highly ritualistic communication system for conflict avoidance and conflict resolution, lifetime employment, and seniority-based wage and promotion systems. It should be noted, however, that all these practices are utilized primarily by large enterprises though they may also be employed in medium and small-size enterprises in Japan.

Decision Making by Consensus[13]

The Japanese decision-making process is that of consensus building, which is known as *Ringisei*, or decision making by consensus. Under this system any changes in procedures and routines, tactics, and even strategies of a firm are originated by those directly concerned with those changes. The final decision is made at the top level after an elaborate examination of the proposal through successively higher levels in management hierarchy, and results in acceptance or rejection of a decision only through consensus at every echelon of the management structure. The decision process is best characterized as bottom-up instead of top-down, which is the essential character of the decision-making process in U.S. corporations. *Ringisei* literally means "a system of reverential inquiry about a superior's intentions." However, the word in this context means obtaining approval on a proposed matter through the vertical, and sometimes horizontal, circulation of documents to the concerned members in the organization. The *Ringi* process may vary from one organization to another, but usually consists of four steps: proposal, circulation, approval, and record. The typical procedure is described in a simplified form below.[14]

When a lower or middle-level manager is confronted with a problem and wishes to present a solution, a meeting is called for in that particular section by its section chief (*Kacho*). The members of a section agree that the idea should be pursued, but they also feel that it needs the overall support of the firm. The section chief reports this to his department head or department manager (*Bucho*) and consults with him. If the department head expresses his support for the section's proposal, the time-consuming activity of getting a general consensus starts.

First, a general consensus among the persons who will be directly and indirectly involved with the implementation in a department is sought. Then an overall but informal consensus in the firm is sought. The department head may arrange a meeting with the other departments concerned. Each department sends one department head, one section chief, and perhaps two subsection chiefs (*Kakaricho*, sometimes called supervisors). They are the ones who would be involved in the implementation stage. Thus, if there are four departments involved, 16 to 20 persons will attend the meeting. If the opinion of specialists or experts on the shop floor is needed, it will be represented. In fact, the meeting is mainly aimed at exchanging information among the involved persons in order to implement the plan. Through the discussion, some other information and materials may prove to be needed. In that case the initiator and his consonant colleagues, under the leadership of the section chief, formally and informally go from section to section and from department to department to collect the necessary information and prepare the documents to be presented at the next meeting. This prior coordination is vital for the *Ringi* system to be effective.

After a number of meetings, the department judges will have attained an informal agreement from all the other departments concerned. The procedure up to this point is called *Nemawashi*, meaning informal discussion and consultation before the formal proposal is presented. This is the moment when the formal procedure starts.

First, the initiator and his colleagues under the section chief's supervision write up a formal document of request, the *Ringi-sho* (*sho* means document), which outlines the problems and the details of the plan for its solution. Also enclosed are supportive information and materials. The *Ringi-sho* is then circulated among various successively higher echelons of management for approval. Each responsible manager or executive affixes his seal of approval. The number of seals can reach ten or twelve. The *Ringi-sho* finally goes up to the top management for formal authorization and the final "go-ahead."

The *Ringi* process can be divided in two broad categories: the *Nemawashi* process, or decision making by consensus, and the *Ringi* process, or formal procedure to obtain authorization.[15] Once a proposal is accepted during the *Nemawashi* stage, it is seldom opposed or rejected at the *Ringi* stage. Scholars have labeled the *Ringi* system "consensual understanding"[16] and a "confirmation-authorization"[17] process of decision making. The *Ringi*, in this context, is used to confirm that all elements of disagreement have been eliminated at the *Nemawashi* stage. It ensures that responsibility is assumed by all who have affixed their seals of approval.

Consensus Decision Making and Internal Communication[18]

The process of communication within a Japanese organization is very much akin to the mating dance of penguins. Communicators resort to both verbal and nonverbal communication. A great deal of ritualistic communication precedes and follows any substantive discussion. Thus participants fully know and understand who is going to say what, and to whom, before it actually takes place. Yet all concerned would be quite offended if the ritualistic behavior were not strictly observed. A great deal is made of forms and gestures, leaving the listener to interpret the meaning of what is being said.

The communication process, however, is quite different when it takes place *within* each group and *between* two or more groups. A group is defined here as members working together in an office, section, or department who have face-to-face, intimate, and less structured relationships among themselves. The size of the group varies according to the company. This is the group that plays an integral role in initiating a proposal, in getting coordination from other groups to promote informal consensus among the concerned groups, and in writing up a *Ringisho*. On the other hand, employees who are in different offices or departments may not have similar relationships and personal ties among themselves, in which case their relationships are more functional, formal, and vertically structured. For ease of analysis, the former group will be referred to as the primary group, and the latter as the secondary group. According to Nakane; "Since the groups are highly institutionalized, it is easy to recognize the hierarchical order among them; and this is the only acceptable pattern among the Japanese for communication that takes place between individuals and between groups who otherwise have weak relations or none at all."[19]

There are many differences between the behavior patterns of the two groups. When a primary group holds a meeting, the subject to be discussed is gone into at length. The atmosphere is highly informal. It engenders a feeling of comaradarie and encourages all primary group members to talk freely. There is an intimate and familiar feeling of participating in a democratic process that makes members of lower status feel comfortable. The first part of the meeting consists of discussing factual information pertaining to the issue. Gradually the members begin to sense the direciton of the group's opinion. Exposition rather than argument is the nature of the discourse. An understanding and acceptance among members is appreciated. After this is done the members try to adjust or arrive at a new line based on the points of disagreement.

Direct conflict is avoided. If a member wishes to present his opinion, he does so by prefacing it with "I happen to know someone who thinks that . . ." or "Let me say this, but I'm just thinking out loud." This gives him a way out, if he sees that what he says is radically opposed by other members. What he wants is a situation of understanding and acceptance by his fellow members before he lets his own opinion be known. Occasionally a heated argument does take place. When this occurs, one of the more prestigious members tries to reconcile the two sides, or one of the middle or lower status members injects some sort of comic relief. After working very hard toward acceptance by all, everyone knows that consensus is near when 70% of the members are in agreement. At that point the minority concedes and is willing to support the decision. Primary groups believe that there must be a basic accord, and that if they try their best they can come to a unanimous conclusion. In this way the decision will be implemented with a high degree of cooperation.

The secondary groups, on the other hand, are the formal, more ritualistic organs of the company. Their meetings are all ceremony and mark the final stage of reaching a decision. Although the members are also members of various primary groups, when they become members of a secondary group they change their behavior accordingly. Each member represents a particular group and must act exactly as that group tells him to. If something unexpected arises, he may have to ask that the meeting be suspended, so he can confer with his members to ascertain how they want him to handle the new situation.

The real power sources at these meetings are the "men of influence," sometimes called the "black presence." They always remain behind the scenes and can be found in every field and every large or influential group in Japan. Over a period of time they have managed to establish effective personal relations with influential members of various groups. The Japanese term for these influential men is *Kuromaku* (a term from the traditional Kabuki drama meaning "black curtain"). Figuratively, it connotes power behind the throne. Perhaps the two most influential behind-the-scenes power brokers in business and politics in Japan are Yoshio Kodama and Kenji Osano, who were involved in the Lockheed scandal in 1976.[20] The relations among such men are established through common bonds of age, school, or workplace. If these contacts are lacking, a relationship can be developed through a mutual compatibility evolving in the course of various meetings attended by both.

These "men of influence" rarely meet together. Instead they use informal channels to relay information among themselves. Their intercommunication is even more open than that among members of a primary group, because they can always say that they are stating their group's opinion. Once they have reached a decision, they write the scenario for the coming meeting. By the time the meeting is convened, each participant has been well rehearsed for the drama to unfold. There is even a scenarist, who sits next to the chairman to make sure that the meeting goes exactly as planned. If a minor figure who wasn't given any lines presents a contrary opinion, it is listened to politely; the chairman thanks him for his opinion and says that it will be considered. This is a face-saving way of ignoring him. Actually, the scenarists want situations like this to arise, so that the process will appear democratic. The dissenter is well aware of his powerlessness in the hierarchy, and his purpose is expressive rather than communicative. And since he is given a chance to express himself, it is easier for him to accept the final decision. In Nakane's words: "It is thus a highly political procedure. The power and the authority of both a formal and an informal hierarchy are effectively employed."[21]

There is some similarity between the decision-making process of secondary groups in Japan and that of the top sector of the U.S. government and the Supreme Court. Highly ritualized activities do occur, but they are restricted to very special areas or occasions. On the other hand, there is no hint in America of a primary group entity. Americans tend to be much less open in disclosing themselves to other members of a group.

Role of Middle Management

Middle managers perform an essential role in the *Ringi* system. However, how well a middle manager performs this role depends largely on his personal relationships and ties with other managers—links that lubricate the flow of information in the organization. Consensus, before a decision is taken, requires a flood of information, and much of the information that is relevant is produced at the place of implementation.[22] Thus the demand for information pulls the decision process down toward the implementation level, while the need for the decision process to be exposed to corporate strategies pushes it upward. The equilibrium point of these two conflicting demands is generally found at the middle-management level. The system is effective only if the middle management is competent in bridging the gap between lower and higher levels of management, so that personal relationships with other people in the organization become critically important. In Japanese firms middle managers acquire these attributes through a system of job rotation from one function to another, lifetime employment, and related training programs. The system also encourages personal ties that enhance the efficiency of information flows.

Japanese middle managers therefore perform the essential role of closing the gap between decision making and its implementation in their activity domains. The closeness of the decision making to the implementation results in a high level of morale and motivation among middle and lower managers.[23]

Role of Top Management

Under the *Ringi* system, presidential or top-management leadership is expected to cope mainly with crisis situations or with abrupt and clear-cut changes in the direction of the firm.[24] Once the general direction of the firm is communicated to the middle and lower-management echelons, both operational decisions and incremental changes are entrusted to the initiatives of the lower and middle-level managers. Most of the working hours of the top management in Japanese corporations are occupied with establishing and maintaining "private" relations with responsible men in policy-making positions in other corporations and government departments. This is accomplished through regular and frequent, formal and informal contacts that are demanded by the sociopolitical environment in Japan, and is evidenced in the close cooperation between the Japanese government and private industry.[25]

Personnel Policies

Most large Japanese corporations employ personnel policies quite distinct from those of the West: lifetime employment; seniority-based wage and promotion systems; in-company training typically on more than one job or function by rotation; and company-oriented rather than trade or profession-oriented unionization.[26] All these practices are found to some degree in Western countries, including the United States—for example, in civil service and public sector employment where salary increments are based largely on years of service; in the recognition of seniority in union contracts; in tenure in judiciary and academic institutions; and in company-based unions. However, in Japan these practices have been used for a very long time and their use is more widespread, culturally accepted, socially encouraged, and officially sanctioned.

Lifetime Employment

Lifetime employment creates a high degree of employee stability and, coupled with other management practices and personnel policies, generates tremendous employee loyalty for the company with all that it entails. A company can invest money in the training of an employee, confident that once he is trained, he will not be hired away by a competitor. A Japanese enters a company right after graduation from school and stays in the company until the official retirement age of 55 to 60. His job security is virtually guaranteed unless he is accused of misconduct.

The origin of lifetime employment can be traced to the family tradition of the old *zaibatsu*, according to which a youth entered the firm as an apprentice and ended up being a trusted manager or founder of a new branch. The bureaucratic structure of government and early state-owned enterprises provided another influence for adapting the old master/vassal system. Furthermore, after World War I socioeconomic conditions made such an arrangement both feasible and desirable. The present form of the lifetime employment system, which includes both blue-collar and white-collar workers, emerged after World War II, when workers and unions tried to improve employment security because of the crisis atmosphere of the postwar period.[27] Then in the 1950s this kind of job security became an established practice primarily in large firms, mainly because it was the most effective means to make employees identify their own interests with those of the corporation. Tsurumi observes that the job security offered by lifetime employment is "not a product of Japanese paternalism, but a necessary economy to all persons in the firm."[28]

The success of the lifetime employment system depends on the fulfillment of a dual set of expectations that are deeply rooted in Japanese traditions and cultural norms. For the worker there is the expectation that he will be able to stay with his chosen firm and intends to do so. This intention is conditioned by the fact that he will be within the norm of Japanese occupational life and that he has a good deal to gain financially by staying on. For the employer there is the expectation that the worker will stay, provided he is offered "standard" wages and conditions of employment. Social conditions and cultural norms impose a sense of obligation on the employer, who is expected to provide work for his employees and take care of them. Moreover, he stands to face a tremendous loss of worker morale, not to mention union resistance, government pressure, and public ill will, if he deviates significantly from the social norm.[29]

Largely because of the ambiguous nature of the employment relationship, there is confusion over what percentage of Japanese employees are covered by lifetime employment. Estimates by various scholars put this number at between 25 and 40% of all employees, with the upper range more representative of employees in manufacturing industries.

The remaining work force is divided into two groups: experienced recruits or midterm employees who did not join the company right after school graduation, and temporary workers. The temporary workers include manual workers and part-time, seasonal, and subcontracted workers. Also, female employees are usually excluded from the lifetime employment privilege, largely because of social norms requiring that they leave their jobs after marriage. An employee who retires at the age of 50 to 60 loses lifetime employment status right after retirement and may be hired as a temporary worker.

Seniority-Based Wage System

Under the seniority-based wage system the remuneration of a worker is determined primarily on the basis of the number of years he has spent with the company. This is subject to age and level of education at the time of entry.[30] The seniority-based wage system is the dominant practice in Japan. Although the income difference between younger and older workers is greater in larger firms, wage differentials according to age and length of service are prevalent in all enterprises, regardless of size.

The seniority-based wage system takes away the often destructive individual competition among employees and promotes a more harmonious group relationship in which each employee works for the benefit of the entire group, secure in the belief that he will prosper with the group and that, in due time, he will acquire the benefits that accrue for long and faithful service. The seniority-based wage system assumes that longer experience makes an employee more valuable. Within the *ie* (group or community) framework, a supervisor must be more than a technically superior worker. He must be able to maintain order in the group and look after its well-being. Thus the older manager acts as the smybol of group strength and continuity. He also functions as the opinion leader and consolidates the community. He acts as the elder statesman and assists group members in all aspects of their lives, including non-job-related activites such as arranging marriages, settling family disputes, and so on. Middle managers contribute to the achievement of community purpose by educating, training, and controlling the young, and by acquainting them with the rules of the community. "These abilities correspond to a skill of seniority. . . . Seniority-based skill is not a simple manual skill, but an overall mental and physical skill originated in a community."[31]

In-Company Training by Rotation

A Japanese employee keeps on training as a regular part of his job until he retires; at the same time he is trained not only in his job, but in all other jobs at his job level. On-the-job training by rotation promotes tremendous flexibility in the Japanese work force, and also helps make of each middle or upper manager a "generalist" with the broader perspective and experience of the company's business and the wider human contacts and friendships that are vital for generating consensus in the *Ringi* decision-making process.

Because of lifetime employment, an employee seldom gets an opportunity to work outside his corporate group. Generally speaking, employment between firms is almost nonexistent, while mobility within a firm is almost unlimited.

Management development by job rotation enables the firm to reassign production and office workers more freely, and makes the employees, because of their job security, more receptive to organizational changes and the introduction of new technology or machinery.[32] Wider experience within the firm tends to nurture the goals of the total firm, rather than those of specific subunits in the firm. It can produce high-quality general managers. The job rotation system allows an employee to build wider interpersonal relationships that may result in freer information exchange. Finally, widespread use of on-the-job training in large companies tends to diminish an employee's capacity to work effectively if he does move on to another corporation. Thus it discourages employee mobility between competing firms.

Company-Oriented Unionization

Japanese unions are usually organized on a company basis, as opposed to the craft or industrywide unions common in the United States. The company union gives both management and workers an identity of purpose and provides an environment in which there is greater cooperation for the achievement of common goals. The system had its origins in pre–World War II Japan. The company-oriented unions (hereinafter referred to as enterprise unions) have certain distinct characteristics. Membership extends to both blue-collar and white-collar workers with regular employee status. White-collar employees up to the level of section chief are included. Union officials consist solely of company employees. The union is regarded as an autonomous organization whose sovereignty is recognized within the whole union power structure.[33] An enterprise union negotiates independently with its own employer, except for a short period of collective bargaining with the affiliated federation, the so-called "Spring Labor Offensive."

Labor disputes are resolved quite differently from in the West. First, there is the societal pressure toward consensus. Second, there is the tendency for industrial action to be taken in a demonstrative form—to make the public aware, as it were, that the workers feel the employer has failed to do what he should to meet their needs. Third, the union, being mindful of the extent to which its members' interests are bound up with the enterprise, is likely to refrain from any action likely to prejudice its long-term future.[34]

THE INFLUENCE OF SOCIOCULTURAL VALUES

It is generally accepted that all Japanese decisions in any organization are a product of group thinking; a consensus among the various groups is demanded in order to preserve a harmonious atmosphere within the organization. Decision-making by consensus is demanded by the value system of Japanese society, which highly prizes *wa*, or spirit of harmony within a group and among groups, and "groupism." At the same time, however, some unique Japanese characteristics—personal, interpersonal, and organizational—help to make consensus decision making work effectively.

The familiar aspects of Japanese social characteristics—variously termed "paternalism," "groupism," and "familyism"—originated in the strong tradition of *ie* or household. Simply put, this is the Japanese sociopsychological tendency that emphasizes (in the sense of protecting, cherishing, finding needs for, and functioning best in) "us" against "them." In social organization, Japanese put far more emphasis on situational frame than on personal attributes. Thus when a Japanese faces an outside group he establishes his point of reference not in terms of who he is but what group he belongs to.[35] At the same time, because of his sense of "groupism," he has a strong emotional need to belong to a group. A Japanese corporate organization can be seen as a large group consisting of all its employees. The *Ringi* system ensures harmony within the organization.

To this, two other social factors should be added. One is the homogeneity of the Japanese people, which makes an identity of outlook and approach possible.

Historically, Japanese companies have practiced strict racial discrimination. Koreans and Formosans were excluded from Japanese companies in Korea and Formosa during Japan's hegemony there before World War II.[36]

The other factor is the Japanese companies' fostering of this conformity of viewpoint even further by hiring only at entry-level positions and training employees not only in the functional aspects of business, but also in the company's philosophy through in-house indoctrination programs. By rigorous pre-employment testing, referral systems, and long interviews, potential employees whose ideas and outlook may be at variance with the company's are screened out and not hired. This practice is accepted to the extent that Japan's Supreme Court upheld the right of an employer to fire a worker because, at the time of his employment, he did not inform the company that as a student he had participated in allegedly peaceful antibusiness demonstrations on campus. The Court stated that, although freedom of thought is protected in the Japanese Constitution, it does not oblige a company to hire someone whose philosophy differs from its own.[37] At the same time, many other students who took part in leftist-oriented demonstrations and allegedly antibusiness activities at the universities obtained positions with large, well-known companies by explaining their participation in student demonstrations "as a matter of going along with fellow classmates" and therefore conforming to the group mores.[38]

The Thread of Personal Relationships

The basic single-unit interpersonal relationship between two Japanese persons is *Oyabun-Kobun* (*Oya*, father; *ko*, child), in vertical system. Members in a work-related group or in any Japanese organization are tied together by this kind of relationship. According to Nakane:

> The extension of this kind of didactic relationship produces a lineage-like organization. The organizational principle of the *Oyabun-Kobun* group differs from that of the Japanese family institution in that the *Oyabun* normally has several *Kobuns* with more or less equal status, not only one as in the case of the household unit . . . [whereas] the Japanese father may discriminate in the treatment of his sons . . . the essential requirement of the *Oyabun* is that he treat his *Kobuns* with equal fairness according to their status within the group, otherwise he would lose his *Kobun* because of the unfair treatment . . .
>
> In this system, while the *Oyabun* may have several *Kobuns*, the *Kobun* can have but one *Oyabun*. This is the feature that determines the structure of the group based on the vertical system. . . .
>
> Within the *Oyabun-Kobun* group, each member is tied into the one-to-one dyadic relation according to the order of and the time of his entry into the group. These dyadic relations themselves form the system of the organi-

zation. Therefore, the relative order of individuals is not changeable. Even the *Oyabun* cannot change the order. It is a very static system in which no one can can creep in between the vertically related individuals.[39]

Interpersonal Relationships and the Group

The Japanese social system is group-oriented. Individuals have a strong need and a tendency to form in groups. When a corporation is small enough to accommodate only one group, their interpersonal relationships are simple: the group members belong to the corporate family and develop relationships like a single-unit family. However, as a corporation grows larger, many subgroups are established within the corporate organization. These groups can be classified broadly into two types: the functional group, and the group united by personal ties.

A Japanese employee spends his daily life as a member in a functional or work group that usually occupies one office. He first develops personal relationships within this group. As he stays in the company, he develops other personal relationships either through his work experience or outside it, but within the corporate organization.

Interpersonal Relationships Within a Work-Related Group

When a class of university graduates enters a corporation, they are assigned to a section of a department for training after a short initial orientation. Under in-house training practices, a person would stay in a section for several years and then would be transferred to another section within the corporation. Because most university graduates obtained degrees in either economics or law, they usually do not have any understanding of business operations. The number of Japanese universities with departments of business administration is very small.[40] Thus most of the training is provided by managers who are at a level above those of the trainees. Moreover, because of the strong Japanese respect for seniority, the *Oyabun-Kobun* relationships will be developed within this work-related group. The concept of *Oyabun-Kobun* or superior-subordinate relationship in the corporate life has the following characteristics:

1. The senior manager is older than his junior, has worked longer for the company, and is in a position of relative power and security. This position enables the senior to assist the junior.
2. The senior is beneficially disposed toward the junior and befriends him.
3. The junior accepts the friendship and assistance of the senior.
4. These acts and related feelings are the basis of the relationship. There is no explicit agreement.

5. Ideally, the junior feels gratitude toward the senior for his beneficence, and this feeling is accompanied by a desire on the part of the senior to become a good elder friend for those younger.[41]

The relationship is developed not only by the social norms, but also through time-comsuming group-centered activities. Members of the work-related or office group do many activities together in order to promote an open and harmonious atmosphere: once or twice-a-year trips, frequent drinking parties after working hours, monthly Saturday afternoon recreation, etc. Most of these activities are usually financed by the company. These activities also serve as a bridge between generations.

The ideal office, from the point of view of smooth relations and thus "better" working environment, has smaller membership and an ascending distribution of ages within its hierarchy. Under the seniority-based promotion system, older men usually occupy the higher position. Figure 26-2 is an example of a seniority-based system in a large bank in Japan.[42] Note that seniority is a major source of control in Japan. Figure 26-3 shows a typical office outlay in a Japanese corporation. One department occupies a large room in which employees of all levels sit at desks arranged in much the same order as the organization chart. It takes about 13 to 15 years to become a section chief. During this period employees may be transferred around in the department and in other departments, so as to make them "generalists" rather than "specialists" in the American or Western sense.[43]

The group is not static but dynamic, in that the members transfer in and out frequently.[44] Whenever a member is transferred to another section, he establishes a similar relationship in the new department. He may still maintain personal relationships with his former senior and junior officials. These personal ties are sometimes transformed into factions, as explained in the next section.

Faction: Its Formation and Personal Ties

The vertical system of personal relationships, coupled with the compulsive Japanese need to belong, are the main bases for the formation of subgroups and cliques (*habatsu*) within a large group. This is true of political parties, government bureaucracies, large corporations, and newly emerging radical and dissent movements. Factions and cliques protect the interests of their members, and also provide a system of checks and balances within the organization. At the same time surface group harmony is maintained and rigidly enforced, until the group in power is challenged by another group. Membership in a faction depends on a combination of such factors as coming from the same university, marriage ties, or assignment in the same section or department of the company.[45]

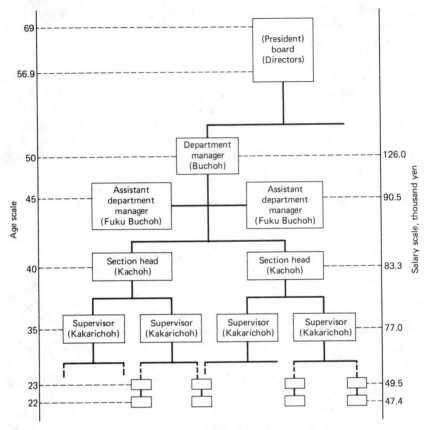

Fig. 26-2. Organizational structure and seniority order.

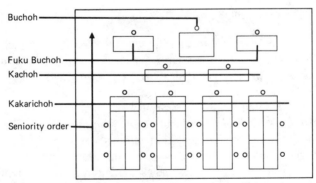

Fig. 26-3 Typical office outley: One department in one room. Source: Naoto Sasaki, *Management and Industrial Structure in Japan* (New York: Pergamon, 1981), pp. 73–74.

Factionism is seen as an evil in Japan, especially by government bureaucrats. But it is tolerated because it has positive function as well, particularly in communications within the organization. A faction promotes and sustains informal and personal relationships between juniors and seniors, and facilitates two-way communication. However, factions also provide for one-to-one relationships between a junior employee and his superior, a relationship that is far more important than larger, more formal group relationships. This is one of the reasons why bottom-up communications are inordinately slow in Japanese bureaucracy: they must go through the entire chain, one link at a time. But it also tremendously facilitates lateral communications and coordination.[46] The needs for coordinaton, easy two-way communication, and one-to-one relationships are some of the reasons why it is extremely difficult for Japanese companies to hire experts from outside the organization at a higher level of management: such experts simply cannot become part of the internal network.

Factions and cliques provide one of the main underpinnings for faster promotions and increased individual rewards in corporate and government bureaucracies. A faster promotion is one of the major rewards for a Japanese employee under the lifetime employment and seniority-based wage and promotion systems, which make rewarding individual excellence difficult, especially in view of widespread official retirement at 55 for employees who are not promoted faster and do not obtain a top or senior management position.

MATRIX MANAGEMENT AND CONSENSUS DECISION MAKING: WHERE SHALL THE TWAIN MEET?

In certain important aspects, matrix management and Japanese-style consensus decision making are remarkably similar. Both stress consensus in the decision-making process. In the role of managers, both emphasize coordination of activities of junior-level employees, and conflict resolution among competing viewpoints instead of primarily implementing decisions made at the top. Both systems push decision making down to the level closest to the implementation point. Both call for a corporate culture and corporate ethos that are conducive to deliberative and consultative decision making, more open and frank communication, a higher degree of mutual trust, and a downplay of hierarchical and top-down authoritative decision making.

These overt similarities between the two systems have caused some scholars to suggest that matrix management and consensus decision making are the same thing in fact, if not in name. According to Davis and Lawrence:

Many aspects of Japanese culture are ideally suited to the matrix format: emphasis on the harmony of group relations, commitment to, and training for, an organization rather than for a specialized role, the subordination of delegated authority to the art of group decisionmaking and the diffused sense of responsibility of each for the actions of others, to name just a few.

If the Japanese culture is so receptive to the tenets of the matrix, why haven't more of their corporations used it? In part, many have, such as in project teams and task forces. But more to the point, the Japanese don't have to create a formal matrix structure and name for "it" the way we do, because matrix structure and behavior is an intrinsic part of their way of being. Moreover, their cultural patterns underscore the major point of this section: patterns of behavior and a culture that support the matrix are not only essential to make the structure work, but they also make the formal structure secondary in importance. The processes are the essence of the form in Japan, as they are in successful matrix organization in the United States.[47]

The surface similarities, however, may be quite misleading. Although the two systems share certain common "process" similarities, they are quite different in terms of their "designed" objectives or outputs, with the result that one can never be a substitute for the other without some serious organizational engineering, which in turn may render both of them structurally weak. It is therefore important that the differences between the two systems be clearly understood and evaluated, before tinkering with either of them.

The essence of matrix management is to deal with specialized information that does not fit conventional organizational modes, and to facilitate work among experts of different orientations so as to direct their expertise and energies toward a common goal. Matrix management facilitates the working together of people who are unfamiliar with each other, come from different academic and professional groups, and have few commonly shared interests.

Japanese-style consensus management facilitates work and communication across departmental lines through a very long and slow nurturing of personal relationships. To the extent that new ideas are not internalized by existing management and professional cadres through in-house training and job rotation, the Japanese management system would have great difficulty in assimilating outside experts.

Another important element of matrix management is the management of uncertainty, where the decision-making mechanism facilitates a speedier and more balanced decision by different specialists. The system originated in response to needs for managing high technology projects where new information and associated high risks had to be absorbed in the decision-making process.

Japanese-style consensus management, on the other hand, is designed to avoid uncertainty and manage continuity. Because of a communication process that is both structured and ritualistic, decision-making is slow and time-consuming. When a fast-changing environment and uncertainty arise, new information is avoided or processed slowly and must be internalized. Japanese-style consensus management is designed to foster group strength, which is often hostile to new informaton as representing a threat to group cohesiveness. Thus while the matrix system and Japanese consensus decision-making both emphasize consensus building, the former applies it in managing new information and risk, the latter in facilitating improved decision-making for conventional and familiar decisions.

The most important difference between the two systems is manifested in the fact that, while Japanese organizations have been known to excel in applications of already developed technologies and in fact to improve upon them, they have not been as successful in developing new technologies, the very forte of the matrix system. As a matter of fact, a number of Japanese companies, notably SEIKO[48] and Matsushita Electric Company,[49] which are facing challenges in world market in high technology fields, have begun to institute matrix systems in their organizations.

REFERENCES

Notes

1. Stanley M. Davis and Paul R. Lawrence, *Matrix* (Reading, Mass.: Addison-Wesley, 1977), pp. 11–18. The discussion in this section is based largely on the material in this source.
2. *Ibid.*, pp. 18–19.
3. Paul R. Lawrence, Harvey F. Kolodny, and Stanley M. Davis, "The Human Side of the Matrix," *Organizational Dynamics* 6 (Summer 1977): 48–49.
4. Davis and Lawrence, *Matrix*, p. 46.
5. Paul Cathey, "Make Profit Centers Work Through Matrix Managing," *Iron Age* (Oct. 15, 1979), pp. 46–47.
6. H. P. Gunz and A. W. Pearson, "Matrix Organization in Research and Development," in Kenneth Knight, ed., *Matrix Management* (New York: PBI-Petrocelli Books, 1977), p. 30.
7. Davis and Lawrence, *Matrix*, pp. 78–81.
8. *Ibid.*, p. 104.
9. Knight, *Matrix Management*, pp. 178–180.
10. Davis and Lawrence, *Matrix*, p. 108.
11. Norman H. Wright, Jr., "Matrix Management: A Primer for the Administrative Manager, Part II: Project Organization," *Management Review*, May 1979, pp. 61–62.
12. *Ibid.*
13. S. Prakash Sethi, *Japanese Business and Social Conflict* (Cambridge Mass.: Ballinger Publishing Co., 1975), pp. 51–56.
14. Naoto Sasaki, *Management & Industrial Structure in Japan* (New York: Pergamon, 1981), pp. 57–58.
15. "Japanese Managers Talk How Their System Works," *Fortune*, Nov. 1977, pp. 131–132.
16. William M. Fox, "Japanese Management Tradition under Strain," *Business Horizons*, Aug. 1977, p. 79.
17. Ichiro Hattori, "A Proposition on Efficient Decision-Making in the Japanese Corporation," *Columbia Journal of World Business*, Summer 1978, p. 12.
18. The discussion in this section is substantially derived from Sethi, *Japanese Business and Social Conflict*, pp. 56–59.

19. Chie Nakane, "An Interpretation on Group Cohesiveness in Japanese Society." Paper presented at the Regional Seminar. Center for Japanese and Korean Studies, Berkeley, University of California, March 1, 1974, p. 12.

20. Richard Halloran, "5 Japanese Had Key Roles in Pushing Lockheed Bids," *New York Times*, March 1, 1976, pp. 1, 8.

21. Nakane, "An Interpretation," pp. 19–20.

22. Sasaki, *Management*, p. 77.

23. *Ibid.*

24. Yoshi Tsurumi, *The Japanese Are Coming: A Multinational Interaction of Firms and Politics* (Cambridge, Mass.: Ballinger Publishing Co., 1976), pp. 228–30.

25. Peter F. Drucker, "Behind Japan's Success," *Harvard Business Review*, Jan.–Feb. 1981, p. 87.

26. The discussion in this section is largely derived from Sethi, *Japanese Business and Social Conflict*, pp. 60–70.

27. Bradley M. Richardson and Taizo Ueda, eds., *Business and Society in Japan: Fundamentals for Businessmen*, (New York: Praeger, 1981), pp. 31–32.

28. Tsurumi, *The Japanese Are Coming*, p. 221.

29. Ronald P. Dore, *British Factory—Japanese Factory: The Origins of National Diversity in Industrial Relations* (Berkeley, Calif.: University of California Press, 1973), p. 35.

30. *The Development of Industrial Relations Systems: Some Implications of Japanese Experience* (Paris: Organization for Economic Cooperation and Development, 1977), p. 19.

31. Masumi Tsuda, "Lifetime Employment and Seniority-Based Wage System," p. 25. Paper presented at International Symposium on Japanese Way of Management and International Business, Tokyo, Nov. 27–28, 1973. Sponsored by Japan Management Conference Board, Tokyo.

32. Arthur M. Whitehill and Shin-ichi Takezawa, "Workplace Harmony: Another Japanese 'Miracle'?," *Columbia Journal of World Business*, Fall 1978, pp. 26–28.

33. Ernest Van Helvoort, *The Japanese Working Man: What Choice? What Reward?* (Vancouver, Canada: University of British Columbia Press, 1979), pp. 130–131.

34. *The Development of Industrial Relations Systems*, pp. 25–26.

35. Chie Nakane, *Japanese Society* (Berkeley, Calif.: University of California Press, 1972), pp. 2–4.

36. Hiroshi Hazama, "Japanese Way of Management—Its Social and Cultural Background," p. 46. Paper presented at International Symposium on Japanese Way of Management and International Business, Tokyo, Nov. 27–28, 1973. Sponsored by Japan Management Conference Board, Tokyo.

37. "Firms May Refuse Hiring on Creed: Supreme Court," *Japan Times*, Dec. 13, 1973, p. 2.

38. Takehiko Yoshihaski, "Linkages Between Social Bonds and Productivity Efforts in Japan," in Allan Taylor, ed., *Perspectives on U.S.-Japan Economic Relations* (Cambridge, Mass.: Ballinger Publishing Co., 1973), p. 78.

39. Nakane, "An Interpretation," pp. 4–5.

40. I. Ueno, "The Situation of Management Education in Japan," *Management Education* (Paris: Organization for Economic Cooperation and Development, 1972), pp. 38–39. 16% (or 62/389) of colleges in Japan had departments of business administration, and only 7% of the total college student population were registered in those departments in 1972.

41. Thomas P. Rohlen, "The Company Work Group," in Ezra F. Vogel, ed., *Modern Japanese Organization and Decision-Making* (Berkeley, Calif.: University of California Press, 1975), p. 197.

42. Sasaki, *Management*, pp. 72–73.

43. Peter F. Drucker, "What We Can Learn from Japanese Management," *Harvard Business Review*, March–April 1971, pp. 117–18.

44. Rohlen, "The Company Work Group," pp. 200–01.
45. Albert M. Craig, "Functional and Dysfunctional Aspects of Government Bureaucracy," in Vogel, *Modern Japanese Organization,* pp. 11–15.
46. *Ibid.,* p. 14.
47. Davis and Lawrence, *Matrix,* pp. 55–56.
48. Hattori, "A Proposition," p. 12.
49. Richard T. Pascale and Anthony G. Athos, *The Art of Japanese Management: Applications for American Executives* (New York: Simon and Schuster, 1981), p. 33.

Bibliography

Butler, Arthur G., Jr. "Project Management: A Study in Organizational Conflict." *Academy of Management Journal* 16 (March 1973): 84–101.

Davis, Stanley M. "Two Models of Organization: Unity of Command Versus Balance of Power." *Sloan Management Review* 16 (Fall 1974): 29–40.

Davis, Stanley M., and Lawrence, Paul R. *Matrix.* Reading, Mass.: Addison-Wesley, 1977.

Davis, Stanley M., and Lawrence, Paul R. "Problems of Matrix Organizations." *Harvard Business Review* 56 (May–June 1978): 131–42.

Goggin, William C. "How the Multidimensional Structure Works at Dow Corning." *Harvard Business Review* 52 (Jan.–Feb. 1974): 54–65.

Huse, Edgar F. *Organization Development and Change,* pp. 188–96. New York: West, 1980.

Lawrence, Paul R., Kolodny, Harvey F., and Davis, Stanley M. "The Human Side of the Matrix." *Organizational Dynamics* 6 (Summer 1977): 43–61.

Sayles, Leonard R. "Matrix Management: The Structure with a Future." *Organizational Dynamics* 5 (Autumn 1976): 2–17.

Section V
Matrix Support Systems

Matrix management is successful in an organization if adequate support systems exist compatibly within the larger organizational system. Supporting systems have to be aligned to support the matrix over and above the existing institutional system. For example, information systems that provide visibility of corporate performance through profit centers and functional staff now have to be realigned to provide visibility to the matrix side of the organization. At the same time an ongoing corporate visibility is needed. This section examines some of the more important support systems.

In Chapter 27 George F. Mechlin and John K. Hulm examine matrix management in the central research laboratory of a large corporation. Their discussion centers around the strategy of the corporate parent, and how the R & D function is carried out in the context of technology transfer, funding, organization, and central laboratory functions.

Hans J. Thamhain in Chapter 28 provides insight into the production function within a matrix environment. He discusses the challenges facing production/matrix management in building and leading a production team. Then he talks about the organization of the matrix-related production function, as well as the planning and control functions within the production matrix. Finally the author puts it all together in terms of "working together across functional lines."

Organizational development (OD) is particularly challenging when undertaken in the matrix environment. In Chapter 29 Milan Moravec discusses what is the most important form of organizational development to use in matrix and how to apply OD techniques effectively. Using a project team as an example, he explains OD as an approach that focuses on issues relating to getting the job done—one that mobilizes resources systematically and regularly to improve project performance.

In Chapter 30 Joe Mumaw writes about control in a project-driven matrix organization where multiproducts are being produced. He examines control through the finance and accounting functions in an aerospace firm. Such topics as budget center management, work breakdown structures, product master plans, design to cost, and progress measurement techniques are examined. He recognizes the inherent contribution that matrix management can provide, if applied properly, to multiproduct control.

Once properly designed, the execution of matrix management depends on the effectiveness of supporting information systems. In Chapter 31 Ivars Avots discusses information systems as they relate to the requirements of the matrix organization, with regard to specific functions such as critical path scheduling and budgeting. He considers both formal and informal information systems in the building of organizational communication networks.

In Chapter 32, Robert L. Brickley discusses the role of marketing in an international corporation. He describes some of the organizational and strategic planning issues to be dealt with in the international matrix ambience. The author provides some guidelines on how marketing can be carried out in the international company as well as some suggestions on how strategic planning can be effected.

27. Matrix Management in the Central Research Laboratory of a Large Corporation

George F. Mechlin and John K. Hulm*

ROLE OF THE CENTRAL RESEARCH LABORATORY

The management strategy and structure of the central research organization of a large corporation must be responsive to the business mission of the corporation and should conform to its inherent strategies. A substantial discussion of the strategy of the corporate parent, and of the mission of the central laboratory and its special problems, is crucial to understanding the structure of the organization.

The parent corporation in this case exhibits extreme diversification in the kind of products and services offered and in the customers served. A very high degree of management decentralization is needed to facilitate optimum performance of the many diverse elements of the corporation's business. A fundamental unit of the corporation's structure is termed a *business unit,* of which roughly 40 make up the operating portion of the whole corporation. Each of these business units is responsible, most importantly, for its strategic planning within its business area as defined by a rather general market-oriented charter.

*Dr. George F. Mechlin has headed the Westinghouse Research & Development Center since 1973. He attended the University of Pittsburgh, where he received his Ph.D. in physics in 1951. Dr. Mechlin joined Westinghouse as a Senior Scientist in 1949. From 1949 through 1957 he worked on the design of submarine propulsion plants at the Bettis Atomic Power Laboratory. He moved to the Westinghouse Marine Division at Sunnyvale, California in 1957. There he participated in the development of the launching systems for the Polaris and Poseidon missiles. His work on the Polaris launcher earned him the U.S. Navy's Meritorious Public Service Award and the Westinghouse Order of Merit. In 1968 Dr. Mechlin became General Manager of the Underseas Division. Four years later, he became Vice-President and General Manager of the Oceanic Divisions. He also was made responsible for the Westinghouse Astronuclear Laboratory.

Dr. John K. Hulm has been Director of Corporate Research and R & D Planning at the Westinghouse Research & Development Center in Pittsburgh since 1980. He attended Cambridge University in England where he received his B.S. and M.A. in mathematics and physics, and a Ph.D. in physics. Dr. Hulm is responsible for planning and developing the Westinghouse R & D Center technical and scientific program, coordinating the technical activities of the Center with the goals of the Westinghouse Companies, and obtaining Corporate and other funding for the projects being conducted by the R & D Center staff. Dr. Hulm joined Westinghouse in 1954 as an Advisory Physicist. Since then he has held a number of management positions. In 1980 he received the Westinghouse Order of Merit for his pioneering efforts in the application of superconductivity to electric power technology. From 1974 to 1976 Dr. Hulm served as Scientific Attaché to the U.S. Embassy in London.

As examples, the charter of a business unit might define the business as the supply and service of equipment for the generation of electrical energy, or the supply of clocks and watches, or the distribution of radio and television entertainment via broadcast and cable. These examples are real and have been chosen to illustrate the diversity mentioned earlier. The strategic plan of a business unit is intended to be a living document, subject to biannual review of the corporation's most senior management, which details the range of products and the markets served within the unit's charter. The strategic plan defines the optimum set of actions for the business unit with reference to the expected condition of the marketplace and the probable actions of competitors. The strategic plan is also expected to forecast and account for expectations in such areas as technological change, government regulatory action, material supply, and the availability of human resources. In short, the strategic plan defines the business and its entire set of operating variables.

As defined then by the strategic plan, the business unit itself normally carries out the full set of actions required by the strategic plan. The business unit builds and operates factories on a worldwide basis consistent with its needs; maintains marketing and product service organizations at its offices and in the field; and plans, develops, designs, and tests its products. One of the informal tests of the appropriateness of the definition and scope of a business unit is the question, "Is this business unit required to cooperate intensively with another business in order to carry out its mission?" A negative answer is needed to create the desired freedom for the business unit's management to concentrate on the business and its all-important externalities.

Many modern corporations are organized along the lines of holding companies, consisting of a set of more or less "orthogonal" business units as described above, plus a small central executive office for overall financial and administrative control. For such fully decentralized companies, the corporation as a whole is essentially the sum of its separate parts. One may pose the question: can a corporation of many independent business units achieve total performance greater than the sum of its separate parts? We believe that the answer is yes, through the judicious use of central corporate services as maintained by the corporation discussed in this chapter.

Experience indicates that there are common functions among the business units that can benefit greatly from economies of scale and are amenable to centralization into a corporate staff function. Obvious examples include the legal function, the accounting of corporate taxes, the management of the investment of corporate cash, and the procurement of the corporation's vehicle fleet. Other functions benefiting greatly from the consistency and coordination available through centralization are accounting principles, relationships with national unions, pension policy and fund management, etc.

Research and development is a special case in that, where certain business units are especially driven by high technology or operate in an area of rapid technological change, the individual unit probably should maintain a strong R & D activity of its own. Such is the case for several business units in our own corporation. However, for the corporation as a whole there are also great benefits to be derived from a strong central R & D activity, which both supplements and complements the technical work of all of the business units, whatever their degree of technical sophistication.

Advantages of Centralized R & D

The benefits of central R & D laboratories for a major corporation can be classified into several categories that we shall now discuss at some length, since they constitute the main argument for maintaining this particular staff function.

Technology Transfer.   While products of the corporation's many business units are diverse and unique to that business unit, the technologies making up the products and their manufacturing processes have much in common. For example, computer-based stress analysis or the materials of electrical insulation or the technology of welding have broad application to many kinds of products. It often happens that a development or capability identifiable at the level of an element of a technology needed by one business unit is an existing capability of another. Where a central research and development center participates in a portion of the technological effort of the whole set of business units, it has a good inventory of knowledge of the capabilities of all. It becomes a simple matter for the business unit to question the central research and development organization, instead of inquiring of all other business units that might conceivably possess the needed answer. No special information retrieval services are needed by the central research and development organization to perform this technology transfer. The fact that there is usually a continuing project relationship between individuals at a business unit and at the research and development center is enough to guide the initial inquiry, and the familiarity of workers at the center with the work of their fellows in technically related areas suffices to develop a reasonably accurate answer. From this point, a computer-based technical report retrieval system can support the informally remembered answer with formal documentation.

Augmentation of Technological Strength.   In an age of increasing technical specialization of people, and where increasingly sophisticated and expensive equipment is required to mount an adequate technical effort, it is obvious that a central technological organization can provide an intellectual climate for specialization and use of shared facilities to a much higher degree than individual business units, particularly those of limited size or that have infrequent need for a particular discipline. The central facility serving many divisions certainly has a better chance of ensuring that expensive instruments such as scanning electron microscopes, electron spectroscopy for chemical analysis, and mass spectrometers are adequately loaded. In addition, the maintenance of a "critical mass" of expert scientists contributes to a higher level of technical excellence in the central group. The program of the central organization can be integrated with the programs of the business units to include the appropriate participation of specialists and the utilization of needed equipment. By such means the business units can have a high and continuing assurance of premium quality in their programs, which would not be feasible otherwise for economic or personnel availability reasons.

Alternative Technological Resource.   The efficient conduct of a business makes the maintenance of personnel levels at some sustainable minimum a highly desirable objective. However, from time to time every business unit needs a pulse of effort or a special project of finite duration. In technological areas this is particularly difficult, since the lead time and cost for recruiting and training people are unattractive. The hiring of consulting firms or outside research organizations sometimes poses difficulties in the control of proprietary information, despite the best of intentions supported by nondisclosure agreements.

The central technological organization can undertake initiatives for business units as an alternative to provision of technological resources from the units' own staffs or by contractors outside the corporation. This type of work can be a valuable source of information and experience on business-unit technology for the central R & D people, but it has some drawbacks. For example, the work is normally quite short-range in nature, hence should not become too large a proportion of the central laboratories' project portfolio. Similarly, there is a danger that business units will depend too much upon the central laboratories for technological capability that they should really acquire for themselves. This usually derives from severe "managed costs reduction" pressure upon individual business-unit managers. Constant vigilance is required at the central laboratories to avoid these pitfalls. However, it seems to be the inevitable fate of many R & D directors to have to compensate for business units intent upon committing technological suicide.

New Technological Initiatives.   Confining the activities of business units to charters and to the limits of approved strategic plans can exclude the corporation from initiatives that might be outside the existing structure of activities. Where the initiative is based on technology, whether new or existing, the central research and development organization offers the means for undertaking these initiatives without creating new organizations or diluting the efforts of existing ones. The identification and authorization of such initiatives must be mediated by interaction between laboratory management and the senior management of the corporation. The technical success of such initiatives, which certainly would not be undertaken without a sanguine belief about the ultimate prospects of success, eventually requires the creation or identification of a business unit to complete and commercially exploit the initiative. This identification must be anticipated in the review and approval process by senior management and in the continuing management of the project by laboratory management.

These four categories seem to us to constitute the four principal benefits that a diversified major corporation should derive from its central R & D operation. Several other dividends might be mentioned in passing: for example, the central laboratories should (though seldom do) serve as a source of highly qualified technical people for the rest of the corporation. The laboratories should act as a "window on the world" of science and technology for the corporation, and in this regard can alert top management to the importance of newly emerging technology that may strongly affect the business.

A not inconsiderable function of the central laboratory is its role as a marketing tool. In sectors of the marketplace serving the industrial customer, an organization vital enough to prosper over its competitors is often willing to take business and technical risks. The nature of a risk is that the unfortunate eventuality envisioned sometimes comes true and the risk taker has a problem requiring solution. If this problem is technological, the central laboratory is often a premium resource for resolving it. A sophisticated industrial customer contemplating a risk being undertaken on his behalf or, worse, being inconvenienced by a problem with a system or product already in his hands, often finds considerable solace in benefiting from the impressive problem-solving capabilities of the central laboratory as expressed in people and equipment. The willingness to accept technological risk, and the ability to expeditiously resolve difficulties for customers, are significant factors in the long-term viability of a business. As before, the central organization must possess the management capability to coordinate customer visits with business units and to represent itself during such occasions.

We will consider now how the central laboratories should be organized and managed to achieve these desirable results. A vital role is played here by the mechanism of funding.

Sources of Funding of Centralized R & D

The central laboratories of most large corporations known to us obtain their research funds from essentially three sources: (1) direct support from the individual business units for specific projects; (2) support from government or private research agencies; and (3) direct support from the corporate headquarters unit. Of course the proportion of these three categories varies enormously from one corporation to another. In the present case, the proportions are approximately 60, 25, and 15%, respectively.

With respect to projects funded by business units, it should be pointed out that research and development are but the initial phases of the life cycle of a product, as illustrated in Figure 27-1. Experience teaches that research and development costs are a minor portion of the investment necessary to bring the product or process under development to a successful (i.e., profitable) status in the marketplace. Obviously, if the whole organizational chain is not committed to making the requisite investment, presuming technical success and adequate market prospects at each successive stage of the life cycle, the work should not be undertaken at all.

When the business unit must (in the accounting sense) pay the costs of the work done at the central laboratory, just as it pays the costs of such work performed by its own people, the chances of utilizing the research and development, as measured by the ultimate appearance of a product in the marketplace, are considerably enhanced. That this should be true is apparent from the nature of the management processes within the business unit. If the work by the central laboratory were "free" (in the accounting sense) to the business unit, the mere plausibility of the ultimate usefulness of the project in question would often be sufficient for its sponsorship. If the financial means to support the work at the central laboratory appear to the business unit as one of many competing needs, it is necessary to have strong advocacy for the project within the unit before the internal resource allocation process yields the necessary sup-

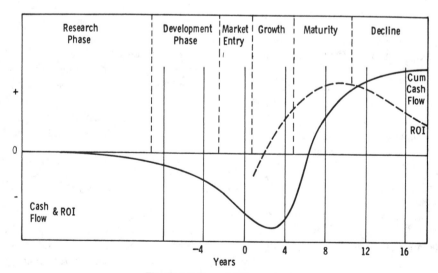

Fig. 27-1. Life cycle of a product.

port. This advocacy becomes a personal commitment on the part of one or a few individual, so that any later project failure (i.e., noncommercialization) must become a lingering unfavorable reflection on their judgment.

The central laboratory in its turn would like to "sell" to business units as many projects as its human and physical resources can support. This immediately implies a project-by-project negotiation between the seller and the now wary buyer, in which each side clearly perceives its interests and seeks to minimize its risks (both reputation and financial). Also implied is the concomitant need for a referee function somewhere in the corporate management structure to assure that the negotiation process is kept within bounds consistent with the corporate interest.

The second important source of funds for the central laboratory constitutes research and development contracts, either from government agencies or from private research agencies such as the Electric Power Research Institute. The usual administrative treatment of such contracts is quite formal, with initiation through a request for proposal based on a sponsor-prepared statement of work. Source selection is based upon a set of formal proposals, with both technical approach and projected cost of performance as important determinants of selection. Following selection, detailed negotiation of price, terms, and conditions usually precedes the formal issuance of a contract, which is itself the prerequisite to creative work on the sponsor's project. The process is intensely competitive: offerers must carefully adjust their proposals to the sponsor's expressed and implied wants and to management's perception of the benefit and motivation for the proposed contract undertaking. Again, the parochial interests of the central laboratory and the business units are mingled on a proposal and contract basis, implying again the need for management capability additional to that needed by either the central laboratory or the business unit taken alone.

The third major source of funds for the central laboratory is the headquarters or top management of the corporation itself. Depending on the subjective inclinations of senior corporate management, this is sometimes identified as funding for "basic or speculative research" with potential for great reward, but at high risk. For less euphoric management philosophies, this direct corporate funding is justified as required to support the central laboratory in keeping up with the body of science emerging from worldwide academic research endeavors, and to fund exploratory research and special initiatives agreed to between the central laboratory and senior corporate management. Any logical relationship of the program of corporately funded work and the remainder of the program implies a management function at the central laboratory capable of relating the highly detailed nature of business-unit and agency-sponsored work with the corporate-sponsored program, so as to avoid duplication and assure their being complementary.

In our own particular case, the corporate research funding constitutes about

15% of the total central laboratories budget, but it plays an especially vital role in the general health of the laboratory. This funding is the only support for research programs where the goal of the program is determined internally within the laboratories. For the remaining 85% of the budget, there are strongly prescribed objectives imposed by outside parties such as government contract officers or business-unit engineering managers. Reasonable success in achieving these objectives is, of course, vital to the future funding of the laboratories, and such programs are prosecuted with maximum vigor. In such circumstances, however, there is little time or opportunity to explore side channels revealed by the ongoing work, or to test out any radically new concepts.

It is in the last-mentioned areas that the corporate funds are so vital. They allow investigators to explore new, untried approaches; they constitute "seed funds" that are used to grow new seedlings, some of which will blossom into brand-new major projects suitable for business-unit or outside-agency support. In other words, the corporate 15% seeds the other 85% funding of our particular central laboratories.

ORGANIZATION OF A LARGE CENTRAL LABORATORY

We turn now to the main question as to how a central laboratory should be organized and operated to achieve the various technical advantages discussed earlier and to ensure the continuous flow of R & D funds. A common choice of large multidisciplinary laboratories is the discipline-based organization. Here the format is to group like subdisciplines and disciplines for which span-of-control considerations are a dominant factor in selection of group size at all levels. For the central research and development center on which this chapter is based, the basic organization is shown in Fig. 27-2.

As with all real-world systems, some lack of fidelity to pure disciplinary logic is apparent from the next lower tier of organization within this structure, as shown in Table 27-1. Among the benefits of the philosophy of a discipline-based organization is that it is familiar to the education and academic personnel who make up the professional staff (roughly 40% hold a Ph.D.). More important, it optimizes the opportunity for peer recognition that is an important element of the motivation pattern of these highly creative people. Modern industrial research and development requires, on an increasing basis, the cooperative efforts of several disciplines. Projects frequently lie in the vaguely defined boundaries where these disciplines interact. Thus the first requirement for matrix management arises from the need for several individuals drawn from different organization units to collaborate on a project. We have not found it profitable to completely recast the basic organization structure of our central laboratories from a "discipline" to a "project"-oriented system. The basic reason for this is that the organization must cope with roughly a thousand major programs each year, with the average program lasting three years. The

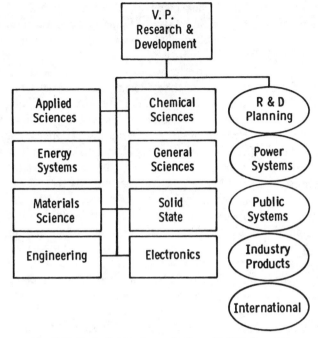

Fig. 27-2. Basic divisional organization with R & D center.

magnitude and frequency of change within a project organization, as it responded to this scale of program complexity and change, would be unacceptable to all concerned.

We employ instead a matrix approach that for each program designates an individual leader, most commonly called the *principal investigator*. For certain sponsors or for very large efforts, this title changes to *program* or *project manager*. In the majority of programs, no formal management organization status is conferred by the appointment to principal investigator. The number of principal investigators and program or project managers is sufficiently large that

Table 27-1.   Department of the Applied Sciences
Division.

APPLIED SCIENCES	
● Optical Technology	● Gas Discharge Physics
● Mathematics	● Nucleonics
● System Modeling	● High Voltage Technology
● Lasers	● Power Interruption
● Lamps	● Superconductivity

no attempt is made to identify them on an organizational chart, except for very large projects, of which there are only a few per year. Each project has a monthly computer-prepared statement of costs, whose distribution includes the principal investigator or project manager.

The responsibility for the performance of the work is shared between the line management organization and the principal investigators or project managers. Principal investigators report only to the line organizations, in that there exists no program organization at the center. This is so because the individual projects are frequently portions of larger programs at the sponsoring business units or agencies, which themselves have adequate program management structures and functions. Program management is therefore limited to situations where it is clearly needed and the costs are directly borne by the sponsor of the work by prior agreement.

Personnel administration matters such as performance reviews of individuals, salary administration, career counseling, and rather rare disciplinary actions are the exclusive purview of line management. Line management is also responsible for the quality of both the work and the work force itself. In addition to these concerns on an individual-by-individual basis, line management is expected to assure an appropriate range of subspecialization for each organizational unit, and to influence the program selection process so as to keep the total program for each work group current with the state-of-the-art of its present scientific discipline. While intangible and difficult to define, motivation and morale are strong factors in the success of any organization and particularly so in research and development. Line management directly carries the responsibility for maintaining the professional pride and performance of its employes.

Line management also carries the budget responsibility for each work unit. Since charges to such budgets consist of charges directly to projects and overhead accounts, line management has a broader cost control responsibility than principal investigators or project/program managers, who are concerned strictly with their individual project.

For the average-size project, which encompasses one to two man years of professional effort, the principal investigators' responsibilities for assuring that adequate effort is applied to their project are relatively simple. The principal investigator, as the name implies, is usually the major contributor to the effort. Other contributors are usually identified in the programming effort to be described later. Since contributors have usually participated in the development of support for the project, they feel a personal commitment to the project that eases problems of personnel availability, except in the case of resource allocation conflicts arising from other sources such as operating or product problems in business units. The resolution of issues of resource allocation is carried out by line management.

The principal investigator is responsible for requisitioning supplies and effort from the laboratory support organizations. Since support effort in most, if not all, laboratories operates in an economy of scarcity, principal investigators usually find the achievement of needed support a constant challenge. In large organizations the lore of "how you get things done around here" seems not to have been uniformly distributed, so that one finds a considerable range of effectiveness in arranging support effort among principal investigators, and a corresponding correlation in personal job satisfaction associated with the position.

The greatest single responsibility of the principal investigator is the creative work of the project, where problems never before encountered must be solved, and new concepts of nature debated and tested. In our laboratory, periodic reporting requirements by the principal investigator have been kept at a minimal level. For projects funded internally by business units or the parent corporation, a paragraph-length report is generated at four-month intervals. Of course full-length technical reports are written whenever appropriate. For agency contracts, progress-reporting requirements are usually monthly and can be an onerous portion of the principal investigator's or project manager's responsibility.

For larger projects ranging usually from 5 to 15 (or rarely more) man years of professional effort, the title *project* or *program manager* is used. The program manager title is usually reserved for activities consisting of several related projects with responsible principal investigators. The choice between the titles *program* or *project manager* frequently depends upon the preference of the sponsor for one term or the other.

The project manager's responsibilities (and shared authority) are broadly similar to those of the principal investigator. However, because of the greater size of their projects, project managers generally experience much greater difficulties in obtaining necessary personnel and service resources. Clearly, they also experience more complexity in financial management and control.

In our particular organization, safety is the direct responsibility of line management. Equipment, materials, and activities in research and development functions are frequently novel and their inadvertent malfunction or ingestion can have effects that the workers themselves are best able to estimate and control. The center has a central safety organization that assists line management in control and education and further supports them in the complex aspects of regulatory compliance.

CENTRAL LABORATORY STAFF FUNCTIONS

The management structure of the subject research and development center also includes several staff functions that operate on a matrix basis with the line management organization. This section focuses on the role of the staff groups in influencing and guiding the technical strategy of the laboratory. In this respect, a major role is played by R & D *Planning* (Figure 27-2).

The function of the Planning Department is to direct and administer the program creation efforts of the center. As previously mentioned, the center executes roughly a thousand significant projects each year, which implies a complex administrative function, particularly since several different funding arrangements are used.

As already mentioned, about 60% of the center's funding orginates directly from the business units, groups, and companies within the corporation as a whole. For simplicity we shall refer to this as "business-unit funding." In practice these business-unit R & D programs fall into two categories: those planned in advance (preplanned), and those funded on an as-needed basis by interworks requisitions.

For preplanned programs, the planning cycle begins in the early spring of the year preceding the calendar year in which the work is to be accomplished. As a first step, research directors from the planning department approach all business units and request the submission of a list of suggested R & D projects for the forthcoming year. These requests are submitted on a standard form (known as an A form) that briefly specifies the research and development result desired and summarizes the benefit expected. Many of the A forms refer to work to be continued from the preceding year. The planning department distributes the A forms to appropriate line organizations within the center, who respond with brief proposals (D forms) outlining the technical approach, schedule, and expected cost. It is not unusual for several radically different technical approaches to be proposed. These are reviewed by the requesting business unit and become the basis for the final program selection. It should be noted that the process of selection and prioritization of the R & D projects is quite a complex one, involving several lengthy review meetings between the technical staff of each operating division and the R & D center technical staff. These meetings are organized by the research directors and their assistants. There is an opportunity to discuss progress on existing R & D programs and the technical strategy of the business unit. The distilled R & D program that emerges from this process contains the best thinking of both the business unit and central R & D.

Before the preplanned program is finalized, the research director for each company reviews the proposed R & D projects with group executives and company presidents for fine-tuning. This process is usually complete by late summer.

Business units also support R & D by means of Interworks requisitions. These are intended to allow short-term response to circumstances unforeseen in the annual preplanned program. The requisitions can be revoked at any time by the sponsoring operating unit. Research directors usually play little role in the creation of these requisitions, except when they are used to accelerate or increase effort on projects in the preplanned sector.

To assist in achieving the central laboratory's goal of about 25% outside-agency funding, the Planning Department operates a centralized marketing function that includes a marketing representative for each line research division. These representatives assist the research division staffs in establishing sources of funding within various outside agencies and preparing preproposal documents and proposals for such agencies. All proposals are assembled in and submitted from the marketing group, using technical material supplied by individual researchers.

In order to coordinate the outside proposal activity with business-unit technical goals, the director of planning operates an outside funding review committee on which all the research directors serve. This committee attempts to ensure that for all major proposals, the intended use of people and equipment is consistent with the business interests of the corporation. A further consideration in the committee decision is whether cost sharing is a feature of the contract. If so, the center must have, or be willing to create, relevant programs that can form the financial basis for the contractually required share. It is the responsibility of the appropriate research director to determine the feasibility of cost sharing.

The planning department also includes a contract administration group that deals primarily with contract terms and conditions. These represent a complex and widely varying ensemble for the numerous agencies normally dealt with. The contract group is also responsible for monitoring compliance on all contracts in operation at the central laboratories.

As noted earlier, about 15% of the central laboratories work is directly supported by corporate headquarters. The program topic selections for this category are entirely the responsibility of the center management, subject only to the fiduciary responsibility of assuring the optimum expenditure in the corporate interest. The programming system for the corporately supported effort accepts proposals from all center organizations in the identical format used for responses to business-unit initiatives. A committee composed of the manager of the planning group and the three research directors make the program selections from the group of proposals, and recommends this selection to the center manager for his approval, which is final. The magnitude of the corporate program is approved annually by senior corporate management.

The corporate program planning is done in the summer of each year, after completion of work on the preplanned business-unit programs; thus it is possible to carry out this particular planning with good up-to-date knowledge of changing emphases in the technology of the business units, and also to anticipate those rapidly emerging external technologies not yet covered in business-unit programs.

Although the corporate program constitutes only 15% of the overall central laboratory program, it plays a more vital role than this proportion would suggest. In some sense the corporate research serves as cement to fill the gaps between the various agency and business-unit programs in a given discipline. As already noted, it permits exploration of new concepts and intriguing possibilities uncovered during the regular program work, which the discipline of the latter would normally prevent.

It is, of course, highly desirable that all the program activities in a given area of technology—that is, business-unit, agency, and corporate programs—be integrated as far as possible into a single technological strategy. Recently we have adopted an additional matrix management technique to improve this process, which will be briefly described.

The technical areas of the laboratories are divided into five Strategic Technology Areas: Science, Materials, Electronics, Energy, and Engineering Science. These five major areas are further subdivided into five or six subareas known as Technical Activity Areas (TAAs), of which there are 29 for the laboratory as a whole (see Table 27-2). These TAAs constitute the basis for the annual corporate program review, which includes an overview of the strategic position in the technical areas, integrating the proposed corporate program with what is to be done for business units and outside agencies. Only one cycle of this process has been carried out to date, and perfection has certainly not yet been attained. However, the process has already provided a better general perspective for the degree of emphasis on various technologies in the total program.

The matrix aspect of this control process is embodied in the appointment of a leader for each Technical Activity Area, whose task is to pull together all groups interested in the particular activity. The TAA leaders are frequently drawn from the technical line management organization, but in many instances are senior technical staff members such as consulting or advisory scientists. The choice of leader is based primarily upon technical stature and experience—the

Table 27-2. Strategic Technological Activities.

1.0.0 SCIENCE	2.9.0 Materials Testing & Analysis
1.1.0 Mathematics	3.0.0 ELECTRONICS
1.3.0 Human & Business Science	3.1.0 Artificial Intelligence
1.4.0 Gas Discharge Science	3.2.0 Data Acquisition & Processing
1.5.0 Superconductivity	3.3.0 Integrated Circuits
1.6.0 Senate-Sponsored Research Activities	3.4.0 Power Electronics
1.7.0 Directed Energy Systems	3.6.0 Electronic Systems
1.9.0 Research Planning	4.0.0 ENERGY
2.0.0 MATERIALS	4.1.0 Advanced Energy Concepts
2.1.0 Metallurgy	4.2.0 Fossil Fuel Systems
2.2.0 Metals Engineering	4.3.0 Energy Systems Analysis
2.3.0 Reactor Materials	4.4.0 Nuclear Energy
2.4.0 Surface Phenomena	5.0.0 ENGINEERING SCIENCE
2.5.0 Physical & Inorganic Chemistry	5.1.0 Computer Assisted Design & Analysis
2.6.0 Ceramics	5.2.0 Experimental Engineering Science
2.7.0 Crystals	5.3.0 Electrotechnology
2.8.0 Organic Materials	5.4.0 Engineering Services

best expert for the job. Clearly, the TAA leader group is matrixed against the line organization, with some overlap.

The final prioritizing of TAA corporate efforts within a given Strategic Technology Area is carried out by a Guidance Council chaired by one of the research directors, with appropriate research division managers as members. The manager of the Planning Department is an automatic member of each of the five STA Guidance Councils, to provide a balance in resources between the five Areas of technology.

By now the matrix character of the research director position at the central laboratories will be reasonably clear. Each director serves in a dotted-line fashion on the staff of one of the company presidents, serving as his principal technical staff advisor and manager of his R & D programs.

Adding significantly to the sensitivity of the responsibility of research directors is the fact that essentially all the work undertaken by the center is on a "best efforts" basis. The inherent nature of research and development is to do what has not been done before. Hence the basis for predetermining costs is, at best, very poor. An organization that commits itself in advance to accomplishing specific research and development goals at a predetermined cost is soon overwhelmed by the necessity of compromising quality of performance or expenditure of funds beyond its budget. The only practical alternative is to eschew such financial commitments. Implicit in this policy is the all-too-frequent duty of the research director to solicit extra support from business-unit clients to overcome some unforeseen difficulty in a research and development project. This requires the essence of matrix management, namely, the confidence and good will of both parties.

Our emphasis on the important matrix function of the planning organization in our laboratory is not intended to diminish the role of other staff functions, which must also operate in a matrix fashion, albeit with less emphasis on technology.

The controller function does not differ much from that of most organizations and will not be treated here. The human resources function also shares authority with the line organization in a relatively conventional fashion. It administers employee benefit programs, employee communications, a labor union contract covering nonprofessional and nonmanagement employes, and the organization's Equal Employment Opportunity program. The human resources function contains a training organization whose primary thrust is enhancing the effectiveness of the management processes of the center.

The matrix management system just described evolved not from any formal effort to optimize a management through matrix management principles, but rather from a long evolution in which the corporate businesses we support became more numerous, complex, and decentralized. It is our conclusion that as an organizational concept it performs very well for us, being well understood and broadly respected by members of our corporation within and outside the center. Nevertheless, the structure evolved concurrently and attuned itself with the culture of our parent corporation, and therefore must be viewed not as a model to be emulated universally, but rather as a source of concepts to be mined by others as they seek to adapt themselves organizationally to their own unique cultures.

28. Production within a Matrix Environment

Hans Thamhain*

THE PRODUCTION FUNCTION WITHIN A MATRIX

Modern production practices are often too multidisciplinary to be structured strictly along functional lines. The project management approach that emerged in the 1960s responds to the need for effective integration and execution of the many operations coming into play during modern production operations. Companies that use the matrix are usually in complex businesses where conventional structures have proven inadequate.

Within the global company matrix the production units have a wide range of options to structure their operations. While these structures take many forms, the production unit typically retains most of its basic functional characteristics. That is, hierarchical relations exist within the production organization with clearly established lines of command, responsibility, and authority. Each of the functional subunits is headed by a department manager responsible for technical implementation of the various projects and tasks that run through production, including the maintenance of personnel, facilities, and technological advancement.

Figure 28-1 presents the organizational construct of a typical production unit within a matrix. In this structure, operational responsibilities are divided along two axes. One axis is concerned with managing the business while the other is responsible for managing the resources. In short, the matrix is built around a cooperative relationship between the business manager, who is responsible for *what* needs to be done, and the resource managers, who are accountable for *how* it is being implemented.

Basic Structural Options

No two production units are structured alike. Fundamentally, however, there are only two choices for organizing production operations within a matrix environment: functional or projectized structure. In practice, dichotomizing between those two options is inappropriate. Many companies are not just large

---

*Dr. Hans Thamhain has over fifteen years of experience in managing and developing project business with International Telephone and Telegraph, Westinghouse Electric, General Electric, and General Telephone and Electronics. He is currently Associate Professor of Management in Worcester Polytechnic Institute, Worcester, Mass., and consults with high technology firms in project management and new product developments.

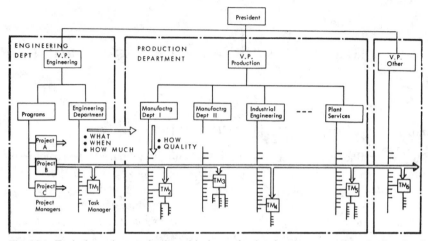

Fig. 28-1. Typical matrix organization with the production department retaining its functional characteristics within the matrix. Project integration is performed by project managers (PM) via production task managers (TM). In this figure, the program management function reports into engineering.

project businesses, but are diversified enough to handle a mixture of large, small, short-, and long-range projects through their productions, which requires the ability to adapt to continuously changing multiproject requirements. Companies that fall into this category have modified the basic concepts, resulting in hybrid structures that overlay the fundamental production organization. What these companies end up with are production units that follow a primary organizational construct, frequently a functional type, but that in addition have organizational subsystems overlays. These overlays may projectize certain production resources or establish miniature matrices within production. All these "in-between" choices must be carefully designed to accommodate the needed intrafunctional integration, while maintaining effective utilization of production resources and overall operational control.

In companies that use the matrix concept, four organizational forms are commonly found in the production units. Rarely is one form used alone; more likely, it is used in combination with other forms or is modified to meet the specific organizational needs. The matrix within the company matrix has become one of the most common constructs, accommodating the widest span of project management requirements with the most flexibility, while retaining most of its organizational stability and resource effectiveness.

Functional Structure within a Matrix.   In this form the production units retain traditional functional characteristics. A clear chain of command exists for all organizational components within the production function. Department managers direct the functional production groups. Task managers or project engineers perform the production's internal coordination and interface with external operating groups along established matrix lines, as shown in Figure 28-1. This is still one of the commonest organizational structures of the production function within a matrix. It provides a high concentration of specialization and production capabilities and assures their economical use. A limitation on this organizational substructure is its dependence on personnel external to production for overall project planning and integration. Especially if the operations require considerable coordination of intraproduction activities, a subordinate matrix structure often evolves and provides the necessary interdisciplinary task integration, control, and initial organizational build-up. This is why an increasing number of production units are introducing an additional matrix layer within their organization, creating a multimatrix environment.

Matrix within a Matrix.   This is essentially an overlay of temporary project organizations on the functional production unit. As illustrated in Figure 28-2, Project manager B is accountable to the program office for managing program x through all production phases. This includes the planning, organizing, and control of all activities necessary to produce the agreed-on results. If the program is large enough, project manager B manages the work via task managers who report to him or her regarding the job requirements, budget, and schedule. That is, the production project manager essentially specifies the work: what the product specifications are and what schedules and budgets must be met, including targets for unit production costs. The task managers essentially have two bosses, being accountable to the project manager for the implementation of the

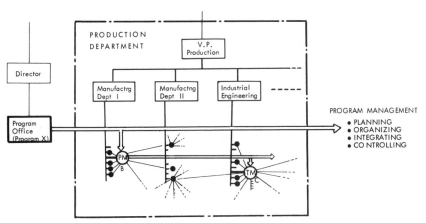

Fig. 28-2. Matrix within a matrix. The overlay of another Matrix to the production department provides a production matrix structure within the global company matrix. Project manager B becomes the central integrator for project X within the production department.

requirements, and to the functional manager for the quality of the work, the availability of resources, and the utility of these resources in accomplishing the desired results.

In describing these multidimensional matrices, executives mostly resort to verbal tools such as charters, directives, and policies, rather than to conventional charts that focus on the primary command channels but often confuse the issues of dual accountability and multidimensional controls.

The advantages of this micromatrix within the larger matrix construct are a more rapid reaction time, better control, and more efficient integration of multidisciplinary activities. The additional integration points, via the production project managers, provide an effective interface with production-external organizations, while at the same time reducing the span of control needed by the program office. On the other hand, the additional organizational overlay is likely to increase cost overheads, and requires a more sophisticated management style to operate smoothly and efficiently within an environment characterized by extensive sharing of resources, power, and responsibility. The multidimensional matrix may not be the panacea for all production organizations, but it can provide a powerful option to the sophisticated executive.

Projectized Structure within a Matrix.   This is an organizational division within the production unit. In its purest forms, the production project manager directs a complete line organization with full authority over the personnel, facilities, and equipment, with the charter to manage one specific project from start to finish. The strongest form of project authority, it encourages performance, schedule, and cost trade-offs; usually represents the best interfaces to production-external organizations; and has the best reaction time. However, by comparison with the matrix, the projectized organization usually requires considerable start-up and phase-out efforts, and offers less opportunity to share production elements, utilize economics of scale, and balance work loads.

Because of these limitations, companies seldom projectize their production units unless the project is large enough to fully utilize the dedicated resources. What is more common is a partially projectized organization where the project manager fully controls certain resources that are particularly critical to the project and/or can be fully utilized over the project life cycle, while other resources remain under the control of functional managers who allocate them to specific projects as needed, based on negotiated agreements with the project managers.

Individual Project Structures within a Matrix.   A very effective method of coordinating smaller multidisciplinary tasks within production is to charter one individual part time or full time with the task integration. This integration usually includes (1) co-ordination with the program office and the resource personnel, (2) front-end planning, (3) progress measurement and reporting, and

(4) directing and controlling the activities toward the agreed-on results. This one-man project organization can be very quickly installed without distributing other ongoing operations or established organizational structures. The individual project organization is a scaled-down derivative of the matrix. It can be viewed as a miniature matrix within the larger construct of the production unit.

Multilayer Matrices

Real-world project organizations are complex, both on paper and in practice; they do not necessarily lead to a single-layer matrix. In fact, so-called multidimensional matrices are quite common for complex businesses. In addition to the project-function matrix with its various levels of integration for each project, there may be overlays of other matrix axes with totally different and unrelated missions, such as a product-function matrix or product-region matrix. Further, the production function may host many minor matrix structures within its functional framework, responsible for a production-internal project such as a new facility development or a quality-control procedure.

The Charter: Defining the Organization

To work effectively, people must understand where they fit into the corporate structure and what their responsibilities are. Given the internally complex workings of the matrix, it is especially important to define the management process, responsibilities, and reporting relations for all organizational components within production and its organizational interfaces. Organizational charters, job descriptions, policy directives, and management guidelines provide the management tools for clarifying the organizational network and directing the business activities coherently. Examples of these management tools are shown in Tables 28-1 through 28-4. However, it should be emphasized that position titles, responsibilities, and matrix structures vary considerably. Therefore these examples provide primarily a typology to management for developing their own tools.

Challenges of the Production/Matrix Manager

To be effective, the new breed of production managers that evolved with the matrix must consider a broad range of issues. They must understand the people, the task, the tools, and the organization. The day of the manager who gets by with technical expertise alone or pure administrative skills is gone. Today, managers in a production matrix must relate socially as well as technically. They must understand the culture and value system of the total organization.

Many of the challenges phased by the matrix manager also arise in more conventional organizations, but they seem to have a stronger impact on oper-

Table 28-1. Project Management Positions and Responsibilities.

PROJECT MANAGEMENT POSITION	TYPICAL RESPONSIBILITY	SKILL REQUIREMENTS
Project administrator Project coordinator Technical assistant	Coordinating and integrating of subsystem tasks. Assisting in determining technical and manpower requirements, schedules, and budgets. Measuring and analyzing project performance regarding technical progress, schedules, and budgets.	Planning Coordinating Analyzing Understanding the organization
Task manager Project engineer Assistant project manager	Same as above, but with a stronger role in establishing and maintaining project requirements. Conducting trade-offs. Directing the technical implementation according to established schedules and budgets.	Technical expertise Assessing trade-offs Managing task implementation Leading task specialists
Project manager Program manager	Same as above, but with a stronger role in project planning and control. Coordinating and negotiating requirements between sponsor and performing organizations. Bid proposal development and pricing. Establishing project organization and staffing. Overall leadership toward implementing project plan. Project profit. New business development.	Overall program leadership Team building Resolving conflict Managing multidisciplinary tasks Planning and allocating resources Interfacing with customers/sponsors
Executive program manager	Title reserved for very large programs relative to host organization. Responsibilities same as above. Focus is on directing overall program toward desired business results. Customer liaison. Profit performance. New business development. Organizational development.	Business leadership Managing overall program businesses Building program organizations Developing personnel Developing new business

PROJECT MANAGEMENT POSITION	TYPICAL RESPONSIBILITY	SKILL REQUIREMENTS
Director of Programs V.P. of Program Development	Responsible for managing multiprogram businesses via various project organizations, each led by a project manager. Focus is on business planning and development; profit performance; technology development; establishing policies and procedures; program management guidelines; personnel development; organizational development.	Leadership Strategic planning Directing and managing program businesses Building organizations Selecting and developing key personnel Identifying and developing new business

ating efficiency within a matrix environment. Seven specific challenges are typically ecountered by the production manager within a matrix organization. The impact of these challenges will be analyzed and suggestions made to increase managerial effectiveness.

Dual Accountability. Unconventional superior-subordinate relationships are common in matrix environment. Both managerial and other work-related responsibilities are often divided along intricate horizontal and vertical organization lines. In order to direct and control activities in such a multiboss environment effectively, the two-matrix axis must be clearly defined in terms of responsibilities, powers, and control. The organizational charter, management directives, policies, and procedures are management tools that can be used effectively. Moreover, the employee performance evaluation and reward system must be consistent with the shared managerial power system. That is, those who distribute the rewards must elicit feedback from all organizations that receive services, and the production managers themselves must be evaluated for the effectiveness with which they provide resources to the many components of their matrix organization.

Power Sharing. No organization has managerial power equally distributed, but the matrix almost encourages managers to jockey for power. Managerial power is vested in the ability to control resources, which can be done in many ways. Both project managers and functional resource managers may try to gain access to resources by (1) influencing the priorities, (2) controlling the budgets,

Table 28-2. Typical Charter of a Program/Project Manager in a Matrix Organization.

Position title: program manager

Authority

The program manager* has the delegated authority from general management to direct all program activities. He or she represents the company in contacts with the customer and all internal and external negotiations. Project personnel have the typical dual-reporting relationship: to functional management for technical performance and to the program manager for contractional performance according to specs, schedules, and budgets. The program manager approves all project personnel assignments and influences their salary and promotional status via formal performance reports to their functional managers. Travel and customer contact activities must be coordinated and approved by the program manager.

Any conflict with functional management or company policy shall be resolved by the general manager or his staff.

Responsibility

The program manager's responsibilities are to the general manager for overall program direction according to established business objectives and contractional requirements regarding technical specifications, schedules, and budgets.

More specifically, the program manager is responsible for (1) establishing and maintaining the program plan, (2) establishing the program organization, (3) managing and controlling the program, and (4) communicating the program status.

1. Establishing and maintaining the program plan.

Prior to authorizing the work, the program manager develops the program plan in concert with all key members of the program team. This includes master schedules, budgets, performance specifications, statements of work (SOW), work breakdown structure (WBS), task and work authorizations. All these documents must be negotiated and agreed upon with both the customer and the performing organizations before they become management tools for controlling the problem. The program manager is further responsible for updating and maintaining the plan during the life cycle of the program, including the issuance of work authorizations and budgets for each work package in accordance with the master plan.

2. Establishing the program organization.

In accordance with company policy, the program manager establishes the necessary program organization by defining the type of each functional group needed, including their charters, specific roles, and authority relationships.

3. Managing and controlling the program.

The program manager is responsible for the effective management and control of the program according to established customer requirements and business objectives. He directs the coordination and integration of the various disciplines for all program phases through the functional organizations and subcontractors. He monitors and controls the work in progress according to the program plan. Potential deficiencies regarding the quality of work, specifications, cost, or schedule must be assessed immediately. It is the responsibility of the program manager to rectify any performance deficiencies.

Responsibility

4. Communicating the program status.

The program manager is responsible for building and maintaining the necessary communication channels between project team members and both the customer community and the firm's management. The type and extent of management tools employed for facilitating communications must be carefully chosen by the program manager; they include status meeting, design reviews, periodic program reviews, schedules, budgets, data banks, progress reports, and team collocation.

---

*Both program and project manager are referred to as program manager in this charter.

(3) interfacing with the sponsor or top management, (4) creating more interest and visibility for their project among the performing personnel, (5) changing reporting relations, or (6) developing a more credible image as sound business managers. For the matrix to work properly a certain balance of power, or matrix parity, must be established via procedures or management guidelines. This should ensure that (1) the basic responsibilities and authorities of each matrix axis is delineated, and (2) the basic process for initiating and executing a project is defined so that it becomes difficult for one or a few individuals to

Table 28-3. Typical Charter of a Functional/Resource Manager in a Matrix Organization.

Position Title: Manager of functional department

Authority

The functional/resource manager has the delegated authority from the general manager or the latter's functional director to establish, develop, and deploy organizational resources according to both long- and short-range business needs. This authority includes the hiring, training, maintaining, and terminating of personnel; the development and maintenance of the physical plant, facilities, and equipment; and the direction of the functional personnel regarding the execution and implementation of the various disciplines according to specific project requirements from the program office.

Responsibility

The functional/resource manager is accountable to the general manager or the latter's functional director for all work within his or her functional area in support of all programs. The functional/resource manager establishes organizational objectives, develops the resources needed for effective program support, and directs these resources to assure effective performance regarding the cost, schedule, and technical requirements established by the program office. This manager also seeks out and develops new methods and technology to prepare the company for future business opportunities.

Table 28-4. Typical Charter of a Task Manager in a Matrix Organization.

Position title: task manager

Authority

Appointed jointly by the functional director and the program manager, the task manager has the authority to direct the implementation of a specific program task within the functional organization according to the contractional requirements.

Responsibility

Reporting to both the functional head and the program manager, the Task Manager is responsible (1) to to the functional/resource manager(s) for the most cost-effective tchnical implementation of the assigned program task, including technical excellence and quality of workmanship, and (2) to the program manager for the most efficient implementation of the assigned program task according to the program plan, including cost, schedule, and specified performance.

Specifically, the task manager
1. Represents the program manager within the functional organization.
2. Develops and implements detailed work package plans from the overall program plan.
3. Acts for the functional manager in directing personnel and other functional resources toward work package implementation.
4. Plans the functional resource requirements for his or her work package over the life cycle of the program and advises functional management accordingly.
5. Directs subfunctional task integration according to the work package plan.
6. Reports to program office and functional manager on the project status.

---

accomplish their activities at the expense of all the others. Matrix parity is based on induced cooperation: the matrix must be designed so that the resource managers have a strong incentive to cooperate with the project managers, and vice versa. This is usually accomplished by tying management performance to both functional and project performance. Proper planning early in the project life cycle will further help to establish these cooperative relations. Proper planning defines the specific work, and the division of responsibilities and controls, in a mutually agreed-on arrangement that encourages involvement and commitment by all project parties.

Product Quality.   The quality of a product must be both specified and designed into it. In a matrix environment many individuals share the responsibility for defining product specifications. Conflict often arises over the technical feasibility and economics of these quality requirements, which should be worked out in concert with all functions that must specify, implement, and live with outcome of the product design. Typically, these functions include (1) research and development, (2) design, (3) manufacturing, (4) field services, (5) product assurance, (6) marketing/sales, and (7) finance. During project execution it is usually the functional resource manager's responsibility to direct the project toward its proper implementation, assuring technical excellence and quality, while the task or project manager directs the overall integration of the multidisciplinary activities in keeping with established requirements, schedules, and budgets. These operating responsibilities must be clearly defined to key personnel in the various matrix axes to assure quality implementation. This can be done orally, but for larger organizations the only consistently effective means are procedural documents, charters, and management directives.

Cost.   In a production environment, cost challenges usually exist in three areas: (1) designing toward a given unit-production-cost (DTUPC) target within an established design budget and schedule; (2) setting up the facilities and personnel for production within an established budget and schedule; and (3) actually producing and delivering the product at the target cost. Since produceability must be designed into the product from its very concept, it is important that key personnel of all product development phases participate throughout the product planning and implementation. Phased management approaches with specific review points, checks, and sign-offs may help to fine-tune the product design toward its lowest possible unit-production cost (UPC).

Conflict.   Conflict is inevitable in managing any complex production task. It is even more pronounced in a matrix environment, with its sharing of power and resources. Conflict can develop over many issues. Project personnel experience particularly intense conflicts over schedules, priorities, resource allocations (especially manpower), and technical issues. It is important for project managers not only to be cognizant of the potential sources of conflict, but also to know when in the life cycle of a project they are most likely to occur. Such knowledge can help the project manager avoid the detrimental aspects of conflict and maximize its beneficial aspects. Conflict can be beneficial when disagreements result in the development of new information that enhances the decision-making process—a topic which is discussed in more detail below.

Resource Loading.   Matching resource requirements with available resources and defining future development needs is a challenge, because of inevitable contingencies that demand deviations from the original plan and fluctuating work loads of the various projects. The impact of both overloads and underloads is usually higher overhead cost, with the obvious additional impact on project schedules for overload situations. *Dynamic project planning* with a minimum of quarterly reviews may provide a realistic up-to-date picture of resource requirements for each project. In this process the best possible timing for resource requirements should be continuously negotiated between the project manager and production resource managers. However, contingencies will happen regardless of how carefully projects are reviewed, resulting in an unfavorable impact on resource loading. *Dynamic resource scheduling* may help offset some of these problems. Such scheduling prompts the production manager to anticipate a certain percentage of sudden work fluctuation on the basis of past experience, and to build these contingencies into the load model. In addition, the resource manager should develop resource alternatives such as relief labor pools, cross-trained employees, temporary employees, additional shifts, and subcontracting.

Developing Future Capabilities.   Production capabilities need to be continuously upgraded to assure optimum product quality, cost, and features, which are necessary prerequisites for unique positioning within the competitive field of the firm. These upgrades require a combination of foresight, creativity, funding, and properly timed actions. It is usually the chartered responsibility of the functional resource manager to identify the needs and develop the plans, including budgets and schedules. Although there is no easy formula, resource managers must think in terms of integrating the long-range requirements for all anticipated projects. To establish meaningful, cost-effective requirements, the manager needs to understand (1) the technologies involved in future products; (2) the production techniques employed; (3) the specific markets, customers, and requirements; (4) the technological trends for both the products and the production means; (5) the relationship among supporting technologies; and last but not least, (6) the production personnel, their skills, needs, and desires.

ORGANIZING THE PRODUCTION PROJECT TEAM

An Effective Team Environment

Building the project team is one of the prime responsibilities of the project/ program manager. Team building involves a whole spectrum of management skills required to identify, commit, and integrate various task groups from traditional functional organizations into a single program management system. This process has been known for centuries. However, it becomes more complex and requires more specialized management skills as bureaucratic hierarchies decline and multidisciplinary teams evolve.

Project managers who are successfully performing their roles in such an environment usually provide an atmosphere that fulfills the needs and leadership expectations of their team members, as shown in Table 28-5. They also recognize the barriers to effective team performance, as summarized in Table 28-6, so that preventive actions can be taken. The effective team builder is usually a social architect who understands the interaction of organizational and behavioral variables and can foster a climate of active participation and minimal conflict. He or she also has carefully developed skills in leadership, admin-

Table 28-5. The Needs of Project Team Members.

TEAM MEMBERS HAVE NEEDS	EXPECTATIONS TEAM EXPECTS FROM PROJECT MANAGER
Sense of belonging	Direction and leadership
Interest in work itself	Assistance in problem solving
Professional achievement	Creation of stimulating environment
Encouragement; pride	Adaptation of new members
Recognition for accomplishment	Capacity to handle conflict
Protection from infighting	Resistance to change
Job security/continuity	Representation at higher management
Potential for career growth	Facilitation of career growth

istration, organization, and technical expertise on the project. The combination of all of these skillful efforts may develop an effective and productive team with the following characteristics:

- Team members committed to the program
- Good interpersonnel relations and team spirit
- Available resources
- Goals and program objectives clearly defined
- Top management involved and supportive
- Good program leadership
- Open communication among team members and the support organization
- A low degree of detrimental interpersonal and intergroup conflict

Organizing the New Production Team

Too often the program manager, under pressure to start producing, rushes into organizing the project team without establishing the proper organizational framework. While initially the prime focus is on staffing, the program manager cannot effectively attract and hold quality people until certain organizational pillars are in place. At a minimum, the basic project organization and various tasks must be defined before the recruiting effort can start.

These pillars are necessary not only to communicate the project requirements, responsibilities, and relationships to new or prospective team members, but also to manage the anxiety that usually develops during the team formation.

Make Functional Ties Work for You.   It is a mistaken belief that strong ties of team members to the functional organization are bad for effective program management and therefore should be eliminated. To the contrary, loyalty to both the project and the functional organizations is natural, desirable, and often very necessary for project success. While the program office gives oper-

Table 28-6. Barriers to Effective Team Building and Suggested Handling Approaches.

BARRIER	SUGGESTIONS FOR EFFECTIVELY MANAGING BARRIERS (HOW TO MINIMIZE OR ELIMINATE BARRIERS)
Differing outlooks, priorities, interests, and judgments of team members	Make effort early in the project life cycle to discover these conflicting differences. Fully explain the scope of the project and the rewards which may be forthcoming upon successful project completion. Sell "team" concept and explain responsibilities. Try to blend individual interests with the overall project objectives.
Role conflicts	As early in a project as feasible, ask team members where they see themselves fitting into the project. Determine how the overall project can best be divided into subsystems and subtasks (e.g., the work breakdown structure). Assign/negotiate roles. Conduct regular status-review meetings to keep team informed on progress and watch for unanticipated role conflicts over the project's life.
Project objectives/outcomes not clear	Assure that all parties understand the overall and interdisciplinary project objectives. Clear and frequent communication with senior management and the client becomes critically important. Status-review meetings can be used for feedback. Finally, a proper team name can help to reinforce the project objectives.
Dynamic project environments	The major challenge is to stabilize external influences. First, key project personnel must work out an agreement on the principal project direction and "sell" this direction to the total team. Also educate senior management and the customer on the detrimental consequences of unwarranted change. It is critically important to forecast the "environment" within which the project will be developed. Develop contingency plans.
Competition over team leadership	Senior management must help establish the project manager's leadership role. On the other hand, the project manager needs to fulfill the leadership expectations of team members. Clear role and responsibility definition often minimizes competition over leadership.
Lack of team definition and structure	Project leaders need to sell the team concept to senior management as well as to their team members. Regular meetings with the team will reinforce the team notion, as will clearly defined tasks, roles, and responsibilities. Also, visibility in

BARRIER	SUGGESTIONS FOR EFFECTIVELY MANAGING BARRIERS (HOW TO MINIMIZE OR ELIMINATE BARRIERS)
	memos and other forms of written media, as well as senior management and client participation, can unify the team.
Project personnel selection	Attempt to negotiate the project assignments with potential team members. Clearly discuss with potential team members the importance of the project, their role in it, what rewards might result upon completion, and the general "rules-of-the-road" of project management. Finally, if team members remain uninterested in the project, then replacement should be considered.
Credibility of project leader	Credibility of the project leader among team members is crucial. It grows with the image of a sound decision maker in both general management and relevant technical expertise. Credibility can be enhanced by the project leader's relationship to other key managers who support the team's efforts.
Lack of team member commitment	Try to determine lack of team member commitment early in the life of the project and attempt to change possible negative views toward the project. Often, insecurity is a major reason for the lack of commitment; try to determine why insecurity exists, then work on reducing the team members' fears. Conflicts with other team members may be another reason for lack of commitment. It is important for the project leader to intervene and mediate the conflict quickly. Finally, if a team member's professional interests lie elsewhere, the project leader should examine ways to satisfy part of the team member's interests or consider replacement.
Communication problems	The project leader should devote considerable time to communicating with individual team members about their needs and concerns. In addition, the leader should provide a vehicle for timely sessions to encourage communications among the individual team contributors. Tools for enhancing communications are status meetings, reviews, schedules, reporting system, and colllocation. Similarly, the project leader should establish regular and thorough communications with the client and senior management. Emphasis is placed on written and oral communications with key issues and agreements in writing.

Table 28-6. Barriers to Effective Team Building and Suggested Handling Approaches. (*continued*)

BARRIER	SUGGESTIONS FOR EFFECTIVELY MANAGING BARRIERS (HOW TO MINIMIZE OR ELIMINATE BARRIERS)
Lack of senior management support	Senior management support is an absolute necessity for dealing effectively with interface groups and proper resource commitment. Therefore a major goal for project leaders is to maintain the continued interest and commitment of senior management in their projects. We suggest that senior management become an integral part of project reviews. Equally important, it is critical for senior management to provide the proper environment for the project to function effectively. Here the project leader needs to tell management at the onset of the program what resources are needed. The project manager's relationship with senior management and ability to develop senior management support is critically affected by his own credibility and the visibility and priority of his project.

Source: H. J. Thamhain and D. L. Wilemon, "Team Building in Project Management," *Proceedings of the 21st Annual Symposium of the Project Management Institute,* Atlanta, Oct. 1979.

ational directions to the program personnel and is normally responsible for the budget and schedule, the functional organization provides technical guidance and is usually responsible for personnel administration. Both the program manager and the functional managers must understand this process and perform accordingly, or severe jurisdictional conflicts can develop.

Structure Your Organization.   The keys to successfully building a new project organization are clearly defined and communicated responsibilities and organizational relationships. The tools for systematically describing the project organization come, in fact, from conventional management practices:

- The *charter of the program/project organization* clearly describes the business mission and scope, broad responsibilities, authorities, organizational structure, interfaces, and reporting relationship of the program organization.
- The *project organization chart,* a simple, conventional chart, defines the major reporting and authority relationships. These relationships should further be clarified in a *policy directive.*
- A *responsibility matrix* defines the interdisciplinary task responsibilities: who is responsible for what. The responsibility matrix covers not only

activities within the project organization but also functional support units, subcontractors, and committees.

- *Job descriptions* are modular building blocks that form the framework for staffing a project organization. A job description includes (1) reporting relationships, (2) responsibilities, (3) duties, and (4) typical qualifications.

Define Your Project.   Both the project organization and the work itself must be defined before staffing can begin. The basic elements typically include:

- Work Breakdown Structure
- Statement of work for all first-level project components
- Overall specifications
- Master schedule
- Cost model and budget

Regardless how vague and preliminary these project components are at the beginning, the initial description will help in recruiting the appropriate personnel and eliciting commitment to the pre-established parameters of technical performance, schedule, and budget. Hopefully, the core production team will be formed prior to finalizing the overall program plan, thus giving it the opportunity to participate in the trade-off discussions and up-front development decisions that will affect producibility and cost during the production phase.

Staff Your Project.   After the project organization and the tasks are defined in principle, the project leader can start to interview candidates for the key project positions. The interview process normally has five facets that are often interrelated:

1. Informing the candidate about the assignments
2. Determining skills and expertise
3. Determining interests and team capability
4. Persuading, or selling the assignment
5. Negotiating terms and commitment

Handling the Newly Formed Team

A major problem faced by many project leaders is managing the anxiety that usually develops when a new team is first formed. This anxiety experienced by team members is normal and predictable. It is a barrier, however, to getting the team quickly focused on the task.

This anxiety may come from several sources. For example, if the team members have never worked with the project leader, the team members may be concerned about his or her leadership style and its affect on them. In a different vein, some team members may be concerned about the nature of the project and whether it will match their professional interests and capabilities. Other team members may be concerned about whether the project will help or hinder their career aspirations. Further, team members can be highly anxious about job expectations, work load and life-style/work-style disruptions.

Certain steps taken early in the project life cycle can be effective in terms of handling the above problems. First, at the start of the project the project leader should talk with each team member on a one-to-one basis about the following:

1. The objectives for the project.
2. Who will be involved and why.
3. Importance of the project to the overall organization or work unit.
4. Why the team member was selected. What role he/she will perform.
5. What rewards might be forthcoming.
6. What problems and constraints are likely to be encountered.
7. The rules of the road in managing the project, such as status-review meetings.
8. The team member's suggestions for achieving success.
9. The team member's professional interests.
10. What challenge will the project present to individual members and the entire team.
11. Why the team concept is so important to project management success and how it should work.

A frank, open discussion with each team member on the above is likely to reduce his/her initial anxiety or identify its source, so that the underlying issues can be dealt with in a timely manner.

Second, the greater the feeling of team membership and the better the information exchange among team members, the more likely the team will be able to develop commitment and effective problem-solving approaches. Third, the team is likely to develop more effective project control procedures. Project control procedures can be divided into two basic areas. The first is the quantitative control procedures traditionally used to monitor project performance, such as, PERT/CPM, networking, Work-Breakdown Structures, etc. The second is the willingness and ability of project team members to give feedback to one another regarding performance. Again, trust among the project team members makes the feedback process easier and more effective.

PROGRAM PLANNING AND CONTROL

Program managers today have available a set of powerful tools with proven capability to plan and control multidisciplinary activities effectively. Most of these tools were originally developed within the aerospace and construction industry. They are, however, equally effective and find increasing applications in other areas, ranging from new product introduction to political campaign management and social programs.

Make Planning Work for You

Effective planning and control techniques are helpful for any undertaking. They are absolutely essential, however, for successful management of large, complex engineering programs. Quality of planning means more than just the generation of paper work. It requires the participation of the entire project team, including support departments, subcontractors, and top management. It leads to a realistic project plan plus involvement, commitment and interest in the project itself. Proper planning fosters an environment conducive to the project goals. Such planning makes everyone's job easier and more effective, because it

- Provides a comprehensive road map of your program
- Pervades and integrates the program, and provides perspective
- Provides a basis for setting objectives and goals
- Defines tasks and responsibilities
- Provides a basis for directing, measuring, and controlling the program
- Builds teams
- Minimizes paper work
- Minimizes confusion and conflict
- Indicates where you are and where you are heading
- Leads to satisfactory program performance
- Helps managers at all levels to achieve optimum results within the limits of available resources, capabilities, environment, and changing conditions

If done properly, the process of project planning must involve all the performing organizations and the sponsors. This involvement creates new insight into the intricacies of a project and its management methods. It also leads to visibility of the project at various organizational levels, and induces management involvement and support. This involvement at all organizational levels stimulates interest in the project and desire for success, and fosters a pervasive reach for excellence that unifies the project team. It leads to commitment toward establishing and attaining the desired project objectives, and to a self-forcing management system where people want to work toward these established objectives.

Managing the Project Through Its Life Cycle

Managing projects from start to finish clearly involves all the functions of traditional business management throughout the project life cycle. To reduce complexity, most project managers use a phased approach for organizing their projects. These phases can follow functional lines such as system phase, development phase, and prototype phase, or business cycles such as preproposal phase, proposal phase, prototype phase, production phase, etc.; or they can follow any other typology that divides the overall project into logical sets of activities with specific outputs.

A generic approach is suggested for dividing the overall project into five managerial phases:

Phase 1. Conceptional

Phase 2. Project definition

Phase 3. Project organization and start-up

Phase 4. Main phase execution

Phase 5. Project phase-out

In each phase certain managerial actions seem to precede others, as shown in Figure 28-3. This permits the development of a framework for planning and controlling projects in a disciplined, systematic way, regardless of project size and complexity.

As the project moves through its life cycle, the focus of managerial activities shifts from planning to controlling. However, many of the activities are interrelated and cannot be confined to only one particular project phase. Plans are managerial tools; they are seldom final and should not be rigged. The purpose of the plan is to provide the basis for organizing the project, defining resource requirements, setting up controls, and eventually guiding the activities. As the various elements of the plan integrate and actual operations begin, modifications of the original program plan may become necessary. Continuous reviews and updates of all components of the program plan are needed throughout the project life cycle, if the plan is to remain a useful reference and guidance document.

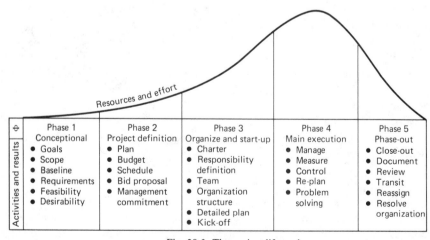

⏀	Phase 1 Conceptional	Phase 2 Project definition	Phase 3 Organize and start-up	Phase 4 Main execution	Phase 5 Phase-out
Activities and results	• Goals • Scope • Baseline • Requirements • Feasibility • Desirability	• Plan • Budget • Schedule • Bid proposal • Management commitment	• Charter • Responsibility definition • Team • Organization structure • Detailed plan • Kick-off	• Manage • Measure • Control • Re-plan • Problem solving	• Close-out • Document • Review • Transit • Reassign • Resolve organization

Resources and effort

Fig. 28-3. The project life cycle

Planning for Measurability and Control

Every program manager takes a different approach to establishing measurable milestones and has a reason for following that particular approach. However, the common theme among those who manage projects successfully is the ability to measure *technical status* at pre-established review points.

While at the time of the review these milestones must be clearly defined down to specific measurable parameters, it is not necessary and often impractical to establish these details at the outset of the program. What is required is a process that predefines the major milestones at the beginning of the program, and allows for incremental detailed development of the measurable parameters as the program progresses.

The process of establishing measurable milestones has two key features: (1) it provides the discipline for establishing detailed, measurable milestones, and (2) it creates involvement at all organizational levels, pervades project planning, and leads to improved communications and commitment.

The five-step procedure summarized in Figure 28-3 relies on conventional project planning and control documents that constitute the program management plan. Project performance data usually evolves from the Work Breakdown Structure (WBS) and the Statement of Work (SOW) into budgets, schedules, and specifications, which form the basis for describing the deliverable items at various points in the project life cycle.

One of the key criteria for establishing measurable milestones is the ability to define deliverable items, not only for the termination point of the program, but also for all critical milestones throughout its life cycle, including reviews and tests. Both the Work Breakdown Structure (WBS) and the Statement of Work (SOW) are usually subdivided into task elements, each associated with a specific (1) budget, (2) schedule, and (3) specification. If done properly, it is usually possible to establish a one-to-one structural relationship among the schedule, budget, specifications, and SOW, all related to the Work Breakdown Structure. With budgets, schedules, and specifications defined for each task element down to the level of actual project control, the basic criteria for measurability are therefore established. The deliverable items can now be defined for groups of task elements or work packages.

Alternatively, deliverables can be defined first for various project phases and then be coupled with those task elements necessary to produce them. In either case the ability to measure the status of the deliverables depends on the ability to determine the underlying task elements and their performance in accordance with the established project specifications, schedules, and budgets.

How to Make It Work

Making the project performance measurement system work requires more than just another plan. It requires the total commitment of the performing organization. Successful companies stress the importance of carefully designing the project planning and control system and the structural and authority relationships that are critical for the implementation of the measurement system. Other organization and management issues, such as leadership style, personnel appraisal and compensation, and intraproject communication, must be carefully considered to make the system self-forcing. That is, project personnel throughout the organization must be convinced that project performance measurements help them to identify potential problems early, get more assistance in solving problems, and in the end achieve higher productivity and efficiency.

There are four pillars on which a workable, self-forcing project performance measurement system is based:

- Management support
- Personnel involvement and pervasiveness
- Program visibility
- Measurable milestones

All four pillars are interrelated. If done properly, the process of establishing measurable milestones must involve both the performing and the customer organization. This involvement creates new insight into the intricacies of a project and its management methods. It also leads to visibility of the project at various organizational levels, and to management involvement and support. It is this involvement at all organizational levels that stimulates interest in the project and desire for success. It fosters a pervasive reach for excellence that unifies the project team and leads to commitment toward reaching each critical milestone. These are the characteristics of self-forcing control.

MANAGING THE PRODUCTION FUNCTION WITHIN A MATRIX

Skill Requirements of Matrix Managers

Production programs are often complex and multifaceted. Managing these programs within a matrix environment represents a challenge requiring technical and administrative skills, and interpersonal skills and leadership. To be effective, managers of these production tasks must consider all facets of the job. They must understand the people, the work, the tools, and the organization. The day of the manager who gets by with technical expertise alone or pure administrative skills is gone. Today, matrix managers[1] must understand not only the technical aspects of the job, but the culture and value system of the organization as well.

Technical Skills.   The project or task manager rarely has all the technical expertise to direct the program single-handed, nor would this be necessary or desirable. Especially in complex problem-solving situations, however, it is essential that the program manager understand the technology, markets, and environment of the business, so as to participate effectively in the search for integrated solutions and technological innovations. More important, without this understanding the consequences of local decisions on the total program, the potential growth ramifications, and the relationships to other business opportunities cannot be foreseen by the manager. Further, technical expertise is necessary to evaluate technical concepts and solutions, to communicate effectively in technical terms with the project team, and to assess risks and to make trade-offs between cost, schedule, and technical issues.

Administrative Skills.   Administrative skills are essential. The project manager must be experienced in planning, staffing, budgeting, scheduling, and other control techniques. In dealing with technical production personnel, the problem is seldom to make people understand administrative techniques such as budgeting and scheduling, but rather to impress upon them that costs and schedules are equally as important as elegant technical solutions.

Understanding the company's operating procedures is important. However, it is often necessary for project managers to free themselves from administrative details regardless of their capabilities to handle them. Particularly on larger programs, managers have to delegate considerable administrative tasks to support groups or hire a project administrator.

Interpersonal Skills and Leadership.   Interpersonal skills and leadership are fundamental to successful matrix management. These skills often involve dealing effectively with managers and support personnel across functional lines while in fact having little or no formal authority. It involves information-processing skills, the ability to collect and filter relevant data, and effective decision-making in a dynamic environment. It involves the ability to integrate individual demands, requirements, and limitations into decisions that benefit overall project performance. Another facet of leadership is the project manager's ability to resolve intergroup conflicts and to build multidisciplinary teams.

The project manager must be a social architect. He or she must understand the culture and value system of the organization—a necessary prerequisite for working effectively with the organization and its interfaces. These organizational skills are particularly important during project formation and start-up, when the manager establishes the project organization and integrates people from many different disciplines into an effective work team. This requires far more than simply constructing a project organization chart. At a minimum, it requires defining the reporting relationships, responsibilities, lines of control, and information needs. Supporting skills in the area of planning, communication, and conflict resolution are particularly helpful. A good program plan and a task-responsibility chart are, of course, useful and recommended tools. In addition, the organizational effort is facilitated by clearly defined program objectives, open communication channels, good program direction, and senior management support.

Leadership Effectiveness in Program Management

With the evolution of modern organizations such as the matrix and their complex environments, a new management style gradually developed. This style complements the traditional, organizationally derived power bases, such as authority and reward and punishment, with bases developed by the individual manager. Examples of managerial power for this so-called Style II management include expertise, credibility, friendship, work challenge, promotional ability, fund allocations, charisma, friendship, project goal identification, recognition, and visibility. Style II management evolved particularly with the matrix, where managers have to elicit support from various organizational units, and where the frequently ambiguous authority definition of the project manager, and the temporary nature of projects, contribute to an intricately complex operating environment.

A number of suggestions may help increase the project manager's effectiveness and ultimately improve overall program performance.

1. Project managers must develop their skills in leading the task team within a relatively unstructured work environment. This involves the ability to

   - Provide clear project direction
   - Assist in problem solving
   - Facilitate group decisions
   - Plan and elicit commitment
   - Communicate clearly and effectively
   - Facilitate the integration of new team members
   - Understand the organization and its interfaces

2. Project managers must foster an environment that is professionally stimulating and challenging, leading to recognition of professional accomplishments. To create such a work environment, the project managers must

   - Understand the needs, wants, and interests of their personnel.
   - Accommodate the interest of their personnel by discussing the assignments with their personnel at the outset and distriouting the work optimally.

- Create visibility and management involvement for their projects via all available means such as meetings, planning, announcements, and news and information channels.

3. Project managers must develop a credible image as experienced, effective task managers. Therefore they should seek out progressively more responsible task management assignments, rather than take on a project that is over their heads. Moreover, an understanding of the technologies involved in managing the project is essential for building a credible image, for integrating technical solutions, and for communicating effectively with the team members.
4. Project managers should build a personnel appraisal and award system that is consistent with the demands on their personnel. The ingredients of such a system include:

- Clearly defined job objectives, rewards, and negotiated commitment to the established project plans
- Clearly defined roles and accountabilities among all project team members
- Shared responsibility for personnel appraisal between functional and project manager
- A nonthreatening work environment
- A minimum of one personnel appraisal per year
- Fair, integrated overall assessment of personal performance
- Strong reliance on nonmonetary rewards such as recognition for accomplishments, visibility, opportunity for professional growth, interesting work, and freedom to act
- Fair monetary rewards, consistent with the level of responsibility and performance

5. Project managers must develop a communication network that facilitates the flow of information and management directions both vertically and horizontally. The characteristics of such an effective communication network are:

- Proper up-front project planning, involving all key personnel from all disiplines needed during the project life cycle
- Established reporting relations among project team members
- Proper project reviews and reports
- Top management involvement, interest, and participation
- Established project management guidelines, directives, policies, and procedures

Effective Conflict Management

Conflict is fundamental to complex task management and is often determined by the interplay of the program organization and its support functions. Complex organizational relationships, dual accountability, and shared managerial powers all contribute to the type and magnitude of the conflict inevitable in such a new frontier environment. Understanding the determinants of conflict is important to a project manager's ability to deal with conflict effectively. Contrary to conventional wisdom, conflict can be beneficial when it produces involvement, new information, and competitive spirit. However, when conflict becomes dysfunctional, it often results in poor program decision making, lengthy delays over operational issues, and a disruption of the team's efforts.

Conflict determinants in project organizations have been investigated by Thamhain and Wilemon, who delineate typical sources of conflict in seven categories.[2] Project managers can prepare for and deal more effectively with operational problems, if they can anticipate these problems and understand their specific sources. The seven conflict categories are:

1. Conflict over schedules
2. Conflict over project priorities
3. Conflict over manpower resources
4. Conflict over technical opinions and performance trade-offs
5. Conflict over administrative procedure
6. Personality conflict
7. Conflict over cost

While it is important to examine some of the principal determinants of conflict from an aggregate perspective, more specific insights can be gained by exploring these conflicts in each project life cycle stage: formation, build-up, main phase, and phase-out. Table 28-7 summarizes the major conflict sources for each stage and suggests specific measures for minimizing conflict that is unproductive.

A number of suggestions may be helpful in increasing the project manager's effectiveness and overall program performance.

1. *Culture and Value System.* Project managers need to understand the interaction of organizational and behavioral elements in order to build an environment conducive to their team's motivational needs. This will enhance active participation and minimize dysfunctional conflict. The effective flow of communication is one of the major factors determining the quality of the organizational environment. Since the project manager must build support teams at various organizational layers, it is important that key decisions be communicated properly to all project-related per-

Table 28-7. Sources of Conflict throughout the Project Life Cycle, and
Recommendations for Minimizing Unproductive Conflict.

PROJECT LIFE-CYCLE PHASE	MAJOR CONFLICT SOURCE AND RECOMMENDATIONS FOR MINIMIZING DYSFUNCTIONAL CONSEQUENCES	
	CONFLICT SOURCE	RECOMMENDATIONS
PROJECT FORMATION	PRIORITIES	— Clearly defined plans. Joint decision making and/or consultation with affected parties. Stress importance of project to goals of the organization.
	PROCEDURES	— Develop detailed administrative operating procedures to be followed in conduct of project. Secure approval from key administrators. Develop statement of understanding or charter.
	SCHEDULES	— Develop schedule commitments in advance of actual project commencement. Forecast other departmental priorities and possible impact on project.
BUILD UP PHASE	PRIORITIES	— Provide effective feedback to support areas on forecasted project plans and needs via status-review sessions.
	SCHEDULES	— Carefully schedule work breakdown packages (project subunits) in cooperation with functional groups.
	PROCEDURES	— Contingency planning on key administrative issues.
MAIN PROGRAM	SCHEDULES	— Continually monitor work in progress. Communicate results to affected parties. Forecast potential problems and consider alternatives. Identify potential "trouble spots" needing closer surveillance.
	TECHNICAL	— Early resolution of technical problems. Communication of schedule and budget restraints to technical personnel. Emphasize adequate, early technical testing. Facilitate early agreement on final designs.

Table 28-7.   Sources of Conflict throughout the Project Life Cycle, and
Recommendations for Minimizing Unproductive Conflict. (*continued*)

PROJECT LIFE-CYCLE PHASE	MAJOR CONFLICT SOURCE AND RECOMMENDATIONS FOR MINIMIZING DYSFUNCTIONAL CONSEQUENCES	
	CONFLICT SOURCE	RECOMMENDATIONS
	MANPOWER	Forecast and communicate manpower requirements early.
		— Establish manpower requirements and priorities with functional and staff groups.
PHASE-OUT	SCHEDULES	— Close monitoring of schedules throughout project life cycle.
		Consider reallocation of available manpower to critical project areas prone to schedule slippages.
		Attain prompt resolution of technical issues which may impact schedules.
	PERSONALITY AND MANPOWER	— Develop plans for reallocation of manpower upon project completion.
		Maintain harmonious working relationships with project team and support groups. Try to loosen up "high-stress" environment.

Source: H. J. Thamhain, "Conflict Management in Project Life Cycles," and D. L. Wilemon, *Sloan Management Review,* Spring, 1975.

sonnel. By openly communicating the project objectives and the subtasks, unproductive conflict can be minimized. Regularly scheduled status-review meetings can be an important vehicle for communicating project-related issues and dealing with problems early in their development.

2. *Leadership.* Because their environment is often temporary, project managers should seek a leadership style that allows them to adapt to the changing requirements of their project organizations. They must learn to "test" the expectations of others by observation and experimentation. Although difficult, they must be ready to alter their leadership style as demanded by both the status of the project and its participants.

3. *Conflict.* The ability to manage conflict is affected by many situational variables. A project manager should (1) recognize the primary determinants of conflict and when they are most likely to occur in the life of the project, and (2) consider the effectiveness of conflict-handling approaches. Proper planning involving all key project participants, lead-

ing to sufficient work detail and measurability, and in the end to commitment by all project personnel, seems to be one key element in minimizing dysfunctional conflict. Another element is a professionally , stimulating work environment.

4. *Interesting Work.* The project manager should try to accommodate the professional interests and desires of supporting personnel. Project effectiveness often depends upon the ability to provide a professionally stimulating work environment. Work challenge can be a catalyst in matching individual goals with objectives of the project and the overall organization. Although the scope of a project may be fixed, the project manager usually has a degree of flexibility in allocating task assignments among various contributors.

5. *Technical Expertise.* Project managers should develop and maintain technical expertise in their fields. Without an understanding of the technology, project managers are unable to participate effectively in the search for integrated solutions, or to win the confidence of team members, or to build credibility within the customer community.

6. *Planning.* Effective planning early in the life cycle of a project is another action which may have a favorable impact on the organizational climate and overall project performance. Insufficient planning may eventually lead to an unprepared work team, confusion, inability to measure and control progress, interdepartmental conflict, and discontinuity in the work flow.

7. *Personal Drive.* Finally, project managers can influence the climate of their work environment by their own actions. Concern for project team members, ability to integrate personal goals and needs of project personnel with project goals, and ability to create personal enthusiasm for the project itself—all these can foster a climate high in motivation, work involvement, open communication, and resulting project performance.

A situational approach to project manager effectiveness is presented in Figure 28-4.[3] Summarizing the effects of managerial influence style on motivation and position power, it indicates that the intrinsic motivation of project personnel increases with the project managers' emphasis on work challenge, their own expertise, and their ability to establish friendship ties. On the other hand, emphasis on penalty measures, authority, and the inability to manage conflict effectively lowers personnel motivation.

The figure further illustrates that the project manager's position power is determined by such variables as formal position within the organization, the scope and nature of the project, earned authority, and the ability to influence promotion and future work assignments. All factors appear to be important for sustained high work effort and organizational commitments over the project life cycle. Moreover, field research shows that project managers who foster a

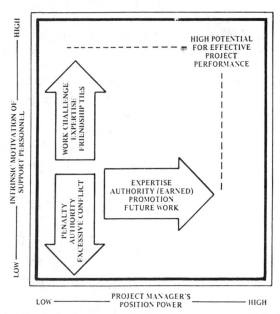

Fig. 28-4. The situational effectiveness of project managers increases with their position power and their ability to motivate personnel. The key variables are show.

climate of highly motivated personnel not only obtain higher project support from their project personnel, but also achieve high overall project performance as perceived by their superiors.

REFERENCES

Notes

1. Various terms are used to define the title and task responsibility of managers in a matrix environment. If used together within the same organization, a certain hierarchy of responsibility is often implied, as for instance program manager, project manager, task manager, and project administrator. In this discussion the term *project manager* is used in a general context to describe any matrix manager with multidisciplinary task responsibility.
2. See H. J. Thamhain and D. L. Wilemon, "Diagnosing Conflict Determinants in Project Management," *IEEE Transactions on Engineering Management*. Feb. 1975.
3. For detailed discussion of the research leading to the situational leadership approach in project management, see G. R. Gemmill and H. J. Thamhain, "Influence Styles of Project Managers: Some Project Performance Correlates," *Academy of Management Journal*, Vol. 17, No. 2 (June 1974): 216–24.

Bibliography

Allen, L. A. *Making Managerial Planning More Effective.* New York: McGraw-Hill, 1982.

Aquilano, J. "Multi-skilled Work Teams: Productivity Benefits." *California Management Review* 19 (Summer 1977): 17–22.

Blake, R. R., and Mouton, J. S. *Productivity: The Human Side.* New York: AMACOM, 1981.

Cleland, D. I. "A Kaleidoscope of Matrix Management Systems." *AMA Management Review.*

The Conference Board, *Matrix Organizations of Complex Businesses,* 1979.

Cook, T. M., and Russell, R. A. *Contemporary Operations Management.* Englewood Cliffs, N.J.: Prentice-Hall, 1980.

Cooper, R. *Job Motivation and Job Design.* London: Institute of Personnel Management, 1976.

David, S. M., and Lawrence, P. R. "Problems of Matrix Organizations." *Harvard Business Review,* May–June 1978, pp. 131–41.

Davis, L. E. "Individuals and the Organization." *California Management Review,* Spring 1980.

Devore, T., McCollum, J. K., and Ledbetter, W. N. "Project Engineering in a Plant Environment." *Project Management Quarterly,* Sept. 1982, pp. 25–30.

Gemmill, G. R., and Thamhain, H. J. "The Effectiveness of Different Power Styles of Project Managers in Gaining Project Support." *IEEE Transactions on Engineering Management,* Vol. 20, No. 2 (May 1973), 38–43.

Gemmill, G. R., and Thamhain, H. J. "Influence Styles of Project Managers: Some Project Performance Correlates." *Academy of Management Journal,* Vol. 17, No. 2 (June 1974): 216–24.

Handbook of Industrial Engineering. New York: Wiley, 1982.

Heckert, R. E. "Productivity: A Matter of Personal Accountability." *AMA Management Reivew,* Aug. and Oct. 1982.

Herzberg, F. "Herzberg on Motivation for the '80's." *Industrial Week,* Oct. 1, 1979, p. 58.

Hopeman, R. J. *Production and Operations Management.* Columbus, Ohio: Merrill, 1980.

Katz, F. E. "Explaining Informal Work Groups in Complex Organizations: the Case for Autonomy in Structure." *Administrative Science Quarterly* 10 (1965): 204–23.

Kerzner, H. *Project Management for Executives.* New York: Van Nostrand Reinhard, 1982.

Kerzner, H., and Thamhain, H. J. *Project Management for Small and Medium-Size Businesses.* New York: Van Nostrand Reinhold, 1983.

Likert, R. "Improving Cost Performance with Cross-Functional Teams." *Conf. Bd. Rec.* 12 (Sept. 1975): 41–9; *Management Review* 65 (March 1976): 36–42.

Marfin, C. C. *Project Management: How to Make it Work.* New York: AMACOM, 1976.

McAfee, R. B., and Poffenberger, W. *Productivity Studies.* Englewood Cliffs, N.J.: Prentice-Hall, 1982.

Monks, J. A. *Operations Management.* New York: McGraw-Hill, 1982.

Moore, F. G., and Hendrick, T. *Production-Operations Management.* Homewood, Ill.: Richard D. Irwin, 1980.

Riggs, J. L. *Production Systems.* New York: Wiley, 1976.

Schroder, R. G. *Operations Management.* New York: McGraw-Hill, 1981.

Slone, H. D., Thamhain, H. J., and Wilemon, D. L. "Managing Change in Project Management." *Proceedings of the 19th Annual Symposium of the Project Management Institute,* Oct. 1977.

Thamhain, H. J. "The Effective Engineering Manager." *Proceedings of the 1981 IEEE Engineering Management Conference,* Dayton, Ohio, 1981.

———. *Engineering Project Management.* New York: Wiley, 1983.

———. "Managing Engineers Effectively." *IEEE Transactions on Engineering Management,* May 1983.

Thamhain, H. J., and Wilemon, D. L. "Conflict Management in Project Life Cycles." *Sloan Management Review,* Summer 1975, pp. 31–50.

Thamhain, H. J., and Wilemon, D. L. "Project Performance Measurement." *Proceedings, PMI-Internet Joint Symposium,* Boston, 1981.

Thamhain, H. J., and Wilemon, D. L. "Skill Requirements of Engineering Program Managers." *Convention Record of the 26th Joint Engineering Management Conference (IEEE),* Oct. 1978.

Thamhain, H. J., and Wilemon, D. L. "Team Building in Project Management." *Proceedings of the 21st Annual Symposium of the Project Management Institute,* Atlanta, Oct. 1979.

29. Organizational Development in Matrix Management

Milan Moravec*

Organization development (OD) as a set of techniques and a concept has undergone many changes of focus and meaning over the years (depending on who was using it and how), has been praised and reviled by approximately equal numbers of managers, and still means different things to different people. The matrix organization is also a concept that has undergone changes in focus and meaning, has its advocates and opponents, and is generally not well understood.

So if you put these two approaches together you will achieve chaos—right? Not necessarily. Although not every business and management thrive on a matrix structure, such a structure is at times the best, if not the only, way to get the job done efficiently and achieve the organization's purpose. And while OD runs the risk of creating more problems than it solves when naively managed, it can be the bulwark or the salvation of matrix management. The trick is to see that the most appropriate *form* of OD is selected, and that this OD is applied at the right time by trained and seasoned specialists.

THE PROJECT MATRIX

There are various types of matrix organizations, but in this chapter I intend to focus on one common matrix element: the project organization that is a temporary alignment of people dedicated to completing a specific project mission on schedule, within budget, and to the satisfaction of a client. (The client may be another company, or the head of another division in the same company.) This project organization is composed of people who have been selected from various ongoing functional groups within the main organization, and are under the leadership of a project manager. The matrix here is the overlaying of pro-

---

*Milan Moravec is Personnel Manager in the Arabian Bechtel Company Limited, Al-Jubail, Saudi Arabia. Formerly, he was Program Manager of Development for the Bechtel Group of Companies in San Francisco. His professional work experiences include management assignments in Australia, North and South America, Europe and the Middle East.

ject responsibilities upon the regular functional responsibilities and reporting relationships.

The Project Manager's Responsibilities

The project manager has one of the most demanding roles that can exist in an organization. For example, in a large organization that has contracted to complete a complex project for a significant client, a project manager would probably have *at least* the following responsibilities:

- Defining with the client organization the goals and objectives of the project
- Establishing and maintaining communication systems with the client and with the project team
- Installing and implementing a project control system, which involves such factors as scheduling, monitoring, assigning responsibility, evaluating the impact of changes, initiating and reviewing forecasts, auditing, reporting, and controlling materials and finances
- Managing the client's involvement in project control, business policies, and daily procedures
- Evaluating and reporting on the progress and performance of the various project elements
- Administering all legal, commercial, and functional aspects of the contract
- Defining the responsibilities and procedures for subcontracts
- Making sure the project quality assurance program is complied with by all project participants
- Maintaining liaison with all functional departments in the company that may be involved in the project

While making it happen, the project manager strives to maintain a balance between human, financial, organizational, management, political, and technical considerations—to achieve cohesion among all those contributing to the project while meeting stringent time, budget, and quality requirements.

As chief point of contract with the client, the project manager also has to be the bearer of bad tidings when, for any of a variety of reasons (some of them beyond his or her control), the job is headed for trouble. In such a situation, it is up to the project manager to see that the client understands the sources of the problems and takes remedial action, even if the primary cause is in the client's own organization.

Difficulties and Pitfalls

Without any traditional line authority, the project manager is deprived of the traditional carrot-and-stick approaches and must find other ways—use of specialized knowledge, force of personality, ability to influence, skill in directing meetings, managing expertize—to ensure commitment by project team members. These people may be a heterogeneous group—from different parts of the business, from different businesses, or even from different geographical areas. They probably have varying technical and experience backgrounds, perspectives, and training, and may have little experience in working with one another.

To complicate matters, the client organization presents its own group of employees. They may themselves be part of a matrix organization, or they may not. Although a good client organization devoted to the purpose can contribute immensely to the cohesion of the work group, a divided or vacillating client group can be disastrous. If the client organization is pyramidal rather than matrix, and each of its departments proceeds to establish its own independent liaison with a counterpart member of the contractor's project team, the project manager has plenty to do.

The project structure is an invitation to constructive conflict. Team members are involved in two channels of responsibility—the functional organization and the project team—and are responding to two sectors of their own business, not to mention the client organization's business. This conflict is not necessarily ineffective. It instantly identifies areas for management decision; balances the technical, administrative, schedule, quality, and financial input of individual members; keeps people on their toes; and may even be viewed by the team members as stimulating. Constructive competition brings out the best individual and team performance, can be a healthy motivator and enhance productivity. However, the power balance does require constant attention. Any imbalance must be caught quickly before it degenerates into a power struggle—"personality conflicts"—and destroys the effectiveness of the matrix.

Despite all these pitfalls and a labyrinth of communications, the project manager must create and maintain a positive project environment receptive to the contributions of each member of the team, while operating in his or her organization's environment in the interests of the client who is paying the bills. To pull it off, the manager may need some organizational development/effectiveness assistance. Many organizations are beginning to introduce organizational development/effectiveness strategies and methods into their project operations. Some of these strategies have been very successful, while others have created more problems than they solved.

ORGANIZATIONAL DEVELOPMENT

What Is OD, Anyway?

Organizational development had its origins back in the late 1940s, when the first T-groups (training groups) and group dynamics processes appeared on the business scene. These phenomena spawned others of similar persuasion and grew into a sort of Business Potential Movement akin to the Human Potential

Movement. Today OD is a large, amorphous mass of varying strategies, methods, and results. It is difficult even to define OD, except in very general terms such as "a process aimed at modifying and improving a work group and moving it in the direction it needs to go."

An infinitely expandable number of techniques will fit into this definition. There are OD "interventions" that involve individuals, groups, values, structure, even the whole organization. They include such disparate methods as sensitivity training, sensing sessions, team-building sessions, conflict resolution meetings, management skill training, communication workshops, situational leadership training, and internal management consulting programs aimed at improving and increasing the effectiveness of the organization's response to internal and external challenges and pressures.

OD interventions are usually directed by facilitators or specialists from outside the organizational unit—external or internal consultants or other professionals who are called in for this specific project. These professionals cover a wide range of disciplines. Some are industrial psychologists, others academics, but a majority are management and business consultants. Nearly all have a pet approach that they think superior to other approaches.

Some business managers are excited by the changes wrought by these OD practitioners, while others are skeptical (or even derisive) about the ability of OD to result in sustained improvement. Many companies have tried a whole palette of OD techniques at various times, and with various results.

It is possible to categorize OD systems according to the type of results they are designed to achieve. One category consists of techniques aimed at increasing the value of an organization's (or a project team's) human resources by making people feel better about themselves, educating them, and providing them with new interpersonal and supervisory skills. Another category is comprised of what might be called "social engineering" techniques; these focus more on the group than on the individual, and include team building and various methods of restructuring organization charts, norms, decision-making methods, management styles, or modes of communication and relating. A third kind is the approach that focuses specifically on bottom-line results—getting things done—such as productivity, timeliness, and cost effectiveness: the work of the project.

Dangers of Misapplied OD

These results can all be beneficial to an organization, if applied at the right time and place. But for a project management matrix, not all of them are practical. In fact, some of them could prove more of a hindrance than a help.

For example, many self-awareness programs (in the human-resource-improvement category) necessitate removing people temporarily from their work setting and involving them in introspective exercises. While this may well be beneficial in the long run for the individual—and by inference the orgaza-tion—it may be detrimental to the project, which is concerned with the short run. An OD intervention whose major short-run impact is to cause delay is simply not going to be time and cost-effective.

Similarly, team-building techniques (the "social engineering" category) tend to rely on a "retreat" setting in which participants step back from their tasks and examine the network of their interpersonal relationships. While a smoothly communicating team with new decision-making modes is vital to a successful project, it is seldom practical for team members to take out large chunks of time to examine the general principles of human and interpersonal relation-ships. And since many project team members will not be working together after the project is completed, many of the new team-building norms attained will have limited application later on.

Not all types of OD processes are appropriate for all settings and cultures. Take the example of an engineering, procurement, and construction firm that manages projects all over the world. On one given project site there may be engineering, construction, and operations specialists from several different nations. A Middle Easterner may not take very well to the self-awareness or confrontation exercises used in many types of OD operations. A Japanese might be completely puzzled by American-style team-building and manage-ment style exercises. A transactional analysis workshop composed of Europe-ans, Americans, Asians, and Middle Easterners might serve as a vehicle for international understanding and education, but not as a boost to pro-ductivity.

In any kind of culture there are people who view OD as a threat. As already mentioned, OD interventions are generally facilitated by a consultant from out-side the work unit who has a particular expertise: running and OD intervention. When this "expert" tells the manager how to behave or "facilitates" the work group on conflict resolution, the result may be anxiety, resistance, or even hos-tility. The project manager may feel that he or she is perceived as an incom-petent, and the team members may feel manipulated.

In a project setting there is already enough conflict arising from the dual responsibilities imposed by the matrix structure. While a certain amount of matrix conflict, when sensitively managed, improves response time and the quality of decisions (as indicated earlier), it is seldom productive to overlay *another* type of conflict on this already delicate situation.

Given all these caveats and dangers, why should OD be used at all in a project setting?

Need for Appropriate OD

The answer is that sometimes this kind of intervention is necessary. As tech-nology advances, money becomes more expensive, and projects grow increasingly complex, the project manager's job—that host of responsibilities referred to earlier—becomes intensely more difficult. Changes in cash flow, environmental and government policies, client objectives, personnel, administrative procedures, and so forth keep the project manager constantly making decisions that balance the crucial "Big Three" project considerations: Time, Budget, and Quality. There is little if any time left for hunting down organizational issues such as unwieldy communications and decision modes or morale problems. This is where OD can help. It is the task of the OD specialist to assist the project manager and his or her team in getting things done and thereby increase project team effectiveness.

There are some kinds of OD interventions that, if properly managed, are *not* dysfunctional to the project but can actually help it stay on track. These belong in the third category of OD types, in which the focus is on productivity, timeliness, and cost effectiveness—the elements that matter most in a project situation.

Does this mean that the results of the other OD mechanisms are of no value to a project team? Far from it. People involved in a project should continue their education, have self-awareness and communication skills, and be able to relate effectively to various management styles, decision modes, and other members of the team. However, these training objectives are secondary to the goal of getting the job done.

If people involved in a project were skilled in communication, satisfied in their career progress, at peace with their management style and methods of conflict resolution, and comfortable with organizational norms and values, we could assume that the project manager possessed an outstanding set of human resources. However, that would not *necessarily* mean that project tasks and programs would be completed faster, with more cost effectiveness and quality. Knowledge acquired from developmental programs does not necessarily translate into visible, short-term, day-to-day improvements in project or task performance. When you have a primary goal (in this case, productivity) and a secondary goal (improvement of human resources), the secondary goal often receives such focus that it is mistaken for the primary goal, and in the end is the only one achieved. Moreover, if the team members did *not* come to the project with all the aforementioned skills and attributes, there would probably not be time and motivation to develop them within the time span allotted to the project and still meet the project objectives.

The project manager, as a rule, is not responsible for the education, training, or development of the team members or the changing of organization structure, management styles, or norms, all of which fall under the purview of the functional manager. The project manager's responsibility is simply to get *this* job accomplished with "given" limited resources, after which the team members return to their functional groups or go on to yet another project.

Characteristics of Effective Project Matrix OD

For the kinds of projects we have been talking about, then, the most effective OD system is one that

- Is primarily aimed at helping the project manager and team achieve productivity rather than developmental goals
- Does not disrupt the work unit or schedule
- Gets quickly to the heart of improving operational effectiveness
- Is understandable by all members of the team
- Is nonthreatening to team members and leaders alike

Such a system would not be concerned with evaluating a manager's performance, altering the project team "climate" or "values," conducting sensitivity training, or making decisions for the project manager on how to behave or how to communicate by using fashionable values. It would focus on specific internal or interface work problems, opportunities, or barriers to attaining project objectives, and on getting the job, task, or program done. Based on the assumption that the project manager is neither weak nor incompetent, it would be an additional resource for the manager and would be managed by, not imposed on, him or her.

We will examine the *how* of such a system shortly. But first, let's look at the *when:* situations in which OD should be applied in a project setting.

When to Use OD

Following are some instances in which OD can increase productivity.

When a project team has just been put together. When project team members come from different functional groups or even different geographical areas, they have probably arrived with different conceptions of the project and their roles in it. OD techniques can aid in planning and mutual goal setting, as well as ensure that all members of the project team are committed to the goals and to their work in achieving these results.

When the project manager seeks to improve contractor-client work, or to make sure they remain healthy and viable. As a member of an organization providing project services to a client, the project manager must not only represent the contractor to the client, but also represent the client's changing objectives and concerns to the contracting company's top management and its project team. Effective and timely communications and decision are essential here, and OD intervention can help identify the work to remove communication, organization, decision, and procedural blocks.

When new people join the project. Newcomers must be quickly integrated into the work of the team and brought up to speed to help on its problems, issues, and operations. OD can help these activities occur quickly and with the necessary quality to accomplish project objectives.

When procedures, responsibilities, or project objectives are readjusted. A significant change in government policy, financial policy, cost of materials, client needs, or even climate conditions can make it necessary for the project manager to take stock of where the project now is and to modify its direction. Such changes affect everyone on the project team, and in different ways. They may also affect the nature of work with the client. OD can help the team members translate the implications of the new parameters into work practices, and define future task expectations.

When there is a problem. If conflicts, morale crises, or mysterious snags suddenly threaten the project, OD is often called in as a troubleshooting mechanism. But this is the *last* sort of condition under which the project manager should have to use OD. Judicious use of OD interventions throughout the life of the project should prevent the accumulation of the problems and tensions that bring about crises in the first place.

Role of the OD Specialist

For OD assistance to be available in these situations, it must be built into the project structure. That is, the OD specialist should be a part of the team, setting up shop right at the project location so as to be available, and to be continuously appraised of the status of the project and of any emerging issues that need action. The specialist should report to the project manager and maintain close communication with other key people designated by the project manager. These people can include members of the client organization, if any of the problems involve client relationships.

Although this communication with team members and client representatives is helpful to the OD specialist in ferreting out operational issues, it should be made clear to everyone from the start that this person is accountable to the project manager *only,* and not to a functional area, the company as a whole, or the client. Matrix management is complicated enough without confusion regarding the OD reporting relationship and its charter. And the project manager's job is difficult enough without having his or her influence and direction eroded by an "outsider."

At the outset, the specialist's role should be clarified by the project manager to the members of the project team. This is the only way to defuse any feelings of threat that the project team may be experiencing. (These feelings are more likely to be present if the OD specialist arrives on the scene *after* the other team members have already become involved in the project.) The specialist can be introduced by memo, or during a project staff meeting, or both. The project manager should point out that this specialist is not a spy for management, an ombudsman to mediate quarrels, or an efficiency expert, but is there *at the request of the project manager* to identify what the members of the project team understand to be their major problems, tasks, and objectives. The specialist, in short, is there to apply his or her resources to progress work.

Once the barriers or opportunities have been identified, the specialist's role is to assist with the identification of work, and to provide impetus for doing something about these barriers. The OD specialist does not, however, solve the problems or implement the action plans; that is the job of the specialists assigned to the project.

OD IN ACTION

The OD process might unfold as follows.

Data Gathering

After discussing the unit's needs with the project manager, the OD specialist begins gathering information about the unit's opportunities and problems, including barriers to performance. The specialist tries to attend all meetings, receives copies of all memos, and conducts brief preliminary interviews with the people who report directly to the project manager. He or she asks questions about roles; about short- and long-term objectives; about what the people are doing and ought to be doing more or less of; about the quality of meetings, control, and planning systems; about the interface with other departments or with the client; and about current blocks or perceived resistances to operational effectiveness.

During this initial set of interviews the OD specialist—in addition to collecting information—assesses both the level of the interviewees' candor on sharing work-related data, and their understanding of the issues facing the unit. This helps in adjusting questions and interview methods in order to obtain quality information easily translated into work assignments.

The specialist then confers again with the project manager, who decides which, if any, issues need to be added to the list identified in the preliminary interviews. The project manager approves a new set of questions for in-depth interviews and joins with the specialist in selecting interviewees. At this point interviewees might include people outside the unit, such as members of the client organization, higher-level managers, or others whose work affects or is affected by the unit's output.

The project manager introduces the OD specialist to the new interviewees and the in-depth interviewing process begins. Sessions with each person vary, depending on the number of issues to be covered, the candor shown by respondents, and of course the skill of the OD specialist. While collecting data, the in-depth interviews focus participants in detail on significant issues and begin to generate alternative work to overcome barriers. Interviews are scheduled so as to interfere with work progress as little as possible.

Since this is a results-oriented approach, personality issues are avoided by persistently turning attention to the tasks themselves. For example, the statement that "Joe and Maria have communication problems" is considered inadequate. The OD specialist will probe: "What *specifically* do these people need to communicate, to resolve their problems and accomplish their tasks?" In other words, what are the specific work issues warranting action? When an OD specialist is told that "the quality of personnel" and "personality conflicts" are interfering with productivity, the problem and conflict usually turn out to be rooted in work, organization, or management practices that can be changed without resorting to long-term education, training, organization, or psychological approaches.

The OD specialist then collates the information as reported by the interviewees, and prepares to feed it back to them and to the project manager for decision.

Feedback / Action Planning Meeting

The project manager sets a date for an intensive business meeting that will involve the manager, those reporting directly to the manager, and the OD specialist. At the meeting, after ground rules are explained and expectations reviewed, the OD specialist feeds back to the group all the work information that has been gathered in an action format. The specialist has organized the data, but does not analyze it or attribute any of the information to individuals, so that no one knows who said what (the objective is to deal with problems, not ascribe blame). Then the project manager and the other members of the group individually review the data related to their own responsibilities and commit themselves to assignments that will address these issues. (They can also decide not to take any action on a particular issue, or to delegate an action item to someone under their direction, or to defer action pending additional available details.) At the meeting's conclusion, the project manager sets a date for the first follow-up session.

Implementation and Follow-up

Each participant in the meeting begins work on his or her action items immediately. At a series of follow-up sessions—perhaps twice a month for several months—progress is reviewed and any new problems are assessed. The focus is always on what results have been achieved, rather than on whether the person carrying out the assignment says, "It's done." These follow-up meetings can involve members of the client organization, if the project manager deems it appropriate. Later the project manager and the OD specialist may meet again to identify succeeding steps to move the project further toward the attainment of its objectives.

Factors that can be used to measure OD results include quality of action items, timely implementation of decisions, improvements in productivity as assessed by the project manager, improvements in performance as measured by the client, and achievements to ameliorate cost, quality, and schedule targets.

Advantages

This pragmatic approach to OD has certain advantages that are not shared by other types. First, it is *time and cost-effective.* Because it does not deal with abstract issues but only with specific, identifiable work items that need action, it gets to the heart of the operational issues, problems, and bottlenecks more quickly than other, more process-oriented training and development techniques. (It aims at better *results* rather than at better process.) Turnaround time, from the day the OD specialist comes on board to the point where action items begin to be acted upon, is usually only several weeks, and could even be measured in days if there are few issues and/or this approach has been used by the team before.

A second benefit is *adaptability.* The system has been used productively in many different industries, at different levels of management, and in all kinds of cultures. A task-oriented OD system is readily understandable by managers and employees in just about any country; whether they are based in Kuwait, Australia, Argentina, Mexico, the United States, the United Kingdom, or France, managers are concerned with getting the job done. It can be used not only within the contractor's project team, but also to improve project/client interfaces. It can be employed at regular intervals or only when changes in project conditions or personnel warrant an objective look at performance.

Third, the process is *nonthreatening.* It does not attempt to change people— managers or subordinates—or to tell them what the experts believe they should do, so neither the project leader nor the other members of the project team resist it. Since the focus is on the task itself, not on issues of personality, style, or belief, the method is acceptable to incumbents of work positions, as the information collected and used for decisions is consistent with getting work accomplished. There is no "blame" assigned to the persons on the team, in the functional organization, or the client organization, as this kind of OD is based on the assumption that the specialists who are doing the work are the best available limited resources.

Fourth, the approach is *nondisruptive;* it takes units, project teams, and individuals where they are and helps them reach higher levels of productivity through the solving of workrelated problems. In many OD interventions the project manager would be overwhelmed by so many unfamiliar recommendations as to end up with a whole new set problems to replace or add to the original ones. In such a case the manager may find himself with employees who are better trained, but cannot apply their training to the project at hand; with new concepts and ways of managing that don't fit this particular work situation; with a group whose new team spirit doesn't help them overcome frustrating operational barriers; with new communication methods that don't contribute to getting the job done. That doesn't happen in the system described here, because input and decisions come from the people responsible for implementation. Sets of work plans and goals for improvement are identified by team members rather than by the OD consultant, and the data on which these plans are based is also collected from those charged with project goals.

Finally, there is a certain amount of *fallout, or indirect benefit*. Although the primary goal, as stated earlier, is getting the job done, this does not mean that secondary goals are never achieved. What usually happens is that when results become evident—the work is getting done on time and within budget and quality standards—people feel a sense of accomplishment and movement, which leads to increased satisfaction with their roles on the job. Appropriate job skills are acquired as action items are identified and implemented. And as people examine the same data and work together in pursuing specific decisions, a realistic, job-related level of cooperation and communication develops: team spirit emerges from accomplishments. The work activities accelerate and intensify the quality of both vertical and horizontal communications. Thus "social engineering" and "increasing the value of human resources" are byproducts of getting the job done.

SUMMARY

Organizational development/effectiveness in a project matrix setting must be carefully managed. The wrong approach can create confusion, delays, and resistance. An appropriate form of OD, on the other hand, can be of immense aid to a project team in meeting its time, budget, and quality objectives and thus improving performance. The most effective type of OD system for a project team is one that focuses on issues relating to getting the tasks accomplished, not on increasing self-awareness, managerial skills, or interpersonal communication (although the later objectives are likely to be achieved as byproducts).

Besides knowing what kind of OD to use, it is important to know when to use it. In a project setting there are many situations in which OD can increase productivity. Although OD can help a project manager out of a tight spot if necessary, it is most effective when mobilized systematically and regularly to improve project performance in matrix management.

30. Control in a Multiproduct Environment

Joe Mumaw*

Today's complex organizational structures, profit competitiveness, and scarcity of resources demand more in-depth financial evaluations of performance—both product profitability and the contribution of each segment of each functional element to profitability and return on investment. Moreover, these performance evaluations need to be made in an overall framework considering all products (regardless of their positioning in the product maturity cycle). Matrix management provides the framework for structuring the Management Information System (MIS) to allow top-down analysis. The MIS contains the traditional four-step closed loop:

Matrix management requires an organizational structure that provides a significant advantage for controlling costs product line by product line, as well as by each functional group specialization within each product line. Establishing interfaces between each product line and each contributing function requires a structured process of planning and performance measurement. The degree of excellence of this process will determine the effectiveness of the total organization to meet the desired performance. It is possible to establish predetermined standards for product nonrecurring and recurring costs as well as design-to-cost objectives for the design activity, and to measure performance against these objectives on a continuing basis. Costs of each functional group

*Joe Mumaw is a twenty-year veteran with the Westinghouse Defense Center in Baltimore. He has been the Business Operations Manager on the two largest Defense Center production programs, radars for the F-4 and F-16 aircraft. Currently, he is Manager of the Purchasing Department. He is a graduate of the University of Baltimore and holds an MBA from the University of Pittsburgh. On summer weekends he is usually involved in racing his sailboat Cold Duck on Chesapeake Bay where, in 1981, he won Race Week and High Point First Place honors.

can be measured by performance versus overall budget objectives, and by comparison with how well each performed to objectives on each product. The cost of each product can be measured by comparison of actual and forecast costs to objective costs. Variances to objective are the basis for management action.

THE FINANCE FUNCTION

Financial management receives significant benefits from the matrix organization's structure, which establishes as mini-entrepreneurs both product line and functional group task managers. This layering of responsibility reaches from top management downward through the entire organization. The participation at each level in all facets of the performance of each task (or grouping of tasks that relate to that level) creates a results-oriented management team. Matrix management imposes controls over all product/functional interfaces that are manifested as forcing functions which demand:

- Full understanding of product line requirements by all involved functional groups
- In-depth, up-front planning
- Early identification of high-risk areas
- Establishment of mutually acceptable cost and schedule objectives
- Regular reviews to highlight problem areas and allow timely corrective actions

This environment fosters the blending of cost and profit awareness into the total output of the functional personnel at a level typically lower than in nonmatrix organizations. This grouping of cost with schedule and performance responsibility at the level where work is actually performed has four significant benefits:

- Credible projections of final costs throughout the product cycle
- Ability to redirect or terminate projects on the basis of full knowledge
- Knowledge by the total organization of the performance to objectives
- Real-time awareness by functional groups of their profit contributions

Matrix management provides an effective control system that furnishes the tools necessary for the total plant to respond to the total requirements. This is particularly true when the functional group contains a vast array of specialization. The manufacturing department of a large-scale electronics producer will contain as a minimum the following types of "hands-on" (recurring) labor specialties, all with their attendant management structure:

- Machine shop
- Printed wiring assembly

- Cable/harness manufacturing
- Specialty component manufacturing
- Final assembly/test

Additionally, factory support specialists are needed in the areas of industrial engineering, manufacturing process engineering, planning and schedule control, production control, and inspection, along with their respective management organization. Other functional groups (engineering, materials, marketing/contracts, finance) have their corresponding unique diversifications.

To achieve this control, the following must be integrated plant-wide for each product line:

- Structure the desired output in a manner fully understandable by all functional contributors
- Define lower-level individually identifiable, deliverable tasks with cost, schedule, and performance requirements
- Measure status of these tasks
- Integrate all tasks with one another and within the overall program, to permit total program assessment
- Feed back the progress in a timely manner (1) on each product, to allow product manager redirection, and (2) on each function, both by product and by total function, to allow functional management proaction

Ideally, the plant will have management control systems in place recognizing responsibility accounting, effective cost-estimating procedures, labor standards, meaningful cost history, and timely feedback of actual costs.

Horngen states: "Responsibility Accounting ... systems recognize various decision centers throughout an organization and track cost ... to the individual managers who are primarily responsible for making decisions about the cost in question."[1]

MATRIX ACCOUNTING

A matrix management responsibility accounting system structurally decentralizes the organization into smaller, more manageable functional elements, each headed by an identifiable functional manager who accepts the responsibility for cost, schedule, and performance control, and has the authority to make relevant decisions. This responsibility accounting functional element receives tasks from many programs and is measured individually for its performance on each task. Moreover, the sum of all its efforts toward objectives form the basis for an effective evaluation system. Similarly, each task within a given product, combining all functions, can be summed together to allow measure of the product's performance (see Figure 30-1).

Costs must be accumulated in a job-order cost-accounting system that pro-

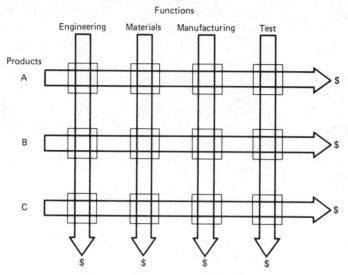

Fig. 30-1. Product/functional interfaces.

vides visibility into categories of labor and material costs within each product. The cost-charging system must be flexible enough to identify all categories of cost by function for each task within each product line. These tasks can then be summarized in a cost matrix both by product line and by function. An example of a cost charging/identification system that is totally responsive to these needs is contained in the following description.

Cost Identifiers

COST CATEGORIES:			IDENTIFIER
Labor	Engineering		A
	Manufacturing Hands-On		B
	Manufacturing Support		C
	Test	Hands-On	D
	Test	Support	E
Material	Proto-Type		F
	Production—Non-Recurring		G
	Production—Recurring		H

Product Identifier

Each product requires an identifier unique for each lot of each product. This will allow the total cost for each product lot to be determined at any point in time.

Task Identifier

A task is an individual detailed job necessary to complete the product requirements. The sum of all tasks equal the total product efforts. Examples of tasks
are: (1) Design Computer Chassis; (2) Manufacture Antenna Array; (3) Test
transmitter Assembly. A task should have the following characteristics:

1. It is clearly distinguished from all other tasks.
2. It is assignable to a single functional organizational element.
3. It has a scheduled start and completion date.
4. It has a budget value assigned in terms of dollars, man-hours, or other
   measurable units which represents the desired final value.
5. It is integrated with other functional tasks and schedules.

Budget Center Identifier

Budget Centers can be established within each cost category for further identification of the type of costs incurred. Manufacturing Support labor can be
divided between recurring and non-recurring by use of Budget Centers. Engineering labor costs can be differentiated between Electrical Design, Mechanical Design, etc. by use of Budget Centers. Material Costs can be identified by
use of a code which shows the value of installed material, attrition material,
tooling, lot test charges, make-to-buy changes, configuration changes and other
one-time charges.

The appropriate grouping of cost type, product, task, and budget center identifiers form the basis for a cost charging system for all incurred direct costs.
(See Figure 30-2).

Hence, a cost charged to A1234BBAAXZ would be identifiable as Engineering labor (A) from budget XZ against task BBAA of Product 1234.

A cost charged to H1234BBAC10 would be identifiable as Production
Material (H) ordered for task BBAC of Product 1234 as installed material
(10).

With the cost charging/identification system outlined it is possible to develop
matrix cost reports for the full width and breadth of the product/functional
activities.

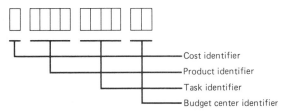

Figure 30-2. Structure of a flexible charge number.

Residual Responsibility

"The Buck Stops Here" is a sign frequently seen on product manager's desks. It is symbolic of the fact the product manager must bear the final responsibility for cost, schedule, and performance of his product line. The product manager is the financial manager of his or her product, assuring an acceptable financial plan mutually agreeable to all functional cost contributors. The product manager is the production manager, because he or she must obtain sufficient capacity commitment to meet promised deliveries. The product manager is the performance manager who assures that the product's quality, reliability, and specification compliance are fit for the intended use.

The product manager must disseminate the product's requirements to all functional contributors to assure their full understanding and obtain commitment on cost, schedule, and performance. Since the product manager has the requirements and the funds, he or she is, in effect, the customer who is procuring services from the functional activities (who can be viewed as suppliers). When considered from this perspective, the product manager's sign could read "The Buck Starts Here."

ORGANIZATION

The first or Organization Phase establishes a firm data base for each product line in a manner consistent and compatible with the total MIS. It is essential that this phase be completed before entering the next phase. The key elements of the Organization Phase are the creation of requirements in the form of a *Work Breakdown Structure* and a *Product Master Plan.* These two documents allow for establishing mutuality between the program manager and the functional contributors.

Work Breakdown Structure (WBS)

To match the decentralized organizational control offered by the responsibility accounting system, the product line requirement needs to be correspondingly defined into smaller, more manageable tasks. The ultimate objective is, of course, to create a structure of work broken down into identifiable tasks each of which is controlled by a single functional manager and can be understood as a deliverable item, can be scheduled with a start and finish, and can be measured from a performance standpoint. Moreover, each task must relate dependently or independently to all other tasks as depicted in an overall framework.

Such structuring has been a requirement of large military programs for over a decade. Department of Defense Military Standard 881 defines their requirements for the structured base that must be utilized by the Armed Services prime Contractor and lower-tiered subcontractors. The current generation of factory business systems that utilize material requirements planning (MRP) techniques use product structuring and bill of materials that are micro-extensions of this theme.[2]

The WBS is a product-oriented family tree consisting of levels of deliverables and their corresponding functional inputs. These inputs are composed of hardware, software services, and other tasks necessary to complete the desired output. The WBS provides a stable framework that is clearly visible and facilitates understanding of where each task fits, in terms of cost and schedule, when combined with the product master schedule (PMS). The WBS provides:

- Defined technical and management responsibility assignment
- Adaptability for both development and production phases
- A framework for controlling the progress and status of cost and performance against objectives

A WBS (Figure 30-3) is a multilevel relationship that defines the product to be developed or produced and displays the elements of work to one another and to the end product. Each element in a WBS is a discrete part of the entire WBS. An element may be either an identifiable hardware product, a service, or technical data.

A WBS for a television antenna rotating system (Figure 30-4) would contain all hardware items necessary for the system and their relationship to their next higher assembly. The functional effort to produce the system—design, material, manufacturing, and test—would be applicable to each item in the WBS.

A WBS is equally applicable throughout the life cycle of a product at the conceptual, definition, cost estimating, development, production, and support phases. The WBS can be conceptual during the definition phase—i.e., through level 3 only—and expand into a final form during the maturation of the product. Hence a WBS is an iterative process throughout a product's life cycle. WBS provides ancillary benefits in matching the functional organization to the

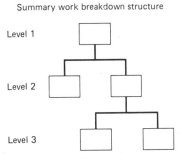

Summary work breakdown structure

Level 1

Level 2

Level 3

Fig. 30-3. Organizational relationships within work breakdown structure.

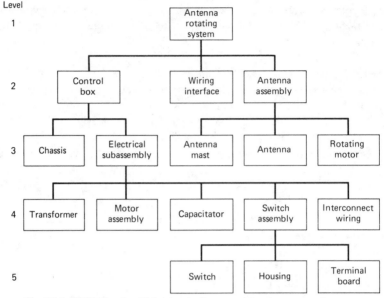

Fig. 30-4. WBS, showing all hardware elements of an antenna rotating system.

hardware elements of the product (Figure 30-5), as well as by creating a realistic base for product cost estimating.

Construction of a WBS and its corresponding levels of elements should consider:

- Scope and nature of the effort
- Accountability to higher levels of management
- Management controls: cost, schedule
- Degree of need to alter (or abolish) original plans

Each WBS task must have a written description of the scope of the task allowing all functions to fully understand the degree of their involvement. Depending on the nature of the task, multiple functions could have an input to the same task. A developing product would require engineering to design, material management to procure parts, and manufacturing to produce. The segregation for control and separation of recurring/nonrecurring costs must be accommodated either by subtasks or by cost tracking by functional spenders. In addition, each task within each product's WBS has a unique identifier (see Figure 30-6).

Once defined, the WBS can be expanded to include functional responsibility and time phasing for schedule purposes. Each element in the WBS is assigned to a functional manager. Each responsible functional manager and the product manager then attempt a negotiated agreement on cost and schedule for each WBS element.

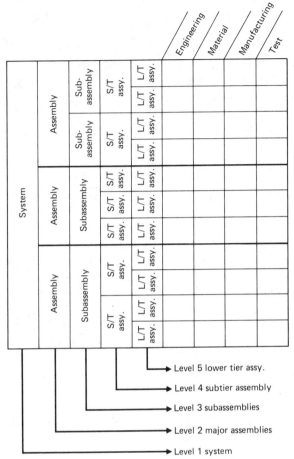

Fig. 30-5. WBS, matching functions to hardware.

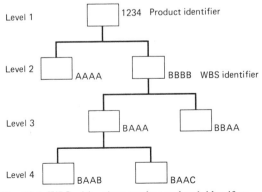

Fig. 30-6. WBS with unique product and task identifiers.

Product Master Plan (PMP)

Since the WBS is an integrated network of relationships, the expansion of each WBS element to include functional inputs (tasks) and time-phasing will yield a Product Master Plan. A master schedule (Figure 30-7) is established for the entire program. The master schedule is driven by the ultimate customers' delivery demands and is met by agreement between the product manager and the top-level functional managers.

The master schedule should differentiate between functional activities:

TASKS

	ENGINEERING	MANUFACTURING
D		
E		
V		
E	DESIGN LAYOUT	
L		
O	BREAD BOARD	PRODUCIBILITY REVIEW
P		
M	ENGINEERING MODEL	MASTER PRODUCTION
E		PLANNING
N		CAPACITY
T		MAKE-OR-BUY
	PROTOTYPE	FACILITIES
		PROCESS
		DEVELOPMENT
-------	-------------------	--------------------
P	PRODUCTIONIZING	MASTER PRODUCTION
R		SCHEDULE
O	DRAWINGS	MFG. ORGANIZATION
D	TEST PROCEDURES	RELEASE SCHEDULE
U		PROCUREMENT PLAN
C	VALUE ENGINEERING	SETTING STANDARDS
T		
I		TRANSITION TO FULL
O		PRODUCTION
N		

Consolidation of all top-level master schedules provides a rought-cut look at total resource requirements over the planning horizon. This is particularly useful when the introduction of (or attempt to obtain) new products is resource-limited. This view also provides a significant precursor of the need for make-or-buy changes, and allows enough time to develop acceptable subcontractors.

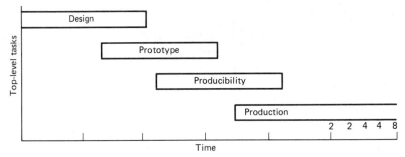

Fig. 30-7. Product master schedule.

Detail plans (Figures 30-8 and 30-9) are constructed within each functional element for each WBS task. A functional organization is assigned to correspond to the product WBS requirements. A functional person at the level that most influences the completion of the task is assigned responsibility for each task. This person then studies the written task description and defines his or her action plan. The plan must include planned start/finish dates, interim milestones, quantity, types, and duration of labor requirements, estimate of material costs, and a recommended measurement technique to monitor completion (cost, schedule, and performance).

Establishing Mutual Agreement

The most important ingredient for a successful performance recipe is establishing consensus on deliverable performance, schedule, and costs between the product manager and the functional task manager. This step assures that the

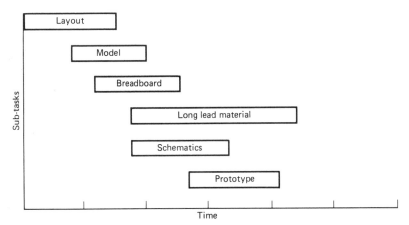

Fig. 30-8. Detail design plan.

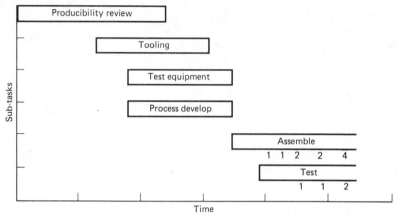

Fig. 30-9. Detail manufacturing plan.

functional task manager fully understands the scope of the task requirements and gives a personal commitment to make it happen, and assures the product manager a reasonable degree of confidence in achieving the overall objectives.

The product manager is ready for this step when his or her product's requirements are defined in a Work Breakdown Structure, each task with a detailed written description of required performance, and an envelope of time within which the task must be started and completed. Additionally, the product manager should have completed a cost budget for each task which, combined with a management reserve, is compatible with the overall product total cost objective. The cost objective value of each task is derived by comparison with prior product's similar tasks, standards, bottom-up estimates, or parametric comparisons. The cost objective value must stress those elements the functional task manager can realistically control, such as direct labor hours, yields, types of special test equipment, material attrition, and labor grade mix. These controllable items are the basis for subsequent performance measurement.

The value of the management reserve should correlate with the degree of risk within the product's requirements. A state-of-the-art design with high packaging density using new technology materials would necessitate a significant reserve, whereas the repackaging of a proven design into the same volume would require a minimal reserve. This reserve must be totally under the sole direction of the product manager. Layering of mini-reserves controlled by functional task managers results in inflated cost estimates (reserves on reserves), decisions on the amount of appropriate reserve made at too low a level, and more difficult interim problem identification. Reserves provide needed flexibility for the product manager either to redirect efforts (make-or-buy changes) or to accommodate valid overruns. For complex product developments, it is unlikely that all tasks will occur on schedule and within objective costs, but with proper visibility of overall performance and available management reserve funds, the product manager has the flexibility to shift to a new course that will still achieve the objective.

The functional task manager is ready for this step after determining how the task will be performed within the resources under his or her control. This manager has to plan and schedule this new task into the total workload of tasks from other products, using projected labor charts to derive any potential labor shortfall and solutions (new hires, transfers, second shifts, overtime, subcontract).

All new resources required (people or equipment) and their projected cost and availability (lead time) must be determined. The projected direct cost to perform the new task must then be estimated. It is imperative that this estimate be based on the "should cost" principle, which reflects a reasonable value in a well-managed operation with known and given performance complexities. No reserves or contingency funds should be included in the cost estimate; rather, the relative risk of meeting the derived estimate should be quantified. The functional task manager defines a method to measure cost, schedule, and performance monitoring that allows for interim evaluation and projections.

In all product phases there is an ordered set of events (subtasks) that are necessary to complete each task. These events can be scheduled and measured. In the design phase, a typical approach is to use "page-and-line" schedules where each event for each task is scheduled and reviewed. A series of design reviews is conducted to assure that all individual tasks are progressing satisfactorily relative to one another. In the material phase, purchased parts must be identified, quantified, scheduled, ordered, and received for each hardware element of the product. These events are conducive to a network schedule to accommodate internal and external lead times, minimal inventory, and material availability. Special facilities, both product and capital, need to be included in this network schedule to assure their availability. The manufacturing/test phase is also conducive to a networking type of schedule of events that relate WBS tasks with time requirements into a master hardware schedule.

The product manager meets with each functional task manager and reaches agreement on all costs, schedule, and deliverable performance parameters. This process is one of negotiation; the ease of agreement is directly related to the perceived complexity of the task and the conditioning of the total organization to operate in the matrix environment. The first factor of perceived complexity is a function of the work to be performed, and differences are usually resolved by including the functional head in the discussions. The second factor of organizational conditioning reflects the prevailing degree of acceptance by management, top to bottom, of the matrix management concept.

Design to Cost (DTC)

Where there is a need to design a product to a given cost, there are techniques to estimate costs successfully throughout the design process for each hardware element, and to compare these costs to objective costs. Unfavorable variances require design iterations or manufacturing process or material selection changes to achieve parity. This technique involves multifunctional involvement early in the design phase, producing synergistic productivity improvement solutions and commitment to a final design that meets production cost objectives as well as technical performance requirements. The design task manager is concerned about the productive cost of his or her design as well as managing the cost of the design. The manufacturing, test, and material task managers are involved in assuring the design is producible within given cost objectives. This technique is widely employed by the Department of Defense in major new system procurements.

A DTC goal is created that defines all variable parameters. The recurring cost (no fee) of the product is to be a given dollar value at the 1000th unit in a continuous production run at a given deliverable rate of systems per month with one purchase of materials. Nonrecurring production costs such as special test equipment and special tooling are excluded from the DTC goal, but are considered on a constant year basis to avoid the impact of inflation. Using experience curves, grouped material procurements, and prototype factors, the goal can be translated back to a dollar value objective for each hardware task at the beginning of the design phase. Reviews are held routinely to measure the success of each design task and of the overall design in meeting their objectives.

The cost of the product in reality will differ from the DTC goal by the impact of variance from the DTC goal-given parameters and by unforeseen and/or performance changes. Inflation changes from the costing rates and material cost estimates in the DTC model and those appropriate in the years when the product is being manufactured cause the greatest amount of variation.

LATER PHASES

The *Implementation Phase* consists of performing the task on the WBS and the accumulating of cost and measurement of progress. The *Analysis Phase* is the payoff of the efforts expended in the Organization Phase; decisions can be optimized by virtue of full knowledge of where you are compared to where you should be.

Once completing the Work Breakdown Structure, master and detail schedules, and concurrence on cost and schedule and performance has been achieved, *progress measurement* is accomplished by periodic reviews of detail and summary statements. Each task is divided into a series of milestones indicating progress toward completion of the task. Each milestone is scheduled and a budgeted value is assigned that represents the estimated cost of the milestone. The cumulative estimate dollar value of the milestones and planned performance achievement is plotted over time (see Figures 30-10 and 30-11). Prog-

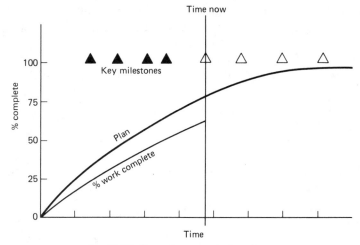

Fig. 30-10. Planned versus actual performance.

ress is measured by adding the estimated value of each completed milestone and comparing the result with the planned objective. This technique provides interim measurement of overall progress. These schedule and cost-controlled milestones may not be rescheduled or altered without approval of the next higher level of functional authority and the product manager.

Experience and lessons learned form the basis for milestone selection, its corresponding schedule and dollar values, as well as the techniques used for

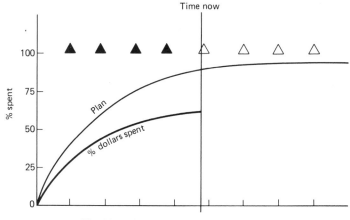

Fig. 30-11. Planned versus actual expenditures.

deriving values for work performed. This methodology is difficult in subjective tasks like design and development, and considerably easier in the quantitative activities such as materials, manufacturing, and testing.

The product manager is concerned that balance be maintained between the value of work actually performed and the planned performance, and the actual costs and the planned expenditures on each task as well as on the overall program. The functional manager is concerned with all tasks, typically from various product lines, within the scope of his or her responsibility. This integrated system provides early identification of problem areas to both the product and the functional manager.

Finally, the *Modification Phase* consists of initiating with full documentation the decisions reached in the Analysis Phase.

SUMMARY

Control in a multiproduct environment is optimized by the organizational structure's recognition of matrix management. The framework of dual controls of both product and functional organization significantly enhances decision making. Disciplines required in the Organizational Phase have effective payoffs throughout the product life cycle.

REFERENCES

1. Charles T. Horngen, *Cost Accounting: A Managerial Emphasis (Englewood Cliffs, N.J.: Prentice Hall, 1972,) p. 158.*
2. The American Production and Inventory Control Society (APICS) has literature and training aids on these subjects.

31. Information Systems for Matrix Organizations

Ivars Avots*

Most managers will readily agree that management information is the key to effective performance of the tasks and functional activities within the matrix organization. Often, however, more time may be spent on the organizational issues, problems of authority and responsibility, and even the subtle issues of conflict resolution, than on the design of the information system.

It is important to note at the outset that project management information involves much more than what is represented by critical path networks, cost charts, and technical progress reports. An effective information system passes a large variety of data, ideas, and concepts in a manner that is clear and does not distort its meaning between the originator and the receiver. In the matrix environment the information system stands at the heart of the organization structure, its planning and decision processes, and all the elements constituting the so-called cultural ambiance. It is critical to the very success or failure of the matrix management system.

In any organization the fundamental tasks of communicating take up a substantial time. People are constantly receiving, generating, and transmitting information and participating in decisions based on this information. Because of its complexity and the broader involvement in the decision-making process, the matrix organization is particularly dependent on the effective operation of the information system.

In this chapter we will discuss the information systems as they relate to the specific requirements of the matrix organization, without regard to specific functions such as critical path scheduling or budgeting. We will consider the

*Ivars Avots is President of Trans-Global Management Systems, Inc., a firm specializing in project-related systems and services throughout the world. Formerly with Arthur D. Little, Inc., he has been responsible for the design and implementation of project management systems in engineering, construction, and research organizations. He has a B.S. degree from Susquehanna University and an M.B.A. degree from the Wharton School. Mr. Avots has received the McKinsey Award for his contribution to management literature, the Arthur D. Little Presidential Award for effective project leadership, and the 1979 Person-of-the-Year Award from the Project Management Institute. He is a vice-president and director of the International Project Management Association.

formal information, which is based on rules, procedures, and calculations, as well as the informal, which is based on face-to-face or telephone-to-telephone contacts, as well as chance encounters in the hallway or cafeteria. An effective information system will consist of a balance of these two modes of communication.

BACKGROUND: THE ECONOMICS OF INFORMATION

A useful way to look at information is as a resource necessary for carrying out the objectives and goals of the organization. As such, one must necessarily consider the implications of its value and cost. The value of any given piece of information is equal to the usefulness that it provides to the organization. Most of the information that we are dealing with will eventually lead to some decision, therefore we cannot entirely separate the value of information from the individual decision maker. Some managers with many years of experience in their particular field can make excellent decisions with a minimum of information. In such a case, the economics of providing the information to the manager will be quite different from the matrix situation where most of the participants may be relatively inexperienced in their roles and require significant amounts of supporting data to make appropriate decisions.

The economics of information are illustrated in Figure 31-1. Typically, the initial cost of information rises at a fairly steep rate, then levels off and begins to rise again as we approach the point where most of the easily available information has been obtained. The value of information, on the other hand, rises slowly and eventually reaches a point where additional information provides only a limited amount of additional value. From a purely economical standpoint, the optimum information system will be the one providing the largest value for a given amount of cost. In practice, however, the cost level will be determined by the budget for the project in question and for maintenance of

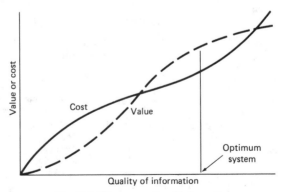

Fig. 31-1. The economics of information.

the information system. Thus it becomes a three-way compromise between the cost and value of the information system and the cost of hiring experienced managers in the first place.

With this as a background, we now will look at the information needed for the matrix organization and discuss the information-related tools available to deal with these requirements. As stated elsewhere in this book, the three principal reasons for turning to the matrix organization are:

1. Pressure for a dual focus
2. Pressure for sharing resources
3. Pressure for high information-processing capacity

We will deal with each of these pressures in turn, exploring them from the viewpoint of the information required and discussing the practical solutions available to management.

INFORMATION ASPECTS OF THE DUAL FOCUS RELATIONSHIP

The matrix organization serves the dual focus requirement by dealing simultaneously with specialized functions, products, services, geographical areas, or sets of customers. It also seeks to prevent arbitrary overruling of one area by another and tries to bring about trade-offs in decisions considering all involved orientations.

Focusing Information on the Issues

There are two principal tools that enable any product-oriented organization to focus more sharply on the issues at hand and on the related functions and responsibilities. These are the Work Breakdown Structure and the Responsibility Matrix. A combination of these two techniques is illustrated in Figure 31-2.

The *Work Breakdown Structure (WBS)* has long been used in project environments to define the product and set up work packages facilitating management and control. Although Figure 31-2 shows only two levels within the structure, in actual practice five to eight levels are frequently used. The WBS is developed by progressively breaking down the project into more and more detailed end items. The key to successful breakdown is observation of end items or product lines rather than functions. This is not always easily done; the process often presents a first test of the effectiveness of communications within the matrix team. A typical question likely to arise is whether development is an end item or a function. In a functional organization, development could be a separate department that carries out developmental work from year to year. In a project, development may be viewed as a specific phase that breaks down into

WORK BREAKDOWN STRUCTURE		RESPONSIBILITY MATRIX							
LEVEL 1	WORK PACKAGES	JK	NJ	DR	FE	NN	ET	X	Y
Product definition	Market survey		PM						
	Product description		PM	S					
Development	Formulation		AP	PM	S				
	Color selection		AP	AP			PM		
	Flavor development		AP	PM					
	Packaging	AP	S			PM			
Production	Samples					PM			
	Pilot line					PM			
	Production facility	AP				PM			
Marketing	Market test		PM		S	S			
	TV campaign		PM		S				

Fig. 31-2. Work Breakdown Structure and related Responsibility Matrix.

specific work packages representing end items. In the development of a cold medicine, such packages could include the formula of ingredients for the product, the color recommendation, the specific flavor, and even the package in which the product will be sold.

In developing the WBS, it is useful to identify the end items as hardware, software, facilities, and services. If any of the elements do not fit into one of these categories, they should be carefully evaluated to make sure they can be broken down into work packages that can be scheduled, budgeted, and controlled in terms of the project objectives.

Figure 31-2 also illustrates how the WBS can be expanded into a *Responsibility Matrix*. The abbreviations used under the individuals' initials indicate whether a particular individual is acting as a prime mover (PM), an approver (AP), or a supporter (S) of a function related to a particular work package. From a communications standpoint, these two techniques have two purposes: (1) to clearly indicate what work is to be done and what work packages completed; (2) to identify which participants of the matrix organization are responsible for what functions, as related to specific work packages. Development of the WBS and the Responsibility Matrix should be one of the first steps after the matrix team has been organized. It is a process that usually involves considerable discussion and reflects much disagreement. Bringing out these points at a very early stage of the matrix activity helps to prevent many problems in later operation.

The Effect of Different Viewpoints

To prevent arbitrary overruling of one area in the project by another, it is necessary for all the involved parties to understand one another's problems. The matrix process requires that an agreement be reached among peers whenever possible; only if there is substantial disagreement should it be necessary to refer to a higher level for a resolution of the problem. To begin with, this process of developing a consensus depends on an effective information system. In addition, it requires an understanding of the fact that the emphasis on what is important in the project changes from one phase of the project to another. Team members with major responsibilities in a later phase of the project may have a viewpoint quite different from team members with current tasks. The reasons for this are illustrated in Figure 31-3. During the early phase of the project schedule is of primary importance, while cost takes second place and quality third. Later in the project cost becomes the controlling interest, with schedule taking a secondary role. After the project has been completed, schedule and cost problems are easily forgotten and quality becomes the key. It is important to recognize these differences and viewpoints and incorporate them in the information system. For example, the sequence shown in Figure 31-3 could be reflected in the priorities for system implementation on a project.

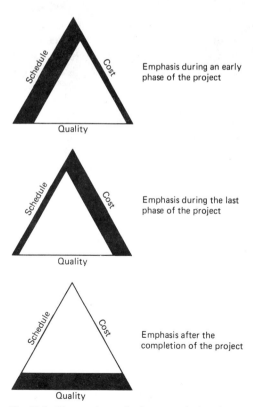

Fig. 31-3. Changes in emphasis on a typical project.

Dealing with Uncertainties

To bring together all the orientations involved in trade-off decisions in the matrix environment, it is necessary to understand the uncertainties of the process and to have visibility over the eventual effects of the decisions. Parametric estimates, various types of trend charts, and extrapolation of current experience often do not provide the desired confidence levels sought by the team. While we have learned to make fairly good forecasts of the cost and schedule for equipment, materials, and even personnel, we have run into a new set of social and political factors that are rapidly changing but that nevertheless have an important effect on the success of our projects.

One approach resulting in better information about these less certain areas of the project involves successive detailing. In this approach, illustrated in Figure 31-4, three typical project components—structure, equipment, and social effects—have each been estimated within a range of probable cost. Statistical techniques have then been applied to calculate the standard deviation and variance, showing the relative uncertainties related to the cost of each of these elements. Thus the figure shows that the structure has a very low uncertainty of five, while the social area has a large uncertainty of 60. In going to the next detailing step, both of the high uncertainty items—equipment and social—have been broken down into further components. This has reduced the uncertainties related to these subelements, so that the highest uncertainty areas are detailed even further. The last item to consider is the political subset of the social element. This has been separated into a fairly insignificant element recognizing that there will be forces trying to stop the plant construction altogether, and a more serious element saying that influences will be exerted to change the design of the plant.

Such successive detailing can be carried out until all the subelements of the project have been reduced to an acceptably low degree of uncertainty, or until it has been determined either that not enough information exists to accomplish further subdividing, or else that the process requires information too expensive or too time-consuming to obtain. In either case the process has resulted in infor-

INITIAL DETAIL	SUCCESSIVE STEP	SUCCESSIVE STEP	SUCCESSIVE STEP
Structure 05	Structure 05	Structure 05	Structure 05
Equipment 40	Heater 40	Feeder 25	Feeder 25
		Burner 10	Burner 10
	Mover 10	Mover 10	Mover 10
Social 60	Environmental 20	Environmental 20	Environmental 20
	Political 45	Political 45	Antiplant 20
			Design 40

Fig. 31-4. Successive detailing to reduce uncertainties.

mation highly useful to the management team. It makes it possible to focus on those issues where the greatest uncertainties remain, and reduces the possibility of arbitrary decisions that have not considered all the points pertinent to the issue.

INFORMATION TOOLS TO FACILITATE SHARING OF RESOURCES

The most critical resource in many matrix organizations is manpower. This means that an information system is required that tells management which personnel resources are available and which to be developed, and how the existing personnel are most effectively allocated.

Companies involved in many projects or product groups requiring assignment of personnel to various matrices often use personnel or skills inventories for this purpose. These can range from computerized data banks to specialized personnel inventory control rooms. One large construction company requiring hundreds of project managers around the world maintains a closely controlled room where the background information, including photographs of potential candidates, are displayed. When a new project team needs to be formed, top management can review the various alternatives and select from the personnel listed in the control room.

Just as important as the resource inventory is the personnel development plan. This plan may be part of a career development program or may be set up as a separate training program. While this is a resource management system quite independent of the overall information system, it nevertheless must be closely related to it. Tracking the personnel needs of the day-to-day management teams helps to identify emerging resource requirements that can be reflected in the personnel development plans.

With the information on available resources conveniently in place, the next step is to link it with the application program that permits comparison of the needs and availabilities. This results in schedules that minimize disruptions of projects as well as help maintain the life styles of the people involved. In recent years companies have begun to recognize that employees no longer accept arbitrary shifts between plants and locations around the world, and that attention must be given to the effect on their life styles. Most companies are now considering the family needs of their employees, and so require more elaborate personnel information systems to support the typically dynamic matrix organization.

Another aspect related to personnel resources is the constant shifting between personnel assignments, authority bases, and interpersonal relationships within the typical matrix organization. The information system must be able to adapt to these changes when they happen. To most people, availability of information means power. If changes are implemented in such a manner that some of the traditional communication channels are cut off, there may be a loss of real or perceived power that can affect the efficient operation of the organization. This situation can become particularly critical when the change involves organizational dislocation. Often, owing to space requirements or other considerations, a department may be moved to another building at some distance from the other departments involved in the matrix. While the actual distance may not be great, the separation can have a pronounced effect on the communications channels. Invariably, new channels get established at the new location and some of the traditional channels are eliminated. This can cause noticeable shifts of power within the matrix organization and affect the way in which the team works together. With regard to sharing resources, it is a well-established fact that people tend to work with personnel who are in their physical proximity, even though others who are not as conveniently located may be better qualified for the assignment.

The selection of membership for the matrix team can also be affected by changes in the company's approach to profit or cost centers within the organization's structure. In one consulting company well-known for effective multidisciplinary teams drawn from various parts of the organization, the selection of team members was drastically narrowed when the corporate management established stringent market development and billability goals. As a result, the new matrix teams tended to come from the one organization that was responsible for selling the project and wanted to keep most of the billings within the group. An effective informational feedback might have alerted management that the advantages of the multidisciplinary approach were being compromised.

INCREASING INFORMATION-PROCESSING CAPACITY

The information-processing function in the matrix organization is related to four separate issues: (1) establishment of effective communications channels among members of the organization; (2) reduction in the human information bottlenecks to support decision makers; (3) facilitating the handling of issues involving uncertainty, complexity, and interdependence of personnel; and (4) ability to have more people take on general management functions. These issues are complicated further by the fact that the information demands of the matrix organization tend to be changing and unpredictable, frequently causing overloads in the system.

In the matrix environment we use *three different types of information.* The most obvious of these is the information required to plan and control the specific project or product that is the key to the matrix's existence. This information does not need to be exact, since it is used for management purposes and not for accounting, but most of all it must be forward-looking. Cash flow forecasts fit into this category. Supporting the matrix activity is the resource-specific information system. This information is much more detailed and often based on operations research algorithms. It is oriented more to the operations people than the manager, and its purpose is to help allocate the various resources required to carry out the work within the context of overall company requirements. An example is the inventory management system. The third type of information is that which is needed to carry on the administrative function of the organization. This information is usually narrow in scope and exact, but it does not always need to be up to date on any given day, and it does not change frequently. It is illustrated by equipment listings, office supply budgets, and personnel files.

One may logically ask whether it would be useful to integrate these three types of information, and what their priorities would be within the matrix concept. We can answer these questions indirectly by turning our attention from the information, as such, to the specific tasks at hand. We should note that the matrix organization may change during the life of the project, but the information system will still be needed to support the ongoing work. Thus emphasis should be on the product and on the role of the team members in the management process, not on the matrix organization as it stands by itself. Staff meetings may effectively bring together members of the team and communicate information among them. Carefully designed distribution lists may assure that important information is forwarded to all concerned. It will not matter which type of information it is, as long as it pertains to the problem and is available when needed. If information integration will facilitate this process, it is justified.

The need for open communications in the matrix organization has been often emphasized. This contrasts with many organizations where information is viewed as a source of power that needs to be protected by the holders of specific offices. Figure 31-5 shows the typical flow of information in a matrix organization. Project-related information moves up through the project side, while operational information goes through the functional departments. By the time project or product-related information reaches top management, it has already passed through the decision-making filter represented by the matrix. When the members of the matrix team agree, they have the authority to proceed without an approval from general management. However, they have an obligation to keep the higher level informed. On the other hand, if there is disagreement, the issues must be sent up the organizational ladder.

Since the response time for decision making often can be very short, and it is important that communications at the matrix level be free and open so that agreements can be readily reached, a problem can arise with new members of the matrix team. A person used to protecting his or her position by controlling information flow may have difficulties in adjusting to the open environment.

It is important to note that in addition to the project and operational information, there is also an informal information flow that does not follow a specific channel but can deal with the department managers, the project personnel, or

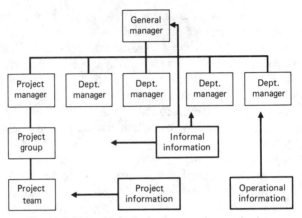

Fig. 31-5. Flow of information in a matrix organization.

with top management directly. In each organization there are certain individuals who serve as informal communication links. Some managers depend on these sources to verify or supplement the information that reaches them through formal channels.

The Role Grid

Considering the importance of communication among members of the matrix team, care must be taken to assure a complete understanding as to the role of each member on the team. Different people often have different interpretations as to what their jobs involve, and this is particularly true in the matrix, which brings together people from different organizations and backgrounds without always formally defining their responsibilities. Earlier in this section we described the Responsibility Matrix as an extension of the Work Breakdown Structure that links the work to be accomplished with those who will accomplish it. A further extension of this technique is the development of the role grid as shown in Figure 31-6. This is a technique for organizational clarification that illustrates to the team members and management their roles in terms of specific responsibilities and relationships to other organizational units.

The role grid is developed from the Work Breakdown Structure, from a critical path activity list, or simply from a checklist laid out chronologically so that various tasks, records, events, and approvals are identified. These are listed in the left-hand column of the chart. The next step is to arrange a meeting for all the key participants of the matrix team (only three are shown in the example) and ask such questions as who is responsible for making each of the key decisions during the life of the project. It is usually very apparent from the answers that the various roles the people ought to be fulfilling on key project tasks and

WORK ELEMENT	DIVISION MGR	PROJECT MGR	DEPT. MGR
Product feasibility	PM/APPR		CONTR
Appoint project mgr	APPR	CONCUR	PM
Develop business obj.	APPR	MSE	PM
Develop tech. objectiv.		PM	APPR
Develop schedules		PM	APPR
Form project team		PM	CONCUR
Initial designs		APPR	PM
Etc.			

Fig. 31-6. The role grid in the decision-making process.

decisions are not clear. As their discussions develop and definitions of the roles are prepared, most of the ambiguities involved in the authority and responsibility of the matrix are resolved. This technique has been particularly useful in research and development environments, where the products are not as clear as in developing a specific hardware. Organizational development specialists point out that the process of developing the role grid is much more important than the eventual result. It should not be viewed as a document, but rather as a tool for effective communication within the group.

The completed role grid assists in identifying the hierarchy of authority and the extent of control exercised by members of the matrix team. It has been emphasized elsewhere in this book that one of the important management approaches that makes the matrix successful is the participative role of its members. The role grid is also useful as a means for testing the extent to which the organization truly practices participative management. To do this, one can quantify the role grid by assigning points to each level of responsibility. For example, the individual who acts as a prime mover is given the highest score and others are given lower scores. This is done for each of the identified functions and then an average is computed. The scores for the matrix team members can then be plotted in a chart as shown in Figures 31-7 and 31-8, to illustrate the amount of control exercised in decision making by each of the team members. In an autocratic organization, the shape of the graph will be trian-

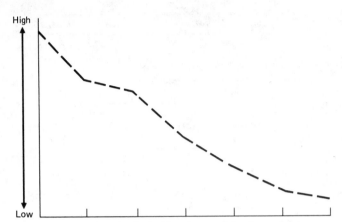

Fig. 31-7. Participation in decisions in a traditional organization.

gular as shown in Figure 31-7. This kind of a distribution would be very undesirable for a matrix organization, which should appear relatively flat as in Figure 31-8.

Facilitating the Communications Flow

Another useful test to determine the amount of flow of communications among the principal participants in a matrix organization is known as the *sociomatrix*. As illustrated in Figure 31-9, this chart shows the number of communications among the members of the matrix team. To develop such a matrix, it is nec-

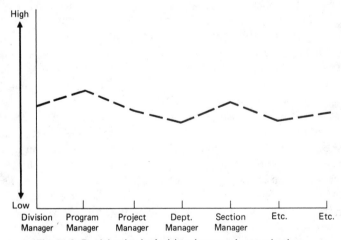

Fig. 31-8. Participation in decisions in a matrix organization.

| INITIATOR | | RECEIVER | | | | Relative |
	A	B	C	D	SUM	frequency
Division manager	0	8	2	1	11	0.25
Project manager	8	0	2	5	15	0.34
Department manager	8	2	0	1	11	0.25
Researcher	4	3	0	0	7	0.16
SUM	20	13	4	7	44	
Relative frequency	0.45	0.30	0.09	0.16		

Fig. 31-9. Sociomatrix, showing flow of communications among principal participants in the matrix organization.

essary for somebody with the appropriate qualifications to identify the initiator and receiver of each communication. In this respect, a communication is defined to include written materials, telephone calls, face-to-face meetings, and even greetings in the hallway. The form or length of the communication is not important for this purpose, only the frequency and direction of the interaction. Analysis of the sociomatrix can show when communication channels are overloaded or not used adequately. In a well-functioning matrix organization the frequency of the interactions should be fairly equal in both the amount and direction of the information.

The problem of short response times in the matrix organization has been greatly facilitated by recent technological developments and the growth of decision support systems, as contrasted to the previous batch systems that generated periodic reports. In the traditional organization the top manager often represents an information bottleneck, because he or she acts as a focus for information coming into and leaving the organizational unit. This executive delegates problems instead of solving them. He or she is the only one who has the big picture. By opening up the information channels and making information available to many members of the team, this bottleneck is to a large degree eliminated and the information serves as a support to decisions that can now be made at a lower level.

Distribution of information by itself, however, does not do away with the information overload problem. Experimental psychologists have shown that humans have definite limits on their information-handling capacity. The human mind identifies and processes simple problems using short-term memory, but for more complex problems information must be transferred to long-term memory. This transfer rate from short-term to long-term memory is quite slow, so one must be selective in the information to be processed. Extensive exception reporting and imaginative visual display terminals can make maximum use of short-term memory, as well as facilitate transfer of information into long-term memory when required.

Developing Better Information

The next major area of information management deals with the issues of uncertainty. Ability to complete the project on time and within budget, and to accomplish the technical objectives in general, are needs familiar to every project manager. During recent years the situation has been further complicated by the emergence of various environmentalist and consumer interest groups, whose activities introduce a significant amount of uncertainty in the accomplishment of projects.

To deal with such uncertainties, it is necessary to consider a whole range of potential scenarios and to use techniques providing a more comprehensive overview of the situation and its potential effects on the project. For example, critical path networks are widely used to plan and schedule projects. In an increasing number of cases project managers assign a range rather than a single activity time estimate, and use Monte Carlo simulations to get a better feel for variations possible in the completion of the project. In other situations, problematic segments of the network may be split out and expanded into decision trees. A technique known as crisis analysis has been developed whereby critical impacts of the project are estimated in advance and the appropriate information organized; then, if the crisis actually occurs, little time is wasted in selecting the appropriate response. These techniques draw on the collective experience of the matrix team members. Thus they not only make the success of the project more likely, but also improve the working relationships of the team.

The TREND Technique

A very important technique to identify the uncertainties and required information flows in a matrix organization was developed in the early 1970s at the Harvard Business School under the name of TREND (Transformed Relationships Evolved from Network Data). While this technique was successfully applied to several situations in Europe, it unfortunately never caught on in the United States. Since the principal purpose of TREND was to forecast problems of coordination and collaboration during the project performance, it should be viewed as a vital tool in designing information systems for matrix management.

The TREND method was designed to show important relationships among working groups such as are found in any matrix organization, and to alert managers to the likelihood of organizational and group interaction problems. At the same time, it was meant to aid in the design of the necessary information systems.

To apply the TREND methodology, it is necessary to know about the participating groups, their interdependencies, the relative degree of uncertainty with which they need to cope, and their relative prestige relationships. Interdependencies between the team members can be determined by analyzing the critical path network charts, if such have been developed. If three-time estimates are used to determine activity durations, they indicate the degree of task uncertainty for the respective organizations. Estimates of relative prestige must be analyzed by reviewing the organization charts and the appropriate organizational background.

Figure 31-10 shows in a highly summarized format the overall relationships involved in a typical project. The figures in the boxes indicate the number of activities in which the organization is involved in the project. Critical path activities are separated from others, since their late completion will adversely affect the downstream organizations. The shading in the boxes designates the amount of uncertainty involved in a particular activity. Typically, design activ-

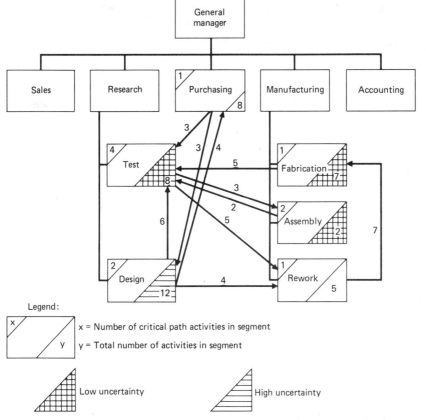

Fig. 31-10. Project functional relationships between members of a matrix organization.

ities reflect high uncertainty while fabrication and assembly activities show low uncertainty. The numbers on the connecting lines indicate the actual number of activity relationships among the organizations.

What this chart does not show is the relative prestige of the organizations involved. In general, research organizations hold higher status than manufacturing organizations. It has been suggested that collaboration is more difficult when relative prestige and relative power relationships are inconsistent. Conflict can be expected to arise when the group having higher prestige depends on input from the group with lower prestige. In this case the less prestigious group may actually have more power, and special coordination procedures may be necessary to make sure that this power is used in a positive manner.

While Figure 31-10 reflects a single point of time in the project, the actual situation and relationships will change from time to time as the program progresses. Thus there will be different needs for information and coordination among the organizations. To make full use of the technique, a network has to be periodically analyzed, with all dependencies noted and mapped on the organization chart for that point in time. The resulting displays will reveal the unique organizational interaction required by the project in the context of the overall organizational design. Difficulties of coordination and problem solving can be predicted, and appropriate coordination or information techniques can be planned to deal with these situations. It should be noted, however, that a TREND analysis will take into account only structured variables, and not issues such as historical working relationships, personality combinations, etc. Some of these may overcome the effect of unfavorable structural variables.

Influence Diagramming

Another highly useful tool to organize project information so as to identify the existing uncertainties, clarify the results of alternative actions, and project their eventual outcomes, is known as influence diagramming. An influence diagram is constructed by working backward from the expected value outcome. In the example shown in Figure 31-11, this outcome is the cost of a particular facility. If we assume that the finance charges and the costs for the capital, labor, and materials involved in the construction process are uncertain, we show them as nodes on the influence diagram. If any of these, however, are clearly known, we exclude them.

Going back step by step, we can then determine the next variables likely to have an effect on the element we are considering. In our example, the method of construction is a summary node that incorporates many of the individual items of uncertainty. In turn, the method of construction depends on the site geology and the eventual quantities or units to be handled. It should be noted that at this point we have moved from the area of responsibility of a construction department into the engineering department. Going back one more step

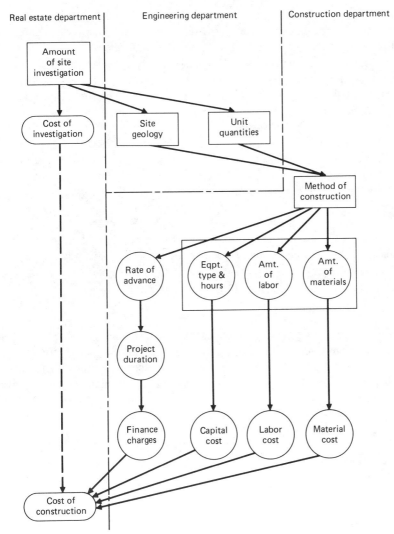

Fig. 31-11. Effects of decision on the work in other departments.

brings us into the real estate department, which is responsible for the initial site investigation.

Through this simple analysis we arrive at the key to the construction costs. The owner, or in this case the real estate department, often may be reluctant to invest much money in the initial site investigation, since the benefits of doing more than a certain amount of preliminary work are not obvious. However, the less known about the site, the greater the possibility that the site geology will present surprises which will increase the cost of excavation by increasing quantities or calling for a different construction method. Following this sequence down the influence diagram, one can readily see that the amount of site investigation will have an important effect on the eventual cost of the construction.

It should be noted that Figure 31-11 shows only a few of the many links involved in the influence diagram for a construction project. Many other considerations such as the extent of preliminary design, negotiation of labor contract, joint venture arrangements, and environmental conditions should also be included. The influence diagram can thus take on formidable proportions. From a theoretical viewpoint, it can then be handled with statistical methods to determine the relative sensitivities of the different factors. Management attention can then be focused on those elements having the greatest potential impact on the outcome. Aside from this use, however, the greatest benefit is in the process of constructing the influence diagram itself. By highlighting the risk consequences and uncertainties involved in the project, much useful information is obtained and the framework is laid for establishing additional information channels and coordination procedures that will contribute to the successful accomplishment of the project.

Effects of Stakeholders

Modern-day projects as varied as nuclear plant construction or the splitting of genes must take into consideration outside parties whose actions can change the course of the project or even destroy it. The need for information on the emergence of such interest groups and the possibilities of managing them have given rise to a technique known variously as stakeholder mapping or environmental scanning. The objective is to give management a tool to analyze the project environment and identify potential problems, so that action plans can be devised to assure successful outcome of the project. While up to this point we have focused largely on information requirements within a project, we must now also look outside and consider virtually everything that can affect us, such as the nature of the product, customer, and competitor, the geographical setting, and the economic, political, and even meteorological climate in which we must operate. These factors and the changes in them will affect the planning, organizing, staffing, and directing of the project, so a continuous flow of information is required to reflect these situations in the matrix operation.

To deal with the outside factors, it is necessary to define the other organizations or subsystems in the project environment. In other words, one must know who the parties are with a stake in the project, who the stakeholders are. Second, one needs to understand the nature of the dependency in this relationship. Why do these parties have a stake in the project and under what situations will they be willing to change? Third, how much uncertainty is there in the situation and how likely is it to change?

In stakeholder mapping the manager identifies the important elements in his

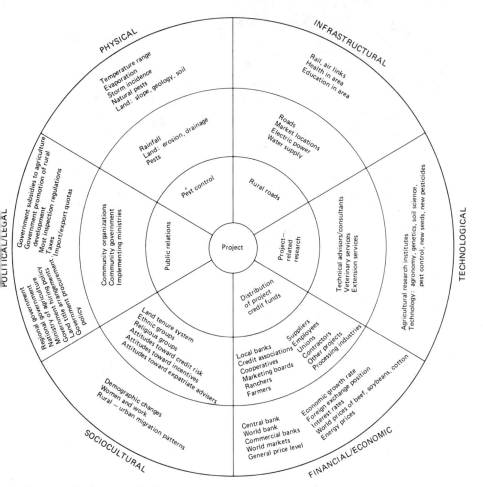

Fig. 31-12. Environmental influences (stakeholders) in a South American livestock and land settlement project.

Source: Reprinted by permission from Nicholas R. Burnett and Robert F. Youker, "Analyzing the Project Environment," CN-848 (Washington, D.C.: Economic Development Institute, World Bank, 1980).

environment and assigns each to one of three categories: controllable, influenceable, or appreciated. Figure 31-12 shows an example of such a stakeholder map, listing the environmental influences in a South American livestock and land settlement project. While this figure attempts to list all the possible stakeholders and influences, in actuality the list changes during the life of the project. A frequently cited example is that of the U.S. Supersonic Transport Program. At the time when this program first started, there were only a half dozen stakeholders who had a serious role in the project. From year to year this group grew in size as various politicians and environmental groups stepped into the picture. In the end there were so many pressures on the project that it became most difficult to carry it further and the whole program collapsed.

The stakeholder map thus serves to identify all the parties having an effect on the project. It also provides information leading to better management of the matrix organization. When properly updated and monitored, the map gives clues as to the necessary management actions to prevent adverse effects on the project. The information gained from the stakeholder analysis can be incorporated in the influence diagrams, so that the sensitivities of the project to the external groups can be better appreciated and information systems designed to help keep track of them.

Participative Planning

A discussion of information processing in a matrix organization would not be complete without consideration of participative planning and the involvement of personnel from various functional organizations who are nevertheless interdependent within the context of the matrix. One tool to facilitate such a process is known as SPEC (Structured Planning Evaluation and Control). This technique was developed by the Canadian Pacific Corporation in Canada to direct the efforts of a group of individuals toward the accomplishment of objectives or the constructive solution of any issues in a matrix environment. The technique is based on the concept of a planning room where information can be developed on a participative basis and visually displayed and manipulated. This process increases understanding of the objectives on the part of the team members and achieves a commitment to action.

The planning mechanism used by Canadian Pacific consists of simple cards that one can attach to the walls with magnets or with any sticky substance permitting them to be removed and relocated. The cards describe the idea set forth by members of the group. The group leader simply extracts more ideas and organizes the cards on the walls until some desired logic is achieved. This procedure, for example, could be used in the development of a Work Breakdown Structure. Once everybody has agreed on the structure, then data can be recorded either through photography or by a transcribing process, and the cards rearranged into a summary level critical path network. When more detail is required, a given card can be replaced by more detailed cards and the process continued.

Canadian Pacific has also established procedures whereby a clerical staff can transfer the information from the cards for distribution and storage. If the project is large, the information can be entered into a computer for analysis, updating, and printing of reports.

In addition to the techniques just described, the matrix organization will also benefit increasingly from the fast-growing inventory of technological aids. Relatively low-cost data bases and decision-support systems will become widely used as microcomputers replace the minis during the 1980s. The wide availability of terminals and desk-top computers should also alleviate the problem of getting more people involved in the decision-making process. Despite all the passwords and regulated access procedures, it is only a matter of time before personnel at lower levels will be able to know as much about a problem as the manager.

While this information explosion will represent a threat to the manager of the traditional hierarchical organization, it will facilitate the open communication and peer-level decision making characteristic of matrix organizations. It has been suggested that organizations develop according to their ability to use information. From this viewpoint, developments in information technology will have a positive influence on organizational developments in the matrix environment.

REFERENCES

Benningson, Lawrence A. "TREND: New Management Information from Networks." Harvard Business School Working Paper 72-19, March 1972.

Burnett, Nicholas R., and Youker, Robert. "Analyzing the Project Environment." The World Bank, EDI Training Materials CN-848, July 1980.

Davis, Stanley M., and Lawrence, Paul R. *Matrix,* Addison-Wesley Publishing Company, 1977.

Emery, James C. "The Economics of Information." *Wharton Quarterly,* Fall 1967.

Handa, V. K., and McLaughlin, W. A. "Planning and Control of Management Teams." *Proceedings of the CIWB-65 Second Symposium,* Haifa, 1978.

Levitt, Raymond E., Ashley, David B., and Logcher, Robert D. "Allocating Risk and Incentive in Construction." *Journal of the Construction Division,* ASCE, Sept. 1980.

Lichtenberg, Steen. "Real World Uncertainties in Project Budgets and Schedules." *Proceedings of the Project Management Institute INTERNET Joint Symposium,* Boston, 1981.

Logcher, Robert D., and Levitt, Raymond E. "Human Information-Handling Capacity in the Design of Project Control Systems." CIB Working Symposium on Organization and Management in Construction, May 1976.

Youker, Robert. "A Participative Process of Group Project Planning." The World Bank, EDI Training Materials CN-317, revised Feb. 1982.

32. International Marketing in a Matrix Ambiance

Robert L. Brickley*

Marketing in the industrial domain may be viewed as an interactive function, inextricably mixed with others, in what constitutes the business enterprise. Some general managers operate the business from the perspective that marketing is the essence of the business. All marketing-oriented concerns must continuously cope with several aspects of the business which we label as "marketing functions." This chapter will develop ideas on some of these functions as they are employed in the ambiance of a matrix organization of the enterprise.

THE SALES AND GEOGRAPHIC ORGANIZATIONS

The Sales Organization

The sales organization is vital to the enterprise. Its function can be viewed as the beginning and end of the business cycle. This function is the day-to-day interface with the marketplace—the interaction with the customer and the competition. It is here that the application of designing, manufacturing, and financial creativity is tested. The sales organization is perhaps the key sensor of the enterprise in continuously reading and reacting to the activity of the marketplace.

In the geographic/product line matrix the role of the sales organization as it carries out its functions of representing the various product lines in given markets becomes a formalization of a dual role associated with the salesperson in any structure. Every salesperson constantly balances the task of representing the interests of the enterprise to the customer with that of representing the interests of the customer within the salesperson's own company. This dichot-

*Robert L. Brickley holds a B.S. from Youngstown University and and an M.S. from Purdue University in mathematics. He completed the program for management development at Harvard Business School. His management experience spans 30 years with Westinghouse in engineering, manufacturing, marketing, and general management in both domestic and international operations.

omy carries over into an identical role carried out by the customer's purchasing organization. It is not infrequent that an effective but perhaps overzealous salesperson may be asked, "Whose interests are you representing?" when conflict in this interface takes place. The question can come from either the customer organization or the supplier organization.

This is but one of many examples of dualities constantly arrising in the day-to-day work of the sales force and its management. It occurs in the product-line organization as well as in the geographically or market-province structured organization. It is not infrequent that the individual salesperson stands alone in this difficult situation, even though other salespersons in the same enterprise may be experiencing similar situations. The individual salesperson may sound like a voice in the wilderness when trying to penetrate the noise barrier at the product-line division.

In international business, the salesperson of the U.S. enterprise must stand alone in representing the firm in competition with other U.S. firms and competitors from other countries. U.S. law is unique in today's international marketplace in the way it prohibits U.S. firms from presenting integrated approaches fashioned in the national interest. Non-U.S. enterprises are not so limited. They often share information and jointly formulate approaches, to the disadvantage of the U.S. firm that stands alone in the face of this action.

The matrix organization cannot solve all these problems, but it can strengthen the co-ordination of resources to get more productive use out of those available. A salesperson talking to his or her product-line division gains clout through use of the matrix. In the product line/geographic matrix the geographic management can identify other salespersons in a given region facing similar problems and can present a broader and more convincing argument supporting corrective action—a new design suggestion, a modified sourcing policy, pricing adjustments, special terms and conditions unique to an industry and a country, and so forth.

The Geographic Organization

The geographic organization has a much greater chance of getting a hearing, when it comes to representing the interest of U.S. business in competition with non-U.S. competitors within the United States. The geographic organization represents a broad spectrum of product-line interests and can make effective use of corporate prestige to bring issues of a larger geographic scale to the attention of the U.S. government—issues like the impact of export-import bank financing policy, the application of foreign corrupt-practices legislation, or government/industry co-operation to match that from government or pseudo-private enterprises that serve as conduits for policy and action by competing governments in influencing trade and capital flows.

That the matrix organization can aid in these matters is widely recognized.

U.S. enterprise generally accepts the sharing of decision making under the matrix. In a study of the organizational structures and operational cultures of European and Asian Pacific firms, it was found that joint or consensus decision making under the matrix had strong ties to the unique cultural heritage of the countries involved. It was also found that many of these countries, having been forced by economic limitations of home markets to serve international markets over centuries of trading, exhibit a natural acceptance of geographic imperatives in product-line decision making. Career planning in these firms characteristically requires long-term international assignments—a practice which facilitates joint product line/geographic decision making.

In the evolving matrix structure, this process evolves slowly. Characteristically, a business unit that historically has operated independently views every shared decision as a degree of freedom or autonomy lost. The challenge, then, is to identify and produce the positive pay-off which justifies this sacrifice of autonomy.

This pay-off does exist. It can be as complex as opening markets not previously available. It can be as simple as reducing independent inputs or complaints from the field. Replacing these singular inputs with better articulated arguments having broader reach and greater depth of substance can add substantially to decision making at the business unit. All the many pay-offs available seem to come only with hard work, and all take time. This highlights the necessity for patience and staying power. It will take a long time—not just a few years, but generations of executives committed to the process—to permit the matrix structure to evolve into a natural and viable process. Matix structure thereby lives a tenuous existence even as it develops and grows.

If we are to make progress on the larger scale of U.S. enterprise competing effectively in global markets, we have to be equally patient. The laws of the United States were established for good reasons over many years. Most of these years have been spent looking inwardly at U.S. enterprise and its impact on the worker and the small businesses of the United States. A more recent driving force in creating legislation has been a concern to protect the U.S. consumer from a perceived threat associated with big business. U.S. business will continue to grow and prosper as domestic enterprise only if it can be competitive in the global marketplace. The geographic matrix can aid the enterprise in forcing a continuous awareness of this dimension of its planning and operations. This too will take time to produce results and to hold the progress made, but it is essential to the future of the enterprise and to the United States as a leading economic power.

BUSINESS DEVELOPMENT

There are many dimensions of this complex subject. The next few paragraphs will address just two: business development of industrial product markets, and business development of the major projects market.

Product/Market Development

Much has been written on this aspect of industrial marketing. It is the focus of many of the marketing case histories of the U.S. business schools. Our concern here is not how it is done, but what happens to aid or impede this process when operating under the matrix.

This, perhaps, is the classic example of how a reasoned approach gives rise to the fundamentals of the matrix system. The very merging of the idea of a market need to be served with the idea of a product that is defined by this need, is matrix management at work. The concept of product/market development is the culmination of trade-offs to arrive at a technoeconomic fit with which the enterprise can live. It requires dual judgment in product design, product costs, market share objective, pricing, and sales methods.

This approach stands in stark contrast to that of a Durant in the early days of General Motors, and the subsequent practices in the evolution of autonomous profit centers and business units within the enterprises. However, one of the characteristics that the "one man, one thought" driving force has in common with matrix decision making is the necessity for total organizational commitment to the application of the decisions once made. Another characteristic common to both in the successful pursuit of the decided course is faith—faith that a successful course has been taken, faith in the decision maker, and faith in the process that led to the decision. On a larger scale, it is this faith that makes the global banking system work; it is the backbone of the free enterprise system.

Without this essential ingredient, none of these systems works. Without faith in the process and commitment to apply its output, the matrix system is impotent. Matrix ideas generated by the involvement and support of both product-line and geographic interests, or product-line and functional interests, have the potential of producing better results than can be achieved through autonomous individualized approaches.

Major Projects Development

Less has been written on business development of major projects, but much has been learned and is practiced in this area that utilizes matrix organization principles. Although a plethora of material has been written on project organizations, the classic application of program management—the application to project selling—is generally bypassed in the literature.

Industrial experience in project or program selling covers many years. It has been used extensively in defense sales, power systems sales, and industrial sales. The more complex the program, the more likely the use of matrix principles in the solution. Figure 32-1 presents a generalized example of major project selling in the industrial marketplace, permitting an examination of some of the contributions of matrix thinking.

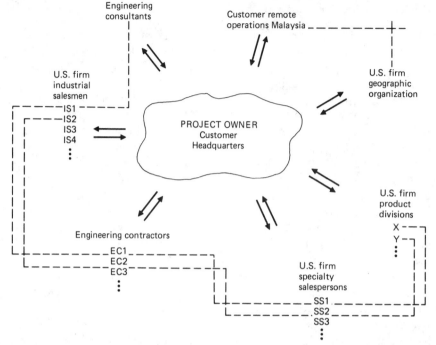

Fig. 32-1. Primary Project Participants (U.S. Firm, Customer, contractors, consultants, and some interrelationships early in the life of the project).

The following scenario depicts a selected subset of events in a typical project-selling situation. This subset presents the order in which the events might occur, and has been created with obvious omissions in order to reflect the typical selling situation, where much less than total coverage and knowledge is available. An asterisk indicates awareness of the situation by someone in the enterprise.

Period 1.   Owners establish preliminary design concepts, scope, timing.
Period 2.   Capital is authorized by owner.
Period 3.   Owner solicits bids from engineering contractors EC1, EC2, EC3.
Period 4.* Industrial salesman IS1 is contacted by EC1 for preliminary product information and pricing estimates
Period 5.* Specialty salesman SS1 learns of possible planned expansion by owner.
Period 6.* Salesman KL1 in Kuala Lumpur learns of new facility being planned for construction in Malaysia.
Period 7.* Industrial salesman IS2 notifies division that EC3 will require preliminary pricing and technical information on product line for bid he is preparing; no details are yet available on location.

Period 8.   Owner selects EC2.

Period 9.* Industrial salesman IS3 discovers that job he has been following with EC2 has been awarded to his account, and at this point first alerts other selling activities in the enterprises.

Period 10.* Specialty salesman SS2 registers negotiation with EC2 and notifies specialty salesman SS1, who calls on the owner. Product line *x* will be purchased by EC2.

Period 11.* Industrial salesman IS3 learns EC2 will place all purchase orders for electrical equipment, then notifies industrial salesman IS4, who calls on owner about job and solicits help.

Period 12.* Product line *y* division marketing manager reads notice of new expansion in international trade news and alerts specialty salesman SS3 to call on EC2.

Other periods, etc., etc.,

Figure 32-2 depicts the sparse nature of the population of communication intersections early in the life of a project.

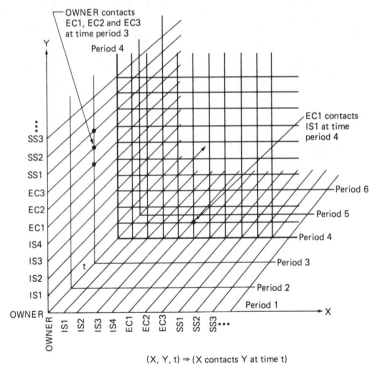

(X, Y, t) ⇒ (X contacts Y at time t)

... typical selling situation, where much less than
total coverage and knowledge is available.

Fig. 32-2. Project awareness within the U.S. firm (early in the life of the project).

A number of observations arise from this series of events:

1. A large number of sales personnel of the enterprise can be involved representing the special interests of various divisions.
2. Individuals can be operating independently for substantial periods during the negotiation phase of the project.
3. There are various levels of awareness of project details by individuals of the enterprise.
4. Many individuals may be active in a given project with no evident awareness of a corporate strategy for winning the order.
5. There is a distinct possibility that an independent posture by one product line may weaken the ability of the enterprise to relate to the project.
6. Individual salespersons and divisions can be working on a specific negotiation and be unaware of its relationship to a more general project opportunity for the enterprise.
7. The effectiveness of any individual sales effort is highly dependent on others involved in the job and can influence that of others involved.
8. The one-on-one situation between the specialty salesman and a given division is clearly part of the total involvement of the enterprise in this sales opportunity.

At this point a key question can be raised by those in the enterprise who are linked in some way to this sequence of events, and it may be expressed in various ways:

1. When does a simple product negotiation become a part of a major project negotiation?
2. What are the parameters that give definition to a project negotiation?
3. What is a project?

To address these questions, it is most convenient to define a project from the perspective of the customer. Yet even this allows ambiguity, since there may be more than one engineering constructor involved, as well as a specialty equipment manufacturer and several operations of the owner/user. So who is the customer? To remove this confusion, it helps to view the project from the point of view of the headquarters of the user firm. This convention is based on the owner creating an operating facility that requires investment in plant and equipment. This investment poses varied opportunities, depending on the capabilities and resources of the selling enterprise. The investment may consist of engineering design, civil work, bricks and mortar, process equipment, mechanical equipment, heating and cooling, electrical distribution equipment, primary substations, perhaps even a captive power plant. It may be as simple as duplicating an existing process line within an existing facility, or revamping an existing manufacturing facility that is now outdated; or as complex as creating a green-field facility such as a totally integrated steel mill. Regardless of whether the job is a $1-million revamp of an existing operation or a $1-billion integrated processing facility, the owner views it as a specific project and generally has in mind a firm start-up date. The owner's goal is to get the new facility working so that the anticipated return can be generated through utilization of the plant.

Accordingly, to give definition to the concept of a project, it is useful to think in terms of that facility being created by the owner. When we speak of the involvement of the enterprise in the project, we have in mind the pursuit of every business opportunity for the enterprise that is to be generated by the owner's investment in the new facility. For an electrical equipment manufacturer this may range from conventional electrical apparatus required for the primary substations, plant distribution systems, process drives and controls to the instrumentation, industrial fans, elevators, and lighting equipment. It could include the early work by distributors of electrical components and supplies in the course of developing a business position with the new operation. It could include the involvement of service divisions from installation through continued involvement in maintenance, repair, and operating requirements of the ongoing facility.

What we see is the investment of the owner in a new facility giving rise to a host of potential business opportunities for the enterprise and for the various selling organizations representing those interests. Many of those interests can be represented adequately through a one-on-one situation with the customer and the customer's agents; however, in many of these situations knowledge about the involvement of others, and accessibility to the knowledge gained by others in terms of their involvement in the project, create likely benefits for all concerned.

A fundamental premise of the enterprise, in its evolution of profit centers and business units, is the requirement that the corporate interests of the enterprise stand above and dominate the specialty interests of any division or business unit. It is imperative, therefore, that a mechanism be established which permits a corporate strategy to be developed; that individual business-unit participation through the various selling agents be integrated; and that a well-orchestrated corporate participation in the project result. It is also essential that at all times the owner and the owner's agents recognize that there is a single interface for relating to the enterprise as a whole.

Matrix management enters this picture in many ways. The following thoughts are noteworthy.

1. Planning principles are absolutely necessary in putting together a meaningful and effective strategy for success. This process requires the development of an overall corporate strategy for winning the order, versus a strategy for successful participation by an individual product line or division. This perspective may be viewed as ominous to the autonomous business unit whose profit,

scope of participation, and independence may to a degree be threatened, in order that corporate scope, profitability, and chances of winning increase. The strategic marketing organization of the enterprise must have the total support of executive management to make this work. That this does not happen automatically is understandable. It is comfortably tempting for an executive to manage several totally autonomous operations by avoiding involvement in decisions directly affecting their profitability and growth. The executive can avoid any personal action that would allow the subordinate profit-center manager to attribute lower profits to a joint decision made in the best interest of the corporation. An executive whose entire career philosophy has been molded through experience in autonomous profit-center operations could hardly allow that to happen: it would be unthinkable! We have all heard the observation, "Show me a profit-center manager who's making his decisions on the basis of what is best for the corporation, and I'll show you a general manager who'll soon be out of a job!" This is the result of a business culture that has evolved in the United States over the last 40 years.

It makes everybody's life a little uneasy to participate and to accept responsibility for participating in a subordinate's decisions. Perhaps this is why matrix management consultants say one should force the joint decision making as far down the organization ladder as possible. This may reflect empathy for the degree of acceptance of matrix and participative management at the executive level, and it just might make that upper management job a little more comfortable.

2. The whole process operates on the principle of employing and allocating the resources of the enterprise most effectively so as to win the order, or at least determine logically whether the enterprise should try to do so. Among these resources, the human resources of the enterprise are perhaps the most critical to its success.

The selection of people to staff the selling team must consider the importance of arriving at balanced judgments. The team needs strong advocates of varying positions. As a matter of experience the selling team seldom produces its best results without a designated devil's advocate to question ideas. Many really bright ideas disintegrate when exposed to the light of reality. Bias—regardless of its source—can seldom stand the test of hard analysis. The process must both generate and question ideas.

3. This brings us to the third observation on selling teams laboring under the matrix system. It is plain, hard work. If the team members do not have the dedication to dig out the basic data that is needed, are not willing to brainstorm until exhausted, or cannot tolerate the rejection of ideas and come back with more, they will add little value to team results.

At some point in this process there has to be time to sit and think individually—to examine what the group thinking has done, and determine where it has become trapped in alleys leading nowhere or walked past an idea that holds great promise. Again, this can only be achieved with general commitment to the effort and hard work. The use of the matrix and participative planning and management are one and the same in these phases of a selling team activity; the most essential ingredient is 100% commitment to the task.

SELECTION OF SALES REPRESENTATIVES

The role of the sales representative or agent in international sales is broad. The sales representative is an individual or firm well positioned to establish an identity as the local representative of specific product lines of the enterprise, or as the representative of several product lines to given local market sectors. The sales representative helps the non-resident enterprise to identify business opportunities, carry out the principal local activities of the selling effort, and provide insight and guidance on local practices and strategic intelligence for the planning effort. The sales representative may perform these services on a commission basis, or in special circumstances may purchase and take title to equipment for resale purposes. His or her responsibilities continue through order follow-up and postsales services.

In all, the sales representative becomes an essential link in the organizational chain that connects a business unit and its marketing unit with the marketplace. This then may help to explain why business units and their marketing units instinctively tend to guard their historical "right" to select and manage the sales representative. But again, an objective assessment of this selection process shows the weaknesses of unilateral action, and the real benefit of participative involvement by other units of the enterprise with local experience or plans to enter the local market.

A collective effort can draw on cumulative experience gained from working in the country: knowledge of local practices, awareness of interconnecting relationships within the country, and exposure to sources of sound counsel.

Not insignificant is the discipline brought by broadened participation and the exposure to other interests which lead to a more substantive look at a larger group of potential candidates. The chance is reduced of accepting the credentials of the sales representative who found the enterprise—a course that is tempting because of the ease with which it can happen.

The geographic input can and should lead to the development of a reasonable number of qualified sales representatives in the country or region who can collectively represent the interest of the enterprise, and who among themselves have compatible assignments.

SOURCING AND PRICING

Experience operating under a matrix system has shown the typical responsibility chart, indicating sole and jointly shared responsibilities of product-line and geographic units, to be of little practical use. Consider the following example. The responsibility for sourcing and pricing is uniformly viewed to be in the realm of the business unit. This is an interesting phenomenon because about the only time the issue of sourcing and pricing occurs is when the in-country sales representative has strong arguments on the competitive situation, or detailed knowledge of unique customer requirements, that can support technical or commercial changes justifying re-examination of the pricing or sourcing. In other words, the reality of the marketplace—not a responsibility chart or management direction—determines that joint decision making is an imperative. Processes, if permitted to flow naturally, will result in joint or matrix solutions to complex problems, much in the manner of Adam Smith's "invisible hand."

FINANCING AND INVESTMENT

This important area requires functional expertise. In today's global interrelationships in trade and capital flows, current functional understanding of availability, rates, and mixing of credit is an essential element of both major project and product/market development. In the earlier sections on project selling, matrix involvement in team selling has been described. This collective pursuit of a strategy for success also assumes the integration of the necessary global expertise on financing.

In the case of product/market development, several interesting avenues are open. The business units with plants manufacturing product x that are operating at less than acceptable capacity will probably adopt the position that they need not add to that capacity. Even when we broaden the vision of the enterprise to include the total industry for product x on a worldwide basis, we generally find an overcapacity situation in mature product areas. Again, a legitimate question can be raised as to why invest in overcrowded industry. Competition is intensive as aggressive companies from industralized nations are attracted to any active international market so as to gain volume and load underutilized plants.

In this case the developing country—the target of this intense competition—may be totally indifferent to the situation of any given enterprise or to the global picture. Indeed, the country's driving force may be no more than a desire to reduce its balance-of-payments deficit. Local or in-country manufacturing of imported goods is often a most important goal. This may be justified not in terms of costs, profitability, or global market share, but simply in terms of trade flow and employment considerations.

In these situations the local market demand may be small, and the scale of local manufacturing must be correspondingly small. This is occasionally offset by the existence of a strong driving force to build an export capability from that country. The complexities created by these interests in mature businesses are obvious. The collective business-unit and geographic viewpoint must be focused on issues such as:

1. Should the penetration mode for the enterprise be through licensing (technology transfer) or through direct investment?
2. Does local law allow foreign ownership and repatriation of profits?
3. Are the potential sales volume and scale of operations compatible with business unit resources? Is the investment worth the effort?
4. Can the enterprise identify suitable local partners?

Clearly, a well-developed understanding of that country and its driving forces is essential to the thought processes of the business unit. The geographic organization can contribute to this understanding and will naturally be drawn into the process through an open system of communications exchange leading to improved country and business-unit strategies.

One interesting outcome of these deliberations is the increased attention the geographic perspective brings to the concept of transnational production. Under this concept, manufacturing resources in two or more regions of the world are employed to improve global competitiveness of a product line. For example, production capabilities in low labor-rate areas like Malaysia or Sri Lanka can be linked with advanced production technologies in Japan or the United States, so as to offer a lower priced "Japanese" or "U.S." product to global markets. This simple example can be misleading. The benefits of this concept reach far beyond the search for cheap labor to improve global competitiveness. Strategically placed production units give rise to a local presence of the enterprise that helps to open local or regional markets. Such units can be part of a global rationalization of manufacturing operations of the enterprise. They can also be placed to gain better access to the supply of basic materials and components.

AFTER-SALES SERVICE

The linking of new business sales with after-sales service tends to be challenging, even though it is universally recognized as essential. We have found that the cyclical concept of negotiation, team selling, order follow-up, and service is the best mechanism for creating a positive climate for negotiating and selling the next job in the marketplace. By the same token, a breakdown in this cycle—say, when order follow-up or prompt and high-quality repair and service are absent—is a recipe for disaster that is difficult to overcome. Once again, re-establishing the dedication and resolve of the enterprise to honor these is so much more costly than doing the job right in the first place, that it seems wasteful to think of nonperformance being allowed. It is the close coupling of order follow-up, repair, and service as integrated functions of the sell-

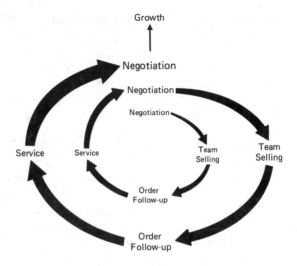

We have found that the cyclical concept of negotiation, team selling, order follow-up, and service is the best mechanism for creating a positive climate for negotiating and selling the next job in the marketplace.

Fig. 32-3. After sales service.

ing effort that lets these functions perform most effectively (see Figure 32-3). Involvement and participation again become watchwords. Creating forward thinking by the selling team makes this a necessity; participation by the service organizations earns their commitment to an overall strategy. Matrix management occurs automatically: it is the product of need—interested response to need, and to satisfaction from value derived.

STRATEGIC BUSINESS PLANNING

Impressive achievements have materialized through application of matrix management in the strategic planning process. This process couples business planning at the country level with business planning in the product-line business unit. The phases of this process carried out by the geographic organization reflecting the interest of key business units for given country or region include:

1. Assessment of the social, political, and economic environment
2. Analysis of the business climate
3. Assessment of selected markets
4. Analysis of the position of the enterprise

5. Assessment of the competition

6. Recommendation of modes of entry

Correspondingly, the business units carry out the development of information including:

- Definition of strategically significant segments
- Analysis of segments from a global perspective
- Alternative strategies for each segment
- A value-based financial analysis of each segment
- Underlying assumptions and recommended strategies

It is the integration process that requires a close coupling of business-unit and geographic perspectives (see Figure 32-4). Adversary modes may be assumed with favorable results. It is essential that theoretically attractive ideas and strategies under consideration be examined from both perspectives. This examination should include both strategic action plans as well as plans not to act.

This mutual examination must take place over an extended period of time, since each business unit must give consideration to several key international areas as well as the home market. Each country organization must give consideration to several key business units. To have an effective interchange of ideas among these groups, it is essential for these to be face-to-face discussions of the issues and strategies.

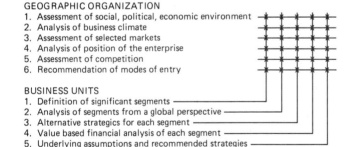

GEOGRAPHIC ORGANIZATION
1. Assessment of social, political, economic environment
2. Analysis of business climate
3. Assessment of selected markets
4. Analysis of position of the enterprise
5. Assessment of competition
6. Recommendation of modes of entry

BUSINESS UNITS
1. Definition of significant segments
2. Analysis of segments from a global perspective
3. Alternative strategies for each segment
4. Value based financial analysis of each segment
5. Underlying assumptions and recommended strategies

The integration of business planning at the country level with that at the product line business unit is compelled by addressing each intersection of information development responsibility. Each business unit strategy or assessment must be viewed as it relates to each assessment of the geographic organization of the country environment, and vice versa. A different family of intersections, each an adjusted assessment resulting from the integration process, will be generated by each business unit in each geographic area.

Fig. 32-4. Strategic business planning in the geographic/product-line matrix.

Both parties must find mutually rewarding approaches within their individually structured priority and value systems. As the interchange between these groups increases, there can be a cumulative effect such that the priorities and value systems become less independent or have greater commonality. A business unit manager's awareness of a country's political imperatives can lead to a substantial change in attractiveness from a commercial or technological perspective. The country managers' understanding of a basic development strategy for a product line can elevate the product-line fit for one country and perhaps reduce it for another.

The benefits of this cross-fertilization carry through to organizational development as well as career development. Having actually worked in another country under an international assignment can substantially benefit the employee and the business unit employing that person in a product-line assignment and vice versa. For the manager participating in the strategic planning function on a global basis, this experience is indispensable.

The long-term development of the matrix organization must rely on the enhancement that results from an interchange of product-line and geographic experience. This is a slow process but it must take place continuously and surely. A compelling argument for permitting this to be a slow process is the need to allow a reasonable time for the incumbent in either assignment to add real value and produce results.

SUMMARY

This chapter has presented observations on several important aspects of the marketing function, and the manner in which activities within these functions can be affected by a matrix ambiance. A few key characteristics and results of the application of matrix structures in marketing have been discussed:

1. The matrix provides the means for getting more productive use of resources of the enterprise. This applies notably to the use of human resources and the resulting value from consideration of varying perspectives which more broadly reflect the total interests of the enterprise.

2. Joint decision making—a concept approached with great apprehension by those managers who have spent their entire careers in autonomous profit centers—occurs quite naturally even in those realms of decision making frequently classified as the sole responsibility of one of the participating groups in the matrix. This occurs when either party openly utilizes the value added from meaningful intelligence contributed by the other.

3. Adding real value generally entails hard work, and the contribution is enhanced through experience and time. Accordingly, the evolution of an effective matrix structure requires both hard work and patience.

4. Many strategic planning principles are absolutely required in utilizing the matrix effectively. Frequently a corporate strategy or action required in the

best interest of the total enterprise must be elevated and applied in lieu of those of one of the parties in situations involving several units or perspectives. This carries with it the implicit awareness that decisions under the matrix need not be totally comforting to all parties; such decisions may not even be viewed as a consensus. However, it is essential that the perspectives and value added by the contributions from the various parties be fully considered and utilized in arriving at the best decision for the enterprise.

Section VI
Matrix Management Techniques

As the concept of matrix management has evolved, corollary specialized techniques have emerged to facilitate the successful implementation of matrix. In this section some of the key managerial techniques are presented. These techniques deal with the determination of organizational roles, conflict management, resource negotiation, management skills, and decision processes.

In Chapter 33 Miles G. Patrick examines project management from the viewpoint of the skills required in a project manager. The chapter discusses the structure of a single matrix chart for use by a project manager to identify and judge the relative importance of the various skills needed by such a manager. The information in the chapter was amassed because of the need of a large research laboratory to expand its capabilities with a project management organization. The author deals with such topics as the functions of project management, the skills and characteristics needed, and the application of these skills in the job.

Anyone who has worked in a matrix realizes how easily conflict comes about as matrix managers compete for scarce resources in the organizations. Conflict is so prevalent that it seems to dominate much of the time of the principal managers. How to manage conflicts in the matrix organization is the subject of Chapter 34 by George H. Labovitz. He looks at why matrix organizations create conflict and the ways that organizations can cope with it.

In Chapter 35 Dundar F. Kocaoglu and David I. Cleland examine the concept and techniques of matrix organizational charting. They describe such charting as a way to determine the specificity of organizational interfaces by establishing defintions of such interfaces through several descriptors centered around the "work packages" in an ongoing organization. The approach they suggest can not only help define relative roles in the organization but also facilitate team building, consensus decision making, and participative management. Sufficient insight is provided into matrix responsibility charting in this chapter to allow a manager to apply the technique.

In Chapter 36 C. E. Leslie describes the process of negotiation in matrix organizations. He uses two common activities found in matrices—*the negotiating process and its aftermath* and *communication in negotiation*—as focal points to present the chapter material. His discussion includes basic negotiating concepts, terminology, guidelines for negotiators, and planning for matrix system negotiations. He also discusses the barriers to communication and some guidelines for more effective communication.

33. Skills Needed by a Project Manager

Miles G. Patrick*

This chapter discusses project management—its elements, functions, needs, and how it works—and explains how a project manager can most effectively use the process of project management.

Project management has been around for a long time. After all, someone ran the job to build the pyramids. Only recently, however, have researchers and others begun to gain a better understanding of the matrix organization and its four major elements, identified by Cleland and King as functional support, project management, routine administration, and research and development (strategic planning).[1] Their studies are bringing into prominence the underlying principles forming the basis of project management. As the use of the matrix concept of organization continues to grow, the element of project management receives increasing study and attention.

This chapter portrays some of the complexities and challenges in the life of the project manager, by showing how the activities or functions necessary in managing a project naturally dictate the skills required in a project manager. Understanding the relationship between functions and skills can be helpful in training and development programs for project managers. For example, the relationship is useful in portraying some of the concepts of project management and describing the project manager's job, including reinforcing the importance of the various skills.

The chapter outline follows the structure of a single matrix chart that can be adapted to reflect any given project or organizational situation. A project manager can use the chart to identify and judge the relative importance of the various skills needed for a specific project management situation. The same chart can be used by management to analyze candidates in selecting the manager for a project or program, or can be useful in evaluating project managers

*Miles G. Patrick has 35 years of experience in engineering, project management, and management systems application. With the General Electric Company and Battelle Memorial Institute, he has managed projects for nuclear research facilities, production plant modifications, research investigations in management sciences, and the development and implementation of a research project management system. He has managed a quality control organization in construction of a major nuclear plant, and a project planning and control support unit. He has also conducted seminars in project management. Mr. Patrick has a B.S. in mechanical engineering from Oregon State University and is a registered professional engineer.

and determining their need for specific counseling, assistance, or training. Project managers can use the matrix to gain additional insight about developing specific solutions to specific problems. The principles, applications, and needs of the skill areas can be taught in classroom situations. Acquiring and polishing the skills, however, requires practice and "coaching" over an extended period of time.

The information in this chapter is an outgrowth of a recent Battelle Memorial Institute (BMI) internal report on project management.[2] BMI, an independent nonprofit multinational organization in the contract research business, has a total staff of more than 7000. At any given time, several hundred applied research projects directed toward new products and processes are under way in Battelle's U.S. Divisions for a wide variety of governmental and industrial sponsors from all parts of the world. Such projects vary in size from a few thousand dollars to more than $500 million each. Because of research market trends for the past several years, Battelle has emphasized increased attention in applying project management skills to the needs and efforts of the projects being conducted.

Not only has attention to project management increased in the Laboratory Divisions, but also the Battelle Project Management Division (BPMD) was formed in 1978 to manage large, highly complex projects. The new division also provides a focus and a resource for the ongoing enhancement of project management effectiveness throughout the institute.

In 1981 the author and Dr. John M. Batch of the BMI Corporate Office prepared an internal report on project management within Battelle for the Institute's Executive Vice-President.[3] A part of this report focuses on what a project manager must know and do to perform the job of project management. The report examines the characteristics of effective project managers, the functions involved in managing a project, and some of the knowledge and skills a project manager needs.

Characteristics and skills are both difficult to define quantitatively; however, we felt that skills are slightly easier to define. We also felt that the major significance of defining personal characteristics is the increased likelihood of a person developing the skills needed. Further, since the term "skills" encompasses the ability to apply knowledge effectively, it provides a distinction between *knowing* how to manage a project and actually *doing* it effectively. Therefore our report focuses on relating the functions to the skills, so as to gain insight into project management and the project manager.

The rest of this chapter, which expands upon the principles expressed in our report, is addressed to you as project manager in a matrix organization.

FUNCTIONS OF PROJECT MANAGEMENT

The usual major functions of project management that you must carry out can be divided into two phases, planning and execution. (For our purposes, proposal preparation is a part of the "planning" phase, and project close-out is part of the "performance" phase.) These usual project management functions are listed below.

The listing is only approximately sequential, because the planning and performance phases of the process usually overlap. Project work generally begins long before planning is completed. In fact, "replanning" could be listed as a function in the performance phase.

Planning functions include the following:

1. Establish objectives.
2. Define the work statement.
3. Establish the work breakdown structure.
4. Establish work package work statements.
5. Develop networks and schedules.
6. Make budget estimates.
7. Define and plan the Quality Assurance (QA) Program.
8. Obtain identified resources and build the team.
9. Establish the review and approval process.
10. Develop make-or-buy decisions.

Performance functions include:

1. Authorize work.
2. Measure progress and status reviews.
3. Track costs.
4. Implement changes.
5. Maintain QA requirements.
6. Conduct project reviews.
7. Maintain the project team.
8. Monitor the use of resources.
9. Maintain external communications.
10. Manage procurement.
11. Complete the project and dissolve the team.

SKILLS AND CHARACTERISTICS NEEDED

The most important factor in successful project management is the project manager. Personal characteristics of effective project managers are listed below. Most of these personal characteristics are mentioned in Russell Archibald's book, *Managing High-Technology Programs and Projects.*[4] As one would expect in any management job, the characteristics and skills needed for working with people predominate.

- Flexibility and adaptability
- A preference for significant initiative and leadership roles

- Aggressiveness, confidence, persuasiveness, verbal fluency
- Ambition, activeness, forcefulness
- Ability to communicate and integrate
- Broad scope of personal interests
- Poise, enthusiasm, imagination
- Ability to balance technical solutions with time, cost, and human factors
- A tendency to be a generalist rather than a specialist
- Ability to identify problems
- Willingness to make decisions
- Ability to handle conflict and confrontations

Effective project management requires using a variety of personal skills, often simultaneously. For instance:

- Communication skills
- Negotiation skills
- Motivational skills
- Leadership and conflict resolution skills
- Knowledge of subject matter
- Skills in structuring work logically
- Sills in network and schedule techniques
- Skills in cost/schedule integration techniques
- Knowledge and skills in information systems
- Skills in using project reviews
- QA conformity skills

The personal skills required in project management can be related to the functions involved in managing a project. The skills required vary considerably, depending upon specific situations. The matrix of functions versus skills (Figure 33-1) gives a general view of these relationships, illustrating how the various skills are frequently utilized by effective project managers. The symbols in the matrix indicate two levels of importance for different skills in different functions. There are no hard-and-fast rules about these relationships, but the ones shown are considered typical. A detailed discussion of each skill area follows.

Throughout this chapter, in the discussions of skills and their relationships to functions, the pronoun "you" refers to you as project manager.

SKILL AREAS

A program manger or project manager must function as a skillful manager to be effective. Your role is frequently viewed principally as that of a business manager; however, work gets done by people, and how well it is done depends

Program/Project Management Functions	Personal Skills Required for Project Management										
	Communication Skills	Negotiation Skills	Motivational Skills	Leadership & Conflict Resolution	Subject Matter Knowledge	Logical Work Structuring Skills	Skills in Network/ Schedule Techniques	Skills in Cost/Schedule Integration Techniques	Knowledge & Skills in Information Systems	Skills in Using Project Reviews	QA Conformance Skills
I Planning Phase											
1. Establish Objectives	(X)	(X)		(X)	(X)						
2. Define Work Statement	(X)	(X)			(X)						X
3. Establish WBS	X	X		(X)	(X)	(X)					
4. Establish Work Pkg. Work Statement	(X)	(X)		(X)	X				X		X
5. Establish Networks/ Schedules	X	X		(X)	(X)	(X)	(X)	X	X	X	X
6. Establish Budget Estimates	X	X		(X)	(X)		X	(X)	X	X	X
7. Define & Plan QA Program	X	(X)		(X)	(X)						(X)
8. Identify Resources & Establish Project Team	(X)	(X)	(X)	(X)	X		X	X	X		
9. Establish Reviews & Approvals	(X)	X		X	X				X	(X)	(X)
10. Establish "Make or Buy" Decisions		(X)			(X)		X	(X)	X		X
II Performance Phase											
1. Authorize Work	(X)	(X)	(X)	(X)				(X)	(X)		X
2. Measure Progress	X				(X)		(X)	(X)	(X)	X	X
3. Track Financial Status								(X)	(X)	X	
4. Conduct Status Reviews	(X)			(X)	X	X	(X)	(X)	(X)	(X)	X
5. Implement Changes	X	(X)	X	(X)	(X)	X	(X)	(X)	X	X	X
6. Maintain QA Requirement	X	X	(X)	X	(X)					X	(X)
7. Conduct Proj. Review	(X)	X		(X)	X		(X)	(X)	(X)	(X)	
8. Maintain Proj. Team	(X)	X	(X)	(X)	X					X	
9. Monitor Use of Resources							X	(X)	(X)		
10. Maintain External Communications	(X)	X		X	(X)		(X)	(X)	(X)	(X)	X
11. Initiate & Monitor Procurements	(X)	(X)	(X)	(X)	X	X	(X)	(X)	(X)	X	(X)
12. Complete the Project	(X)	X			X				X	(X)	(X)
13. Dissolve Proj. Team	(X)	(X)		(X)							

(X) – Primary Importance

X – Secondary importance

Fig. 33-1. Program/project management functions and personal skills matrix. Source: J. M. Batch, and M. G. Patrick, *Project Management within Battelle,* Battelle Memorial Institute internal report, Columbus, Ohio, July 1981.

heavily on your skills in managing people and your knowledge of the subject matter along with your administrative skills.

One major difference between a project manager and a functional manager stems from the mission-oriented thrust of a project. Schedule and cost factors are important to all managers, but they are critical to the project manager in accomplishing the mission. It is especially important that costs and schedules be integrated and balanced with the third factor, work scope or technical content, to measure project progress effectively. The project manager must be informed continuously of project status and must identify trouble spots as early as possible, in order to take appropriate action and maintain the drive toward the project's mission.

Another major difference between the project manager and the functional manager in the matrix organization is the authority or power base relative to people working on the project. The project manager does not, in general, have the traditional "boss-worker" authority that the functional manager usually has. The distinction is not clear-cut because project managers usually contribute either formally or informally to performance appraisals of team members. The authority delegated by higher management to you as project manager also adds to your power base as perceived by the project team. This kind of authority, however, generally remains less than that conventionally viewed as vested in functional managers. Consequently, as a project manager you must rely heavily on the power you earn by your expertise or other attributes. Essentially, your power comes in a large part from how you are perceived by the project team, as has been well documented in the work of Gemmill and Thamhain in 1973[5] and later by Thamhain and Wilemon.[6]

All management jobs are complex. The differences in this regard are in the degree of complexity. The project manager's job in the matrix organization ranks very high in complexity as compared with most functional managers. For example, the manager of the engineering function faces a multidisciplinary situation in many organizations. This function may include eight or ten distinct engineering disciplines. However, the project manager may be exposed not only to all those disciplines, but also to several other disciplines in research and development, construction, procurement, and business, not to mention the political situation and public opinion. The project manager must face multiple interest as well as multidisciplinary issues. One analogy is, "Like the juggler . . . , the project manager must use all his skills to keep the many pieces of his project 'in the air.' The effective project manager has learned to live with complexity."[7]

The matrix in Figure 33-1 provides a way to examine the complexity of project management by breaking the task into small, discrete elements as an aid in management analyses of specific situations. An even wider range of the spectrum of project management issues and problems can be presented succinctly, when this matrix is coupled with the Cleland-Kocaoglu Linear Responsibility Chart[8] to relate responsibility assignments to functions.

Not only is the process of carrying out the functions shown in Figure 33-1 highly iterative, but frequently several of the functions are being worked concurrently. In addition, many of the skills need to be applied simultaneously, also as shown in Figure 33-1, indicating how complex the project manager's job is. Looking merely at functions and skills does not directly address the added complexity of the multiplicity of disciplines and interests that must constantly be juggled. Each area of skill and knowledge is broad, hard to define, and difficult to measure. The next sections look at each area, including subsets of skills, techniques, styles, and methods of practice.

Communication Skills

The effective project manager must be a good communicator in a very broad sense, in order to both understand information and convey it. Good two-way communications are essential in applying most of the skill and knowledge areas in managing a project. Volumes of literature exist on the subject of communication skills and how to develop them. It is important for any manager, but essential for the project manager, to develop good communication skills, as both a receiver and a transmitter of information. In addition to being articulate in transmitting information to others, it is important to learn to listen for understanding. Easy to say, and all too often very hard to do! It is easy to "hear what we want to hear," make assumptions about the meaning behind the message, listen for rebuttal, or too quickly place our own value judgment on a message. All these tendencies interfere with gaining a true understanding of what someone else is trying to convey.

The manager of any except the very smallest and simplest of projects has to communicate (both ways) with a wide variety of people with different specialties, disciplines, and interests, most of whom use their own jargon or specialized languages. Scientists, engineers, builders, cost and schedule specialists, customers, government employees, regulators, the public, environmentalists, and buyers all express themselves differently. To the extent that you are conversant in all these "languages," your effectiveness as a communicator is enhanced. Since most project managers are limited in this regard, you also need to use the simplest, commonest terms possible and to maximize your use of feedback, paraphrasing, and questioning to strive constantly for the best true understanding that can be reached.

It can be helpful to identify individuals who are good interpreters of some of these languages and in whom you have a lot of confidence and trust. These persons can be effective aides in maintaining the upward, downward, and sideways flow of understood information so essential to project success.

Negotiation Skills

In general, you must be a skilled negotiator if you are to be an effective project manager. The situations requiring these skills frequently cover a broad spectrum. Some instances require a formalized negotiating process, such as establishing the contract with the customer, labor agreements, or subcontracts with suppliers. In many organizations skilled assistance is available from service groups such as legal counsel, contract specialists, labor relation experts, or procurement officers. However, it is important for you as project manager to understand the negotiation process, if you are to properly represent your project.

In addition to such formalized situations, there are informal, ad hoc negotiations with functional managers and project participants to "cut the deals" and reach agreements on who will do what and when and for how much. Basically, for the project manager the term "negotiate" implies a process where two parties come to an agreement with a common understanding of that agreement's content and meaning. In a project situation the skilled negotiator can be guided with respect to the other party by remembering two concepts: (1) what you need or want from the other person to get your job done; and (2) what you are willing to give to the other person to meet his or her needs to get the job done.

Preparation is a key element in successful negotiating. Preparation includes going in with a firm objective and a strategy in mind. It is important to know as much as possible about the other party and what to expect in the way of demands, wishes, or needs, and also to be careful not to underestimate that person.

The skilled negotiator goes in to win, but recognizes that winning does not require that someone else lose. As reported by Royce Coffin, "It is the very hallmark of successful negotiations that both sides emerge as winners."[9] Particularly when the negotiation is the beginning of an ongoing relationship, it is important that both parties be satisfied with the outcome and with each other. Once both parties are satisfied, the considerable amount of energy channeled into forming the relationship can be rechanneled and directed to accomplishing the work.

Keeping in mind that the outcome is usually determined more by how the negotiation process is conducted than by the terms of the agreement or the subject matter, you should study some of the wealth of published material on this subject and sharpen your own negotiating skills.

Motivational Skills

Motivating others is, of course, an important skill for any manager; much has been published on the subject, primarily by behavioral scientists. For the project manager, however, the specific need is for people to feel highly motivated toward accomplishing a specific mission as team participants, whether the project manager has the traditional boss-worker relationship or not.

In the matrix organization the question "What will working on this project do for me/us?" is always present, but not always articulated. To be an effective project manager, you need to find the best answer to this question, even if it is not directly asked. One way is to determine the objectives and needs of the organizations and key individuals involved with the project, and find ways to show the maximum possible congruency between those and the objectives and needs of the project.

Obviously, no single motivational factor, method, or combination of techniques will work in all situations. The three principal elements affecting motivation (the project manager, others, and the situation) are closely interrelated, unique to each project, and constantly changing. If you skillfully watch the interactions of these elements, and apply basic principles accordingly, it can pay tremendous dividends in terms of project success.

Leadership and Conflict Resolution Skills

Leadership and conflict resolution can be treated separately, but are combined here because they have much in common when applied to project management. These skills are applied jointly to various functions in managing a project.

Leadership in this sense means having the ability to exercise authority or influence. It is a quality or skill that defies quantification, yet we all know when it is there and when it is not. One way to examine leadership as a skill is provided by the Tannenbaum-Schmidt model of leadership styles (Figure 33-2) as

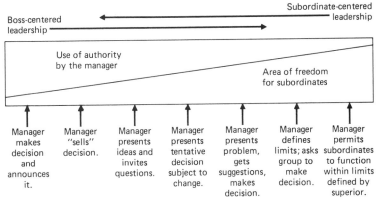

Fig. 33-2. Continuum of leadership behavior. Source: R. Tannenbaum and W. H. Schmidt, "How to Choose a Leadership Pattern (HBR Classic)," *Harvard Business Review*, May–June 1973, p. 164.

applicable in management. The key for the project manager is to develop the skills to function across the complete spectrum of these styles or behaviors, depending upon the combination of his characteristics, the follower's characteristics, and the particular situation. A single style of behavior is not enough.

The ability to lead or influence others is a key factor in resolving conflicts on projects. It is important to recognize that conflict is inevitable and not necessarily bad. Well-managed conflicts can contribute significantly to project success.

Thamhain and Wilemon list the following bases for a project manager's ability to influence others:

- Expertise
- Authority
- Work challenge
- Friendship
- Future work assignments

- Fund allocation
- Promotion
- Salary
- Penalty

Their studies related these bases to five ways of handling conflicts as applied in several potential conflict areas in project management: confronting or problem solving, compromising, smoothing, forcing, and withdrawing. After developing correlations with an assessment of overall project management effectiveness, they concluded in part:

- "It appears that project managers do not select a management style which minimizes the overall conflict experienced in managing their project."
- "The effectiveness of conflict-resolution modes in minimizing conflict is largely determined by the situation."
- "Project Managers generally have more flexibility in altering their conflict-resolution modes than in modifying their influence styles."
- "The less a project manager emphasizes organizationally derived influence bases—such as authority, salary and penalty—and the more he relies on work challenge and expertise, the higher he is rated in his ability to effectively resolve conflict and manage projects."[10]

In essence, the effective project manager strives for a "win-win" resolution of conflicts not just to make people feel good, but because over the long haul, it is most effective in getting the job done.

Knowledge of Subject Matter

One of the advantages of the matrix organization is that it provides reservoirs of expert technical knowledge and skills that can be applied to several projects simultaneously. Therefore you can look to the functional organizations for available expert knowledge. It is not necessary for the project manager to be an outstanding, in-depth expert in the technical subject the project addresses. It is important, however, that the project manager be sufficiently knowledgeable in the overall subject of the project, to be able to understand what the experts are doing or not doing and to properly integrate their work. You must be able to communicate in a knowledgeable way with the customer and the project participants. Since most projects are multidisciplinary, it is very helpful to have at least a general working knowledge of as many of these disciplines as possible and to be thoroughly conversant with the dominant one.

On very large, complex projects you may supplement your knowlege with that of key technical advisers on your own staff. However, keep in mind that studies show that expertise is most often the major basis for a project manager's ability to influence or exercise authority with respect to project participants. Particularly in a matrix organization, take care not to undermine this important authority base by giving the impression that you rely unduly on other experts.

A project manager needs general subject matter knowledge as opposed to knowledge in depth. In fact, one expert in project management observes that effective project managers are generalists rather than specialists.[11]

Skills in Structuring Work Logically

An overall sense of how the various elements of a project can be divided logically into smaller parcels or packages while maintaining project integrity is very helpful. The logic involved is important to assure completeness and enhance the proper flow and summarization of information on the status of the project.

The Work Breakdown Structure (WBS), an outline of the program structure, provides the common denominator for integrating schedules and cost data. The WBS also facilitates the assignment of responsibility to project participants in a clear, unambiguous manner, and is helpful in developing good clear statements of work across all levels of detail.

Fortunately, there are now specialists available to assit and advise in preparing Work Breakdown Structures, properly subdividing the work in meaningful ways. However, as with other areas of expertise, you need to be sufficiently knowledgeable in this area to work effectively with these specialists, be satisfied with the results, and have a good understanding of how the WBS will assist you.

Skills in Network and Schedule Techniques

Networks and schedules are important tools for the effective project manager. It is therefore necessary that you be sufficiently knowledgeable to understand them and the information they convey. With the advent of sophisticated systems such as CPM and PERT, specialists in these techniques have evolved and are very helpful.

Well-developed and carefully thought-out networks and schedules are invaluable in balancing resources, developing alternatives, and tracking project progress. When changes occur or problems develop, good networks and schedules can help you determine how other parts of the project and the overall project may be influenced or affected by the changes.

You do not have to be an expert in computer software or know fine points of detail about networks. However, you must be able to review them, relate them to the work on the project, and make your own judgments as to accuracy, adequacy, and impact analyses. You also need to understand the jargon and terminology, so you can interact effectively with network and schedule specialists.

Skills in Cost/Schedule Integration Techniques

The ability to combine cost and schedule data in a meaningful, accurate manner is necessary for good project control. However, neither schedule data nor cost data by themselves provide a satisfactory picture of where the project really stands. One of the purposes of a Work Breakdown Structure is to provide a common denominator for both schedules and costs for the various elements of the project effort. The "earned-value" system provides an excellent technique for combining schedule and cost data in a meaningful way. Basically, this method assigns an estimated dollar value to identifiable, scheduled increments of work.

With the earned value method, both schedule and cost status, as information for control purposes, are indicated by the combination of Budgeted Cost of Work Scheduled (BCWS), Budgeted Cost of Work Performed (BCWP or "earned value"), and Actual Cost of Work Performed (ACWP). An analysis of these three elements identifies quantified variances from plans of both schedule and cost at any desired level of the WBS. This therefore represents a valuable tool in project management; it is important for you to understand it, so you can use it effectively. The use of this method is now mandatory on most large projects for the U.S. government. It is a basic requirement in conforming to the government's Cost/Schedule Control System Criteria.[12] Further, it is such a helpful tool that its use in industrial projects is increasing, particularly for large projects.

Knowledge and Skills in Information Systems

An excellent way to find out what is going on and the true status of your project is to go out and look for yourself. This includes sitting down and talking with functional managers and other participants. Such first-hand observations are very valuable; however, large projects are frequently geographically dispersed and the project manager's time is limited and valuable. It is often necessary to rely heavily on the automated information system reports and formal reports from others on technical accomplishments and problems.

Computerized project management information systems are capable of producing very large quantities of information. Remember that quantity does not necessarily mean quality. Take care in addressing the issue of the quality of information and working out the right kind of outputs with the information specialists. Your goal is to receive the information you need in the most useful form, instead of being "buried in paper." Some project managers also prefer having a computer terminal in the office so they can call up the stored information on a selective basis, whenever they want it.

Information about the project is so vital to the project manager that investing in the effort to understand the system is essential. This does not mean that you must have the intimate knowledge of the programmer, just a good knowledge of the system's basic functions. The emphasis is on understanding enough principles to be able to use the system effectively. Intelligent information is essential in taking intelligent action.

Skills in Using Project Reviews

For the project manager, there are two kinds of project reviews: (1) those that are conducted primarily for your benefit by the project team, and (2) those that are conducted for the sponsor, outsiders, and/or your management. In both instances the effective project manager will use the project review as an opportunity for a two-way flow of information. Project reviews should be planned and conducted carefully. Give consideration both to what the other parties need to know, and to what you want from the review. This will help to maintain that important two-way flow. You will need to draw upon your communication and negotiating skills in this area also.

Most people try to present things in as positive and optimistic a manner as possible. However, ignored or concealed problems rarely go away or get better; even though it is usually painful to face them openly, delaying such confrontations will probably only increase the pain.

Project reviews are meetings, and there is a wealth of literature on how to conduct meetings. Use this to enhance your skills and upgrade the project reviews for your own benefit as well as that of others. For example, develop your skills in preparing and using visual aids. Visual presentations are powerful when well done, and can produce significant positive impacts.

Carefully consider who will present which material to further the development of your staff and at the same time provide the strongest presentations of material and information. Expert responses to questions, without being defensive or apologetic, help generate the needed feeling of trust and confidence in the team.

QA Conformity Skills

Nothing is more important than the quality of the finished product. At least a modest level of schedule and cost sins can be overlooked, forgiven, and forgotten, if the product of the project is right, sound, and of value. However, even organizations with highly sophisticated and extensive quality assurance programs have gotten into serious trouble on issues of quality.

The importance of quality in all phases of a project must not be underestimated. For example, in a design and construction project an effective QA and QC program in construction will result in a plant conforming to the design. If the lack of effective QA work in the design phase results in a poor or bad design, then the net result of the very good QA program in construction will be to assure an unsatisfactory facility.

To be skillful in the use of QA, be sure that in it fact focuses on quality and does not fall into mere lip service. The project manager is responsible for quality and the responsibility cannot be delegated. It can, however, be shared by being sure that others understand their responsibility for quality.

The right system of checks and balances must be in place, and a sense of the importance of quality must be instilled. Ideally, every participant should feel responsible for the quality of his or her work rather than adopt the view that quality is the responsibility of the QA organization (those other guys).

Quality levels or requirements should be established as realistically as possible, neither too low nor too high. All participants need to be aware of the QA requirements and guided accordingly. Few of us are perfect. The right kind of product is most likely to result when the appropriate checks and balances are followed. The intent is to have QA conformity for the sake of quality, not just as a procedural process.

APPLICATION OF SKILLS

With this brief discussion of the skill areas generally associated with project management in mind, let us examine their applications to some of the functions or activities involved in managing a project. The matrix of skills and functions shown in Figure 33-1 represents a composite picture of some of the complexities in project management. To examine these issues further, the following sections discuss each management function in turn. Remember that the process is highly iterative, and several functions often take place simultaneously.

Establishing Objectives

The cornerstone of all project planning is the statement of the project's objective. Ideally, this is a simple, concise statement of precisely what the project is to accomplish. One often quoted as a classic is: "This nation should commit itself to achieving the goal, before this decade is out, of landing a man on the moon and returning him safely to earth."[13]

In the world of project management, such statements are generally not arrived at easily or quickly, yet nothing is more critical to project success. Until the project manager, the customer, and the project team reach a state of common understanding (a "meeting of the minds") on the project's basic objectives, unification of work and direction is impossible.

Establishing the overall project objective can result from direct face-to-face discussion between project manager and customer. This process is then repeated, particularly in a matrix organization, between the project manager and key members of the project team for two purposes: to assure understanding of the overall objective, and to establish appropriate subsidiary objectives for the various supporting efforts required.

The process is basically one of communication about a particular subject area; therefore the process requires that the project manager be knowledgeable about the subject and be a skillful communicator in the sense of communicating for understanding. Since the objective must be realistically attainable, some balance is frequently required between desires, needs, and available resources, including capabilities. Reaching this balance equitably is usually the result of a negotiating process, hence such skills on the part of the project manager are very important.

In the matrix organization the program or project manager must participate with the functional managers in establishing the subsidiary or subproject objectives. This process begins to develop the project team, in addition to reaching a common agreement on what is to be accomplished. It is at this point that you as project manager can establish yourself as the team leader in the eyes of the team members.

Further, it is not uncommon to experience some level of conflict between the objectives of the project on the one hand, and on the other the objectives of the functional organization or even the personal objectives of the individuals involved. A complicating factor in resolving such conflicts is that they may be concealed, either consciously or unconsciously. Relying on the laws of chance to resolve unrevealed conflicts is very risky. Your skills in communication, leadership, and conflict resolution, along with your knowledge of the subject matter, are key factors in this objective-setting and conflict-resolving process. Although these skills are largely intuitive, difficult to define, and nearly impossible to meansure, they are critical not only in establishing the correct set of objectives, but in establishing the "tone" or "cultural ambience"[14] of the project team.

Defining the Work Statement

Once the project objective is established, defining what is to be accomplished, you can address what work must be done to reach that objective. Initially, the work is defined in general terms and becomes more specific as the plans are developed in more detail. The skills required are similar to those applied in the objective-setting process, with increased emphasis on the importance of subject matter knowledge and consideration of QA. Leadership and conflict resolution skills are not indicated in Figure 33-1 because the overall work statement may not require the depth of detail that involves the project team. If it does, then these skills are also needed.

Establishing the Work Breakdown Structure

Another key element in laying the foundation for a successful project is developing a good, workable WBS—one that structures the various product-oriented tasks or work packages in a logical, straightforward manner so as to reflect all the planned work and subsequently facilitate project control.

The effective project manager applies the same skills to this work as those used in establishing objectives, plus skills in "thinking through" the project component products in a logical, structured sense to assure that the work flow, in at least one dimension, is appropriate. A second dimension of work flow can be shown in the network diagrams.

One thing involved in this process is giving due consideration to project organization and structure independently of the functional organization structure, which usually gives rise to some conflict and requires leadership ability for resolution. This provides another opportunity to enhance team building, since team members probably have different views about how the component products or work should be structured.

Establishing Work Package Work Statements

Defining the work to be done in the individual work packages assigned to various functional organizations continues the process of establishing subprojects within the larger project and involves communicating needs, constraints, and approaches about the work to be done. Agreements must be negotiated consistent with project needs and resources available. Quality assurance requirements must be considered and the work statements must be technically sound, hence skills in those areas apply.

As project manager, you must provide the leadership in this basic team effort and must be prepared to assure that the inevitable conflicts are appropriately resolved. You must also be cognizant of the information system capabilities, because writing work package work statements will establish a major part of the basis for information that will later flow to you. The larger the project, the more important it becomes for information on individual work packages to be in a form that can be readily aggregated for summarized reporting.

Developing Networks and Schedules

Developing and reviewing good networks and/or schedules usually will require you to draw upon nearly all the skills and knowledge at your command, as indicated in Figure 33-1. In particular, you must be familiar with the network/ schedule technique being used, whether it is bar charting, PERT, CPM, or one of the variations, and whether it is computerized or not. There are very few supermen (or superwomen) around who are truly highly skilled in all these areas. You can assess your own strengths and weaknesses and take steps to augment those areas that you feel should be strengthened. For example, it can be very helpful to utilize support staff such as good cost/scheduling specialists with strong skills in setting up logical work flow diagrams, Work Breakdown Structures, and networks.

As project manager you could use a matrix such as the one shown in Figure 33-1 to assess your own strengths and weaknesses and their relative importance to the project being managed. Such an assessment can have a bearing on your actions to augment your own strengths appropriate to the needs of the particular project. It is especially important to apply good strengths in developing good networks and schedules, because these are critical in establishing key reference points for many future decisions.

Making Budget Estimates

The process of developing budget estimates is quite similar to the process of developing schedules. Both are more art than science and both rely heavily on accumulated experiential data and information.

Budget estimates must be integrated with the schedule to provide management with a sound basis for later making project control assessments and decisions. A common practice is to develop the schedule first, then estimate the budget, then maintain the relationship through the inevitable iterations and trade-offs to produce an integrated cost/schedule portrayal of the project plan.

If the system used conforms to the Cost/Schedule Control Systems Criteria developed by the U.S. Department of Defense and since adopted by other government agencies, then the earned-value technique will be utilized. This technique can provide the project manager with a very valuable method for measuring progress or performance on the project, when objective indicators of accomplishments are properly utilized as the basis for progress measurement. When other techniques are used, such as comparing percent complete to percent spent, it is frequently difficult to obtain meaningful results, and misleading status information can result.

Whatever technique is used, the essential requirement is that budget planning and schedule planning be integrated in order to establish a good basis for later tracking the project progress, so that necessary corrective action can be effective. Therefore it is very important that good skills in integrating budget estimates and schedules be applied in the planning phase.

Defining and Planning the Quality Assurance (QA) Program

Establishing an effective QA program involves much more than setting up a QA component in the organization. Whether the QA group is a part of the project organization, or is designated as a component in the functional part of the corporate structure or as a separate entity reporting to higher management, the success of the QA work on a given project depends heavily upon the skills and actions of the project manager.

Particularly on a high technology program, the project manager's leadership skills are of paramount importance in instilling a positive attitude toward QA. Engineers and scientists frequently resist, even resent, the application of QA to their work. A common attitude is "We need it for those other guys, but not for me." This results in conflicts and requires effective leadership, combined with subject-matter knowledge and a good understanding of QA plus communication and negotiating skills, to establish the appropriate attitudes and perceptions. If as project manager you find yourself caught between very zealous QA persons and a resisting technical team, you must bring about the right kind of balance and accord, if QA is to be effective. Effectiveness in this sense means not only maintaining QA procedural conformity but also producing the right level of quality in the project products.

Obtaining Identified Resources and Building the Team

In the matrix organization, project managers may have to compete with one another for resources (people, facilities, etc.) available in the functional organizations. Your skills in communicating, negotiating, and resolving conflicts are key factors in your degree of success in this area. You must clearly communicate your project's needs and you usually must negotiate with functional managers for the assignment of appropriate resources to meet these needs.

Once the right kinds of deals have been made with the functional managers and at least the best available lead people have been identified and assigned, the project team-building process is under way. Assembling the team is one thing, but getting it to function as a team is something else again. The importance of team building by the project manager is well documented. For example, Archibald states:

Whether a person is assigned to the project office or remains in a functional department or staff, all persons holding identifiable responsibilities for direct contributions to the project are considered to be members of the project team. Creating awareness of membership in the project team is a primary task of the project manager, and development of a good project team spirit has proven to be a powerful means for accomplishing difficult objectives under tight time schedules.[15]

Creating this "awareness of membership in the project team" and laying the groundwork for focusing its efforts requires you to utilize sveral skills. You must communicate well with the team members, and they must feel motivated to a team effort and be willing to follow the team leader. Assigning tasks and responsibilities appropriately requires the use of skills in scheduling, cost/ schedule integration, and the information system to integrate activities and foster the notion of a team effort. When team members know and understand their assignments (including what is expected of them and how their efforts contribute to the overall project), and when they feel a desire to participate as team members, with confidence in the leader and other participants, then energies are likely to be focused, and productive output can be astounding. A true team is a thing of beauty and can be awesome in its accomplishments.

Establishing the Review and Approval Process

Determining who has to reveiw what and who has to formally approve what is important throughout the life of the project. It is necessary that those who need to be involved are kept informed and that formal authorities are defined. Appropriate reviews of work in progress are important to assure quality. Further, timely approvals of key documents can be essential in maintaining project progress, especially in today's world where there seem to be so many different interests and agencies involved in any given project.

One important factor in reviews and approvals is related to the QA program. Performing the quality checks and documenting that they have been properly performed are key to assuring a quality project. Pay careful attention to the QA reviews and approvals, so they will be positive and supportive rather than punitive and disruptive during the project's life.

The skillful project manager will focus the review and approval process to help maintain communications and minimize extraneous interference as the project proceeds. You must strive to negotiate review and approval requirements to assure that the appropriate level of management control is maintained. However, you must also provide yourself and other project participants with freedom of action and decision commensurate with responsibilities. Establish in advance that periodic project reviews will be conducted to help build confidence regarding your management of the project.

Developing Make-or-Buy Decisions

Another key issue in planning a project is deciding what will be done in-house and what will be subcontracted. These decisions are usually based on balancing a number of factors: (1) capabilities, (2) resource availability, (3) cost, (4) schedule, (5) QA conformity, (6) level of direct management control, (7) contractual constraints, and (8) corporate policies or business strategies.

In many instances such decisions are easily reached and present few problems. These include the use of standard products such as basic construction materials, routine equipment items, and many specialty items produced by established industrial organizations. However, even in these areas reaching the make-or-buy decisions sometimes can present complex problems. Examples include projects that are large in scope, technologically advanced or difficult, complex in terms of range of disciplines or specialties required, or very sensitive from a political, institutional, and/or business standpoint.

With relative ease, you can make the make-or-buy decisions simply by going where you judge you can get the job done properly in the least time and for least cost. Although this is a good attitude to start with, actual cases are frequently more complicated. For one thing, the capability of an organization is not always easy to determine, and schedules and costs are estimated with some level of uncertainty and sometimes bias. Further, as indicated above, the decisions may be greatly influenced by other factors and constraints, thus requiring the project manager to negotiate make-or-buy issues with several interests: the customer, the funcitonal organizations, and corporate management. In addition to your negotiating skills, you will probably rely on your technical knowledge, your skills in balancing cost and schedule factors, and your QA cognizance in establishing the preferred choices.

Authorizing Work

Authorizing people and organizations to perform the actual work of the project is usually the finalization of negotiations that began in the planning phase. Depending upon how far the planning has been carried, this authorization may be only a formality. However, last-minute differences frequently arise in finalizing agreements.

For the project manager, communications, negotiation, motivation, and resolving conflicts are important in this step. Others must clearly understand what is to be done, and must resolve differences and feel motivated to perform. It is also important that the authorizations include requirements for periodic status information, so that cost and schedule performance can be integrated and properly assimilated in the information system for summarization and correlation with other work elements. Quality and QA requirements are finalized at the same time.

Measuring Progress and Status Reviews

Is work on schedule? Are costs in line with the estimated budget? Is the work itself acceptable from the standpoint of quality and subject matter? Answers to these questions provide a measure of progress. Even with the best of schedules and milestones, there are inevitably areas where management and/or subject-matter judgments are required in measuring progress. Constant vigilance is also required in overseeing quality as the work proceeds. The project manager must assure that reported progress is real progress.

This caution does not imply an element of dishonesty, but is needed because the perspective of the project manager is usually different from that of those doing the work. Such differences in perspective can produce amazingly different conlusions as to where the work really stands. Good communication skills can help in this area, particularly good questioning, paraphrasing, and listening for understanding.

Tracking Costs

The project manager is always concerned about costs as one of the major elements in determining project success. To be meaningful, cost data must be related to schedule progress and must be compared to plans as the best available reference point.

It is important for you to understand and have confidence in the cost information reported to you. The cost to date and the expenditure rate are significant factors in estimating the cost at completion, and such analyses need to begin very early in the project life cycle.

The terrible problem of the last 5% of the work costing 25 or 30% of the total can be avoided by careful attention to integrating cost and schedule progress information right from the start of the work. In the early phases of the project, the project manager has the maximum flexibility for corrective action. As the project progresses, the extent of this flexibility diminishes to near zero in the final stages. Doing work over and trying to accelerate schedules in the later phases of a project can be extremely costly.

Different cost-accounting information systems may use different ways to categorize cost items, such as direct, indirect, overheads, commitments, obligations, reserves, and even "actual." Therefore you need to understand the accounting information system being used well enough to use the information with confidence. In addition, you may have to be concerned about cash flow and the cost of money, and make sure that funds are available when needed. Even when you have the help of skilled finance specialists, it is important that you be familiar with budgeting processes, understand funding or cost categories, and be able to analyze for trend forecasting.

Implementing Changes

Although the following advice was originally targeted to managers of design and construction projects, it is applicable to any project. A good slogan for all project managers is:

> The project you estimate and propose is never the project you finally execute and complete. The largest single contributor to cost overruns, schedule delays, loss of confidence, and deteriorating relations between parties on a project is failure to recognize, accept, and act on the above principle. The responsibility for remembering this (at least every 3 hours) rests with the P.M. He also has the responsibility for acting intelligently on it; choosing techniques, procedures, actions, personnel, etc., that acknowledge and respond to the problem; and above all, monitoring his project for its effectiveness.[16]

Changes are inevitable; the key is to recognize when they are happening and manage them effectively. No matter how good a plan is, it cannot anticipate everything that will occur. Changes from plans will occur that vary widely in magnitude and consequences. It is suprising how often changes can slip in almost undetected. The project manager must be alert in identifying and consciously managing changes.

As indicated in Figure 33-1, effectively managing the implementation of changes in the aggregate will cause you to call upon all the skills at your command. Proposed changes and approved changes must be communicated, negotiated, and analyzed for their impact on project cost, schedule, and subject content. Changes frequently generate conflicts; they have strong impact upon the motivations of those involved or influenced; and they require close attention to QA conformity.

Maintaining QA Requirements

As indicated in the discussion of QA conformity skills, maintaining the right level of quality involves much more than mechanically following QA procedures. The people doing the work on the project ultimately have the most influence on the work quality. Therefore the project manager needs to draw upon people-influencing skills, knowledge of the subject matter, and QA conformity skills and knowledge to be effective in controlling quality to the appropriate level. Clearly communicating the quality requirements, negotiating to reach agreement with project participants on these understood requirements, and maintaining a healthy, motivated environment relative to quality work are some of the skills needed.

The process of others' reviewing and judging the quality of work done assures that there will be conflicts. Resolving these conflicts with the focus on the work, not the worker, is strongly recommended. The goal is to simultaneously maintain quality, productivity, and a highly motivated team.

Conducting Project Reviews

Well-conducted project reviews can be very beneficial in maintaining effective relations and communication with the customer, higher management, and the project team. As indicated in Figure 33-1, you need to utilize most of your skills to maximize the benefits from these occasions.

Conducting a project review is very similar to negotiating in at least one respect: how the review is conducted may be more important than the subject matter covered. Generating the right kind of understandings and perceptions of the project and its team are important objectives.

Another element in project success is the level of confidence and satisfaction the customer and higher management feel toward the project team and its work. Reviews present excellent opportunities to communicate with such people in a way that helps generate those kinds of good feelings. Project reviews can function in a two-way mode: in addition to having an interchange with higher-level management and the customer about things internal to the project, you can learn about developments and events external to your project that may affect it in some way. Skillfully conducted project reviews help to elicit this kind of external information.

Project reviews frequently generate the need for specific actions or some changing of project plans. Assignments should be realistic and be consistent with other scheduled activities and needs or negotiated changes. In other words, the P.M. must "keep his cool" and avoid overcommitting in the excitement of the moment.

Good reviews communicate crisply and concisely. Apply communication and leadership skills in planning and conducting these reviews by using such techniques as the following: ask experts in the subject matter to make presentations, keep the review on track, focus on the objectives and the exception items, and confront the problems.

Maintaining the Project Team

Defining how or when a collection of individuals becomes a team is very difficult. It is also very difficult to pinpoint when or why a team begins to "come unglued." The effective project manager, however, will be sensitive to this and utilize interpersonal skills to forestall such an event by maintaining that ephemeral thing called "team spirit" or "high morale."

Focusing attention on the "people" issues throughout the life of the project is obviously preferred to rebuilding a project team. Keeping up to date on key participants' personal goals and objectives, resolving conflicts skillfully, and maintaining a healthy work-challenging environment seem to be important. In short, the project manager should continue to give attention to those factors and activities discussed in the earlier section on "Building the Team," in order to maintain it as the project develops and changes.

Even in the best of families, relations can deteriorate. Here are some symptoms to watch for:

- The level of conflicts escalates.
- The nature of conflicts changes: i.e., persons rather than problems are attacked.
- Productivity falls off.
- Communication that should take place between participants increasingly goes through managers.
- Complaints increase and/or become more "person"-oriented rather than "situational."

If relations deteriorate, then rebuilding is in order, which requires increased attention to the "people issues." It may be necessary to renegotiate roles, to focus on resolving conflicts in a problem-solving mode (strive for win-win), or to bring in a third-party facilitator to help get to the roots of the problem.

Monitoring the Use of Resources

As used here, "resources" include funds, people, equipment, facilities, and suppliers. Good use of these resources leads to project progress. The effective project manager will use schedule progress/cost information to review whether the right kind and quantity of resources are being applied in the right places at the right times. Trend analysis is an important part of this process, in order to anticipate and plan for future needs. Continuous monitoring is important as a basis for timely adjustments as required.

Maintaining External Communications

As a minimum, external communications about the project must include the customer. Such communications take several forms, both formal and informal. Generally, informal communication is as important as providing formal reports and project reviews. As shown in Figure 33-1, a number of the skill and knowledge areas have a bearing on the effectiveness of these communications.

Particularly in large projects, the external community (such as multiple government bodies, both domestic and foreign) may include a wide scope of varying interests. The general public, numerous special interest groups, the news media, technical communities, associations, and business groups may also be actively interested in the project.

Communications in this broad arena must be geared to the ability (or inability) of the listener to understand and interpret (or misinterpret) what is presented or said. Strategic planning with the assistance of communications specialists can be very helpful in getting the desired message across. Remember that "how you say it" may be as important as "what you say." Maintaining credibility with widely varying interests and levels of expertise or sophistication can be very difficult, but is essential in minimizing opposition or, better yet, building support.

The area of external communications represents another major difference between managing a project and managing an internal function. When well done, external communication can reap major benefits for the project manager, the corporation, and the customer.

Managing Procurement

Except for obtaining the simplest kinds of standard materials, supplies, and equipment, the process of effectively obtaining needed outside resources can require the project manager to apply all the skills discussed here. Even with the help of highly skilled procurement specialists, project managers have had to agonize their way through late deliveries, poor quality, cost overruns, and even lawsuits involving suppliers.

In the whole procurement process, it is necessary to follow good business practices, abide by corporate policies, and if the project is for the U.S. government, follow the Federal Procurement Regulations. Other constraints may also apply. The project manager usually does not have to be intimately familiar with all such requirements and constraints, but a general knowledge is very helpful. You should at least understand today's numerous types of contracts and their applicability.

Since the project manager in this arena functions both as a manager and a businessman, you need the skills of both. Every skill and knowledge area discussed here may have to be applied, since a procurement action may really be a subproject. Skillfully done, the whole process can go smoothly, obtaining the right quality on time and within budget.

Completing the Project and Dissolving the Team

Completing the project and closing it out is easy to say and frequently very hard to do. Sometimes you literally have to work yourself—and others—out of a job. It is helpful to approach closing out as an opportunity to cement good relations with the customer and the various other participants in the project. Primary skills applied include communications, negotiation, and leadership.

Archibald recommends preparing a project termination plan covering at least these major items:

- Contract—Delivery and customer acceptance of products and/or service, and completion of all other contractual requirements

- Work authorization—Close out work orders and assure completion of all subcontracts
- Financial—Collection from customer and closing of project books
- Personnel—Reassignment or termination of people assigned to project office or project team
- Facilities—Close office and other facilities occupied by the project office and team
- Records—Deliver the project file and other records to the appropriate responsible manager.[17]

It can also be very beneficial to conduct a penetrating post-completion evaluation of the project. Each project is a learning experience and the value of such learning can be enhanced with a good review. In general, the points to be covered are:

- Determine the original and final objectives in terms of performance (end product), cost, and schedule.
- Determine whether these objectives were met.
- Where things went right, determine what factors contributed to the success.
- Where things went wrong, determine the basic causes.
- Develop policy and procedure changes to eliminate causes of missed objectives or other problems.
- Implement the changes.[18]

A well-handled project close-out may pay major dividends on the next or another project; it deserves to be handled with skill and care.

REFERENCES

1. D. I. Cleland and W. R. King, *Systems Analysis and Project Management,* (New York: McGraw-Hill, 1975), p. 195.
2. J. M. Batch and M. G. Patrick, *Project Managment within Battelle,* Battelle Memorial Institute internal report, Columbus, Ohio, July 1981.
3. *Ibid.*
4. R. D. Archibald, *Managing High-Technology Programs and Projects* (New York: John Wiley and Sons, 1976), p. 55.
5. G. R. Gemmill, and H. J. Thamhain, "The Effectiveness of Different Power Styles of Project Managers in Gaining Project Support," *IEEE Transactions on Engineering Management,* May 1973, pp. 21–28.
6. H. J. Thamhain, and D. L. Wilemon, "Leadership, Conflict, and Program Management Effectiveness," *Sloan Management Review,* Fall 1977, pp. 69–89.
7. L. C. Stuckenbruck, "The Ten Attributes of the Proficient Project Manager," *Proceedings of the Project Management Institute Symposium,* 1976, p. 40.
8. D. I. Cleland, and D. F. Kocaoglu, *Engineering Management* (New York: McGraw-Hill, 1981), pp. 46–51.

9. R. Coffin, *The Negotiator,* AMACOM, Division of American Management Association, 1973, p. viii.
10. Thamhain and Wilemon, "Leadership, Conflict, and Program Management Effectiveness," pp. 83–85.
11. Archibald, *Managing High-Technology Programs and Projects,* p. 55.
12. *Cost/Schedule Control Systems Criteria Joint Implementation Guide,* Departments of the Air Force, the Army, the Navy, and the Defense Logistics Agency. AFSCP/AFLCP 173-5, DARCOM-P 715-5, NAVMAT P5240, DLAH 8315.2, Oct. 1, 1980.
13. President John F. Kennedy, May 25, 1961.
14. Cleland and Kocaoglu, *Engineering Management,* p. 51.
15. Archibald, p. 116.
16. E. Jenett, "Guidelines for Successful Project Management," *Chemical Engineering,* July 9, 1973, p. 73.
17. Archibald, p. 235.
18. *Ibid.,* pp. 237–38.

34. Managing Conflicts in Matrix Organizations

George H. Labovitz*

Conflict is the mother's milk of matrix organizations. It is as natural as breathing, as predictable as taxes, and as healthy as exercise. Conflict occurs in matrix organizations for all the reasons it occurs in nonmatrix organizations, and for other reasons as well. So essential is conflict to the proper functioning of a matrix organization, in fact, that the management of conflict is often a key factor in the success or failure of such an organization.

In this chapter we will look at why matrix organizations create conflict, and ways in which organizations can prepare themselves and their people to better cope with it. We will draw on what surprisingly little literature there is on how to deal with conflict in matrix organizations, and present some general theories and then some specific guidelines based on our research and experience as consultants to matrix organizations and in training managers to operate within them.

THE NEED FOR CONFLICT

First, though, let us examine the common notion that conflict is a dirty word to be avoided at all costs. Because so much of our education has been influenced by a belief in harmony and good will (e.g., "love thy neighbor as thyself"), most of us have a built-in bias against accepting conflict as a necessary and perhaps even desirable fact of life. Abraham Maslow noted this bias when he described our society as one characterized by a general "fear of conflict, of disagreement, of hostility, antagonism, enmity. There is much stress on getting along with other people even if you don't like them."[1]

Today it is recognized that conflict is an inevitable part of organizational life, and that any organization in which it doesn't occur, or occurs only rarely, is one in which people either don't feel free to think for themselves or don't feel

*George H. Labovitz, Ph.D., is a Professor of Organizational Behavior and Management at the Boston University School of Management. An active consultant and researcher in organizational development and analysis as well as conflict management, he is the President of Organizational Dynamics, Inc., Burlington, Massachusetts.

free to express their thoughts and ideas. Either way, such an organization will find it hard to adapt to new challenges. As Stephen Robbins has noted, conflict leads to change, change leads to adaptation, and adaptation leads to survival.[2]

All change represents a challenge to established patterns of operating or behavior. Matrix management is an attempt to institutionalize change within an established organization. Whether its specific purpose is to increase flexibility or responsiveness, strengthen coordination and team-work, overcome parochialism, or a combination of all of these, matrix management brings a new dimension to bear on decision making, challenges established ways of thinking and operating, and subjects conventional wisdom to the test of open scrutiny and reasoned argument. Thus the critical need in matrix organizations is not to avoid conflict, but to recognize its potential for productivity and manage it in such a way as to squeeze maximum gain from that potential.

CONFLICT IN MATRIX ORGANIZATIONS

To understand the special role of conflict in matrix organizations, let us first review the sources of conflict in organizations generally. As management theorists Paul Lawrence and Jay Lorsch have pointed out,[3] there are three main causes of organizational conflict.

Structure. Whenever people compete for resources, pursue different goals, or operate in different time frames, conflict is likely to occur. The more that people perform distinct and specialized functions within an organization, the greater the tendency to parochialism and "tunnel vision." If differentiation is the hallmark of complex organizations, then conflict is an inevitable byproduct.

Communication. Each of us speaks the language of our background and discipline. As we adopt the protective coloration of our functions, we become reluctant to share information, to let others know what we know (after all, "knowledge is power"). Our communications tend to be self-serving, our views of other functions stereotyped. Accountants are perceived as penny pinchers, planners as ivory-tower intellectuals, salesmen as expense-happy boozers, etc.

Interpersonal relationships and behavior patterns. Differences in management style, personality traits, personal ambition, or policy issues all play a part in producing organizational conflicts. If I want to market Product X and my boss wants to market Product Y, we are in conflict, even if we represent the same function and enjoy good communications.

In matrix organizations these three universal sources of organizational conflict are overlaid by three special conflict sources.

Dual command structure. By setting up a dual chain of command (i.e., a matrix team member reports to both a functional boss and a matrix boss, as seen in Figure 34-1), matrix relies on a belief in the productive force of contending interests. The system can only flourish when every interest participating in the matrix believes that its viewpoint can prevail if he or she makes a

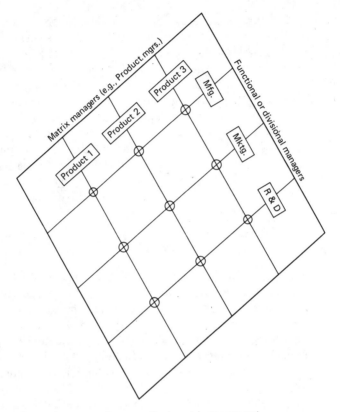

0 = matrix team members, reporting to matrix manager and
functional or divisional manager.

Fig. 34-1. A typical matrix system.

good enough case for it, and that no one has the inside track. Maintaining this
system of checks and balances is the central challenge of managing conflict in
a matrix system.

Ambiguous authority. The system requires one set of managers (functional,
divisional, etc.) to share some of their power with a new set of managers
(matrix), but it leaves the precise level of authority of the two groups ambiguous. The two groups over time will work out their own relationships, based on
the distinct kinds of issues they face, the competence level of the parties, and
their skill in working out conflicts.

Cultural differences. Matrix management is committed to values sometimes
at variance with established organizational culture. A matrix organization
tends to be more horizontal than vertical in structure, collegial rather than
hierarchical in relationships, more mindful of teamwork and individual com-

petence than titles and status, and more intent on cutting red tape than following established procedures. To the extent that these values represent a departure from those prevailing, matrix creates a language, culture, and style all its own.

Given these distinct challenges, it is clear that a power struggle in a matrix organization is, as Stanley Davis and Paul Lawrence have observed, "qualitatively different from that in a traditionally structured hierarchy because in the latter, it is clearly illegitimate." The matrix is designed to encourage conflict, in the belief that the organizational welfare will be advanced by the clash of contending viewpoints and interests. Thus, note Davis and Lawrence, in matrix "the boundaries of authority and responsibility overlap, prompting people to maximize their own advantage."[4]

CONFRONTATION, THE WIN-WIN SOLUTION

As noted by John E. Jones and J. William Pfeiffer, there are five basic ways to deal with organizational conflict:

- Denial or withdrawal (i.e., pretend it doesn't exist, and maybe it will go away)
- Suppression or smoothing over (pretend it's no big deal, and maybe it will fizzle)
- Forcing or power (impose a solution, and don't worry about bruised feelings)
- Compromise (give everyone something, but not what they're asking for)
- Confrontation or integration (try to find a way to meet everyone's goals or needs)[5]

While each of these options is appropriate in some circumstances, confrontation has been found to be the most effective method of dealing with most conflicts. For example, examining the use of three conflict management methods in six organizations, Lawrence and Lorsch concluded that the two highest-performing organizations used confrontation to a greater degree than the other four, and the next two organizations in order of performance used confrontation more often than the lowest two. (The other two methods studied were forcing and smoothing over.)[6]

Similarly, another study asked 74 managers to describe how they and their immediate superiors dealt with conflict. Of the five types of conflict-management techniques noted, participants reported the best results from confrontation, followed in order by smoothing, compromise, forcing, and denial/withdrawal.[7]

We believe that confrontation is especially appropriate for managing most conflicts in matrix organizations, for two reasons. First, confrontation requires facing conflicts squarely rather than deflecting them. In this regard, confrontation differs from two of the four other conflict-management techniques, denial and suppression. If conflict is, as we have seen, a way of life in matrix organizations, it must be managed in a way that surfaces and deals forthrightly with the issues standing in the way of resolution, or the raison d'etre for the matrix is defeated.

Second, confrontation can embody the principle of win-win, which is essential to any negotiation in which the parties need to retain a co-operative relationship over the long run. In this respect, confrontation differs from the other two conflict-resolution techniques, forcing and compromise, which are at best win-lose and at worst lose-lose solutions.

In contrast to win-lose and lose-lose strategies, win-win strategies are typically the result of participative management techniques in which all parties to a conflict collaborate in establishing common or superordinate goals, and also engage in joint problem-solving efforts to determine how those goals can best be met. As Allan Filley has pointed out, managers operating in a win-win mode are establishing a basis for dialogue in which each participant agrees: "I want a solution which achieves your goals and my goals and is acceptable to both of us. . . . I would like to find a solution in which you get what you want and I get what I want—that is, neither your solution nor my solution but a strategy which satisfies both of us."[8]

Handling conflicts with a view to confronting the differences while seeking ways in which all parties can win is no easy challenge, of course. Note, however, that we are not saying all parties have to see themselves as winning every battle. Rather, all parties must feel that the process of decision making is fair, that they'll have another turn at bat, and that each decision has been, and will be, made on the merits (i.e., that no one's thumb is tipping the scales).

The importance of win-win in matrix conflicts stems from the fact that the matrix represents a delicate balance of forces in which no party can afford to—in the phrase of Davis and Lawrence—achieve "total victory" over another. "To win power absolutely is to lose it ultimately," they point out, because absolute victory "finishes the duality of command and destroys the matrix."[9]

WAYS OF DEALING WITH CONFLICT

In our own work with companies, we draw a clear distinction between those actions that can be taken in advance of conflict and those that are helpful in dealing with conflict when it occurs. Here are some steps to follow in setting up a matrix system to help raise the likelihood that conflicts will be effectively managed:

1. Chose matrix managers with a view of how much talent they show for managing conflicts. Among the qualities to look for are communication skills,

tact, persuasive powers, flexibility, resourcefulness, commitment to teamwork and participative management, and creative problem-solving ability.

2. Make clear, to both the matrix managers and the functional or divisional managers they interface with, what responsibilities they do and don't share. Make clear, too, what kinds of issues the boss

- wants to be consulted about before a decision is made;
- wants to be informed about after a decision is made;
- doesn't want to be involved in.

3. Provide a training/team-building experience to matrix managers, functional managers, and matrix team members in how to operate in a matrix system. This training/team building should include the following elements:

- Background on the role and rationale, the problems and pitfalls of matrix management, preferably with some case histories to generate discussion.
- Joint identification and analysis of factors helping and factors hindering a matrix system in the organization, followed by a collaborative problem solving and action planning to determine how to strengthen the helping factors and overcome or minimize the hindering factors.
- Clarification—in advance of any specific conflicts—of which function should play what role in the basic kinds of decision the managers are likely to face in the course of their new relationships. For example, which functions should be involved in recommending pricing changes? In approving pricing decisions? In implementing pricing changes? If such questions are not worked out in advance, they can become a major—and unnecessary—source of unproductive conflict.
- Instruction and practice in techniques of managing conflict, using exercises and case histories involving role plays, and including group feedback on how effectively each conflict was handled and how its handling could be improved.

4. Tailor the performance appraisal system to support matrix management in general and managing conflicts in particular. A performance appraisal system that doesn't reward the conduct the organization wants to encourage won't obtain that conduct for very long.

Actions for Dealing with Actual Conflict

In dealing with actual on-the-job conflicts, the role of facilitator normally falls to the matrix manager. Unlike an impartial mediator, the manager is not strictly neutral. Chosen for his or her ability to find ways of helping competing interest groups recognize a larger interest than their own, the manager needs to gain and retain the confidence of all sides in a conflict, to fashion consensus out of controversy, agreement out of acrimony.

To cite an extreme example: to optimize efficiency, a manufacturing manager in a blue jeans factory may want to produce one size and one color. To please customers, a marketing manager in the same company may want to produce all sizes and all colors. The matrix manager's role is to help the rival managers agree on what sizes and colors best serve the company's interests.

A format for managing conflict that some clients have found helpful is as follows:

1. Get the issues on all sides on the table.
2. Get agreement on a process to follow in arriving at a solution.
3. Seek agreement on a common or superordinate goal (in the above example, this might be "the most profitable mix of sizes and colors").
4. Take the least controversial issue and facilitate a dialogue by the parties, asking questions to clarify thinking, working for ways to widen areas of agreement within an issue and to narrow areas of disagreement.
5. Take the next issue and do likewise.
6. Summarize what has and what has not been agreed to, and create an action plan spelling out what each party will do to implement agreements.
7. Arrange for another meeting, if appropriate.

As Richard Walton has observed, the essence of successful mediation lies in making the warring parties recognize they are dependent on each other for success.[10] In a sense, it can be argued that matrix management is a lot like marriage. To be successful, the parties have to be aware of their mutual interdependence, of the inevitability of conflict, and of the necessity to find win-win solutions to those conflicts. Effective conflict management is an essential ingredient in making a "matrix marriage" work.

NOTES

1. Abraham Maslow, *Eupsychian Movement* (Homewood, Ill.: Richard D. Irwin, Inc., 1965), p. 185. Reprinted in Stephen P. Robbins, *Managing Organizational Conflict* (Englewood Cliffs, N.J.: Prentice-Hall, 1974), pp. 17–18.
2. Robbins, *Managing Organizational Conflict*, p. 20.
3. Paul Lawrence and Jay Lorsch, *Organization and Environment* (Homewood, Ill.: Richard D. Irwin, Inc., 1969), pp. 8–11.
4. Stanley M. Davis and Paul Lawrence, "Problems of Matrix Organizations," *Harvard Business Review*, May–June 1978, p. 134.
5. Paraphrased from: John E. Jones and J. William Pfeiffer, Eds., *The 1977 Annual Handbook for Group Facilitators*, San Diego, CA: University Associates, Inc., 1977. Used with permission.
6. Lawrence and Lorsch, p. 152.

7. Allan C. Filley, *Interpersonal Conflict Resolution* (Glenview, Ill.: Scott, Foresman, 1965), p. 31.
8. *Ibid.*, pp. 27, 30.
9. Davis and Lawrence, p. 134. ·
10. Richard Walton, *Interpersonal Peacemaking: Confrontations and Third Party Consultations* (Reading, Mass.: Addison-Wesley, 1969), pp. 117–27.

35. The RIM Process in Participative Management

Dundar F. Kocaoglu and David I. Cleland*

This chapter describes a participative approach to the development of organizational roles and interactions. It is a process through which the responsibility interfaces are defined, negotiated, and resolved. The participants in the process are the individuals who ultimately carry on those responsibilities in the organization. The output is the Responsibility Interface Matrix or RIM (also called Grid, Linear Responsibility Chart or LRC, etc.), a pictorial representation of "who does what for whom" in the conduct of business throughout the organization. It is a blueprint of the responsibility flows for the completion of organizational activities and for the coordination of people in their organizational roles. An analogy will illustrate this point. In a classical concert, a transformation takes place as soon as the conductor moves his baton. It is a transformation that blends the scores of seemingly independent instruments into a unified body of musical output. The audience is delighted with the delicate balance, precision, and accuracy of the performance. When the program is over, the orchestra receives an ovation. The audience enjoys the synergistic effect of the interactions among the instruments, and the precision with which the individual roles are integrated and performed throughout the concert. What the audience sees is the climax of a long and deliberate process that has molded the individual roles of the orchestra into a unified body in the final performance. Regardless of how expert each performer is, the success of the concert depends on one thing: how well the individual experts play supportive roles for the rest of the orchestra. If they cannot play those roles, it does not matter how good each one is, or how capable the conductor is. The baton will not perform the transformation, and the performance will prove a disaster.

*Dundar F. Kocaoglu is the Director of the Engineering Management Program at the University of Pittsburgh. He is the editor of the McGraw-Hill Series on Engineering Management, and coauthor of a book and author of several articles on the same subject.

Dr. Cleland is a Professor of Engineering Management in the Industrial Engineering Department at the University of Pittsburgh. He is author or coauthor of ten books and has published numerous articles in leading national and internationally distributed technological, business management, and educational periodicals. Dr. Cleland has had extensive experience in management consultation, lecturing, seminars, and research.

A manager is just like a conductor, having a responsibility to present to the "customers" the output of the performance of the entire team, controlled by the movement of his or her "baton." The baton symbolizes the means of communication available. The movements of the baton are determined by management decisions. As in the case of the orchestra, the manager's success depends on one factor: how well the individual roles of the organizational unit interact with each other, and how well they support each other as directed by the manager's baton.

Each manager has a specific responsibility to carry out elements of work that lead to the creation of something of value for the organization. The something of value may be a product, service, or some combination thereof. A production worker assembles parts on the production line leading to a subassembly or a total product. A programmer develops a software package. An accountant assembles financial records into a balance sheet. The manager integrates the various functions of the organization to satisfy his or her obligation to higher management levels.

THE MOGSA HIERARCHY

Throughout all work—whether performed by workers, professionals, or managers, several common characteristics are found:

- Specific *actions* are taken to perform the discrete tasks in an organization.
- Actions are a reflection of *strategies*. Each strategy is supported to varying degrees by one or more actions.
- Strategies are the building blocks of organizational *goals*.
- Goals support broad *objectives*.
- Actions, strategies, goals, and objectives lead to the organizational *mission*.

Organizational mission is the reason for an organization's existence. It defines the business of the organization. Objectives are broad statements specifying the destinations that should be sought in achieving the mission. Goals are the specific targets or milestones, usually expressed in measurable terms. Strategies are the means through which the goals are satisfied. Actions are the specific activities that, when taken together, form strategies. These components form a hierarchy, such that the elements at each level support one or more elements at the next higher level, going from the actions all the way up to the mission as depicted in Figure 35-1.

The MOGSA (Mission-Objectives-Goals-Strategies-Actions) hierarchy is a general model applicable to multinational corporations, companies within corporations, business units, divisions, departments, product lines, programs, even projects. Each level of the hierarchy has a different type of work requirement.

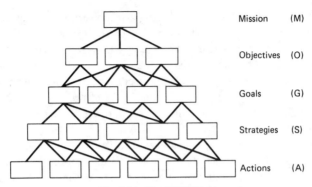

Fig. 35-1. The MOGSA hierarchy.

The definition of the work requirements depends upon the type of organization to which the model is applied. For example, in a company in systems business the hierarchical work requirements can be summarized as follows:

Mission level: Design, development, and installation of industrial process control systems for the domestic U.S. market.

Objective level: Become a leading systems supplier.
Develop maintenance and service capability.
Capture the OEM market.

Goal level: Reduce costs by 10% within three years.
Reduce lead time to three weeks by the end of 1982.
Establish a viable service organization by 1984.
Increase market share to 20% in the eastern U.S. markets by 1984.

Strategy level: Explore and develop new technologies.
Increase quality assurance efforts.
Emphasize service of all current and future products.
Improve inventory control capability.
Improve productivity.

Action level: Implement cost-control procedures.
Design new products in each product line.
Acquire new technology through licensing.
Initiate a quality-control program.
Standardize existing products.
Hire and train service technicians.
Establish an R & D department.

Work Breakdown Structure

The MOGSA model can also be applied to projects when a project is broken down into hierarchical work elements following the MOGSA pattern. The result is the Work Breakdown Structure (WBS) for that project. The WBS is a mission-oriented hierarchy composed of discrete work elements at each level. For example, the WBS for the construction of a sports complex can be developed as shown in Figure 35-2.

Work Packages

The work elements at the hierarchical levels of the Work Breakdown Structure are called the *work packages*. They are used to identify and control work flows in the organization, and have the following characteristics:

1. A work package represents a discrete unit of work at the appropriate level of the organization where work is assigned.
2. Each work package is clearly distinguished from all other work packages.
3. The primary responsibility of completing the work package on schedule and within budget can always be assigned to an organizational unit, and never to more than one unit.
4. A work package can be integrated with other work packages at the same level of the WBS to support the work packages at a higher level of the hierarchy.

Work packages are level-dependent, becoming increasingly more general at each higher level, and increasingly more specific at each lower level. In the construction project, "sports arena" is a work package near the top of the hierarchy, while "electrical work" is at the bottom. In both cases, the primary responsibility can be assigned to an individual or a group in the organization, and the supporting roles of the other organizational units can be identified. Primary responsibility for the "sports arena" rests at the project manager level. In order to deliver that work package on time and within budget, the project manager has to integrate many lower-level work packages through his interactions with a variety of organizational units, including the electrical subcontractor. That subcontractor in turn has the primary responsibility for the work package "electrical work," and has to deliver it on time, within budget and according to specifications, if the entire project is to remain under control. In carrying out this responsibility, the electrical subcontractor interacts with the concrete, masonry, and mechanical subcontractors, the carpenters, and the steel erectors, coordinating his or her own work with the work done by all those units in order to place the wiring, the conduits, and the switches at the proper location at the proper time. The tasks are very specific and very detailed at that level.

On the other hand, the project manager responsible for the "sports arena" work package has a much more general perspective when looking at the project from the level of his or her work package. The manager has to have the showers, basketball court, indoor tennis courts, and indoor swimming pool com-

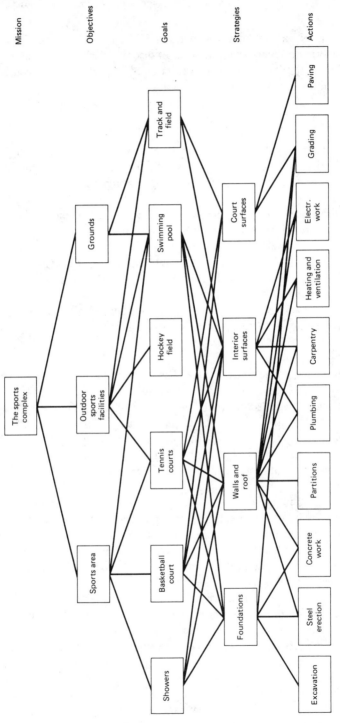

Mission   Objectives   Goals   Strategies   Actions

Fig. 35-2. The Work Breakdown Structure for construction of a sports complex.

pleted and delivered to him on time, within budget and according to specifications, so as to deliver the work package without violating the time, cost, and performance requirements. Under normal conditions, he or she does not become directly involved with whether or not the electrical conduits have been placed properly. The project manager controls the detailed work of the electrical and other subcontractors by making sure that the showers, indoor courts, and indoor pool are completed as specified. If those components of the manager's own work package are indeed on time, within budget and within the technical specifications, then clearly all the supporting work has been completed properly. If, however, some of the lower-level work packages fail to meet their criteria, the delays are carried all the way up to the top levels of the hierarchy. In that case, the manager responsible for the sports arena sees the effects of those delays in his or her own work package and takes corrective actions that are transmitted through the hierarchy down to the lowest levels where the specific changes are made.

THE RESPONSIBILITY INTERFACE MATRIX (RIM)

The organizational positions and the responsibilities assigned to them in carrying out the work package requirements constitute the basis for the Responsibility Interface Matrix. It is developed by specifically identifying responsibilities for each of the work packages. The responsibilities are defined at the work package/organizational position interfaces, using symbols or letters to depict the work/position relationships (WPRs). Thus RIM has three key elements:

1. Organizational positions taken from the traditional organizational chart and used as the column headings in the RIM.
2. Work packages defined at all levels of the MOGSA hierarchy and used as the rows in the RIM.
3. WPRs specified at the interface points and located at the intersections of the organizational positions and the work packages. These are the points where activities are carried out to make the organization function effectively. The major WPRs are as follows.

P—Primary Responsibility. Prime authority, responsibility, and accountability for the accomplishment of the work package rests in this position. It implies leadership for the work package to see that the work is accomplished on time, within budget and meeting the performance standards. The person who has primary responsibility is "it" for doing what needs to be done. Each work package has one and only one P associated with it. No work can be conducted without assigning the primary responsibility to an organizational position, and the primary responsibility for no work package can be assigned to more than one position at a time. This follows the definition of a work package as a discrete element of work at a given organizational level, for which the primary responsibility is uniquely identified.

A—Approval. This organizational position approves the work package. *A* is used when the authentication of a work package is required, as in the case of the approval of a capital investment project by the chief executive officer of a corporation.

R—Review. The individual who has an *R* reviews the output of the work package to decide if the predetermined quality and/or quantity standards have been met. For example, a financial executive reviews the financial credibility of the bid price on a proposal submitted to the customer, and a quality-control manager reviews the outputs of the production line. Typically, *A* includes *R* unless otherwise specified, since if the review of a work package is not explicitly assigned to an organizational unit, the approval authority carries with it the review function.

N—Notification. The individual occupying the position with an *N* is notified of the output of the work package. As a result of this notification, he or she then makes a judgment as to whether or not any action should be taken. For example, when the sales department bids on a request for proposal, the contract administrator is notified of the decision and takes the necessary contracts-related legal precautions.

O—Output. This position receives the output of the work package and integrates it into the work being accomplished. For example, the contract administrator receives a copy of the engineering change orders, so that the effects of changes on the terms and conditions of a project contract can be determined. There is a subtle difference between *N* and *O*. *N* (Notification can be a telephone call, a memo, or a short verbal notice, while *O* (Output) implies that the final product or service resulting from the work package is delivered to the organizational position thus identified.

I—Input. This position provides input to the work package. Input is generally required from various positions in the organization. For example, a bid/ no bid decision on a contract cannot be made by a company, unless inputs are received from the manufacturing manager, financial manager, contract administrator, marketing manager, and the manager of the profit center division that will eventually be responsible for that contract.

W—Work Is Done. The actual labor of the work package is accomplished by the position with a W. This legend is used when the work is done by a position other than where the primary responsibility resides. For example, the primary responsibility for the preparation of a project proposal normally resides with the sales manager, but the preparation of the document itself is accomplished by the publications manager in the organization.

I*—Initiation. The work package is initiated by the position marked with *I**. For example, new product development is the responsibility of the R & D manager, but the process is generally initiated with a request from either the profit center manager or the marketing manager. For this work package the R & D manager has a *P*, the profit center manager or the marketing manager an *I**.

If *W, A, R,* and *I** are not separately identified, then *P* is assumed to include them.

Figure 35-3 portrays a sample RIM in an actual industrial organization. The management council shown as one of the organizational positions on top of Figure 35-3 is included in the RIM as the "plural executive" composed of the key managers of the organization, dealing with key operational and strategic issues.

Each column of the RIM describes how a certain organizational position operates in the company. It summarizes the line, staff, and support roles and the information dissemination characteristics of that position. For example, in the sample RIM the R & D manager has primary responsibility for the development of new products, but plays only a support role for the other work packages. The R & D manager provides input to the financial manager for annual budget and another input to the manufacturing manager for master schedules, and in turn receives the corporate objectives from the management council and the final budget from the financial manager. The R & D manager is also notified by the manufacturing manager when the master schedule is completed.

In a similar way the rows contain information about how certain jobs are accomplished, and who does what for whom in carrying out the work package requirements. For example, bid strategies in the sample RIM are developed by the marketing manager with inputs from the contract administrator, the manufacturing manager, and the profit center division manager. After the strategies are developed, each of those managers receives a document from the marketing manager, describing the policies on how the bids are to be prepared and submitted.

Taken as a whole, RIM is a blueprint of the activity and information flows that take place in the organizational interfaces in a company. The authority/responsibility patterns and the organizational interdependencies can be read directly from the chart. Once the RIM is developed, it can be sorted for each organizational position, first with all the *P*s, then with *I*s, *O*s, etc. Looking at the sorted work packages related to his or her organizational unit, the manager immediately sees a listing of the activities for which the manager has direct responsibility and those for which he or she supports the other units in the organization. A study of the WPRs (work/position relationships) vertically gives the manager's position description: basically, *what* the manager does for each work package. Looking at the WPRs horizontally, the manager sees *whom* he or she interacts with and *how*. RIM identifies the manager's contact points in the organization, and the nature of the contacts to be maintained.

RIM is a valuable tool as a succinct description of organizational interfaces. It conveys more information than several pages of job descriptions and policy

WORK PACKAGES	CONTRACT ADMIN.	MFG. MANAGER	FINANCIAL MANAGER	PROFIT CTR. DIVISION MANAGER	R & D MANAGER	MARKETING MANAGER	MGMT. COUNCIL	COMMENTS
Setting Corporate Objectives	I, O	I, O	I, O	I, O	I, O	I, O	P	
Developing Corp. Maint. Agreements	P	N	I, N	I, A		I, N		
Negotiating Customer Contracts	I, R	I, N	R	P			A	
New Product Development	I, O	I, O	I	I*, O	P	I*, R		
Developing Bid Strategies	I, O	I,O		I, O		P		
Preparing the Annual Budget	I, O	I, O	P	I, O	I, O	I, O	A	
Developing Master Sched. for Opers.	N	P	N	I, N	I, N			
Establishing Standard Costs		W	P	I, O				

Legend: (P) *Primary Responsibility*—The prime authority and responsibility for accomplishment of work package.
(R) *Review*—Reviews output of work package.
(N) *Notification*—Is notified of output of work package.
(A) *Approval*—Approves work package.
(O) *Output*—Receives output of work package.
(I*) *Initiation*—Initiates work package.
(I) *Input*—Provides input to work package.
(W) *Work Is Done*—Accomplishes actual labor of work package.

P includes W
P includes A
P includes I*
A includes R unless otherwise specified.

Fig. 35-3. A sample Responsibility Interface Matrix (RIM).

documents, by delineating the authority/responsibility relationships and specifying the accountability of each organizational position. However, by far the most important aspect of RIM is the process through which the people in the organization prepare it. If RIM is developed in an autocratic fashion, it becomes simply a document portraying the organizational relationships. On the other hand, if it is prepared through a participative process, the final output becomes secondary to the impacts of the process itself. The open communications, broad discussions, resolution of conflicts, and achievement of consensus through participation provide a solid basis for organizational development and managerial harmony. By the time RIM is developed this way, the organization goes through such an "education" that the chart becomes secondary. Organizational units fully subscribe to the ideas behind RIM, and protect its integrity because of their vested interests amplified by the participative process.

THE RIM PROCESS

The RIM process can be introduced into an organization in four interrelated stages:

1. MOGSA development
2. Role and work package definitions
3. RIM meetings
4. RIM implementation

Stage 1 is the introduction of the RIM process into the company. At this stage the organizational units get together to lay a common ground and to develop a conceptual framework for the critical issues. Key managers define the mission, objectives, goals, strategies, and actions of the company. They discuss each level, modify the definitions as necessary, and argue about the hierarchical elements until they reach a consensus. This process may take two or three iterations until the MOGSA hierarchy is developed and agreed upon by all the key managers involved. Once developed, MOGSA becomes the basic framework for the Work Breakdown Structure in the entire organization.

In *Stage 2* each key manager defines the work packages under his or her responsibility and identifies the responsibility interfaces with other organizational units which contribute to those work packages. Each manager does this individually, or with discussions within his or her unit, but independently of the managers of other units. It is important that the managers should define their work packages at a level of the MOGSA hierarchy that is relevant and meaningful for their roles in the organization. Although no rigid rules can be stipulated about the levels, the relevant work packages for the vice-presidents are usually at the objective and goal levels; the division managers typically find their work packages at the strategy level, and the department managers at the action level.

Stage 3 brings the key managers together again. Having first established a common baseline as a group, then defined their work packages and interactions individually, they are ready to discuss the organizational interfaces and resolve any differences that may exist. The meeting preferably takes place at an off-site location, away from the daily routine and potential interruptions. Each manager presents his or her work packages to the others. The responsibility interfaces are discussed by the whole group. Some of them are redefined, some are deleted, and some new ones are added as necessary. As the presentations are made, the overlaps and inconsistencies become apparent and are corrected. Issues are discussed until agreement is reached for all the work packages and the responsibility interfaces. Once that agreement is reached, the RIM meeting completes its mission. It may be necessary to have more than one meeting to achieve this result. The visible output of the RIM meetings is the RIM chart itself. But more important, they provide a forum for the participative management process in the organization. Because the work packages and the responsibility interfaces are discussed openly, key managers fully accept and subscribe to the roles of the organizational units and the interactions among them. Usually, there is a high level of enthusiasm about RIM at this point, partly because of a feeling of achievement, and partly because of the pressure of the concentrated effort during the meetings. However, if RIM is set aside after the managers return to their daily activities, the enthusiasm can rapidly die and the whole process can prove a futile exercise. To prevent this from happening, it is important that Stage 4 be approached very carefully, with full support and commitment of the top management.

In *Stage 4* several tasks are performed simultaneously. The key managers who attended the RIM meetings hold regular meetings within their organizational units to explain the RIM and to obtain the commitment of their subordinates to it. After the entire organization becomes familiar with the RIM, the organizational units start developing additional RIM's at a more detailed level, reflecting on the work package responsibilities within each unit. At that level the interactions are inside the unit, very rarely going beyond the unit manager. While the lower level RIMs are developed, it is not uncommon for the key managers to redefine some of the original work packages and to develop new responsibility interfaces through group discussions among themselves as needed. As these discussions become a way of life in the company, the participative management style is gradually embedded into the organization, and a team spirit prevails throughout. As the work packages are implemented, the organizational relationships are developed as depicted in the RIM, and the information flows start taking shape as agreed upon. These activities lead directly into detailed position descriptions, manpower planning, activity planning, and resource scheduling in the company.

VALUES OF THE RIM

Value to the Individual Manager

To do a good job, a manager must have a clear understanding of his or her own role in the organization, vis-à-vis role relationships with peers, subordinates, and supervisors. Figure 35-4 illustrates the interdependencies that exist among the individuals. To be effective, everyone must understand how to work with the others. These role interrelationships come to focus through the work packages and are held together by accepted authority, responsibility, and accountability of the managerial team.

The RIM process helps each manager to understand fully the specific responsibility assigned to him or her in the organization. An individual is designated as having primary responsibility for each work package. Others who have collateral responsibility involving the work package are also identified. Once these collective roles have been identified, there remains "no place to hide in the organization." If the work package is not completed on time or does not meet the performance standards, someone can readily be identified and held responsible and accountable for that work. Responsibility, authority, and accountability—the triad of personal performance in organized life—are not left in doubt when the RIM process is used in organizations.

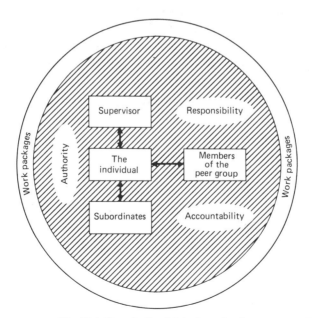

Fig. 35-4. Organizational interdependencies.

Building the Managerial Team

The use of the RIM process in determining relative roles in the organization creates a very high degree of the awareness of teamwork that goes on in the day-to-day activities of the people. RIM shows that interdependence is a fact of life in modern organizations. This interdependence operates both vertically and horizontally; superiors and subordinates depend on each other and on their peers in the conduct of the organization's affairs. Subordinates relay information and comply with the manager's orders. The boss has information that the subordinate needs concerning his or her own job. When an individual has to coordinate and work with peers to obtain information or compliance with a procedural requirement, an interdependency relationship exists. For a well-functioning organization these relationships come together in effective teamwork.

Such team work is subtle; formal teams do not exist per se. Yet as managers work together on reciprocal work packages, many of the cultural characteristics found in a well-functioning team become visible:

1. Managers work at integrating work packages through information sharing and decision making.
2. Subordinates share information with others and build a communication network centered around the work packages to be accomplished.
3. An ambiance of trust, loyalty, confidence, and commitment develops as managers share resources, views, results, and decisions.
4. Managers become better problem solvers thanks to more communication and peer support. A collective strength is formed that is far greater than would be realized with one manager working alone.
5. More creativity and innovation emerge as managers and professionals share ideas (and problems) with others. Having a small peer group off whom one can "bounce ideas" helps to develop ideas into worthwhile concepts.

When the team culture becomes established, conflict becomes socially acceptable. Conflict is a natural form of integration among the managers when they have different primary responsibilities, interests, perceptions, and attitudes on how to handle issues. Since the managers have had an opportunity to participate in developing the specificity of their organizational roles through charting of the RIM, and know what is expected of them, conflict can be accepted and worked through for a solution.

Value to the Organization

Many problems in organizations can be traced to an important oversight. No one takes the time to clarify roles of the managerial team, to ask who does what, who works for whom. As a result of this failure to clarify roles, much psychological and social energy is tied up by individuals trying to seize more power or territory in the organization. Conflicts that are not confronted in the organization are played out in indirect and destructive ways. In the typical organization there are challenges in developing a cultural ambiance supportive of interpersonal confidence, trust, loyalty, and commitment. Territorial imperatives often drive a manager to think only in terms of an individual role and to neglect how that role interfaces with others in the organization.

However, a powerful movement under way in America today is testing traditional managerial styles. A new family of approches is coming into prominence that promises to break with the rigidity and formality of the past. It is difficult to apply a single name to what is happening. Some call this movement "Quality Circles" or "quality of life programs"; others call it a "work innovation movement." Whatever the name, the end result is a greater degree of manager-employee participation than many old-timers can recall. The underlying principle behind this movement is that all employees can truly participate in helping to manage the affairs of the organization.

The movement highlights the much neglected matter of improving the way people work together. Some organizations are good at "people managing." Attention to people in every aspect of business is a fundamental part of the culture at well-run companies as diverse as IBM, Delta Airlines, Hewlett-Packard, and Walt Disney Productions.

The growing popularity of employee participation programs makes it clear that the managers at all levels will have to discover new ways to assert leadership—to become a leader in an environment where people can work together with economic and social satisfaction. Those managers who cannot deal with in-depth discussions of their individual and collective roles in the organization in a context of sharing decisions, rewards, and results will find themselves losing the race between modernity and retirement. The employee participation movement is making it clear that people want to manage themselves and cooperate in the management of the peer group. American management is finally beginning to recognize that the individual doing a job probably knows more about how that job is done than anyone else. Yet many managers (fortunately a diminishing number) still believe their principal job is to exercise "supervision" over the people who "work for" them.

The word "supervision"—the act of overseeing subordinates—is an interesting word. Breaking the word into two parts gives "super" and "vision." Presumably one who is a supervisor exercises "super vision" over subordinates. In today's complex and dynamic organizations there are limits to the amount of supervision that one can exercise. Supervisors depend on the technical and detailed knowledge of the specialists working for them. While the supervisor is able to judge the effectiveness with which organizational goals and objectives are accomplished, much of the detailed intricacies of how those goals and objectives are accomplished are simply beyond the time and depth of the manager's understanding. Managers who knows their limitations in fully exercising "supervision," and their dependence on the people working for them, can develop a broader organizational philosophy in which employees plan and schedule their work, participate in the design of their jobs, and assist in the management of the peer group with whom they work to produce results.

If the manager does all these things well, success is still not ensured, but its chances are greatly enhanced. The development of a cultural ambiance that supports participative management requires months, if not years, to develop. Like a favorite rose garden such a culture requires frequent tending.

RIM AS A CONTINUOUS PROCESS

The RIM process affords managers an opportunity to discuss their role as well as the roles of the key people with whom they work: supervisors, subordinates, and members of the peer group. Those discussions are carried out in the context of authority, responsibility, and accountability. As these relative roles are discussed, appropriate questions of "territory" emerge. Some of these questions can be resolved at the RIM meeting; the more difficult ones are identified for future resolution. Throughout the RIM process, the managers should be aware that the discussions about relative roles will "unclothe" the organization. Overlaps and gaps between authority and responsibility are laid out for all to see. Conflict, much of which may have remained hidden, tends to come out into the open. Things that have "fallen through the cracks" start surfacing. The net result of the process is a clear understanding of organizational roles and an agreement on who works for whom and under what circumstances.

Because the organizational roles change as different peers and peer teams become involved, a continuous redefinition and reinforcement of those roles are required. The RIM process cannot be viewed as a one-time activity of finite duration. It should be a continuous process, if it is to be an effective mechanism for participative management in the organization. As the new work packages are identified, the organizational units affected by those work packages should communicate with one another and develop new entries into the RIM. As the organization's goals and objectives change, the RIM should be modified to reflect those changes. As the projects go through their own life cycles in the organization, new responsibility interfaces will emerge. They should be incorporated into the RIM.

In short, a commitment should be made to maintain the continuity of the RIM process in order to assure its success. If such a commitment is made, RIM becomes the focal point of participative management in the organization, and provides the proper forum for the understanding of work flows, the delineation of individual and collective roles, the identification of supportive mechanisms, and the formation of participative teams leading to group discussions, conflict resolution, and consensus in the organization. If that commitment is not made, however, RIM could end up being simply an additional exercise of paper shuffling, and an expensive and time-consuming alternative to the development of position descriptions.

CONCLUSION

The Responsibility Interface Matrix is a tool for participative management. If used properly, it provides valuable assistance in getting the managerial team organized, and in pushing the management responsibility to the lowest levels of the organization. It helps to develop open communications, mutual trust, and confidence among the organizational units working toward common goals and objectives.

The resulting matrix itself is a simple pictorial representation of the work flows and responsibility interfaces in the organization. But the process through which the matrix is developed is much more critical. Through this continuous process the organizational interactions become visible; the territorial fears are subdued; the roles, responsibilities, authorities, and accountabilities are crystallized; and the entire organization is pulled together toward achieving its mission.

REFERENCES

Allen, Louis A. *Charting the Company Organization Structure.* Studies in Personnel Policy, No. 168, National Industrial Conference Board, Inc., New York, 1959.

"Changing the Company Organization Chart." *Management Record,* Nov. 1959.

Cleland, David I. "The Professional in Matrix Management." Paper presented at the Project Management Institute 1982 Seminar/Symposium, Toronto, Oct. 1982.

Cleland, David I., and King, William R. *Systems Analysis and Project Management,* 3rd ed. New York: McGraw-Hill Book Co., 1983.

Cleland, David I., and Munsey, Wallace. "The Organization Chart: A Systems View." *University of Washington Business Review,* Autumn 1967.

Cleland, David I., and Munsey, Wallace. "Who Works with Whom." *Harvard Business Review,* Sept.–Oct. 1967.

Donnely, John F. "Participative Management at Work." *Harvard Business Review,* Jan.–Feb. 1977.

Duncan, Robert. "What is the Right Organization Structure?" *Organizational Dynamics,* Winter 1979.

Higgans, Carter C. "The Organization Chart: Its Theory and Practice." *Management Review,* Oct. 1956.

Janger, Allen R. "Charting Authority Relationships." *The Conference Board Record,* Dec. 1964.

Kocaoglu, D. F. "A Participative Approach to Program Evaluation." To appear in *IEE Transactions on Engineering Management,* 1983.

Kocaoglu, D. F., and Prokay, J. *Participative Management.* Paper presented at the TIMS/ORSA Meeting, Houston, Oct. 1981.

"Linear Responsibility Charting." *Factory* 121 (March 1963).

McConkey, Dale D. "Participative Management: What it Really Means in Practice." *Business Horizons,* Oct. 1980.

Mintzberg, Henry. *The Structuring of Organizations.* Englewood Cliffs, N.J.: Prentice-Hall, Inc. 1979.

Youker, Robert. *A New Look at WBS (Work Breakdown Structure).* CN-851 Course Note Series, July 1980, International Bank for Reconstruction and Development.

This chapter appeared in the article "RIM Process in Participative Management," published in *Management Review,* October 1983. Copyrighted by AMACOM, a Division of the American Management Association. Used by permission.

36. Negotiating in Matrix Management Systems

C. E. Leslie*

INTRODUCTION

Matrix management systems are diverse in structure, composition, and in the complexity of detail required to achieve specified objectives. The opening chapter by Cleland detailed this diversity ranging from Quality Circles to project management to joint venture operations. All matrices, however, share two common activities: *the negotiating process and its aftermath,* and *communication in negotiation.* This chapter will discuss some basic factors that enter into these processes.

The discussion will include basic negotiating concepts, terminology, guidelines for negotiators, and planning for matrix system negotiations. In addition, there will be a brief discussion of interpersonal communication, barriers to effective communication, and guidelines for more effective communication.

The application of the material in this chapter to specific matrix negotiations is left to the reader, for each negotiating session is unique in terms of characters, situation, and time. Readers interested in further exploration of some of the material, briefly exposed, are referred to the reading list at the end of the chapter.

Matrix Negotiating Problems

In a matrix management system of whatever form, there is shared input, responsibility, authority, and accountability. Each matrix participant shares these to a greater or lesser degree depending on functional charge, organizational status, and latitude of decision-making authority. Additionally, the participants, when drawn from different operating, administrative, or professional sectors of the organization (or different enterprises), have individually assumed

*C. E. "Gus" Leslie is President of C. E. Leslie & Associates, a management consulting firm established in 1965. He holds degrees in management (BBA) and industrial psychology (MBA) and is Adjunct Associate Professor at the Roth Graduate School of Business of Long Island University's C. W. Post Center. He is also associated with the New York Institute of Technology Center for Labor and Industrial Relations. Mr. Leslie has over 35 years of experience in management, management training, sales, personnel, and production. He is the author of many books in his areas of interest.

responsibilities for the efficient administration of their own sections, departments, or enterprises and attainment of specific ongoing objectives. This *duality* can and often does create *differences in perceptions* as to modalities to be employed; effort, material, resources, and time to be expended; and assignment of personnel for achieving matrix system objectives while meeting other individual responsibilities. In some companies there may be more than one matrix in operation, and these matrices may involve a particular individual in more than one project. This would call for matrix managers of different systems not only to negotiate with the specialists required for their own project, but also to settle problems of priority and responsibility among themselves. To further compound the matrix negotiating problem, there may be instances of conflict engendered by past relationships of the participants, group loyalties, clashes of ideas with attendant emotionality, individual psychologies, and organizational dynamics and history.

This chapter does not address the variety of methods used to intensify, reduce, or otherwise manage conflict. Methods of conflict management are found in the reference readings. The possible conflicts will be borne in mind, however, and are referred to in the words "past or present relationships."

Because of its structure, successful matrix management cannot rely solely on executive order, fiat, or authoritarian control. While still necessary on occasion, *direct control* over the matrix participants must give way to a more democratic approach. Intermediaries who effect conciliation, rather than arbitrate disputes, should be considered as a form of democratic approach. The final decisions of what the matrix will do, how it will function, and the degree of tolerance of impeded progress, are of course in the purview of the chief operating or executive officer and some cases—joint ventures, for example—in the control of a board of directors, the entrepreneur, shareholders, or on occasion appropriate legislative or judicial interests.

If a matrix system is to operate effectively, the following conditions should exist:

1. Measures for easing the participant burden created by the duality (or dualities) of responsibility for attaining short-range and long-range objectives.
2. Conciliation of possible adverse past relationships and differing ideas of participants, to avoid the blocking of negotiations.
3. Acceptance of the superordinate objective of the matrix.
4. Development of harmonious, willing cooperation and effective coordination among participants drawn from diverse departments, during the negotiation process and the postnegotiation period.
5. Development of a problem-solving approach, rather than the ascendancy of matrix participants over one another.

Achieving these conditions represents the crux of the matrix negotiating problem and is easier said than done.

Recent literature in management has focused on several proposed solutions to the stated matrix problems. The use of mediators, facilitators, or intragroup and intergroup bargaining representatives has been proposed. Schemes such as a mixing of persons or groups within bargaining groups, intergroup therapy sessions, and others have been promoted. Differentiation of bargaining procedures—distributive, attitudinal, integrative, and intraorganizational—has also been proposed (Walton and McKersie, 1965). These proposals have much merit. Which-ever of these approaches is chosen by matrix participants, the implementation has as its foundation the participants' development and application of effective personal *negotiating* and *communicating skills.*

Negotiating Style

Negotiating is a pervasive, lifelong, interactive process. While people engage in this process, many of them are unaware that they are "negotiating" except when making a business deal of some kind. In reality, the process of negotiation transcends business deals and is an important part of social living. Since it is a pervasive and lifelong activity, many persons develop negotiating habits or styles, reflective of personality factors and successes or failures, which they exhibit in negotiating situations. Success achieved by a person in negotiation is attributed to his or her exercise of will, mind (including deviousness), foresight, research, personality, and other "inborn" or developed factors. Failures are often laid at the doorstep of another person's obstinacy, ineffective communication, inability to compromise or conciliate or accommodate, and lack of foresight and sensitivity.

Negotiating style or habits that may have been beneficial in particular situations, when carried over into the variety and uniqueness of other negotiations including matrix systems, may produce random successes and failures. Yet one of the hallmarks of an effective negotiator is not to be fixated on a particular style, but to exercise *flexibility* in thought, striving for goals, approach, and action. To achieve this flexibility requires a critical study of the factors involved in each negotiation, including the personalities (if possible) and the group dynamics. The effectiveness of communication, proper planning, a heightened self-awareness, and knowledge of the negotiating process also can help.

Because negotiating style or habit is evidenced by behavior, if that behavior can be changed by the individual to adapt to specific negotiating situations in organizations or business or personal life, that individual may demonstrate a greater degree of responsive flexibility. A simplified view of behavior indicates that it has four major components: motivations, attitudes (beliefs, opinions, and attitudes are often combined), intelligence, and physiological states that include effects of emotion and energy arousal. If any of these components are changed in an individual, the perception of situations and behavior will also change. Acquisition, retention, and application of knowledge can affect change in certain behavioral components. As a consequence, if new, flexible behavior is prompted and results in satisfaction, that new behavior hopefully may supplant the old. In this chapter our concern is largely with the knowledge factor, since it can influence the development of flexibility in different negotiating situations.

Negotiating: A Working Definition

Management has often been defined as getting things done by (or through) other people. Virtually every managerial or commercial activity, with the possible exception of a single entrepreneur setting objectives or manipulating funds, is a matter of interaction with people. This interactive process, with effective communication as its core, is one of the main ingredients of negotiating. It is extremely difficult—and at times not possible—to differentiate the social from the business aspects of an interpersonal interaction taking place within an enterprise or between representatives of different enterprises.

Whenever one individual (or group) depends on another to satisfy particular wants or needs, and there exist *options of action* and/or *differences in the perceptions* and expectations of the parties, the contact and subsequent interaction for possible resolution of these differences or options is a negotiation. Since many conditions affect a particular negotiation, resolution of these differences, or concordance, may or may not result.

In business negotiations (trading, bartering, merging, buying, selling) usually there are tangibles and intangibles at stake. The same holds for negotiations in a matrix system, where the tangibles—which may or may not be spelled out—include capital, budgetary allocations, financial rewards, time, and resources assignment, and the intangible s include such matters as recognition, achievement, and continuing effective relationships among matrix participants.

A working definition of negotiating might be: *Negotiating is an interpersonal or intergroup verbal or nonverbal interaction incorporating objective and subjective factors, acting alone or in combination, in connection with a contemplated future transaction involving the exchange of goods, services, resources, knowledge, behavior, money, terms of exchange, and/or expectations and satisfactions.*

Implied in the definition are the following conditions:

1. One or all the parties is interested in either obtaining or delivering one or more of the items detailed in the definition.

2. One or all of the parties is willing to give up something for the item(s) wanted or needed. There is no complete giving or taking of any item without a trade-off—something conceded for something accepted.
3. The negotiation may terminate (discontinue) at the option of one or all parties.
4. There must be a recognition or expression of difference regarding the worth of the item being contemplated for exchange.

The origin of a negotiation lies in desiring satisfaction of needs or wants, either personal or organizational and often both, expressed by a *request, demand,* or *offer* by one party to another and requiring the second party's action for fulfillment. Further, there must be implied options of action based on differential perceptions of the satisfactions to be derived from the *contemplated transaction,* as previously stated. What is truly negotiated are *differences which should be into agreements,* rather than items on which all parties agree.

Condition 3 above indicated that a negotiation can be terminated at the option of one or both parties. This is variously called a deadlock, stalemate, or discontinuance. A discontinuance (temporary or permanent) of negotiation may be perceived as a success or a failure by one or all negotiating parties. When a party other than the original negotiators is asked or required by prior agreement, by implied imposition, or by law to mediate, conciliate, or arbitrate a negotiation, the original negotiation can be considered discontinued.

In a matrix management framework, for example, if the original negotiating parties cannot satisfactorily resolve differences, for whatever underlying reasons, and the nonresolution interferes with progress toward achieving established matrix goals, the responsible executives may impose a solution or mediate the disputants. Intrusion, arbitration, or mediation by a higher authority may not necessarily improve the relationship of the negotiators that failed to reach concord on their own.

Negotiating Terminology

Recent literature on negotiating has focused on the "winning" aspects of a negotiation. A few writers have tempered the striving for a "win" by promoting a "win-win" rather than a "win-lose" orientation for negotiators. Others have approached negotiation by using game theory techniques for summing results such as positive sum, zero sum, negative sum, variable sum, and so on. The winning and losing aspects of negotiating require some elaboration. This will follow after a discussion of terminology used by negotiators. To understand baseball or other games, the player or spectator must not only know the rules but also the terms that describe a certain action or condition; the same holds for negotiating.

A *win* in the negotiating process is an outcome in which a negotiator has achieved a result better than nominal.

A *loss* in the negotiating process is an outcome in which a party has retreated from nominal.

Nominal is a position in the negotiation process that a party assumes as a valued estimate of the worth of what is being offered, requested, or demanded either prior or subsequent to the start of the negotiation. A nominal is bracketed by a maximum and minimum for either buyer or seller. Gradients of advance or retreat from nominal to maximum and minimum positions should be established. For a buyer, the maximum payment would be that point beyond which there will be no agreement. The theoretical minimum for a buyer would be zero, but practicality tempers this and a buyer normally should have several gradients below nominal. A seller's theoretical maximum would be "everything," but again this is tempered by the establishment of gradients upward from the seller's nominal to what it is perceived the market or prospect will bear. A seller's minimum, however, can vary in accord with many extenuating circumstances, as will be shown later. A seller, just like a buyer, should set gradients of retreat from nominal to minimum.

A win for a seller is any point agreed upon above nominal (getting more), and a loss any point below (getting less). A win for a buyer is any point agreed upon below his or her nominal (paying less), and a loss any point of settlement above nominal (paying more). A *draw* for either buyer or seller is a settlement at their respective nominals.

A *buyer* in a negotiation is a party who accepts an item and offers something in return. A *seller* is a party who offers something and accepts something in return.

Parties to a negotiation may at times be buyers and at other times sellers. This role switch has as its fulcrum a trade-off, as previously described. Buyers and sellers in a negotiation need not be purchasing agents or salespersons, and people in those positions may even switch roles when negotiating.

A common term used by negotiators is *power*. Power in negotiating can be viewed from two different perspectives. The first is the ability of the negotiator (or more precisely, whoever the negotiator represents) to grant or withhold benefit from the other negotiator. The second perspective is the power of the negotiator to exercise personal influence and persuasion over the other negotiator. This latter power of persuasion or influence is what should be used more often in matrix system negotiations, rather than the granting or withholding of benefit. In joint venture operations, mergers, or acquisitions, granting or withholding of benefit by the parties may be of major importance.

Strategy refers to the long-range objectives to be achieved by the negotiation. *Tactics* are those devices used to gain some subtle psychological advantage over another. For example, where the meeting is held can often play an important role in negotiations. *Ploys* are deliberate acts designed to influence the other negotiator to change position or to hasten agreement. Some ploys may be two-edged swords. For example, the ploy "take it or leave it" may be used to hasten or force an agreement, but the other negotiator may "leave it," which may not be in the best interest of the first negotiator, who believed the other party would accede.

THE TWO PARTS TO A NEGOTIATION: PROCESS OUTCOME AND AFTERMATH

There are two parts to a negotiation. The first concerns the process and its outcome, starting from initial request, demand, or offer, the parties reaching either agreement or a discontinuance. Most of the popular literature in the field of negotiation has been devoted to this process and its outcome. The second aspect of a negotiation—in many instances the most important, and frequently neglected in the literature—is the postnegotiation period or aftermath. It is in this period that implementation or abrogation of an agreement takes place; that benefits or penalties accrue from agreement or discontinuance owing to unforeseen events; that misunderstandings of the agreement arise; and relationships may get better or worse.

Regarding the process and its outcome, most negotiators strive to win. As previously stated, recent literature has promoted the "win" aspect, tempering it to a "win-win" orientation. An effective negotiator, however, recognizes that winning in the immediate negotiation process is not the totality of negotiating. The effective negotiator considers the type of negotiation (about which more later) in which he or she is engaged. Short-range and long-range implications—the aftermath—and the risks to be taken are carefully weighed against the instant win. There is also a recognition that foresight is limited and that possible reversals of negotiation outcomes may take place after the negotiation owing to unforeseen events. It is possible that an astute negotiator, after evaluating all the *foreseen* variables, may have a "lose" rather than a "win" objective for the instant negotiation. He or she may take a *calculated risk* to recoup that loss in the future.

Persons responsible for matrix systems, who must negotiate with others, should bear this idea of a "loss" possibility in mind just as much as the orientation to win. When matrix managers operate under the pressure of time, the resultant stress may obscure the acceptance of a short-term loss. The need to win in the short term may not be as advantageous as maintaining beneficial relationships for the long haul.

Winning and Losing: A Perspective

To provide a basis for the elaboration of winning and losing in the negotiation process, as defined earlier, negotiating for a tangible x will be used for illustrative purposes. When the illustration is concluded, a parallel of negotiating for something in a matrix system will be discussed.

Except for the possible trade-off of volume in the illustration, other variables that could influence the negotiation will be kept in abeyance. The assumption is made that the negotiation is a bona fide negotiation.

The seller of x has set its nominal value at $2.50, predicated on a $0.50 profit. If x were to be sold in volume, however, the seller might accept a lesser profit for each x. The seller's minimum acceptable price (meaning no profit or just breaking even) is thus $2.00, while the practical maximum has been set at $3.00. Any agreement as to price above $2.50 would be considered a win, and less than $2.50, a loss. Mention has been made in the terminology discussion of unusual circumstances that would impel a seller to accept less than normal minimum or the break-even point. The purpose might be to gain a foothold in a new market, to shut out competition, to maintain skilled productive labor for a given period, to reduce excessive inventory to improve cash flow, to help a valued customer in time of crisis (maintaining the relationship), and the like. In doing any of these things, the negotiator takes a calculated risk to recoup the loss in the future. Futures, however, are not guaranteed. Except for the unusual circumstances, the seller may set as a practical minimum, let us say, $2.10 for x.

Negotiating maxim: A seller initially should ask for more than is expected to be settled upon. The initial asking price should be consonant with held power to grant or withhold benefit, or with the other negotiator's perception of that power.

The prospective buyer of x also sets a nominal value for x as a single purchase or in volume. If the seller's initial asking price is agreed to by the buyer, there is no negotiation per se (provided again all other possible intrusive factors such as a payment schedule are held in check), since there is *no difference expressed.* If the buyer's nominal bracketed by maximum and minimum are different, a negotiation may ensue.

Negotiating maxim: A buyer initially should propose less than is expected to be settled upon. This initial proposal should be consonant with power to grant or withhold benefit.

We just said that a negotiation *may* ensue because, if the buyer has *other options* and the initial asking price is not in the buyer's range, the buyer may not choose to pursue the negotiation. Or the buyer may ask for another price, to see whether the seller will bring it within the buyer's market range. If, on the other hand, the buyer perceives the seller's price to be within the buyer's acceptable limits, the buyer can continue the negotiation by offering a trade-off of volume to achieve a price below his or her nominal value of x.

Negotiating maxim: Maximums, minimums, or nominals are not revealed in the initial stages of the negotiation. Normally, neither buyer nor seller will know the other's nominals, maximums, or minimums.

When buyer and seller in a bona fide negotiation attempt to reach an agreement, the following outcomes may obtain:

1. When, by coincidence, nominals of buyer and are the *same:*
   a. If the buyer wins, the seller loses, and vice versa, or
   b. If the buyer pays nominal—a draw—the seller also draws.
2. When the seller's nominal is *higher* than the buyer's:
   a. If the seller wins or draws, the buyer loses, or
   b. If the seller agrees to a price that is less than selling nominal, the seller loses and the buyer may win, lose, or draw, depending on the magnitude of the seller's retreat from nominal and the disparity of buyer and seller nominals. Example: In the instance cited, if the buyer's nominal is $2.25, and the seller agrees to $2.25 the buyer draws, the seller loses. If the seller and buyer agree to $2.40 both buyer and seller lose. If both buyer and seller agree to $2.15 for "x" then the seller loses and the buyer wins.
3. When the seller's nominal is *lower* than the buyer's:
   a. If the seller draws, the buyer wins, or
   b. If the seller wins, the buyer may win, lose, or draw, depending again on the magnitude of the disparity of the nominals and at what price agreement is reached. Example: If the seller obtains $2.75, the seller wins, and the buyer wins if the buyer's nominal was $2.80. If the seller and buyer settle at $2.80, the seller wins and the buyer draws. If the agreed-upon price is greater than $2.80, the seller wins and the buyer loses.

It can be seen from the given example that a negotiation outcome, exclusive of a discontinuance, may be any of a variety of win/lose/draw combinations. Only in a very special case can both negotiators, having a certain magnitude of disparity in their nominals, come out winners when they settle within a particular range.

It should not be inferred that winning, losing, or drawing in the negotiation process has to do with *actual wins* or *losses*. It is merely a way of gauging the degree of satisfaction with the agreement obtained, assuming that both parties intended to arrive at that agreement. The degrees of satisfaction obtained by the negotiators need not be equated. Thus the negotiator that wins may feel very elated, while the negotiator that drew may be very satisfied, and the negotiator that retreated from nominal may be minimally satisfied. These feelings of satisfaction are subjective and cannot be put on a balance scale. On the other hand, executives who may be in a position to evaluate the negotiator's achievement may not be favorably impressed. To safeguard against this possible repercussion, *an effective negotiator insists on performance criteria from the party he or she represents.* These performance criteria may be in quantifiable as well as qualitative areas. Whatever criteria are set, there must be an acceptance, by both the negotiator and the represented party, that agreements reached will be in an acceptable range, from maximal to minimal levels.

In the example just discussed of winning, losing, and drawing in the negotiating process, intrusive variables and parameters were held in check. These variable factors can be of paramount importance to a real negotiation. Such factors include the desire for possible continuing effective relationship between the negotiators; the effectiveness of communication; personality factors (likes, dislikes, prejudice, persuasion, obstinacy, and others); introduction of trade-offs; quality of antecedent interactions of the negotiators; uniqueness of the item; balancing the importance of the instant negotiation against long-range objectives; parameters of law; payment policies; a discontinuance due to alternative courses of action available to buyer or seller; and the like.

Matrix managers and participants are concerned with costs, some of them tangible and capable of being put on books of account or budgets, and some intangible and not quantifiable. Estimating tangible costs for project accomplishment is not always an easy task. To draw an analogy between the example of negotiating for x and negotiating in the matrix, an engineering effort will be used.

The engineering department representative is asked to design a "widget" needed for project purposes. It will be assumed that there exists no friction from previous contacts between engineering and the project or program manager. Preliminary investigation, feasibility studies, and possible shifting of engineers from previous to new project assignments has to be done by the engineering group. Time and cost records for this preliminary analysis may or may not kept, depending on the accounting controls in effect. Nevertheless, there is a cost incurred by the engineering group and consequently by the company for the preliminary work.

The request by the project manager for the development of the widget is made. If the project is on a "profit center" mode, the project manager of the matrix has also estimated, albeit roughly, the budgetary allocation for the widget development. The estimates made by both the project manager and the engineering group may or may not be the same. This parallels the nominals for x made by the seller and buyer in the example. Estimates of the maximals and minimals bracketing the nominal cost for the widget development also should have been made by both matrix negotiators. The engineering representative also should have made an estimate of "lost time" or "delay in schedule" for other engineering projects, and what it will take to make that lost time up due to reassignment of engineers to the new project. This also represents a bargaining point for the engineering representative in the negotiation with the project manager. Thus the engineering representative has two items or "chips" to be used in arriving at an agreement with the project manager. This is closely akin to the volume trade-off for x by the buyer.

In response to the request by the project manager for the development of the widget, the engineering representative would initially ask for more budgetary allocation than the engineering estimate. How much more will depend on the cost for the preliminary work, the lost time to make up, and a safety factor of cost over nominal to make up for unforeseen events such as absence of assigned personnel, "rush" demands from customers or executives, breakdown of engineering equipment, test time, liaison time, negotiating time, and other specific factors. This follows the seller's negotiating maxim.

If the project manager (the buyer) accepts the initial demand for budgetary allocation by the engineering representative, there will be no negotiation since there is no expressed difference. If the engineering demand differs widely from the project manager's nominal estimate, then the give-and-take of negotiating will follow the pattern to arrive at win, loss, or draw as for x in the example. The engineering representative will offer evidence in support of his or her demand. The project manager will also present arguments in support of his or her case for obtaining an agreement within an acceptable range of cost. What trade-offs and concessions are made and accepted, and how this comes about, will determine the outcome of the negotiating process for both sides.

The manner in which the negotiating is conducted—that is, the communication effects, the tactics or ploys used, the emotions displayed, the force of persuasion or power, the perceived rewards and other recognition, the relations of the engineering representative to the engineering group and the project manager to his or her superior in terms of winning or losing in the process—are the intangibles involved. Implementation and cost overruns in the aftermath may also be affected by what occurred in the negotiation.

It can be seen from the foregoing brief exposition that what may happen in a matrix negotiation involving two parties is quite analogous to the illustration of bargaining for x. There is also the possibility that the project manager and the engineering representative may reach a temporary deadlock or discontinuance, so that a third party is needed to effect agreement by imposition, mediation, or conciliation.

Even in a situation where a genuine negotiation has been discontinued, there may be feelings of satisfaction if one or both parties have avoided entering into too onerous an agreement or one where future relationships were suspect. Avoiding a detrimental agreement may be as beneficial as making a satisfactory one.

Aftermath

Initial satisfaction with a win, draw, or loss in negotiating, or the discontinuance of a negotiation, is only part of the total scenario. The second part of the negotiation is concerned with the *real win or loss*—the real success or failure—which occurs during the negotiation aftermath. The real win or loss may result from many factors: reflection by one negotiator on the tactics and ploys used by the other; how power was perceived to be used or abused by the other; a re-evaluation of the relationship of the negotiators; the possible discovery of deceit, manipulation, or misrepresentation; a broken promise or failure to implement the agreement; a misunderstanding of oral or written communication; a turn of unforeseen events militating against deriving the apparent benefit from either an agreement or a discontinuance; or the feelings about a third party that may have been called in by one negotiator to undo the temporary discontinuance.

The second part of the negotiation, as previously stated, is critical. Depending on the purpose of the matrix, abrogation of agreements can result in drawn-out and costly litigations. Actual monetary or physical damage may ensue from misrepresentation or "cutting corners," and relationships may be broken beyond easy repair. Avoidance of such pitfalls requires that during the negotiating process adequate safeguards be observed and incorporated into agreements, and that credibility be maintained and harmonious relationships established and continued, so that neither party feels injured in any way. Negotiating is a thinking process requiring constant awareness of the factors involved, and effort on both sides to achieve effective communication.

In addition to the problems previously enumerated, that create negotiating opportunities in a matrix system, the ideas of winning and losing in the negotiating process must be carefully balanced against what may happen in the post-negotiation period.

TYPES OF NEGOTIATIONS

Negotiations can be classified in a variety of ways. An effective negotiator is constantly aware of the type of negotiation in which he or she is engaged. The classifications are made under these categories:

1. Periodicity of contact or length of relationship continuity
2. Number of parties involved, integrally and peripherally
3. Product or service specialty
4. Perceptual aspects along a spectrum ranging from open hostility to collusion
5. Ethical character, real or spurious
6. Formality of the interaction
7. Combined forms

Many of these types may be characteristic of matrix management systems, because of the diverse ventures with which matrices are involved.

In a *one-time or singular negotiation* the probability of future contact by the parties, except for execution of the agreement, is extremely low and may be nonexistent. In this type of negotiation the negotiators may employ a wide variety of strategies, tactics, or ploys, some even bordering on what some persons may consider as unethical or even illegal. For obvious and practical reasons, this type of negotiation should not occur in matrix management.

Periodic negotiations are those in which the parties meet at specified times to review progress, clear up possible misunderstandings, and perhaps extend previous agreements and solve problems in implementation. Relationships of the participants, attitudes, motivations, and interim communications strongly affect how these negotiations will proceed. The tactics and ploys used will be somewhat limited, as maintenance of a good working climate between participants is necessary if the matrix is to reach objectives.

Continuing negotiations are those in which the negotiators are in frequent contact. Often the participants are not aware that they are negotiating, accepting the interaction as an everyday communication. Continuing negotiations are found in close associations of people such as superior-subordinate relations, peer groups, service operations, matrix management sytems, labor-management contacts, and family groups. Here the tactics and ploys used are or should be extremely limited, since each communication and interaction affects the nature of present and future negotiations.

Periodic and continuing types of negotiations ought to be the constant concern of responsible executives and participants in matrix management systems.

Another way of classifying negotiations is by the number of parties involved: *two-party* or *multiparty*. A party can be either an individual or a team representing one department or enterprise. A multiparty negotiation is one in which three or more departments, enterprises, or other agencies with interest in the negotiations are involved at the same or at separate times.

Matrix management systems fall into both categories. Two-party negotiations are frequently found in day-to-day contacts for resolving differences, clarifying prior agreements, and solving problems not envisioned during the original negotiation. Multiparty matrix negotiations are characteristic of project or product management, international ventures, task forces, and other operations where several departments are involved in coordinated efforts. Matrix system participants should be aware of group dynamics when engaged in either two-party or multiparty negotiations. For example, if a participant in a matrix system is "worked into a corner" by the use of either logic or power, an apparent win may be in the offing. However, it is not uncommon for other participants in the multiparty negotiation to rally behind the "defeated" party and close ranks against the apparent winner.

Many ramifications and parameters surround specific *product or service negotiations*. To cite a few: technological aspects, regulations, law, accounting practices, hard policies, competitiveness, uniqueness, etc. Negotiating to buy a fleet of new trucks can be quite different from a real-estate development transaction, because of the many technicalities and legalities involved in real-estate negotiations. The reader is left to assess the many specific factors of service or product that impinge on a particular contemplated transaction in his or her matrix system.

The *perceptual aspects* classification of negotiations relates to how individuals or organizations view their posture, orientation, or intent in a particular negotiation. They may view the negotiations as

1. Adversarial
2. Compromising
3. Accommodating
4. Collusive
5. Manipulative
6. Spurious
7. Real or bona fide

Adversarial negotiations occur when participants view each other as "enemies" or "opponents." The negotiation climate may range from concealed to open hostility. The perception of enmity may arise from any of a variety of reasons or causes, real or imagined. Such factors as incompatible attitudes or ideas, stereotyping, clash of motivations or aspirations, antecedent relationships or events, departmental cohesiveness, time factors, and differing objectives of enterprises may give rise to adversarial negotiations in matrices. Other examples of adversarial negotiations are labor-management contract negotiations and administration, hostage negotiations, defendant-plaintiff actions both in and out of court, and the like.

Adversarial negotiations also may erupt in matrix systems for a variety of causes. In many organizations, departmental or group cohesiveness may have built "walls" between departments. Negotiators in matrices, especially when starting a new project, should be alert to signs of resistance stemming from this type of organizational dynamic. Perhaps one of the first tasks for the matrix negotiators or facilitators or responsible executives is to build a climate (which may take patience and time) conducive to smooth negotiating to reach matrix objectives.

Compromising negotiations occur when each party is willing to give up reaching a total objective in order to get on to other things. Usually this happens in more or less equal power situations where the impeding of progress cannot be tolerated by either side.

Accommodating negotiations are those in which one party, for whatever reason or motive, concedes more than what would normally have been conceded from a nominal position.

Collusive negotiations are only an ostensible form of negotiation, since all parties know in advance what agreement will be reached. Differences are slight or nonexistent.

Manipulative negotiations are those in which one party seeks to gain a particular end by maneuvering the other into a situation or agreement through deviousness or deceit, without that party's being aware of it. By normal societal standards this type of negotiation is considered unethical, but it does occur. This kind of negotiation also falls into the spurious category.

Spurious negotiations are those in which one party has no intent of reaching an agreement that will commit it to a future action of exchange. In everyday living this negotiation takes place when people are "shopping the market" or "picking the brain" of others. For example, when negotiating to buy a car (new or used), a person may go to one car dealer to gain knowledge. The "prospect" gives every indication of interest and has the salesperson go through a "sales pitch" and figure a deal, yet the shopper may really mean to make a deal with someone else. Having obtained information about accessories, performance, and price, the shopper will be in a better position to bargain harder with another dealer.

Spurious negotiations are sometimes resorted to by a manipulative organization involved in merger or acquisition talks. The *real* negotiation may be taking place among other parties elsewhere, though the negotiators involved in the spurious negotiation may not realize that they are being used in this way.

Real or bona fide negotiations are those in which the parties originally intend to reach an agreement. Since what is under consideration is a *contemplated* future transaction, a real or bona fide negotiation does not imply that the parties must or will reach an agreement or resolve differences. A real negotiation can result in a discontinuance (deadlock or stalemate) in which one or both parties fail to agree, as well as in agreement concerning a future transaction.

It can be easily concluded from the preceding discussion that not all negotiations are ethical and open and above board, nor do all promise concordance. At the same time, the variety of the "not so nice" negotiations should not evoke cynicism, but rather alert negotiators to be constantly aware of the type of negotiation and the intent of the other party. It is hoped that most matrix managers will be involved with bona fide rather than "not so nice" negotiations.

Many negotiations are also characterized by their structure, ranging from "informal," "off-the-cuff," "vestibule," or "in the street" types to those with high protocol. In most matrix systems the prevalent forms are the informal or semiformal. In international negotiations such as joint ventures, there is a greater observance of structure and protocol. Agendas, seating arrangements, turn of speaking, and status are extremely important. The customs of the host country should be scrupulously observed, to preclude degeneration of the negotiating climate.

In domestic matrix systems some structure may be engendered by the introduction of agendas, opening statements of position, and agreement to parameters and ground rules. Some structure is necessary to keep the discussions "on track." Sufficient advance notice of the agenda is a courtesy extended to matrix participants and allows time for preparation for the negotiation meeting.

Often several types of negotiations may be combined. For example, in project management matrix systems there may be a combination of multiparty and continuing, formal, bona fide, and compromise-seeking negotiation. It is important that a negotiator properly identify the type of negotiation, so as to be effective in not only the process but the aftermath. Proper identification can also heighten awareness of effective communication and development of mutually beneficial relations.

MANAGERIAL SKILLS AND NEGOTIATING

Management is often analyzed as being composed of several functions. Prime among these is the *setting of objectives*. Attendant to this are planning, organizing, directing, and controlling or evaluating, which are interwoven to reach objectives. The managerial skills developed in these areas can also be applied to intra- and interorganization negotiations.

Setting objectives consists in determining what is needed; establishing nominals, maximals, and minimals *for each item* to be negotiated; relating the objectives of the negotiation to maintaining a good working relationship with participants; and meeting matrix objectives consonant with corporate strategy.

Planning for a negotiation is a major contributory factor to achieving negotiating process and aftermath objectives. Any planning involves foresight and decision making in the light of available facts, in choosing prime and alternative courses of action, "Available facts" implies not only homework or research with regard to the specifics (real estate, technology, marketing programs, etc.) relating to the negotiation, but also an examination of the current attitudes, motivations and antecedent relationships of the matrix participants. Further, it includes preparation of agendas, an astute selection of tactics and ploys (if warranted), and advance notification to participants of date, time, place, others attending, and the purpose of the meeting.

Planning should consider opening statements; economical use of time for preparation; lead time to the actual negotiation; how initial requests, demands, or offers will be made; how trade-offs will be handled; and contingencies, should a negotiation be temporarily discontinued. If the negotiation will involve a team of persons representing a party, the chief negotiator should prepare the team as thoroughly as possible as part of the planning process. In addition to briefing the team concerning latitude of decision-making authority, roles to be played during negotiation, tactics and ploys available, signals to be used by the chief negotiator, and communication pitfalls, the chief negotiator may involve the team and others in "dry runs" of the negotiation.

One element of planning frequently overlooked is the prenegotiation stage of communicating effectively with all persons who will be involved in implementing negotiation outcomes. It has been the experience of the author that, when this activity is neglected, many problems arise in the postnegotiation period that otherwise could have been avoided. Two other aspects of a negotiation that are also slighted is the need for a method of controlling progress during the process, and the debriefing of the negotiator or negotiating team in the post negotiation period or during a delay of negotiations.

Adequate prenegotiation planning is critical and cannot be overstressed. "Murphy's law" can go into effect, when adequate planing is lacking, at almost any time during a negotiation. Fortunately, the tactic of using a caucus the moment things start to go wrong can save many a negotiation from producing unfortunate results. Planning for matrix system negotiations is further detailed at the close of this chapter.

Organizing involves the selection of the negotiator or negotiating team members and the task assignment of these persons for the prenegotiation planning and the negotiation proper.

Directing involves communicating tasks to others. In management, policies and procedures may be planned, persons selected, and tasks assigned, but it remains for people to be directed to action to accomplish objectives. Not only is communication the central feature of getting things done through or by people, but it also lies at the core of interpersonal actions and consequently of negotiations.

Controlling or evaluating consists of appraising progress toward the attainment of objectives and suggesting that deviations from prime or alternate plans be corrected. Control of negotiation progress is a bit more difficult, but there is sufficient literature in management theory to enable effective negotiators to plan a control system for certain negotiations.

COMMUNICATION IN NEGOTIATION

At the outset of this chapter it was stated that the concern would be with both negotiation and communication. Throughout the chapter the effectiveness of communication has been stressed as a variable affecting the negotiation process outcome and the postnegotiation period. Communication was also mentioned as part of the directing function in the brief exposition of managerial skills applicable to negotiating. Attention will now be focused on this vital area of negotiating and everyday living. Communication, the central activity of humans, is at the same time one of the most complex subjects to "communicate" about by whatever single medium is chosen. Informed readers probably know that there are several communication sciences: kinesics (body language), proxemics (distance or spatial language), linguistics and paralinguistics, semantics, semiotics, and pragmatics, plus graphic arts, theater, and other art forms. To discuss or illustrate any of them in depth would require volumes of writing and displays appealing to the several senses. Because of their complexity, the nonverbal aspects of the communication process are deliberately bypassed in this chapter, but in real life they cannot be overlooked. The brief discussion on communication that follows will be limited to definitions of inter-personal and effective communication, a few axioms, some common barriers to effective communication, and a few guidelines for more effective personal, managerial, and negotiating communications.

All the objectives set for a negotiation, the plans made, the organization or team developed, antecedent relationships of the negotiating parties, negotiator orientations, motivations, and attitudes—all these come to a head during the communication sequences on both sides of a negotiation. Communication is the vehicle by which a negotiator determines what the other party wants or needs. By means of it a negotiator probes for information, grasps the goals, motives, attitudes, and problems of the other negotiator, and establishes and maintains good or poor negotiating and working relationships.

Axioms of Communication

Communication has been defined in many ways. According to one of the most inclusive definitions, *communication is any behavior that results in an exchange of meaning.* By extension, it encompasses what is said and left unsaid, what is done and not done, and how it is said and done. Since it concerns behavior, and there are no nonbehaviors available to people, it becomes apparent that every-one communicates.

Axiom 1: Everyone communicates. But not everyone communicates effectively.

Interpersonal communication requires that at minimum there be a person (the sender) who emits a verbal or nonverbal signal, message, or behavior, and another (the receiver) who has an awareness of the message through sensory perception and interprets it for significance or meaning. In face-to-face com-munication the sender and receiver are interchanging roles constantly. While the original sender is emitting signals, the receiver is also reacting and sending signals back to the sender. The original receiver's reaction may be nonverbal (change in facial expression or movement, body movement, distance change, etc.) or oral (O.K., Whoa! Uh, huh!). The significance of these events to a communicator is that communication proceeds in a straight time line and is never to be repeated; the interaction constantly changes, and each communi-cator's actions affect the other in some way. Also, in either role, a communi-cator should be alert to the need to assess carefully the possible impact of the exchanged behaviors.

In the working definition of communication, the salient word is *meaning.* Meaning is developed in a receiver of a communication and is indicative of how the receiver's psychology and physiology affects the interpretation of the send-er's message. Further, the developed meaning can be a trigger for the receiver's response, or it may be stored in the receiver's memory.

Effective communication implies that a receiver's developed meaning be equivalent to the sender's intent. This definition does not hold if there is a conscious intent of the sender to deceive the receiver. In that case the sender wants the receiver to develop a meaning other than the sender's intent to deceive. If the receiver detects the deception through verbal or nonverbal "leakage," the sender will have been ineffective.

Achieving effective communication is not an easy task, nor is it a task solely in the province of a sender, but it is incumbent on both sender and receiver *if indeed they have decided to be effective with each other.* Individual psychologies and behaviors vary, so that senders and receivers can easily distort messages. For want of a better term, and because it is in wide usage, the word "barrier" will be used to indicate something that distorts meaning or assessment of intent. A *barrier* in the sense used here is not to be construed as a blockage of communication, but rather as some personal or exterior interference to be avoided on the path to effective communication.

Axiom 2: Two or more different meanings (or inferences) can be drawn from any one communication act.

This axiom has much significance for communicators and negotiators. As a receiver, a negotiator cannot rely on a first reaction to a message, for to do so may result in misunderstanding, mistakes in counteroffers, discontinuance on an emotional basis, or possible damage to an ongoing relationship with original sender. In a negotiation as in their private lives, individuals have different frames of reference and consequently must understand a communication from not only their own but from the other negotiator's as well, in order to be effective. In an earnest communication, both senders and receivers can safeguard against misinterpretation by using *feedback.* Feedback in communication is a method used to help clarify either intent or meaning. In the interest of face-saving or maintaining good relations, the burden of understanding should be assumed by either sender or receiver, as the case may be. For example, instead of a sender saying, "Did you understand what I said?" he or she may better say, "I don't know if I made myself clear. What was your understanding of what I said?" A receiver, using the same principle, may say something such as, "Correct me I'm wrong, but I understood you to mean . . ." Failure to use feedback in communication is one of the prime communication barriers, reflecting an assumption that what is heard or seen is clearly understood, which the second axiom belies.

Axiom 3: In interpersonal communication there are two aspects to a message: content and command.

When people communicate face to face, they send two separate messages at the same time. The first aspect of the message is the *content* (technically, digital), that portion of the total message about which two or more impartial observers or listeners can agree. The second portion of the message is the *command* (technically, analogic), that portion that imposes behavior and defines the relationship between the communicants as to how that message is to be received. This is done either through the paralinguistics (voice tone, volume inflection, timbre, etc.) incorporated consciously or unconsciously by the sender, or through the sender's observable action. This is so much of a human problem in face-to-face communication that frequently a message may be prefaced by such remarks as "Now don't get me wrong," or "I'd like to be perfectly frank," or "What I really mean is . . ." and others that in and of themselves convey fear of a distortion of the sender's intent. Here again feedback—careful listening and observation with self-awareness and analysis on the part of both communicants—can frequently avoid distortion.

More Barriers to Effective Communication

In addition to the failure to use feedback for clarification purposes, there are many other barriers that create distortion. One important one is the failure to listen effectively. This failure can arise from any of a multiplicity of causes: inattention, lack of interest, self-preoccupation, physical impairment of hearing, noise and other external interferences, selectivity, retention span, and other causes of psychological origin. The failure to listen effectively creates problems not only in private lives but also in negotiating in a business environment.

The failure to listen effectively in a negotiation not only precludes adequate response to offers, demands, requests, or concessions, but also affects the social aspects involved. It may evoke feelings about the receivers' lack of attention or courtesy, their indifference or apathy, or their competence. Part of listening effectively is the mechanical process of hearing *content,* as well as the paralinguistic conveyance of *command.* For example, a negotiator who has a hearing impairment should state it to the other negotiator at the outset. Failure to do so may prompt negative feelings and so alter the interaction.

The failure to listen effectively can also be coupled with the receiver's failure to observe the sender, since actions also communicate. The combination of these barriers, along with the failure to analyze and use feedback before reacting, can create problems in negotiating. These problems often cannot be quickly rectified and may create attitudes that will affect future negotiations.

The making or holding of invalid assumptions about people and situations can prompt mistakes not only in communicating but also in attempting to arrive at agreement in the negotiation process. While we live in a highly assumptive world and need assumptions for psychological economy, it is the quality or validity of our assumptions that create problems. *Communicating and negotiating effectively require constant testing and evaluation of assumptions.*

Tunnel vision inhibits the effectiveness of negotiating and communicating by lowering sensitivity to others. Tunnel vision occurs when a communicator becomes task-oriented to the extent that he or she neglects, ignores, or overrides the communication of others with respect to their ideas and feelings. While the instant negotiation process outcome may be perceived as a win or success by the highly task-oriented negotiator, future relationships and negotiations may be adversely affected.

Misunderstandings also arise between sender and receiver when there is a *lack of specificity* in the communication. One frequently encountered example will illustrate this barrier as it occurs in both oral and written communication. Memos often end with "ASAP," but "as soon as possible" can mean any time from "instantly" to whenever a person "gets around to it," which may be a week or a month later. "By 3 p.m. Tuesday, January 12, 1983" is much more specific and less likely to be misunderstood.

There are many other conditions that distort or inhibit effective communication. Whether in the role of sender or receiver, negotiators should avoid them whenever possible.

1. *Interrupting senders* by completing sentences or superimposing expression of thoughts and feelings may often engender hostility.
2. The use of semantic ordinates or *absolute words* such as, "none, all, every, best, worst" may trap a communicator.
3. To understand others it is necessary on occasion to appreciate the *differences in frames of reference*. The popular expression "where he/she is coming from" refers to the frame of reference of the individual, comprising knowledge, experience, motivations, and attitudes.
4. The implications of *circularity of behavior* must be understood. In a negotiation, if one person becomes hostile or cold, similar behavior will probably be prompted in the other person and the hostility may be escalated. If warmth is exhibited (friendliness, admiration, respect, etc.), the other communicator will probably return that warmth. There are exceptions, but the probability of coldness being returned with warmth is slight. Smooth negotiations and resolution of issues are more likely to occur when there is a "warm atmosphere" between negotiators.
5. *Lack of credibility or trust* by either communicator in the other can inhibit effective communication.
6. *Ignorance* of the substance of the topic being discussed can create problems.
7. Holding *stereotypes* can create distortions of communication. "Lawyers are shysters," "Engineers are not human relations–oriented," "Salespersons talk too much" are typical sterotypes.
8. *Talking too much or not saying enough* about topics under consideration can cause communication problems.
9. *Perceived status differentials* also may inhibit effective communication. Often a subordinate will tell a superior what he or she thinks the superior would like to hear, rather than what the superior should be told.

Guidelines to Effective Communication

The possibilities of distortions that can be created in communicating with others are numerous. It is almost miraculous that, despite these barriers, people are still able to "reach" or "get through" to others. Communicating effectively is not an easy task. Learning to communicate effectively is not the same as learning how to build a bird house or knit a scarf. Communication reflects the personality of the individual, whether sender or receiver. Constant awareness of the axioms and possible barriers will help avoid ineffective communication in part, but much more than that is necessary. Some of what is required is far beyond the scope of a written article and has to do with modification of personality.

Nevertheless, some improvement in communication can be effected by observing some simple guidelines proposed by many managers in the 1940s. These guidelines appear in many places as the "Ten Commandments of Communication." They are as helpful today as when first formulated.

1. *Clarify your ideas* before communicating.
2. Consider the *total physical and human setting* whenever you communicate.
3. Examine the *true purpose* of your communication.
4. *Consult with others* in planning communications.
5. While you communicate, *watch for the overtones* as well as the basic content of your message.
6. Be sure your *actions* support your communications.
7. Be sure each message includes *help for the receiver* among its objectives.
8. Always seek to *understand* as well as to be understood; be a *good listener*.
9. *Follow up* your communication to find out what it meant to the receiver.
10. *Communicate for tomorrow* as well as today.

THE NEGOTIATION PROCESS

Goals of Negotiators

Negotiators in matrix system and other situations seek to achieve personal goals as well as organizational objectives. Motivations are complex, vary in intensity, change in intensity over time, and may be conflictive. The psychology of motivations is beyond the scope of this chapter. It is important, however, for negotiators to recognize some of the goals relating to motivations.

Some persons view a negotiating situation as an opportunity to gain power over others. Others may have such goals as achievement, recognition or gaining status, establishing or protecting affiliation with others, security, financial rewards, and the sheer excitement of the give-and-take in negotiations.

Matrix managers, as well as matrix system participants, should remain aware of these goals as they affect a negotiation. Information concerning an individual's goals can be obtained by an examination of prior communication with a particular individual. Some persons may express their goals during the negotiation process. A blockage of goal attainment can create frustration, evoke emotional behavior, and alter effective communication for the resolution of negotiating issues.

Recognizing a goal of a particular negotiator can assist the other negotiator in the persuasion effort. One way to show appreciation of a possible perceived blockage in attainment of an individual's goal is by a technique called *face-saving*. One negotiator may say to the other, "Joe, I can understand and appreciate your need to have this . . ., as it reflects on your . . ., however, . . ."—at which point he or she presents the other side of the issue.

Awareness of one's own goals and sensitivity to those of others in a negotiation will also help curb tunnel vision.

Tactics and Ploys

Many tactics and ploys have been developed and used over the years. Special labels have been given to some ploys such as "bad guy/good guy" "backdown," and others. Tactics are actions taken to gain a slight psychological advantage over the other negotiator. Some tactics, while providing a slight "edge," are also calculated to benefit both negotiators in holding negotiations on course.

Tactics and ploys must be carefully chosen to fit a particular negotiating circumstance. They are not magic answers to achieving negotiating ends. In most negotiations there are no "magic answers." As seen from the previous discussion, only hard work is required in dealing with the specifics of the negotiation, maintaining effective communication and relationships with others, and achieving negotiating objectives during both the process and the postnegotiation period.

One tactic mentioned earlier is *determining the site of the negotiation*. Whenever possible, a negotiator should hold the negotiation on his or her home ground. Failing that, a neutral site is next best. In some negotiating situations there may be no option of choice regarding site. Home ground provides comfort and availability of resources for some negotiators, while it may make others a little bit uneasy.

Preparation of an agenda is a tactic designed to provide structure to a negotiating session that can help both negotiators. It also provides a measure of control of the meeting for the agenda preparer. The agenda communicates to others that the preparer has done the necessary homework. Providing the agenda to other negotiators in advance of the meeting not only is a courtesy, but helps others to prepare as well. On occasion, another negotiator may also prepare an agenda, in which case a negotiation over the acceptance of the agenda and modifications thereto may be the first order of business.

Arranging the seating of the negotiators is also a tactic. An effective negotiator will occupy (when possible) a "seat of prominence." This position is one permitting the negotiator to have the best view of his or her team members and of the other negotiators. The arrangement of the seating at the negotiating table can have a direct effect on communication. Occupying the seat of prominence should be done as a natural action rather than as a display of dominance over the situation. Some negotiators try to get light behind their positions at the table, so that others will have difficulty seeing their facial reactions.

A set of probing questions, prepared in advance of the meeting, is a tactic helpful in determining the negotiating positions, attitudes, motivations, etc., of the other negotiator. Questions that probe feelings, attitudes, and the like should be open-ended.

Ending each meeting by *summarizing points of agreement* conveys a positive tone. If at all possible, a summary of the meeting marking the progress made should be written and sent to each participant. In certain instances where the negotiation is highly informal, a short written memo can serve as a record of the agreements reached.

The *number of participants* in a negotiation should be kept to a manageable size. This tactic prevents runaway conversations and accelerates the progress being made.

As a tactic, a negotiator should be prepared to *support oral statements with action.* Overpromising and nonfulfillment of commitments not only lower credibility but also create problems in the negotiation aftermath.

There are several other tactics that can be used, but attention is now turned to *ploys.* Ploys, as said before, are deliberate actions taken in attempt to get negotiators to change a position. Here are a few.

1. *Bad guy/good guy.* In a team negotiation, one member of the team is appointed to bear down on issues while another plays the role of rational conciliator. Care must be taken that the "bad guy" doesn't overdo the role, which might create hostility.
2. *Trial balloons.* This ploy takes the form of asking, "If we did this, would you then do that?" The trial balloon does not indicate that a concession will be made; rather, it is designed to determine whether there can be movement by the other negotiator from the stalemated position.
3. *Caucus.* A "time out" is called for any of a variety of reasons: issues are getting sticky, team regrouping, consideration of a concession, etc. A caucus can also make another negotiator uneasy, if he or she doesn't know what is being discussed.
4. *"You can do better than that."* Combined with sincerity and a little play acting, this statement, when judiciously timed, can prompt a reconsideration of an offer.

Maxims for Negotiators

A few negotiator maxims or guidelines have been previously mentioned. These had to do with buyer and seller entry positions and with not revealing nominals, maximals, and minimals. Additional guidelines to keep in mind include:

1. Assume that your opposite is always smarter than you. This helps to keep a negotiator alert to concessions, requests, demands or offers, and surprise tactics or ploys.
2. Test all assumptions made about the other negotiator's objectives, personal goals, and position.
3. Patience is a necessity in negotiating, yet many matrix managers are under time pressures. During a negotiation these time pressures may cause a precipitous concession or offer.
4. Silence is a powerful tool. Use it to get more communication from the other. Use it to listen and analyze.
5. Nothing is settled until everything is settled. All decisions should remain tentative until all associated issues are resolved.
6. Reduce agreements to writing whenever possible. The written paper won't implement an agreement, but it's hard for someone to retract or deny a commitment when it appears in print.
7. Take steps to avoid externally imposed solutions to resolve differences.

A Personal Planning Guide for Matrix Negotiators

To close the chapter, a planning guide for matrix negotiators is presented in a personalized way. Here are some steps you can take in preparing an effective negotiation.

1. Determine what you need.
2. Determine what you would like.
3. Establish trade-offs of what you would like for what you need.
4. Establish gradients to maximals and minimals from your nominals of what you need and want.
5. Substantiate the rationale for your requests or demands (at whatever gradient levels you choose) by supporting facts or data. Documentation helps.
6. Examine your prior relationships with other matrix participants. Review their behavior toward you, and vice versa. Ask yourself about possible conflicts you may have had in the past. Plan to "mend fences" prior to negotiating.
7. Examine your personal goals and the personal goals of others with whom you will negotiate. Ask yourself which of your personal goals you will delay achieving to accomplish prime matrix objectives.

8. Research (if possible) problems that other matrix participants may have.
9. Check with all parties you believe will be affected by negotiation outcomes. Determine whether the possible agreement will be a viable one.
10. Prepare an agenda for the negotiation. Notify other negotiators of time, place of meeting, others attending, and the agenda items. Prepare an opening statement of position, objectives, ground rules, and specific parameters.
11. Plan to make a record of the negotiations. A secretary or a tape recorder will do. If the negotiations are informal, plan to summarize in a memo.
12. Plan to summarize negotiation progress at the end of the meeting.
13. Choose tactics or ploys, if warranted, in a judicious way.
14. Plan to create and maintain a warm climate for the negotiation. Remember the implications of the circularity of behavior.
15. Review some of the major barriers to effective communication.
16. Try to provide an environment free from interruption and noise for the negotiation.
17. Prepare sets of probing questions to get answers that you need from others.
18. Be aware of the need for patience.
19. Remember to use the face-saving technique, effective listening, caucus, silence, and feedback. Focus on issues. Use a problem-solving approach to resolve differences.
20. If you require a team for negotiations, make certain they are adequately prepared.
21. Make sure you have an agreement as to acceptable performance criteria from you superiors.
22. Go after the whole package, but be prepared for trade-offs.

Negotiating in a matrix system is a complex affair. There are many different personalities with varying perceptions with whom the matrix manager and the other participants will have to work. Managerial and effective communication skills should be developed and used. Knowledge of behavior, motivations, and attitudes help one to negotiate effectively. The negotiator must also be a low-key persuasive salesperson so as to influence, rather than dominate, matrix participants. The effective negotiator should maintain awareness of the group dynamics that may be involved in any negotiation. Also, the matrix negotiator should be as flexible as organization structure and demands will permit.

Finally, since the people who negotiate in a matrix system will be doing so time after time, maintaining organizational cooperation and harmony must be high on the list of priorities for the matrix negotiator. Who knows who will need whom, and when?

REFERENCES

Bateson, G. *Steps to an Ecology of Mind* (Communication Section). New York: Chandler Publishing Co., 1972.

Blake, R. R., and Mouton, J. S. "Reactions to Intergroup Competition Under Win-Lose Conditions." *Management Science,* Vol. 4, No. 4 (July 1961).

Coffin, R. A. *The Negotiator—A Manual for Winners.* New York: American Management Associations, 1973.

Davis, Flora. *Inside Intuition.* New York: McGraw-Hill Book Co., 1971.

Fast, Julius. *Body Language.* New York: M. Evans and Co., Inc., 1970.

Hampton, D. R., Summer, C. E., and Webber, R. A. *Organizational Behavior and the Practice of Management, rev. ed.* Glenview, Ill.: Foresman and Co., 1973.

Illich, John. *Art and Skill of Successful Negotiation.* Englewood Cliffs, N.J.: Prentice-Hall, Inc., 1973.

Kahn, R. L., and Cannell, C. F. *The Dynamics of Interviewing. Theory, Technique and Cases.* New York: John Wiley & Sons, Inc., 1957.

Karrass, Chester L. *The Negotiating Game.* New York: World Publishing Co., 1970.

Montensen, C. David. *Communication: The Study of Human Interaction.* New York: McGraw-Hill Book Co., 1972.

Nierenberg, Gerard I. *Fundamentals of Negotiating.* New York: Hawthorn Books, 1974.

Wall, James A. "Effective Negotiation." *Business Horizons,* Oct. 1977.

Walton, R. E., and McKersie, R. B. *A Behavioral Theory of Labor Negotiations. An Analysis of a Social Interaction System.* New York: McGraw-Hill Book Co., 1965.

Watzlawick, P., Beavin, J. H., and Jackson, D. D. *Pragmatics of Human Communication. A Study of Interactional Patterns, Pathologies and Paradoxes.* New York: W. W. Norton & Co., 1967.

Weitz, S., ed. *Non-Verbal Communication.* New York: Oxford University Press, 1974.

Section VII
Organizational Strategy in Matrix
Management

In Section VII the subject of developing organizational strategy for matrix management is presented. This chapters in this section should prove useful to the manager who is considering the introduction of matrix into the organization. The topics dealt with include organizational development, employee participation, worker problem-solving groups, and innovation.

Andrew O. Manzini in Chapter 37 uses a brief discussion of the evolution and nature of matrix management to set the stage for a discussion of organization development. He uses the experience of many major engineering construction firms and several of their clients in his examination of organization development "interventions" and their contributions to matrix development and effectiveness. The author assumes that individuals can and do learn to behave in new ways, if help is given in the early phases of a matrix before "critical mass" in the change process has been achieved. He concludes that "unless individuals and groups receive support and help in adjusting to a matrix at this point, the matrix will probably fail."

In Chapter 38 Gary D. Kissler presents material on the role of employee participation in the context of matrix management. He believes we are witnessing a fundamental change in the way employees view the legitimacy of having greater influence over their work lives. He discusses some of the historical precedents in traditional organizations where employee schemas have developed. Then a strategy is given for beginning the matrix participation process. Finally, the author presents some "real-life" information about matrix participation.

Chapter 39 by Joe Kelly describes how employee participation can be made productive. He poses the need, as we move through the 1980s, to evaluate and ascertain how participation has stood the test. Then the author examines the factors arguing for participative management. The "Japanese experiment" is reviewed, followed by a brief description of "Theory Z" and the Swedish Volvo experiment. He examines other participative attempts that include Glacier Metal Company in England, Quality Circles, the Scanlon Plan, General Motors at Tarrytown, and General Foods at Topeka. Finally, some of the problems with participation are cited. In concluding, the author looks at the role of the Chief Executive as a philosopher-king.

In Chapter 40 Gervase Bushe, Drew P. Danko, and Kathleen J. Long talk about a structure for successful worker problem-solving groups. They begin by examining the deficiencies with the most popular Quality Circle programs currently "on the market." These programs are examined within the context of the typical structure and process of manufacturing organizations. Then they present a structure model they have developed and discuss how it deals with the deficiencies of current Quality Circle programs. Finally, the use of co-ordinators in implementing their structural model is treated.

Joseph G. P. Paolillo in Chapter 41 describes the use of project teams as an alternative to enhancing innovation. He discusses the major types of innovation in organizational settings, then provides the results of a study to gain insight into the design of innovative R & D subsystems using project teams to enhance innovation. He concludes by raising a key question: what is the optimum size of research project teams, assuming such teams do indeed enhance innovativeness?

The concluding chapter of this handbook by David I. Cleland presents an approach that can be used as a model to introduce matrix management into an organization. A large transnational corporation is discussed as an example of how an international matrix management system is introduced into an organization. Several models are presented to depict what strategy was used to introduce and integrate the matrix into the corporation culture. The author emphasizes the role of communication in implementing matrix management, and concludes by citing some caveats to consider when initiating and using the matrix design.

37. Enhancing the Effectiveness of Matrix Management with Organization Development Interventions

Andrew O. Manzini*

THE EVOLUTION OF MATRIX MANAGEMENT

Differentiation and Integration

Traditionally, engineering and construction organizations engaged in major project work and in a high technology field have always developed and maintained technical expertise by centralizing resources into well-defined areas such as electrical, civil, mechanical, and chemical engineering disciplines. Each discipline is differentiated from the other by virtue of the unique work it has to do. In the words of Lawrence and Lorsch, this concentration of similar resources is called *differentiation*.[1] However, the demands of an ever increasing technology, and a project time span that for large facilities such as nuclear plants can run as high as 12 years, mean that high interdiscipline cooperation and communication are almost impossible without some form of central coordination. The efforts of differentiated disciplines have to be *integrated* so that they can work together toward a common goal. This need for coordinated control is what led to the establishment of integrative mechanisms such as project management.

The Traditional Organization

It is fair to say that traditional organizations have been effective for a considerable period of time. In this kind of organization, each major organizational unit reports to a top manager (see Figure 37-1). There were good reasons for

*Andrew O. Manzini is presently Vice-President of Human Resources for Ebasco Services Inc. of New York, a large engineering, construction, and consulting organization in the energy generation field. Mr. Manzini's primary interests are in the design and implementation of integrated strategic human resources planning systems, large-scale organizational diagnostic processes, strategic planning and management, OD interventions for team development, and matrix and project management design and implementation. Mr. Manzini has an M.A. in international relations from Boston University, a B.A. in political science from Widener University, and is a graduate of the Columbia University Advanced Program in Organization Development and Human Resource Management.

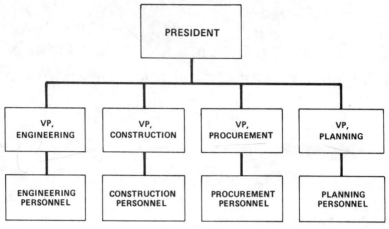

Fig. 37-1. Traditional organization.

this structure. Some of the strengths of traditional organizations are easy to identify:

1. It is considered easier to manage engineers and other technical personnel if they are grouped together and are responsible to a department head who has training and experience in that particular discipline.
2. The strength of the functional discipline organization lies in its centralization of similar resources.
3. A functional department, such as an engineering discipline, provides a reasonably secure and comfortable management with well-defined career paths for a young engineer. Also, mutual support and identification is provided by physical proximity.

On the other hand, management under the traditional organization became increasingly difficult owing to progressive increases in scope and complexity of modern technical projects.

1. When the traditional discipline-oriented organization is involved in numerous large projects or a single very complex project, conflicts arise over relative project priorities in competition for resources.
2. The functional discipline department often places more emphasis on its own work than on the goals of the project.
3. Large projects are difficult to design and build today without central project direction and coordination.

Since the limitations of traditional management structure could not be overcome, several initial experiments with alternative organization design led to the establishment of a function now known as project management. There are essentially two approaches to project management, each with its own advantages and disadvantages.

The Projectized Organization

In this organization, all resources necessary to perform a complex project are separated from their regular functional disciplines. These resources are then set up as a self-contained unit headed by a project manager. Advantages of such an approach are obvious:

1. The project manager is given considerable authority over the project, and all personnel on the project are under his or her direct authority for the duration of the project. Essentially, a projectized organization gives a project manager all the key resources to complete the project.
2. In effect, a large organization sets up a smaller, relatively temporary special-purpose structure with a specific objective—the completion of a project.
3. The internal structure of a projectized organization, however, is functional. The project team is divided into various functional areas (see Figure 37-2).

Disadvantages of this approach are also evident:

1. Setting up a new, highly visible temporary structure may upset the regular organization.
2. Duplication of facilities and inefficient use of resources are inevitable. As long as personnel are assigned to a project, they are not available elsewhere. Since resources are seldom plentiful, this constraint is of major concern.
3. A serious problem in a projectized organization is the question of job security upon termination of the project. Technical personnel are usually concerned lest they lose their "home" in the disciplines while assigned to the project.
4. Technical disciplines usually perceive a loss of control, of power, and of personnel loyalty. In some cases lessened control over quality of work is manifested.

The Matrix Organization

The other alternative structure in project management is the matrix organization. Matrix organization maintains a balance between technical disciplines and project management to assure in-depth support of the project. A matrix organization may be considered as one that falls roughly between a functional

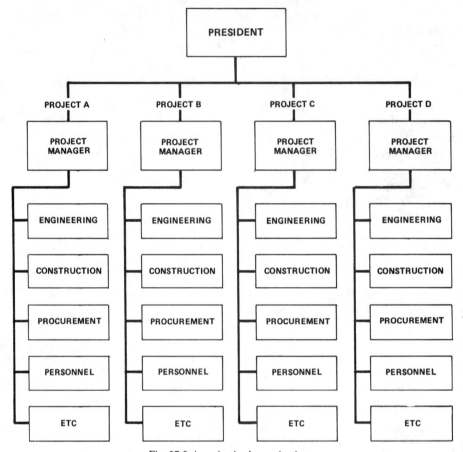

Fig. 37-2 A projectized organization.

and a projectized organization. It is a system that adds horizontal reporting requirements to the traditional vertical chain of command (see Figure 37-3).

There are many reasons why large technical organizations utilize matrix management for controlling complex major project work. Not only are the limitations of both traditional and projectized organizations avoided, but also many important advantages are gained. The engineer or technical worker is still part of his or her department and is also a member of the project group. As a result:

1. There is a real balancing of objectives.
2. There is improved coordination across functional department lines.
3. A steady visibility of project objectives is maintained through the person of the Project Manager.

Organizationally, a matrix organization would be as shown in Figure 37-4. The implementation of a project management matrix structure essentially provides for an independent organization whose activities, by definition, cut across departmental lines in the overall organization. For the project management function to be effective, its role and operating parameters need to be carefully defined.

In such a framework, each engineering discipline such as civil, mechanical, or electrical is headed by a lead discipline engineer. It is his or her responsibility to ensure that all work for that particular discipline is done according to specifications, within budget, and on time. The overall responsibility for coordinating activities of all engineering disciplines is in the hands of a higher-level manager, generally called project engineer.

As previously discussed, reporting relationships within the above framework can be established in two ways, depending on whether the project is set up in a projectized organization or in a matrix organization. In a projectized organization, all personnel report on a direct line to the project manager. On a dotted line, they report back to their functional managers and supervisors at the technical discipline level. The project team essentially belongs to the project until its completion, at which point they revert back to their home technical department. However, if the project is organized along a matrix, the key people would then report on a solid line to their discipline supervisor, from whom they receive technical direction. On the other hand, they are also directly responsible to the project manager, who in turn is responsible for project direction and primarily for cost and schedule management. In the matrix, key people are asked to report to two bosses and receive evaluations from both of them. Should a conflict arise that impacts on the priorities of the matrixed individual, it should be resolved at the project manager/technical manager level.

Essentially, a matrix organization by its very nature demands a power balance between the technical groups and project management. It thus introduces the concept of dual loyalties to both technical disciplines and projects, and the difficult concept of working for two bosses.

It is essential that the position of project manager be defined in such a way

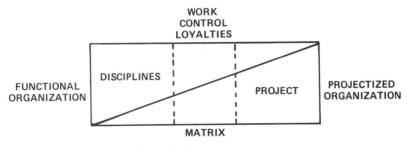

Fig. 37-3. The matrix concept.

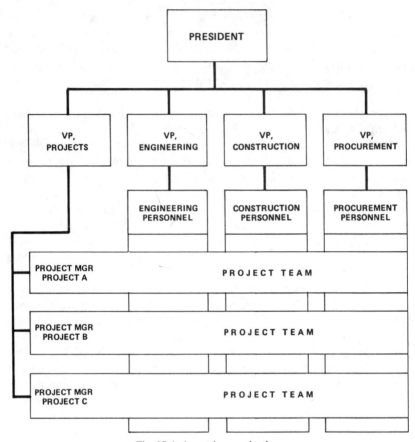

Fig. 37-4. A matrix organization.

that he or she is the person responsible for overall coordination and management of an interdepartmental project. The project manager needs to be at a level higher than the rest of the project team managers. In an effective matrix, the key discipline managers needs to be accountable not only to their functional supervisor, but also to the project manager. The project manager is responsible for coordinating activities of the project team, and for managing project budget and schedule.

Obviously, an organizational change to matrix is, at best, difficult to manage. Moreover, care must be exercised to plan properly for full implementation of such a system. The predictable flaw in most matrixes is one of implementation rather than design. One cannot set up a basically sound structure without adequate consideration of other elements such as processes, reward systems, and the readiness of the people themselves. Quite often individuals placed in these new roles are not adequately counseled and coached about what their new roles demand. Thus they have a tendency to interpret the job in accordance with their personal inclinations.[2] Furthermore, there is a tendency to assume that the project manager can perform his difficult role without formal preparation. However, without careful preparation, significant problems are likely to surface. Inadequate understanding of a matrix can aggravate fears that the matrix system would weaken other areas within the larger organization.

Problems and conflicts that develop in a malfunctioning matrix are caused by frustration. This frustration results from inadequate efforts to systematically bring project team members together to help them sort out their roles in this new context. Ambiguity of roles confirms fears regarding perceived reduction of traditional roles, and the ability to perform quality work. This frustration can lead managers to engage in political ploys to maintain power within the organization. Since one aspect of power is the ability to control what is going on, managers might employ such tactics as:

1. Withholding information from the project manager when he or she needs it
2. Challenging the "newcomer"
3. Working to the rules (which will stop any organization cold)

On their part, the new project managers in their new role are likely to exercise behavior more appropriate to the projectized organization than the matrix organization, further alienating other groups.

Experience in many organizations clearly indicates that it is not sufficient to tell people that they are being shifted from a conventional line manager to a manager in the matrix. To be effective, they must quickly build effective working relationships with others in their matrix. It is too risky to let chance events in their contact form the character and process of the group. If left to chance alone, the group or individuals in it may develop patterns that destroy trust and reduce the potential effectiveness of their relationship. Furthermore, individuals left to their own devices do not benefit from what is already known about issues that are generally faced in the process of matrix development.

A matrix organization is at best a difficult way to manage. Furthermore, it is not a management structure that can be implemented in an instant, and rarely does it operate smoothly. The best that can be done is to make it work at an acceptable level.

ORGANIZATION DEVELOPMENT INTERVENTIONS

What can be done, then, to design and implement a matrix so that its chances of success are maximized? An answer to this question is provided by techniques that have evolved from traditional organization development concepts. Experience with major engineering construction firms and several of their clients provides a framework for examining organization development interventions and their contribution to matrix development and effectiveness. An analysis of

70 such interventions makes possible the elaboration of a process that can be effective in most organizations contemplating implementation of a matrix.

Planned interventions in support of matrix systems can be used for the following purposes:

1. Organization diagnosis leading to a decision to establish a matrix
2. Change strategy and tactics
3. Organization design using open systems planning and development criteria to ensure a proper balance of factors crucial to the effectiveness of the new organization
4. Identification of key matrix managers
5. Training in the matrix concept
6. Role analysis sessions for key personnel
7. Project team building
8. Intergroup team building
9. Conflict resolution sessions, as necessary

The remainder of this discussion describes the various interventions listed above. However, emphasis is on interventions directly impacting on matrix implementation.

Organization Diagnosis

This is perhaps the most powerful intervention available, and one of the most difficult to sell to management. The reason for this is that a properly conducted diagnosis will examine an organization to its core. The diagnosis is analogous to a complete physical for an individual, including a psychoanalytic examination. The most comprehensive model of organization diagnosis was developed by Harry Levinson.[3] This diagnosis examines an organization from the day of its establishment and traces its development through the years. It identifies the various challenges and crises with which it has dealt, and correlates past coping behavior with present practice and processes. This past behavior may or may not be appropriate for the present environment. Thus an explanation for current organizational behavior is obtained. In addition to this historical perspective, this diagnostic model provides analysis of current organizational functioning. The net effect is a comprehensive evaluation of an organization's past, its present strengths and weaknesses, and its capability to face present and future challenges.

Other models, in various degress of sophistication, attempt to do similar things with varying degrees of effectiveness—Mahler[4] and Weisbord,[5] for example. Regardless of the model used, and organizational diagnosis is highly desirable prior to making any changes, because changing to a matrix management approach involves a radical restructuring of an organization's norms, behavior, and processes. If other means of organization are adequate, they should definitely be considered. Matrix management should be adopted only if it is the only alternative available to the organization, given its situation and resources. A matrix is expensive, cumbersome, and hard to manage. Furthermore, it will never be smooth. If nothing else, a diagnosis should help in making the decision.

If the benefits of a diagnosis are easily ascertained, why is this methodology difficult to sell to management? Primarily, because it reveals the weaknesses of the organization and what led to them, which is often threatening to managers who may have been part of the problem. Furthermore, Levinson states that managers who are forced to ask for help from a consultant feel subconsciously guilty for not being able to solve the problem themselves, and thus resist outside involvement or consultation.[6]

One way to minimize resistance on the part of management is to make them part of the process. In one successful example, the top twelve officers of a major engineering comany were designated as an Organization Development Council.[7] The chief internal consultant spent the necessary time to explain to the council the concepts of organization development and presented them with various choices of diagnostic models. The council eventually selected a suitable model and authorized the study. In this case, the council was the client. To minimize resistance in the ranks, the internal consultants conducted a preliminary "power diagnosis." By asking people who they thought had power and influence in the organization, and by verifying these perceptions with the nominees, they were able to identify 24 managers who apparently had power to help or hinder the process of development. To minimize potential opposition to the effort, these managers were incorporated as part of the diagnostic team and made a substantial contribution to the comprehensive diagnosis that was produced 14 months later.

Change Strategies and Tactics

Upon completion of a diagnosis, several key changes in the organization may be necessary. If implementation of matrix management is justified, a strategy needs to be developed to ensure its proper design. In addition, plans must be formulated to manage the transition period between the present organization and the matrix. Key players and affected departments and managers must be identified. Organizational reaction to the changes has to be anticipated, and plans to deal with it must be prepared. In addition, resources required to effect the change need to be identified and funded.

Organization Design

Design of a matrix involves some reformation of the functional organization. At the very least, it involves establishing a project function that cuts across various organizational units. Organization design criteria are crucial and must be an integral part of the new system. Galbraith[8] suggests the incorporation of several mutually reinforcing parameters:

1. *Task,* its definition in terms of diversity, difficulty and variability.
2. *People,* or a means of properly handling human resources through training, selection-recruiting and promotion-transfer.
3. *Reward systems* designed to reinforce behavior appropriate for a matrix through compensation, promotion, development of appropriate leadership styles, and job design.
4. *Information* and decision processes that provide for an adequate decision mechanism, identify frequencies of communication, define the degree of formality in communication and decision making, and establish a centralized data base for all to use.

All these factors facilitate the design of the ultimate structure, with a viable mode of organization in terms of division of labor, departmentalization, configuration, and distribution of power.

Identification of Key Matrix Managers

In most cases key matrix managers other than those who are assigned to the project management function should already be in place. These are the heads of the functional disciplines who heretofore have been the principal players in the organization. With the establishment of a project management function, these managers must accept the new requirement for sharing power and influence. They must also accept the fact that their engineers and scientists are no longer responsible to them alone, but to the project function as well. This is the time to ascertain their ability and willingness to do so. Some people may be temperamentally unable to do so. Others, with guidance and training, may effectively make the transition. The important point is to assure that all the key managers make a commitment to the new organization. Otherwise, destructive power struggles can be anticipated.

As far as the candidates for project management positions are concerned, their selection is crucial. The initial implementation of a matrix organization is highly dependent on their ability to achieve synergy among the various contributors. A major pitfall into which many organizations have fallen is to identify appropriate and qualified engineers as project managers without considering leadership and interpersonal skills crucial to effective management performance. The result in many cases has been that the organization lost good engineers and got poor managers. Just as damaging, the first few projects implemented under those conditions were less than successful, thus giving the opponents of the matrix fuel to vent criticism of this method. During the transition to matrix, extreme care must be taken to select the right key managers and to ensure their commitment to the difficult task ahead.

Training in the Matrix Concept

Before embarking on a matrix project, it is essential to train every member of the organization to ensure that they not only understand the concept of the matrix but accept it. It is particularly important that they recognize the legitimacy of reporting to more than one manager and understand and accept the concepts of dual authority, dual responsibility, and dual accountability.

Role Analysis Sessions for Key Personnel

This is one of the more important interventions to ensure success of a new matrix. It is not sufficient merely to tell people that they will be shifted from being a conventional line manager to a manager in a matrix. A formal definition of their role does not suffice, because human perceptions lead to misunderstandings and ultimately conflict, no matter what the procedures say. Pfeiffer and Jones describe a common way of starting a role analysis session by pointing out that a *role definition,* as specified in the procedures, is rarely sufficient as an explanation of respective duties, responsibilities, and interfaces.[9] This is because each key individual in a matrix has a unique *perception* or *understanding* of his or her own role and that of colleagues. This role perception may or may not match the intent of the official definition. Furthermore, there is the issue of *role expectations.* These are the expectations and degrees of understanding of a person's role on the part of others in the matrix. These may diverge from both the official role and the individual's own perception of the role. Last, we have to deal with the problem of *role acceptance.* Individuals may understand the official role definition well enough; they may be clear on what others expect of them in their role; but they may not accept it either overtly or covertly. A role analysis session is invaluable because it enables a matrix project team to deal with these issues in a collaborative manner.

The role analysis process is relatively simple. A facilitator asks the participants to write on a piece of paper their perception of their own role and to do the same for the roles of each of their colleagues. Once this is done, each individual is asked to transfer the perceptions of his or her own role to a flip chart and to explain this perspective to the others. The latter may point out differences and disagreements. Thus each person's role is thoroughly discussed and defined to everyone's satisfaction. Each member of the team takes turns in publishing and discussing his or her role. The output is typed and distributed to all participants for continuous reference.

Project Team Building

Once the key members of a project team have been assembled, it is helpful to meet regularly away from the daily pressures of the job. This gives them the opportunity to examine how they work together and to find ways to maximize their effectiveness as a team.

To illustrate the procedures and techniques employed in project team building, one series of interventions will be described. These interventions were conducted to improve the situation affecting the management of a large power plant project. When these interventions were proposed, the purpose was listed as follows:

1. Clarify responsibilities
2. Identify information needs/outputs
3. Determine more effective ways of interfacing
4. Develop specific action plans to correct and/or avoid schedule problems

The OD facilitators proposed a two-phase approach to team development to solve problems of company-client interface. The first phase was based on the assumption that working out the interface relationship would be easier and more effective if the company team were strengthened first.

A two-hour meeting of the Senior Vice-President of Projects, the Vice-President of Engineering, the Manager of Projects, the project manager, and the project engineer was held. This was followed by a two-day team-building session with key members of the project: the project manager, assistant project manager, project engineer, lead discipline engineers, project expediter, and project purchasing agent.

Collection of Interview Data.   After the first meeting was held, the OD facilitators conducted interviews with all members of the project team selected as participants. All data obtained from the interviews were grouped into major categories:

- Project team effectiveness
- Roles and responsibilities
- Work relationships
- Work processes
- Problem solving

A data base was prepared that presented a compilation of the interview data divided into the above categories, some of which were further broken down to allow for more detailed differentiation. The interview data clearly reflected the pressures, concerns, and problems encountered by the project team, which was trying to meet a very tight project schedule. Some respondents characterized the schedule as "impossible" and "totally unreasonable."

Data Feedback Session.   A half-day data feedback meeting was then scheduled. The specific purpose was outlined as follows:

1. To present a summary of problems or issues identified in the interviews that affect the smooth operation of the project team.
2. To gain a common understanding of these problems/issues.
3. To arrange the issues in order of priority and importance to the members as a team.
4. To determine how, as a team, the participants are going to address the problems/issues in need of resolution.

Following a brief introduction to the concept and purpose of team building, the project team was given a group problem-solving exercise. The purpose of the exercise was to highlight the differences between facts and perceptions, and to point out the actions required of the group to overcome those differences when engaged in a problem-solving mode.

The data about the team, collected via individual interviews and categorized, were presented to the team members. Questions resulting in further clarification and understanding of the issues were encouraged and considered. The project team was divided into three groups and asked to:

- Indicate their feelings about this team-building effort
- Identify any other issues that had not been addressed
- Rank the issues in order of priority

They then ranked the issues in order of priority.

1. Communication
2. Interfaces between disciplines
3. Design changes and out-of-scope changes
4. Project personnel turnover
5. Paper work
6. Insufficient technicians

It was acknowledged, however, that the interface relationship between various disciplines in the project team is central to the team's ability to function effectively. If those relationships were improved, many of the other identified issues would become less of a concern and the team could manage them more effectively.

Following receipt of the report of this meeting, the project team agreed to:

1. Have representatives of the various disciplines participate in an intergroup team-building session

2. Establish action committees to work toward resolving each of the other prioritized issues
3. Establish a special committee to investigate how to optimize the balance between meeting the project schedule while maintaining high quality standards. Team members were chosen for this committee.

Project Interface Meeting.   As the first item agreed upon by the project team, a project interface meeting was held two months later, involving the project's supervising engineers and two project engineers, to deal with the issues relating to the interfaces between disciplines. Each supervising engineer and project engineer was again briefly interviewed prior to the meeting.

Following a review of the major steps taken thus far in the project's team-building effort, the participants were asked to voice their expectations of what they hoped to accomplish at this meeting. The expectations were posted in flip chart paper for all to see. They were then given a short lecture on the importance of continually examining the process by which work on the project gets done. The purpose of the "lecturette" was to sensitize the particpants to the skills needed to resolve the interfacing issues.

Next, statements were presented regarding interfacing elements collected during interviews. Participants were asked to consider how this data compared with concerns voiced at the start of the meeting. The group then attempted to resolve two of the interfacing issues that had been raised. Resolution of each of the issues was facilitated by having the two "sides" of the issue describe to each other the effects of the issue on them. Resulting agreements and commitments were then reduced to writing.

Follow-Up Project Team Building Meeting.   Six months later a follow-up project team-building meeting attended by 14 members of the team was held. The purpose of this session was to:

1. Identify the progress made to date in the project's team-building activites
2. Review the original list of project team concerns/issues and determine if those issues were still relevant
3. Identify any new project team concerns/issues that had surfaced since the last meeting
4. Set new priorities for resolving relevant issues
5. Agree upon a set of action plans for each priority concern noted.

Before the meeting started, each of the seven prioritized issues that were identified in the feedback session as being of primary importance to the project had been listed on separate sheets of flip chart paper. These papers were posted around the room. Members of the project team were asked to describe specific changes they perceived had taken place within the last six months. Their comments, both positive and negative, were noted on the appropriate flip chart paper.

The participants were then provided a summary of the formal project team-building activities and their outcomes to date. Specific resolutions were reviewed and participants were asked to identify any additional changes in the issues as a result of these resolutions. These changes were also noted under the appropriate issue on the flip chart page.

Each of the original seven prioritized issues was then discussed further and judged as to whether or not it was still an issue worthy of further resolution. If an issue was still considered a concern, clarification was sought to highlight the specific aspect that required further resolution. New issues and concerns that had surfaced since the last meeting were then identified and recorded.

Each participant was asked to choose those four or five issues from the entire identified list of concerns whose resolution was of criticial importance to the project. These choices were then combined to form a single priority listing. Redundant or overlapping statements were stricken from the list of issues requiring resolution. Participants then identified who would be responsible for investigating each concern and developing a recommendation or plan for resolving that issue. The project team members chose 11 issues as being critical to the operational effectiveness of the project. They also assigned resolution responsibilities and deadlines for each issue.

Thereafter a form was distributed to those individuals who had participated in the team building. It was also distributed to team members who had not attended the meeting but who could offer constructive insight. This form contained a list of the issues and a status code to be used for evaluation and comment.

Intergroup Team Building

In addition to strengthening individual project teams, effective transition to matrix often necessitates interventions between interfacing groups. Intergroup team building attempts to reduce stress, gain cooperation and harmony, and provide effective communication and collaboration between two interfacing groups. In the beginning stages, it is helpful to ask management to provide a list of priority intergroup sessions for this purpose to be held regularly over a period of time. In one specific organization, department heads selected five representatives to participate in the intergroup sessions according to the following guidelines. It was stressed that participants must have:

1. Ability to respresent fairly the views and interest of the participants' departments
2. Authority to commit their departments to mutually negotiated agreements

3. A high degree of influence among members of their departments to assure implementation of agreements

Each session was totally independent and dealt only with the issues of the two participating departments. Sessions were held away from the workplace and lasted two days. The meetings were facilitated by three internal OD consultants: one for each group, and the third as a session leader and coordinator. Facilitators followed Argyris's guidelines for intervention:[10]

1. Generation of valid and useful data
2. Preservation of free choice by participants to maintain their discreteness and autonomy
3. Generation of internal commitment to follow through on the participants' choice and its implications

Facilitators prepared extensive documentation that included all proceedings of meetings. These records were later used by participants for follow-up.

Sessions were preceded by one short preplanning meeting involving designated representatives and facilitators. This meeting was held so that participants could understand the intergroup process and express concerns. In addition, facilitators were able to identify the main issues of concern expressed by the participants. This ensured that the intervention would be appropriate. Another purpose of the presession was to ensure participant commitment.

The actual intergroup sessions started with five participants from each group. Three facilitators reviewed the purpose of the intervention and anticipated outcome. Then the groups were placed in two separate rooms and were asked by their facilitators to list answers to the following questions:

1. What does the other group do that is helpful and valuable to this group?
2. What really "bugs" you about the other group? What does it do that hinders this group?
3. What will the other group say about things that are valuable or not helpful about this group?

The facilitators recorded all comments on flip charts. The coordinator went back and forth between the two groups to ensure consistency of approach. Following this listing of comments, participants reconvened in a common room and exchanged lists. With the guidance of the facilitators, they then clarified misconceptions, identified areas of agreement, and prepared a prioritized list of issues. In addition to the listing of issues, resolutions were negotiated by both groups and specific action plans were formulated to implement solutions.

The following are the basic steps utilized by the group in addressing each item on the problem list and in developing action to resolve each issue.

Problem-Solving Structure and Process. The five problem-solving steps are:

1. Define and clarify unclear terms for each item on the problem list.
2. Establish the nature and extent of each problem.
   a. What is happening
   b. To what extent
   c. When
   d. Who is involved
   e. Where
3. Identify where you want to be, that is, the objective or outcome desired. (State what conditions would exist if the problem were resolved.)
4. Determine alternative solutions or steps for moving from the existing condition to the desired outcome.
5. Establish an action plan.

Illustration of the Intergroup Team Building Process. To illustrate the intervention process, a session between the construction and engineering departments of a firm is used as an example of two interfacing groups. The work of these departments is internally complicated by the coordination required for the many and varied engineering and construction disciplines. Further, external complications involve changing government regulation, environmental issues, varying client's needs, etc. The issues raised by the participants of the construction-engineering teams reflected the individual and particular difficulties encountered by each group. Among the problems raised were:

Technical Issues Concerns were expressed by construction about project schedule development and implementation. Complaints were made about the quality of specifications and drawings in certain projects. Suggestions were made to reexamine the constructibility review process, so that drawings would consider physical aspects of construction.

Interpersonal Issues Construction engineering felt that they were not appreciated nor utilized effectively by central engineering, particularly regarding the level of responsibility assigned to site personnel. The question was also raised as to how engineering and projects could enhance their relationship.

Structural and Territorial Disputes There was an expression of the need to clarify the role of project engineer in such areas as technical decisions and scope of work change. It was suggested that the role specification of the project engineer needed review. Also, the quality assurance and construction departments wanted to reexamine overlapping activities and responsibilities, so they could establish clear and mutually acceptable roles.

Communication Problems   One of the problems raised was the need for projects to communicate more effectively to engineering. Specifics included early definitions of contractual obligations, schedules, and work sequences.

Within two weeks of each intergroup session, the facilitators submitted to participants a preliminary report for review and comments. Following adjustment and modification of this draft, a formal report signed by the participants was sent to department heads for written approval of recommendations and action plans.

Follow-Up Study.   Six months after completing the final intergroup intervention, a formal follow-up study was conducted and the findings submitted to management.

Purpose   The purpose of the follow-up study was to:

- Determine the current relevancy of the identified interface issues.
- Analyze the extent of implementation of agreements and results achieved.
- Evaluate the overall effectiveness of the intergroup process.
- Develop recommendations for future intergroup processes.

Procedure   The study was conducted with the original participants of the intergroup sessions. Continuity of representation was maintained since the same participants identified issues, negotiated agreements, and ensured implementation of action items. These participants were given a questionnaire survey and worksheets containing a record of each session typed on word-processing equipment. These work sheets listed the issues, original action plans, and agreements. In addition, the work sheets provided space for action plan follow-up and current status of issues. For a sample of word-processing output of the follow-up study, see Table 37-1.

Participants were asked to rate and comment on the status of each issue according to the following scale:

1. Issue has been resolved.
2. Issue unresolved; action has been taken.
3. Issue unresolved; no action taken.
4. Issue has been expanded.
5. Issue no longer of concern.

After all responses had been collected, they were added to the word-processing data. The completed material was then ready for analysis.

Analysis of follow-up study results has led to the following observations regarding the value of intergroup process and technique in resolving interface issues:

1. Intergroup interventions do provide a significant forum for identifying and diagnosing problems affecting the interface between organizational units.
2. Intergroup interventions are significant in addressing general role definitions, communication, and policy issues.
3. Commitments and agreements reached during intergroup meetings tend to be implemented more often when senior members of management who have authority are present and participating.
4. In cases where participants are not personally affected by, responsible for, or assigned responsiblity for resolution of problems raised in the intergroup meetings, they perceive action plans as only recommendations to senior management. As a result, they await senior management's endorsement and responses before taking any initiative.
5. Resolution of interface issues is, in general, most effectively accomplished on a project-by-project basis. Such an approach ensures that participants raise problems that, for the most part, can be addressed, managed, and evaluated by them.

OD Interventions: Summary

The usefulness of the intergroup technique for major project-type organizations is intrinsically related to the complexities of the matrix organization. This is highlighted by the continuous need simultaneously to balance organizational, project, and departmental goals when solving problems or conflicts.

From the findings of this study, the intergroup technique serves as a strategy for increasing interaction and communication among groups, developing a superordinate goal (a goal that both groups desire to achieve), and establishing collaboration among groups for problem solving and conflict resolution.

1. In general, the intergroup process is likely to be most effective in the following situations:

   - Plans indicate two organizations or departments will be working closely on a new project or series of projects.
   - Management is seeking creative solutions and recognizes that the thinking of the two departments will be essential for optimum results.
   - Two departments or units on a project have (or should have) frequent contact with each other; there is an interest in achieving a more effective working relationship.

Table 37-1. Follow-up Study Format Example of Word-Processing Output.

PROJECT: INTERGROUP FOLLOW-UP   DATE PRINTED: 11/15/79   Page 3 of 4
DATE 11/15/79, TYPIST: Page? n DISKETTE NO.: MISC D-653 PAGE 3

Reference: J Silvert Memo—September 15, 1978   CONSTRUCTION—ENGINEERING INTERGROUP
                                                          (FOLLOW-UP)
            To: E B Golden F O Schartz   LAKE HILLS PROJECT
            Subject CONSTRUCTION—ENGINEERING
                     TEAM-BUILDING MEETING
                     AGREEMENTS AND ACTION PLANS
            A B Mann Memo (confidential)—December 5, 1978
            To: R C Jones
            Subject: INTERGROUP FOLLOW-UP

ISSUE	ORIGINAL ACTION PLANS/AGREEMENTS	ACTION PLAN FOLLOW-UP	CURRENT STATUS OF ISSUE
3. Schedule (Cont'd)			
b. The need to modify construction sequence to accommodate what engineering can provide within limits of maintaining completion date.	b. Modify construction sequence to accommodate Engineering	As of 12/5/78 both Construction and Engineering had agreed to the dates and were working toward their attainment	Implemented and working.

674

c. The need to change strategy— sequence of installation in accordance with 3.b.

d. The need to identify priority items.

e. The need to maximize Level 3 and 4 schedules and marry them to engineering drawing schedules.

f. The need to implement bill of materials program.

Additional Action Plans Related to Schedule

— J Carter and R Roland are to work on resolving conflict between field's schedule and project milestone schedule.
— The field has agreed to revise CPM logic to accommodate engineering.

c. Change strategy—sequence of installation.

OK and working.

Accomplished manually.

d. Identify priority items; J W Hall, and R Smith were to resolve drawing priorities

e. Construction was to provide planners to assist R Smith in examining Level 3 and 4 schedules to marry them to engineering drawing schedule.

f. Bill of materials program was to be implemented.

As of 12/5/78, this was being implemented.

As of 12/5/78, these were in process of being implemented.

B/M program continuing as required. Most items which require a long lead time or bulk quantities have been identified by Engineering prior to drawing completion i.e. rebar support steel.

675

2. In cases where broad organizational interface issues are to be addressed, the following involvement of senior management is likely to be helpful:

- Senior Management identifies the critical areas to be discussed.
- Senior Managers from represented organizations jointly determine areas to be addressed in situation of conflict.
- Senior Management chooses key individuals with appropriate experience, expertise, and authority to represent their organizations.
- Senior management, in a "kick-off" session, communicates its perspective of critical issues to be addressed and the authority to be exercised by individuals participating in the meetings.
- Senior management openly and systematically endorses or implements the recommendations and agreements reached in the meetings.

3. For intergroup interventions to accomplish more effectively the basic goals of increasing communication, collaboration, and problem solving between groups, interface issues should be addressed on a project, rather than a generic, basis. Such an approach meets the critical criteria for resolving problems and conflict by ensuring that participants:

- Share a mutual interest and incentive for reaching resolution
- Are "close" to the problem and possess the necessary experience and knowledge
- Have the responsibility, authority, and necessary control for the implementation and follow-up steps to be taken
- Have the optimum commitment

The precedent set by all successful interventions in the organization supports this approach. Specifically, in each case there have been action plans implemented by the participants as well as ongoing follow-ups conducted by the participants themselves.

The effectiveness of intergroup interventions in project-type matrix organizations depends on:

1. Senior management's initiative, direction, and visible involvement.
2. Selection of participants based on their expertise, knowledge of the situation, ability to represent their organization fairly, and ability to implement action plans and agreements developed at the sessions.
3. Implementation of the intergroup process on a project-by-project basis rather than on a generic basis.

The findings in the follow-up study demonstrate that intergroup meetings provide a forum for identifying and diagnosing problems affecting the interface between two groups. Intergroup meetings are significant in addressing general role definitions and communication and policy issues. The findings also show that the intergroup process does not guarantee resolution of all interface issues. However, they do demonstrate the part the intergroup technique plays in ensuring that problems and conflicts are openly discussed within a structured context and mutually resolved. Properly used, intergroup interventions can indeed make a major contribution to the effectiveness of complex matrix organizations.

CONCLUSIONS

In reviewing the general characteristics of an effective matrix, the quality of flexibility has been emphasized. By examining in some detail the key roles unique to the matrix, one can deduce where the flexibility comes from those individuals in key roles challenged by the matrix to respond to each new situation in a fresh and flexible manner. This approach is admittedly at odds with the present engineering tendency to rely more and more on written procedures to make things work. In point of fact, a matrix organization was never meant to rely heavily on procedures; it was meant to rely on the flexibility of personnel within it.

Experience with matrix in various organizations suggests that its implementation is not an impossible dream. Individuals can and do learn how to behave in new ways. What is clear is that they need help in doing it, and that it is particularly difficult in the early phases of a matrix before "critical mass" in the change process has been achieved. In this early phase individuals feel that they have been asked to change their ways of doing things, although it seems to them as if no one else has. Unless individuals and groups receive support and help in adjusting to a matrix at this point, the matrix will probably fail.

REFERENCES

1. Paul R. Lawrence and Jay W. Lorsch, *Developing Organizations: Diagnosis and Action* (Reading, Mass.: Addison-Wesley Publishing Co., 1969.
2. Stanley M. Davis and Paul R. Lawrence, *Matrix* (Reading, Mass.: Addition-Wesley Publishing Co., 1977).
3. Harry Levinson, *Organizational Diagnosis* (Cambridge, Mass.: Harvard University Press, 1972).
4. Walter R. Mahler, *Diagnostic Studies* (Reading, Mass.: Addison-Wesley Publishing Co., 1974).
5. Marvin R. Weisbord, *Organizational Diagnosis: A Workbook of Theory and Practice* (Reading, Mass.: Addison-Wesley Publishing Co., 1978).
6. Harry Levinson, *Psychological Man* (Cambridge, Mass.: The Levinson Institute, 1976), p. 89.
7. "When Bosses Look Back to See Ahead," *Business Week,* Jan. 15, 1979, pp. 60–61.

8. Jay R. Galbraith, *Organization Design* (Reading, Mass.: Addison-Wesley Publishing Co., 1977), pp. 28–32.
9. John E. Jones, "Role Clarification: A Team-Buidling Activity," in *A Handbook of Structured Experiences for Human Relation Training,* Vol. 5 (San Diego: University Associates, 1975), p. 136.
10. Chris Argyris, *Intervention Theory and Method: A Behavioral Science View* (Reading, Mass.: Addison-Wesley Publishing Co., 1973), pp. 17–20.

38. A Matrix Management Approach to Employee Participation

Gary D. Kissler*

There exists today an historic and slowly diminishing conflict in most organizations. Those responsible for managing organizations would have other members—the employees—become more closely aligned with the organization's purpose, products, and services. They believe that others do not appreciate the "big picture" of the work to be done. These other members view their membership as but a part of a larger scene, a scene encompassing all life activities. Unlike these other activities, however, those related to work appear largely outside one's personal control. We are witnessing a fundamental change in the way employees view the legitimacy of having greater influence over their work lives. Conflict arises when management resists relinquishing what it views as its traditional decision-making prerequisites, while other employees (including other managers) insist that further commitment on their part should be exchanged for a relaxation of these prerequisites.

One method to foster more involvement of multiple organizational segments in accomplishing overall goals envisions a restructuring of the working assignments and responsibilities. One such structure is called a "matrix organization" (Mee, 1964). Davis and Lawrence (1977) define a matrix organization as one having a "multiple command system"—in other words, more than one boss per person. Such an approach was found predominantly in the aerospace industry in the 1960s, where relatively short-term projects had to be completed by teams of technical experts whose functions were guided by multiple management groups (Haberstroh, 1965). As it became clear that this violation of the age-old "unity of command" principle not only was harmless but actually resulted in more effective work output, the prospect of such a change in other areas grew. The application of matrix management to permanent projects was occurring with increasing frequency in the 1970s.

*Dr. Kissler received doctoral training from the University of Tennessee prior to beginning work as a personnel research psychologist for the federal government. His work included an examination of productivity incentives and impediments among civilian employees in the Navy's industrial community. He is currently employed by the General Electric Lighting Business Group in Cleveland, Ohio, as Program Manager of Human Resources Systems.

As one steps back to look at the phenomenon of the matrix management approach, it can be seen that such a structure succeeds because the working circumstances demand a more streamlined communication network, a more responsive support system, a management team with organizational views extending beyond their immediate functional areas, and a work force that is responsive to such a changed environment. It is to this last point that we now turn.

MATRIX PARTICIPATION

The matrix management approach is a response to meet working goals that would probably have overwhelmed managers operating in a traditional hierarchical manner. Over a quarter of a century earlier, another demanding circumstance—the Depression—had been the catalyst for a change in management operation that resulted in the rescue of a failing steel business. This change was called the Scanlon Plan, after the man who introduced it. For those interested in a review of the history of the Scanlon Plan, broader treatment of it can be found in Davenport, 1950; Goodman, Wakely, and Ruh, 1972, and White, 1979. The present discussion of the Scanlon Plan will be restricted to one of its basic elements: productivity committees.

To those familiar with the Scanlon Plan, the main feature is the bonus paid to workers who produce more goods through reduced labor. The bonus formula takes the labor input and dollar value of completed products into account. There is another ingredient in this process, however, that has not received comparable attention. In place of a traditional suggestion program, the Scanlon Plan typically introduces the concept of *productivity committees*. These committees consist of groups of workers, led generally by their immediate supervisor, who all discuss ways to improve the quantity *and* quality of the products they make. In short, the Scanlon Plan introduces a more streamlined communication network and the other features associated with a matrix organization.

Douglas McGregor's work (1960) in the area of participatory management has stressed the importance of employee involvement. According to March and Simon (1958), "So long as an individual desires independence in decision-making, the more authoritarian the supervisory practices, the greater the dissatisfaction aroused and the greater the pressure to withdraw" (p. 95). These statements suggest that management should adopt a philosophy with high regard for employees' input on how the work place can function more efficiently.

Unfortunately, these suggestions were not acceptable to most managers because, as mentioned earlier, they were seen as encroaching upon the historic responsibilities and privileges of management. There has been a suggestion that the adoption of matrix management can be traced, in part at least, to cultural differences within which managers operate (Davis and Lawrence, 1977). The

British, for example, apparently do not look favorably upon matrix management because of the strong cultural identification they have with hierarchical forms of management. Managers in the United States are seen as less tied to such history and more willing to attempt such a structural change. Japan is viewed as a culture with elements of matrix management within it and a general willingness to put group goals above one's own.

It is possible to make too much of cultural differences in management styles and philosophies, but such differences to some degree exist. The Japanese, more so than their U.S. counterparts, listened to McGregor in the 1960s and made deliberate efforts to encourage employees to contribute their ideas and become more closely identified with the quality of the products they were producing. As Cole (1979a; 1979b) has pointed out, the Japanese combined the managerial teachings of McGregor, Abraham Maslow, and Frederick Herzberg with the quality control techniques of Juran (1967) and Dr. W. Edwards Deming. The result was something the Japanese called *Quality Circles*. Cleland (1981) cites Quality Circles as one form of matrix management. It should be clear, however, that our discussion of matrix management and employee participation has now approached an interesting point. It can be seen that matrix management is used as a vehicle for increasing employee participation. Quality Circles serve as a useful example in the present discussion, because they combine a management-sanctioned interaction across functions with a well-publicized intent to foster broader employee input.

DEMANDS FOR MATRIX PARTICIPATION

All the foregoing might make for interesting and somewhat casual reading, were it not for the nagging question, "Why bother?" Even Davis and Lawrence (1977) issue the caveat that the matrix organization is not for everyone and that there are easier ways to manage an organization. There are two important changes occurring that make decisions regarding matrix organizations/ employee participation far more important than was the case just a few years ago. The first change is the increasing desire on the part of employees to gain more control over their lives at work. March and Simon (1958) state, "The greater the amount of felt participation, the greater the control of the organization over the evocation of alternatives" (p. 54). The "alternatives" referred to include the positive as well as negative ones. The second change is the increasing economic pressure being brought to bear upon U.S. employers by the international marketplace. Seeing this pressure as threatening job security, employees are transmitting their concerns through their willingness to compromise on labor-management issues that had been held sacrosanct until recently.

Such concessions are not being made without bitterness and frustration. There is deep resentment among workers that their management has allowed the economic situation to deteriorate. In many instances the issue can be summed up as follows. Management has asked employees to identify with the products they produce and to take pride in their work. The international marketplace delivers the message that the products aren't worth buying. The workers resent the implication that *they* are to blame. Instead, they point to management's emphasis on short-term goals and a willingness to accept quantity over quality. The employees believe they could help improve the acceptability of the products they produce, but management won't listen.

MATRIX CLIMATE EVALUATION

Naturally the picture is not always this clear-cut. Other issues such as deciding on product mix, product pricing, market strategies, and technology shifts play important roles in this area as well. Many employees do participate on the job and many managers are as frustrated as their employees over how to improve their standing in the marketplace, national as well as international. Many other employees and managers are looking for ways to improve the working environment. The popular press touts what managers fear are fads and panaceas. They would prefer changes addressing *their* needs, rather than global prescriptions that do not consider their particular working climates.

Cleland 1981 lists several criteria for examining management systems properties. These criteria are helpful in deciding if matrix management already exists in some form, or would be likely to succeed if management attempted to develop such a structure. These criteria include the existence of a team culture, the sharing of results and rewards, consensus decision making, and the realignment of support systems (e.g., human resources, finance) to support a matrix management design. Cleland concludes that the existence of several of these criteria accompanies some type of matrix management system. These systems can include, among others, project management, task force management, the plural executive, and Quality Circles.

The General Electric Lighting Business Group has identified criteria related to management practices that are supportive of a particular type of productivity program. Through the efforts of line and staff employees, two major vehicles have been developed to "test" the climate of an organization. The first is called a *productivity program inventory* and is intended to focus on specific management policies and practices at a given location. The second is an *attitude and management practices survey* that is circulated to all employees on an anonymous basis.

The productivity program inventory consists of questions related to eight major categories of managerial policy and practice. The eight areas and a brief definition of each follow:

- *Accepted and quantified measurement base:* an explainable, acceptable, and comprehensive performance measurement system with at least monthly reports available by natural groupings (shift, unit, process, etc.).

- *Employee involvement experience:* employee group experience at all levels in structured decision-making and problem-solving processes affecting areas such a quality, processes, and practices.
- *Effective management response systems:* established and effective structured responses to employee complaints, suggestions, ideas, and concerns in areas of manufacturing, engineering, quality, and materials. An exempt employee team that is sensitive to employee inputs and capable of timely and creative resolutions.
- *Business awareness communication:* business, economic, and productivity communication that has resulted in increased employee interest and participation.
- *Employee-generated productivity potential:* significant and sustainable productivity improvement over a period of several years solely through employee effort, ingenuity, and attrition—improvement that is not dependent upon increased sales volume or improved process and/or technology.
- *Employment stability:* five-year forecasts that do not include significant work-force additions or reductions, other than attrition, due to business decline, transfer of work, or automation.
- *Suitable relations climate:* a labor climate receptive to sustained productivity improvements.
- *Program administration and resource individuals:* identified management and clerical employees available for program leadership and administration, including financial and data-processing support.

This inventory is intended to serve two major purposes. First, it provides an open dialogue between upper-level management and staff on the areas outlined. Such introspection helps evaluate the current level of effort being carried out in each category. Second, the individual presenting the material (usually a human resource expert) provides information to management and staff about how other management groups are approaching these areas, thus providing cross-fertilization in a nonthreatening manner. It should be noted that only salaried employees are involved in these discussions.

In the development of the inventory, it was decided that a survey of baseline information was needed. Therefore the survey was administered to a number of management teams, dispersed over a wide geographic area, that represent numerous manufacturing specialties. Although this device was designed with the manufacturing community in mind, it has since been used successfully by nonmanufacturing managers as well. The baseline data demonstrates profiles of management policy and practice in many diverse settings, and therefore have allowed subsequent management teams to match their resulting profiles to those that most frequently involve specific employee involvement programs (e.g., Quality Circles, the Scanlon Plan).

In addition to the inventory, a survey of employee attitudes and views of management practices is also used to gauge the climate within an organizational component. The survey is intended to be administered to 100% of the work force and is completed anonymously on company time. The results are placed into a computer file, thus providing the management team with information on how their location views their practices and policies, and how their location responses compare to the remainder of the data base. The survey is divided into nine areas, each having a number of items drawn from prior surveys submitted by either staff experts or operational managers. The areas and sample items from each follow:

- *Productivity*
  —I have the resources (e.g., equipment, tools, etc.) required to do my job.
  —Management encourages productivity by recognizing good performance.
- *Work management*
  —Management establishes a clear priority of work to be done.
  —There is an effective match between people and resources in my organization.
- *Supervision*
  —My supervisor tries to get the employees involved in decision making affecting their work.
  —My supervisor encourages employees to see "the big picture" of what our work is about.
- *Pay and benefits*
  —My pay is fair compared to what other employees earn in similar jobs at GE.
  —Overall, wage and salary administration is fair and equitable.
- *Communication*
  —I receive sufficient information about how the business is doing.
  —Management is aware of the opinions and attitudes of the employees here.
- *Equal Employment Opportunity*
  —A person's age is no barrier to getting a promotion or better job in my immediate work area.
  —I have a clear understanding of GE's commitment to Equal Employment Opportunity/Affirmative Action.
- *Safety and health*
  —Regular safety inspections are conducted to assure safe conditions within my work area.
  —Trained coworkers are available to administer first aid in case of injury or illness on the job.
- *Personal development*
  —I know whom to contact for information about available training.
  —I can get help in finding out what job opportunities are available to me.

● *Work climate*
—Management consults employees before decisions affecting their work are made.
—My work group holds meetings to plan and coordinate our efforts.

These two tools provide information that allows a management team to decide where they stand regarding practices and policies, as well as how they are perceived by those whose working lives are affected by them. This combination gives management information to be considered when deciding upon a restructuring of the management system (e.g., matrix) so as to encourage employee involvement in a way that best fits the existing climate.

THE GROWING OF MATRIX PARTICIPATION

Davis and Lawrence (1977) state that a matrix "demands new behavior, attitudes, skills, and knowledge" (p. 103). They also warn that a matrix structure is more likely to succeed if it is "grown" rather than merely introduced. This same advice pertains to employee involvement programs as well. There are two aspects to growing matrix participation that deserve more attention: management's responsibilities, and the pitfalls of such an approach.

Primary among management's responsibilities is to provide an example to the rest of the organization by adopting matrix participation practices at the highest level possible. More will be said about this later.

A second major management responsibility is to provide resources. One might naturally consider this to be primarily facilities for meetings, equipment, etc. However, a point raised by Davis and Lawrence (1977) regarding the acceptance of a matrix organization is pertinent here. They indicate that there will always be a "dinosaur" in an organization who cannot adapt to a new management structure. Among other resources, therefore, management must include the training of these dinosaurs and other employees struggling to adapt to this new environment. A few points can be made regarding this training.

It was mentioned earlier that one of the Scanlon Plan's basic elements was a greater emphasis on employee participation. Quality Circles use this as their central theme. There are many who regard the Scanlon Plan as Quality Circles with a bonus. Achieving greater participation among employees requires that management provide a vehicle to take advantage of these approaches. One vehicle is the training offered to the leaders and participants of what can be called matrix participation groups (e.g., Quality Circles, the Scanlon Plan's productivity committees). This training contains three ingredients: problem-solving techniques, group dynamics, and the addition of structure to the overall involvement process.

In most Quality Circle training sessions (e.g., Dewar, 1980) one finds that the leaders and the participants are instructed in several problem-solving techniques. These include brainstorming, the use of Pareto analysis, cause-and-effect ("fishbone") diagrams, charting techniques (e.g., histograms), in addition to ways to combine and display the logic of the group's solutions. For the supervisors, the use of these techniques is usually less of a problem compared with their having to shift their working role somewhat. They are asked to relinquish temporarily their assigned role and shift to a role requiring that their influence be channeled through group participation. In other words, the supervisor now has only one vote and must work on a peer level to solve problems common to all group members. This shift is not an easy one to make, particularly to those who know that the historical trend has been one of declining first-line supervisor authority. Not only must a quid pro quo be made clear in terms of management's view of this shift, but also the supervisors must be made aware of some of the idiosyncrasies that occur in group settings. These dynamics must be mastered by the group's leaders if a focused and efficient group effort is to be maintained. This implies that a structure must be given to the whole effort. As Davis and Lawrence (1977) indicate, a lack of structure can lead to "group anarchy."

A final management responsibility is to provide support for matrix participation. This can take several forms such as communication, restructuring of reporting channels, giving of budgetary authority, etc. One particular example will be discussed here: contingencies on performance.

Among management's control systems, perhaps one that is overlooked when considering such issues as a change in organizational structure or employee involvement, is a formal evaluation of employee performance. The importance of combining the results of such an evaluation with pay decisions has been stated by Lawler (1981): "the more entreprenurial, achievement-oriented individuals seem to be attracted to those organizations where rewards are based on competency and performance. Since most organizations desire this type of individual, it makes sense for them to try to relate pay to performance" (p. 81). The elements to be considered in an evaluation of employee performance include those whom management believes contribute most to the success of the organization. Therefore, by focusing attention on an individual's support of and participation in a matrix environment, management can establish a clear link between such behavior and potential reward. This discussion assumes that multiple forms of organizational rewards exist and that these rewards, as well as negative sanctions, can also be made contingent upon an individual's performance. However, by including matrix participation among other elements considered formally by an organization in its "pay for performance" effort, a clear signal of the organization's regard for such behavior can be given.

There are pitfalls involved when considering matrix participation. Davis and Lawrence (1977) provide a detailed discussion of several of them. They point out that the balance of power must be considered, since an imbalance can result in reduced group performance. There is the risk of assuming that *all* decisions must be made by a group. Internal disputes can take up all the group's time and exhaust the energies needed to solve external issues. One of the pitfalls— or "pathologies," as Davis and Lawrence term them—will be discussed in more detail here. They call it "sinking" and describe it as follows (p. 140):

> There seems to be some difficulty in keeping the matrix viable at the corporate or institutional level, and a corresponding tendency for it to sink down to lower levels in the organization . . . where it survives and thrives. Sinking may occur for two reasons. Senior management has either not understood or been unable to implement the concept as well as those beneath them. Or the matrix has found its appropriate place.

There is a third reason why sinking may occur, and it is not nearly as "natural" a phenomenon as those described by Davis and Lawrence. This writer has observed that the intensity of a manager's support for matrix participation may be inversely related to the distance of such participation from the manager's level. Facetious as it may sound, there is the real problem of management not "practicing what it preaches": ironically, an organization's most vocal proponents of matrix participation may themselves be faulted for failing to take advantage of such opportunities at their own level. It was stated that the most important management responsibility was to provide an example. Failure of matrix participation is often due to inconsistent signals sent from upper management regarding their support. In short, their *own* involvement will speak louder than their words of encouragement to lower levels in the organization.

BEGINNING THE MATRIX PARTICIPATION PROCESS

Earlier it was stated that an important consideration was whether the environment contains elements conducive to a matrix structure or whether such a structure, even in rudimentary forms, already exists. It cannot be stressed too strongly that the starting point is to "know thyself" by whatever techniques are available and have utility. The flow of data should come from at least two of the following sources:

- Interviews with individuals who will be affected by the changes being considered
- A review of Cleland's (1981) criteria and the inventory content by the management team involved
- A survey of existing management practices that includes all employees' views

To echo somewhat the caveat by Davis and Lawrence (1977), there is a need to examine current practices to be able to justify a change to matrix participation. The following example shows how such an examination was made prior to introducing Quality Circles. The basic problem was to decide whether or not to introduce Quality Circles and whether such a program was different from the existing suggestion plan.

Table 38-1 lists the points describing how each program fared when compared with a common set of criteria.

CRITERIA	SUGGESTION PLAN	QUALITY CIRCLES
Employees involved	Very few.	More employees have expressed an interest than there are resources to accommodate them.
Management support	Minimal communication. Individual completes suggestion on own time and with own effort.	Extension communication with employees, union. Facilities provided. Outside sources (e.g., engineering, quality) made available.
Management visibility	Minimal.	Management shows initial verbal support and provides supervisor as leader. Later receives group input through presentations.
Program sanctions	Employee must obtain information on own time and from sources who have no formal justification to provide it.	Groups work on company time and have access to sources that are encouraged by management to support this effort.
Feedback/timeliness	Generally, feedback is given in written form, is brief, and can be delayed for several months.	The groups will receive their feedback verbally from management when their presentations are made.
Focus of benefits	Suggestions tend to focus on individual benefits and are oriented in most cases to cost savings.	Groups have broad focus. They solve issues affecting many employees and are a mix of issues to improve working conditions, material efficiency, cost savings, and others.
Motivation to participate	Money, recognition, and influencing change.	Money, recognition, and influencing change.

The Quality Circle program offers management an opportunity to take advantage of employee input in ways that are quite different from those associated with the suggestion program. It might appear that the decision to introduce Quality Circles is an obvious one, but this may not be the case. The above table also points out several areas in which the suggestion program could be improved to encourage more participation. There are intermediate steps that can be taken to make such improvements. Several are mentioned by Ouchi (1981) in his examination of Japanese and American organizations and how the United States can adopt changes to become more competitive in the international marketplace.

Much attention has been focused on American industry's need to improve productivity. This chapter has already indicated that an inventory of management practices was developed initially for this very purpose. Quality Circles are seen predominantly as a "manufacturing" program by many. While such emphasis is certainly justifiable, management should turn its attention to other areas often overlooked when searching for increased productivity. Although outside the scope of the present discussion, white-collar productivity has been receiving more attention lately in the press. It has been suggested that a typical office setting is essentially similar to those existing shortly after the turn of the century, and that the capital investment made to support blue-collar workers since that time is far greater than that made to support white-collar workers. One aspect of the issue is the increasing need to process information and the lack of person/computer interface in the office setting. Another aspect, however, involves reexamining the way office systems are structured and finding ways to involve employees in improving these systems.

SUMMARY AND CONCLUSION

There exists in today's organizations an historic conflict between management and other employees regarding the legitimacy of each group's influence on the working environment. Management would like other employees to accept, and work harder toward achieving, the organization's goals. The other employees believe that such involvement requires management to allow greater input by them in how those goals are achieved. The types of goals required to compete successfully in today's marketplace have put greater pressure on management to resolve this conflict rapidly.

Certain structural changes have occurred in organizations to achieve working goals that would not have been met using traditional hierarchical forms of management. One of these changes is a matrix organization that acknowledges the need for a broader use of skills and abilities. Not only do these skills and abilities come from multiple organizational functions, they also are required to respond to multiple sources of managerial direction. The matrix management approach was first used in situations requiring cross-functional expertise on short-term projects. However, the success of this structure led to the adoption of matrix management for permanent as well as project management efforts.

Another change that occurred in response to economic demands was the introduction of the Scanlon Plan, which combined a bonus payment for improved productivity with greater opportunities for employees to contribute their ideas for such improvements. The ingredient of employee involvement was suggested by researchers in this area as an alternative style of management—participative management. For historical and cultural reasons, the United States did not make large strides in participative management. The Japanese, for these same reasons, did. They developed a vehicle for greater employee input into the improvement of product quality and called it "Quality Circles," a group-oriented and highly structured method of focusing attention on specific problem areas. Quality Circles have been described as a form of matrix management. This categorization takes into account that the solutions of problems often require multifunctional input, of which Quality Circles make considerable use.

Increased pressure from the international marketplace, as well as social pressures from work groups to take greater control over their work lives, have accelerated changes in managerial structure and practice. Both management and other employees are beginning to agree that greater cooperation and broader input from all parties will be needed to strengthen the organization, increase competitiveness, and ensure job security. Management still faces a problem in deciding if the working climate is ready for the changes needed to take advantage of this cooperative effort. It needs to know what criteria should be taken into account before deciding on a course of action, and what method can be used to help evaluate these criteria.

A number of criteria were listed that suggest whether an environment can support or already is supporting a matrix management approach. It was pointed out that the participation of employees under such circumstances, termed matrix participation, will also require that management examine the working environment. Several methods were discussed that lend themselves to such an evaluation, including interviews, an inventory of management practices and policies as seen by the management team, and a survey of management practices seeking input from all employees in the organization.

A review of management responsibilities was conducted to suggest some specific actions that can lead to greater acceptance of matrix participation, as for instance setting an example, providing physical resources and training, and including such participation among elements considered during formal performance evaluations. A brief discussion of pitfalls focused attention on the importance of management's "practicing what it preaches" with regard to employee participation.

Finally, some guidelines were offered to help management begin self-examination and to suggest areas where productivity might be increased. The overall message is that management must have a thorough knowledge of the working climate prior to attempting to develop matrix participation. It is important that this knowledge be broadened to include white-collar as well as blue-collar employees.

The overall goal of this discussion is to provide some "real-life" information about why matrix participation is needed, why there is an increased demand for greater employee control over their work lives, and some specific criteria that can be used to examine the working environment. Some examples of tools used to conduct such an examination have also been provided. I believe that the success of American business organizations lies in their ability to adapt to the challenge of change. One example may be for them to take advantage of greater matrix participation and to regard such a step as important, just as they do strategic planning in other areas. Booker T. Washington was quoted as saying that an ounce of application is worth a ton of abstraction. The survival of many American business organizations will depend upon whether the management teams can recognize the need for making changes as something other than academic exercises.

REFERENCES

Cleland, D. I. "A Kaleidoscope of Matrix Management Systems." In *Management Review* October, 1981

Cole, R. E. *Work, Mobility and Participation.* Berkeley: University of California Press, 1979.

————. "Diffusion of New Work Structures in Japan." *IAQC International Conference Transactions,* 59–65 (1979). (b)

Davenport, R. "Enterprise for Every Man. *Fortune* 41 (Jan. 1950): 51–58.

Davis, S. M., and Lawrence, P. R. *Matrix.* Reading, Mass.: Addison-Wesley Publishing Co., © 1977. Reprinted with permission.

Dewar, D. L. *Quality Circle Leader Manual and Instructional Guide.* Red Bluff, Calif.: Quality Circle Institute, 1980.

Goodman, R. K., Wakely, J. H., and Ruh, R. H. "What Employees Think about the Scanlon Plan." *Personnel* 49 (1972): 22–29.

Haberstroh, C. J. "Organization Design and Systems Analysis." In J. G. March, ed., *Handbook of Organizations.* Chicago: Rand McNally & Company, 1965.

Juran, J. M. "The QC Phenomenon." *Industrial Quality Control* (Jan. 1967), pp. 329–36.

Lawler, E. E., III. *Pay and Organizational Development.* Reading, Mass.: Addison-Wesley Publishing Co., 1981. Reprinted with permission.

March, J. G., and Simon, H. A. *Organizations.* New York: John Wiley & Sons, Inc., 1958.

McGregor, D. *The Human Side of Enterprise.* New York: McGraw Hill, 1960.

Mee, J. F. "Matrix Organization." *Business Horizons* (Summer 1964).

Ouchi, W. G. *Theory Z: How American Business Can Meet the Japanese Challenge.* Reading, Mass.: Addison-Wesley Publishing Co., 1981.

White, J. K. "The Scanlon Plan: Causes and Correlates of Success." *Academy of Management Journal* 22 (1979): 292–312.

39. Making Participation Productive

Joe Kelly*

A. Baker Ibrahim**

Flint, Mich.
A year ago, Eddo Brantley's biggest responsibility was blowing dust off equipment with an air hose. Now, she can shut down one of the principal production lines of an automobile factory whenever she wants.

And she does. On a recent day, the slender, 49-year-old laborer strode assertively to a control panel and pushed a fat maroon button. A rolling line of metallic saucers, destined for a spot in every new front-wheel-drive car produced by the General Motors Corporation, jerked to a halt.

"I could never do this before," said Mrs. Brantley, one of 2,500 Buick workers who has been trained in what are termed "Quality of Work Life" principles. "Now, I stop the line whenever there is a problem."

She and hundreds of other production line workers at Buick's huge assembly plant here make decisions on adjusting machine settings, rejecting faulty raw materials and moving machinery into step-saving positions overlooked by engineers. In return for taking on this added responsibility, they no longer have to answer to foremen, punch timeclocks or work continually at the same monotonous job. And they can get more money.

"I think I've found myself a home," said Fred Bailey, 41, a welder and another work team member. "People get along better here and they let you put out a better product."[1]

An increasing effort is being made to involve shop-floor people in decision making that affects them. Unfortunately, U.S. companies have allowed themselves to fall behind both European and Japanese firms, which have made

*Dr. Joe Kelly is a Professor of Organizational Behavior in the Department of Management at Concordia University, Montreal, since 1968. Dr. Kelly has considerable experience as an organizational behavior consultant in the fields of restructuring organization and structured sensitivity training. He has worked with Canadian companies and a federal government agency mainly in the fields of supervisory and executive training. Dr. Kelly is the author of over 25 articles and six books.

**A. Baker Ibrahim, Ph.D., Assistant Professor of Management at Concordia University, is specially interested in the areas of executive selection, policy, and the interface of management and accountancy.

immense progress in utilizing participation. Firms such as the Glacier Metal Company in Britain, Volvo in Sweden, and many Japanese companies have been experimenting with new job and organizational designs.

In the United States companies such as GM, IBM, GF, and many others are experimenting with new designs. The aim of this chapter is fourfold:

1. To review the arguments favoring participation.
2. To specify what we in North America can gain from looking at the Swedish experiments at Volvo, the Japanese effort, and the British effort at Glacier.
3. To evaluate the American experience with participation.
4. To develop a contingency view of participation.

In the 1980s there has been a marked change in corporate climate, facilitating a new move toward employee participation. Dynamic companies such as IBM, General Motors, Xerox, and GF are investigating participation. Participation has been with us now for some 30 years. As we move into the 1980s, the time seems to be appropriate to ascertain how well participation has stood the test.

THE CASE FOR PARTICIPATION

Most seasoned executives are usually willing to admit that the classical form of management does generate some problems. Within rigidly structured organizations shop-floor people feel suffocated, experience alienation, and have a general feeling that the organization is not making the best use of their resources. To overcome these problems and make the best use of the potentials that people bring to work, managers ought to employ participative management. Six factors arguing for participative management will be considered.

1. The Basic Assumptions of Participative Managers

The assumptions of human relations management are that people are spontaneous, responsive, and responsible. People function better in a democratic work environment with bosses who are supportive. They work better in groups where they feel they belong. If people participate in the decision making that affects their lives, they are more likely to give of their best.

2. Changing Employee Values

Among both academics and executives a growing consensus is emerging that there has been a shift in employee values that reveals increasing dissatisfaction with many aspects of work. A frequently cited example is the Lordstown strike of 1972 in the Vega plant in Ohio. The general expectation is that these trends are continuing and that this gap in satisfaction between managers and shop-floor people is growing. Cooper, Morgan, Foley, and Kaplan have presented a new synthesis of employee attitudes gathered over a 25-year period. Their argument is that there is a hierarchy gap, and that discontent among hourly employees and managers seems to be growing. Further, most employees increasingly expect their company to do something about this job dissatisfaction. The argument is that while most shop-floor people feel happy about security and pay, most are unhappy in their attitudes toward management and supervision. They feel there is little opportunity for advancement. The basic problem is rising expectations that companies will find it difficult to meet. As the authors point out: "The changes reported here are ubiquitous, pervasive, and nontransient; any reversal is unlikely in the foreseeable future. The goal for management is to be aware of and prepared for new and surfacing employee needs, before it is forced to take reactive, ignorant, and resistive postures."[2]

There is abundant evidence that dissatisfaction at work is increasing. This dissatisfaction is greater among the younger, more educated, and less authoritarian workers. Despite increasing unemployment, the reaction against dehumanization of work seem to be growing. When the workers are simply asked to say how satisfied they are with their work, the vast majority report themselves satisfied. Yet basically, because of improved education and rising levels of expectation, dissatisfaction and alienation are increasing.

3. Communication

The key to employee participation is communication. If a manager and his or her subordinates do not communicate and try to solve the problems that they jointly face, alienation may arise. This alienation may become more acute because the workers feel that the working situation is very different from the democratic environment in which they grew up. In other words, employees in our society come to work with the expectation that they can participate in problem solving and decision making, and that their work can be made challenging and not dehumanizing. For this participation to be effective, it must be related to the decision-making structure of the organization.

A major argument favoring participation is that it will reduce power differences between levels. Research evidence suggests that this is not always the case, and where there are substantial differences in skill levels, participation increases the power differences.

4. The Spread of Industrial Democracy

Industrial democracy is quite widespread in Europe, particularly in Germany, Sweden, Yugoslavia, and Britain. But when people talk about industrial democracy, they are liable to lump together the management of the kibbutz in Israel, a works council in Yugoslavia, a supervisory board in the steel industry in Germany, and the management of the new General Foods pet food plant in Topeka, Kansas. Putting all these different examples of industrial democracy together fudges rather than clarifies the issue. In the United States there is clear evidence that informal interpersonal participation can produce good effects independent of formal forms of industrial democracy. Participative styles of leadership are very extensively used in the United States and are highly appropriate to both the psychological and managerial styles of American managers.

5. Productivity and Participation

There is abundant evidence that American productivity growth has slipped from positive to negative—a problem that will be overcome only if companies develop more effective participation. It is widely believed that this productivity slippage is due to a breakdown in the work ethic.

Japanese industry has been able to maintain high levels of productivity. A major reason for this is that the workers do not resist technological innovation. Since Japanese workers have a guaranteed lifetime job and bonuses based on productivity and profitability, they do not resent innovation.

In addition, productivity in Japan is not exclusively a management function. The problem in North America is that management's objectives are too short-term and tied to earnings-per-share indexes. North American MBAs are taught quantitative techniques that focus the manager's mind on short-term gains. Japanese top executives, on the other hand, take a longer view. Also, they do not run their businesses on a strictly adversary basis, and try to think in terms of the national interest.

6. Moving Beyond Classical Organizations

Classical Organizing.   Classical organizing makes extensive use of organizational charts, role descriptions, rule books, and makes certain assumptions about organizations. Basically, it assumes that organizations are like pyramids, with the Chief Executive as the conductor of an orchestra who runs the whole show, because only that executive knows the whole score. The classical type of organization is still the most widely used.

For example, the typical North American car assembly plant is organized along classical lines with maximal task breakdown and with the individual effort tied to the speed of the assembly line. On the other hand, Volvo in Sweden has come up with an auto assembly line where people work in teams that allow job enrichment and job exchange. But so far, neither U.S. managers nor shop-floor workers are particularly attracted by this approach, which they feel will require them to make a major additional effort to cooperate.

The Systematics of Organizing.   The first step is to define the job in terms of duties, physical and mental requirements, and tools, which means job analysis. The next step is to organize the jobs into groups according to some principle. For example, the jobs can be organized according to function or geographical area. The old type of organization is called bureaucracy; the new type of organization is called adhocracy.

The Democratic Style of Organizing.   Realizing the limitations of the classical approach, behavioral scientists switched their efforts to defining the effect of supervisory style. The most important research on the effect of supervisory style was carried out by Rensis Likert, who compared employee-centered to job-centered supervisory style. Likert's research suggested that supervisors who were employee-centered and democratically oriented had more effect on work groups, in terms of both higher production and employee satisfaction. Likert went on to argue that the supervisor provided a key link between his own work group and other work groups that he had to interact with.

The superior advantage of the democratic approach is that is rejects the organization model built on the accounting and industrial engineering analogy and gives pride of place to the individual and the primary working group. But in fact, human relations are usually grafted onto the classical organization after it has been developed. Thus while the rules, roles, and relations have already been specified, the human relations manager seeks to encourage individual participation by facilitating consultation. The presumption is that employees will make mistakes, but it does not assume that this will inevitably maximize error. But unfortunately, consultation can easily degenerate into pseudo-participation.

THE JAPANESE EXPERIMENT

In the age of the steamship, locomotive, and telegraph, economics was defined as the allocation of scarce resources to achieve what society wanted. What society wanted could be summed up in a four-letter word—more. As we moved out of steamships and into jets and computers, the scarce resource became management, which, in the case of Japan, Inc. (a vast army of disciplined, devoted executives), could start with virtually nothing in terms of physical resources such as iron ore, oil, and rubber, and yet end up with the double-digit growth in GNP called the Japanese miracle. And what was the rational explanation for this miracle? When academics were finished explaining how the Japanese stole their ideas on shipbuilding from the Clyde, on cars from Detroit, and on radios from RCA, they turned to Japanese ancestor worship and their penchant for corporate calisthenics. Another seminal idea brought here by visitors from outer space who had used their 21-day excursion ticket to visit Japan, was that Japan contained only Japanese. "Not like the good old U.S. of A., which is just one large melting pot of ethnics."

The Zen Classical Manager

Americans carry around in their heads a stereotype of the Japanese executive as the "guy who brought us Pearl Harbor" and who is now disguised as a transistor salesman with a Kamikaze pilot's scarf. This stereotype is unfortunate, for the Japanese classical, but existential, manager has established "the greater coprosperity sphere" in Asia in terms of achievement and affluence, but without alienation and anxiety. And of course, in Japan an employee has a job for life. Further, in Japanese businesses a consensus is established by allowing direction to float upward from the junior executive level to the board of directors. If we discount for the moment the impact of Zen Buddhism (full of kinky koans and a major source of existential ideas), it seems reasonable to argue that the Japanese economic miracle was facilitated by a traditional managerial elite who worked systematically by the book, with a homogeneous population of workers who had a deep and pervasive respect for authority.

Western executives have much to learn from the way authority is exercised in Japan. The Japanese have a positive genius for exercising authority in a way that retains ritual as a saving grace. Frank Gibney has pointed out "the importance of ceremony as art in keeping a civilization together. Consensus and collectiveness are more than a virtue. They have almost the quality of religion."[3] Drawing on the central value of sincerity, the Japanese have established "a great steel web of contract and commitment," which is the basis of the Japanese system called *amaeru,* which means to presume upon the affections of someone close to you. Out of *amaeru* has developed a tremendous sense of team work, of esprit de corps, which generates tremendous efficiency. And to make the managerial miracle into an economic exponential, the Japanese were able to buy and build on Western technology, instead of paying the exorbitant price of inventing their own technical know-how.

Participation: The Japanese Example

Japanese companies typically outproduce their American counterparts. The Japanese have proven that significant gains can be made in productivity if shop-floor people can be motivated to take an interest in raising productivity. And we know that American productivity is declining, while Japanese productivity is rising at a rate of approximately 10% per year.

To make an automobile in Japan requires only about 90 hours of labor at an hourly cost of about $11, whereas in the United States 130 hours are required at a total hourly cost of $20. American managers often treat their subordinates as adversaries. Japanese managers treat their subordinates as family in the hope that productivity will "bubble up" from the workplace. Japanese companies let workers make more decisions.

Japanese firms work with less hierarchy than American firms. For example, one Ford plant has 11 levels of organization and 200 worker classifications, while a Toyota plant got by with six levels of organization and seven worker classifications. According to James O'Toole, Japanese auto firms operate with one manager/inspector for every 200 workers, while American firms have one manager/inspector for every 10 workers.[4] And the quality of Japanese products is often stunning: 96% of their automobiles leave the assembly line in fit shape for delivery, versus 75% of U.S. cars. Japanese managers give workers a genuine stake in decision making through participation. And Japanese managers have also proved that they can stimulate productivity via participation in their American plants. O'Toole describes how the old Motorola plant in Chicago was put on the map by Matsushita through participation. Sony achieves high productivity in its San Diego color-television plant.

In many Japanese plants, employees meet two or three times a week on their own time to discuss how productivity and quality control can be improved. General Motors is trying out this idea of "Quality Control Circles" to improve productivity.

But Motorola asks in its advertisements, "Are Japanese Executives better people managers than American Executives?" Motorola argues:

> Much has been written about Japanese management style.
>
> Quality circles, lifetime employment, corporate anthems, exercise programs, etc.
>
> The implication is that the style of Japanese management is superior to that of American management. And that for Japanese companies, this particular style works both at home and abroad.
>
> But did you know that this style is not universally practiced in Japan? In many Japanese companies, employment tenure varies, quality circles don't exist, the boss is the boss.
>
> And did you know that many American companies have an even better record of management than some Japanese companies? We believe that one of these companies is Motorola. Why? It all starts with our respect for the dignity of the individual employee. We apply this philosophy in many employee-related programs.
>
> Our Participative Management Program brings our people together in work teams that regularly, openly and effectively communicate ideas and solutions that help improve quality and productivity. In the process, many employees tell us that the program also enhances their job satisfaction.
>
> Motorola's Technology Ladder provides opportunities for technical people, such as design engineers, to progress in professional esteem, rank and compensation in a way comparable to administrators and officers.

Our ten-year service club rewards employee dedication and loyalty with special protection for continued employment and benefits. And in an industry noted for both explosive growth and high mobility, almost one quarter of our U.S. employees have been with us for more than a decade.

Our open door policy enables employees to voice a grievance all the way up to the Chairman. It's rarely needed, but it's there. And it works. These and other programs reflect our respect for, and commitment to, the individual and the team. Their continuing effectiveness is reflected in the direct and open, non-union relationship among all our people, whether production workers, engineers or office workers. All of this works for us as we work for you.

Meeting Japan's Challenge requires an enlightened management style—demonstrated by our participative management style that respects the dignity of the individual.[5]

Theory Z Organizations in Action in America

Theory Z organization as proposed by William Ouchi captures the best in management methods from Japanese and American approaches that are egalitarian and participative, and stress good interpersonal relations. It is market by shop-floor cooperation and commitment to the objectives of the company. The main idea is to coordinate people, not technology.

Theory Z builds on the following principles:[6]

1. A manager of integrity must lead the effort.
2. Make an audit of your company's philosophy.
3. Get the CEO on board.
4. Create structures and incentives that will facilitate participation.
5. Develop interpersonal skills, particularly in problem solving, by splitting decision-making groups into participants and observers.
6. Test yourself and the system by facilitating feedback. This may not be fun.
7. Get the union involved.
8. Stabilize employment.
9. Slow down evaluation and promotion, thus demonstrating to employees the importance of long-run performance.
10. Broaden career paths through a program of career circulation.
11. Begin at the top, but implement at lower levels.
12. Solicit suggestions from the shop floor and implement them.
13. Develop holistic relationships, so that employees feel that they are part of an important institution.

Using Theory Z approaches, Japanese companies have achieved dramatic improvements in quality. For example, in the Toyota Motor Company, at the close of each day employees discuss performance quality. As J. M. Juran points out:

> The West is in serious trouble with respect to product quality. A major reason is the immediate threat posed by the Japanese revolution in quality.
>
> During the early 1950s Western product quality was regarded as the best. The label "Made in USA" (or Germany or Switzerland) was a distinct asset to product salability. In the subsequent decades Western quality has continued to get better, as it had been doing for centuries. (Those who contend that Western quality has deteriorated are usually generalizing from isolated cases.) But its quality position has not kept pace with quality improvements made by the Japanese.
>
> Prior to World War II Japanese product quality was poor—it was widely regarded as among the worst. Such products could be sold, but only at ridiculously low prices, and even then it was difficult to secure repeat sales.
>
> Soon after World War II, Japanese quality began to improve, and the pace of improvement has been remarkable. By the mid-1970s this pace had brought the Japanese to a state of equality with the West. Within this broad state of equality the West retained quality leadership in some product lines. In other product lines the Japanese had attained quality superiority. A well-publicized example is color T.V. In that case, Japanese quality superiority was a major reason for a dramatic shift in share of market from the West to Japan. During the 1980s the situation will get worse, much worse. The scenario of the color T.V. set is being rerun in many product lines, including such essentials as automobiles and large-scale integrated circuits. The magnitude of this treat has yet to be grasped by the West.[7]

THE SWEDISH EXPERIMENT: VOLVO

More Human Assembly on a New Kind of Line

Instead of the traditional high-speed conveyor line, the new Volvo plant in Kalmar, Sweden, uses 250 "carriers" (18-foot-long computer-guided platforms) that deliver the frame for a single Volvo to one of the plant's work teams. One team will put in the Volvo's electrical system; another will complete the interior finish.

The man who brought us this type of organization is Pehr Gyllenhammer, the dynamic forty-year-old managing director of Volvo. Gyllenhammer's meteoric career took off in 1969. At the age of 34, the well-groomed, good-looking lawyer took over from his father as head of Skandia, Sweden's largest insurance company. In 1971 he succeeded his father-in-law as managing director of Volvo, Sweden's biggest industrial concern. Under his leadership, Volvo has substantially increased its sales. Soon after taking over, he replaced a centralized management structure with four semiautonomous divisions, each one a profit center. He also expanded production of trucks and marine and industrial engines.

These work groups at Kalmar organized themselves according to their own preferences. While the worker in a traditional assembly line might spend a whole shift of eight hours a day putting on head lamps, members of a Kalmar work team rotate from tail lights to head lights, from fuse boxes to signal lights, from the electronically controlled fuel injection system to the dynamo in a starter motor.

Working conditions are good in the roomy, airy, uncluttered workshop with its brightly colored walls and huge picture windows that let in the sunlight. (Curiously, many American factories either have no windows or else have the glass blacked out.) The new plant, which cost $23 million to build, includes the most modern devices to control production and maintain quality control.

American Realism

Oddly enough, American auto workers are not too impressed by plants such as Volvo's at Kalmar. Recently, six auto workers, all employees of G.M., Ford, or Chrysler, spent four weeks working at a Saab engine plant in Sweden. The Americans were integrated into groups of 18, working an eight-and-a-half-hour shift. In work teams the task of each worker was determined by the autonomous decisions of the assembly team.

One of the most important conclusions to be drawn by the six U.S. workers was that they found team work boring. A majority of the group said they preferred the traditional assembly-line method of working, and claimed that the Swedish and Finnish workers whom they met did, too. Group working was generally regarded as undesirable by the Swedes. One reason expressed was that it was looked upon as "women's work." American workers were surprised to find that those involved in the group engine assembly were women. According to the management of the Saab plant, this was because a four-day week allowed them to spend more time with their families.

The American workers recognized that group working offered several advantages to management: for example, if one member of the team was sick or on leave production need not be interrupted because other members of the work group could of course do his job. But according to the report by Arthur S. Wineberg, co-ordinator of the workers' exchange programs at Cornell and an observer of the team, the Americans were critical of the work councils, which to them sounded like a "mixture of a shareholders meeting and a general sales meeting." In general, the Americans found the respectful relationship between management and labor disturbing; they hankered for a good grievance procedure.

On the other hand, the Americans did like the working conditions at the Saab plant, which was superior to those in Detroit in working environment, noise levels, lighting, and quality of the air. For example, early in the exchange program a Swedish worker asked one of the American visitors why he always seemed to be shouting, to which he replied that at Pontiac you had to shout in order to be heard.

Team Assembly Is Not for Everybody

In 1976 Volvo was planning to open a plant in Chesapeake, Virginia. But oddly enough, Volvo's new U.S. plant would have used traditional methods. The Swedish car company did not intend to use team assembly, which involves small groups of workers who completely assemble a single component such as an engine.

But it's not for America, Volvo says. "The team assembly system was designed to operate in a unique sociological environment," says William Baker, a Volvo spokesman. "Those conditions simply do not exist here." The American worker "seems to prefer to learn one task and stick to it."[8]

Volvo based this judgment on the views of American executives and union leaders who went to Sweden to see the Volvo system in action. What this Volvo example illustrates is that work systems cannot be lifted out of their sociological context; the attitudes that workers bring to work are as important as those they find there.

Volvo: Postscript

Volvo announced in 1973 that it would set up an automobile assembly plant in the United States, but the company has been forced to postpone the startup indefinitely. Volvo was unable to maintain its sales in the United States, because the company was hit by two major problems. First, labor costs due to increases in wages rose very quickly in the mid-1970s. But an even greater problem was absenteeism, which in some of their plants ran as high as 20% daily, which meant Volvo had to pay five employees to do the work of four. The system of sick pay benefits was so generous that it was widely abused, causing some workers to be absent 65 days a year. Even the Kalmar plant experienced a 15% absenteeism.

THE BRITISH EXPERIMENT: GLACIER

Getting Beyond Human Relations:

One of the best examples of the application of the human relations approach to bring about changes in the attitudes of both managers and workers, was the famous research at the Glacier Metal Company in Britain, which began just as World War II was ending and is still going on.

The Glacier was made famous as a place to work by the Tavistock Institute of Human Relations, which set out to examine in a scientific way the ongoing relations in one organization. This project, which was supported by the government, was described by Elliot Jacques in *Changing Culture of a Factory.*[9] And the factory management change the culture. The shop-floor people didn't clock in; there were no foremen, just section managers; and the business was run by an elected works council. It was Plato's republic writ small, with four systems determining the quantity and quality of life. The legislative system, made up of the various work councils, hammered out the policy; the executive system, consisting of all white- and blue-collar employees, did the actual work; the representative system served as the shop steward movement for both workers and executives; and the appeals system allowed subordinates to appeal decisions right up to the hierarchy.

To get a glimmering of Glacier, it is useful to have some notion of the product: plain or "big-end" bearings that are used in automobile engines to reduce friction on the crankshaft where it connects with the piston rod. The manufacture of these big-end bearings begins with the baking on of copper-lead-tin alloy on a bimetal strip by a continuous process using sintering methods— rather like a huge tape of steel that has an exact coating of smooth metal baked on as the tape is continuously pulled through an oven. The Thin Wall Shop cuts and bends these very large rolls of steel with their backing of low-friction linings into semicircular cylinders that are subjected to a series of operations, including drilling an oil hole, cutting an oil groove, boring the surface of the bearings to a specified standard, and cutting a locating nick. These complicated and tricky operations are performed at high speed on semiautomatic equipment with transfer machines.

Certainly it was a very exciting place to work, with all those people from the Tavistock Institute of Human Relations floating around as social science consultants, working through organizational problems. The top Tavistock consultant was Elliott Jacques with his M.D. from Johns Hopkins, and a Ph.D. from Harvard's Department of Social Relations awarded for his dissertation "Changing Culture of a Factory." Jacques, who was a founding member of the Tavistock Institute of Human Relations and a qualified psychoanalyst of the British Psychoanalytical Society, worked half time as an analyst in private and devoted the other half to the Glacier project as a social analyst. In those days he "lived" for part of the working day in a small building near the main gate of the London plant of the Glacier in Alperton, and was theoretically available to anyone in the company who wanted to consult him.

Elliott Jacques was perfect for his part—the medical psychiatrist who had turned his hand to industrial anthropology. Those were the days when psychiatrists were still in good standing. The Tavistock had helped win the war, conducting officer selection via the WOSBIE or group selection, and psychiatry interviews to select paratroopers and eliminate the bravado boys (I often wondered who that left), while after the war it provided therapeutic communities for returned POWs, work groups for coal miners facing mechanization, T-groups for executives, and air-crew sociology for British Airways. In addition, the Tavistock was one half of the famous interdisciplinary journal *Human Relations,* the other half being the Center for Group Dynamics at Ann Arbor, University of Michigan, which gave them an "American connection."

Fusing Psychoanalysis and Field Theory

Elliott Jacques, with his Tavistock knack for fusing Freudian psychoanalysis and Lewin's field theory ("a group was like a dilute electrolyte which could be polarized into positive and negative ions or roles, forming wonderful subgroups which attacted and repelled each other while the charges were built up or discharged"), was an excellent partner for Wildred Banks Duncan Brown, who was made a baron in 1964.

Brown, born in 1908 in Greenock, Scotland, was educated at a minor public school, joined Glacier in the 1930s as a sales manager, and rose to become the managing Director and Chairman of Glacier in 1939. Brown, a union member and supporter of the Labour Party, was a philosopher-king who wrote a variety of interesting articles, as for instance "Can There be Industrial Democracy?," included in *Exploration in Management,* a Penguin best seller in the United States. Brown and Jacques were the perfect pair: the tough, analytical executive who had made it to the top, and the human relations tough-but-nice analyst who helped him make it a beautiful social experiment that everybody wanted to read about and which turned the Glacier into a sort of show plant.

After all, the plants were run by soviets—elected work councils that decided policy. The Kilmarnock factory being within the socialist Red Clyde, the shop stewards insisted on the term "work committee." The committee (always pronounced coMMittee) is always a big thing in working-class Scotland.

The works committee was made up of seven shop stewards who sat on one side of the table, and seven manager-representatives who sat opposite, including the general manager of the plant. On one occasion a personnel officer who arrived early took a seat in the middle of the shop stewards' bench; when the stewards arrived, they excluded him by all sitting to his left.

Good-Bye Human Relations

By 1960 Glacier was getting out of human relations and into what Eric Trist of Tavistock called the task or systems approach, where the work to be done was a function of the resources (technological, human, and economic) available. "Organization was not a function of personality," said Brown. Brown was, of course, largely right. Glacier had become a victim of the British economy, sometimes at war set up in case the main plant was knocked out by bombs, and other times at peace (hence boom, which was quite good, and bust, which was very bad); a victim too of bearing technology (originally American-Clevite, a U.S. company that had given a Glacier competitor, Vandervell, its license in Britain, and the Glacier Metal Co. had been allowed to use this technology by the Ministry of Supply during World War II); and a victim as well of British workers (mostly union men) and managers (mostly up from the shop floor). Glacier policy was pure Tavistock and took the form of Great Britain writ small with its four systems. What knocked Glacier sideways was the stop-go British economy, the tricky technology of marrying semiautomatic machines and transfer equipment, awkward programming problems, and the good old British shop stewards who thought they could have done the job singlehanded.

THE AMERICAN EXPERIMENT

A high-quality working life is one that an employee finds interesting, challenging, and responsible. Increasingly, managers are setting out to improve the quality of working life (QWL). It is argued that engineers and social scientists must co-operate to develop new organizational designs that somehow get efficiency and QWL together.

Quality of Working Life (QWL)

QWL is an umbrella concept embracing studies dealing with job enrichment, organizational development, and industrial democracy. It tries to humanize work without sacrificing productivity. QWL is a new label that has won international recognition because it focuses organizational efforts on improving work life. It has pulled together a variety of efforts including "people people" from Tavistock working on sociotechnical systems, researchers in the United States working on job enrichment, and managers in Sweden writing on industrial democracy.

A close tie exists between QWL and productivity, at least in theoretical terms, in the minds of the proponents of such programs. The aim is to give workers an enriched job and increased participation in decision making. There are also ergonomic and safety aspects, and implications for the workers' nonworking life.

QWL interventions are quite complex in nature. Improvements in productivity and morale, usually measured at the actual time of the intervention, tend to be positive but temporary. Which raises the question, "Is QWL just a bubble on the surface of organizational life?" Maybe so, but perhaps organizations need an occasional shake.

QWL is attempted at a technical level involving diagnosis, design, and implementation, and also at a value level. QWL can also produce irreversible value changes that produce new optics, perspectives, and processes. But this is not the case with union leaders, who apparently are strongly resistant to such efforts. In Canada Steinbergs, Esso, and Alcan are especially interested in QWL programs.

Improved QWL through Participation

A good number of both blue- and white-collar workers feel underutilized, experience job frustration, and lack motivation. This is because the educational level of the employee is going up while work is being increasingly simplified. In a nut shell, smarter people are being asked to do dumber things. This unhealthy situation will not change until it is realized that the ultimate responsibility for the quality of working life lies with both management and the unions.

In an attempt to upgrade the QWL, Rockwell International and the U.A.W. together planned a new production facility at Battlecreek, Michigan. The new organizational design required job procedures that gave employees a complete unit of work to do and also made them responsible for the quality of the product, but gave them as much autonomy as possible on the job. Following the Scanlon Plan, teams of labor and management were formed to plan not only their own work but also their own vacation and overtime policies.

The Scanlon Plan is based on the principles developed by the late Joseph Scanlon, a former union official later associated with M.I.T. The object of the Scanlon Plan is to get management and labor to work together so as to be more productive, reduce costs, and produce a monthly bonus for all. To get this plan into action, production committees made up of a manager and two elected shop-floor people review written or verbal suggestions. Base labor costs as a percentage of sale dollars are set up on the basis of experience. Savings over the base labor costs are divided: 75% to the workers and 25% to the company. But the Scanlon Plan cannot work without the support of the union. Where it has been employed, union officials generally see the Scanlon Plan as hopefully providing something extra beyond the essentials of the collective agreement.

QWL at GM, Tarrytown

QWL programs are systematic attempts to make employees more creative and get them involved in discussions affecting their work. These programs are concerned not only with extrinsic goals such as improving productivity and efficiency, but also with intrinsic ones such as improving the quality of the working environment so that employees can achieve greater self-fulfillment. A good example of a QWL program is the one undertaken at the General Motors car assembly plant in Tarrytown, New York. When the program began, the plant was in a fairly bad economic state. Operating costs were high and there was much absenteeism and labor turnover. For example, at one time 2000 labor grievances were on the docket. In such a state of chaos, management was essentially on the defensive. As Robert H. Guest recounted:

> Union officers and committeemen battled constantly with management. As one union officer describes it, "We were always trying to solve yesterday's problems. There was no trust and everybody was putting out fires. The company's attitude was to employ a stupid robot with hands and no face." The union committee chairman describes the situation the way he saw it: "When I walked in each morning I was out to get the personnel director, the committeeman was shooting for the foreman, and the zone committeeman was shooting for the general foreman. Every time a foreman notified a worker that there would be a job change, it resulted in an instant '78 (work standards grievance). It was not unusual to have a hundred '78s hanging fire, more than 300 discipline cases, and many others."[10]

To overcome some of these difficulties, in 1973 the Quality of Work Life and General Motors negotiated a national contract that included a letter of agreement providing for recognition of the QWL concept. In April 1974 a professional consultant was brought in to involve supervisors and workers in a joint training program for problem solving. But in November 1974, at the height of the OPEC oil crisis, General Motors had to shut down the second shift in the plant and lay off half the workers. Men on the second shift with high seniority "bumped" hundreds of workers on the first shift. A shock wave went through the plant. In spite of these difficulties, the Quality of Work Life team decided to press on. Early in 1977 the plant made the big commitment and the Quality of Work Life effort was launched on a plant-wide scale.

The trainees who took part in these programs were to learn three things: first, the concept of Quality of Work Life; second, the functions of management and union; third, problem-solving skills. The problem-solving skills helped the workers to diagnose themselves, their own behavior, how they appeared in competitive situations, and how they solved problems.

The QWL program was a significant success, in terms not only of productivity but also of quality of the output. Further, in 1979 the Tarrytown plant became involved in the production of a radically new line of cars that achieved considerable success in spite of many technical difficulties.

Participative Management at Donnelly Mirrors

A brilliant example of participative management in action is provided by a quick look at Donnelly Mirrors, Inc., located in Holland, Michigan. DMI, which holds 70% of the domestic market in automobile mirrors and is a major manufacturer in other glass products, saw its sales rise in 1975 from $3 million to more than $18 million in 1977.

What is interesting, from our point of view, is that DMI makes use of many of the concepts developed by either the Tavistock people or Rensis Likert. DMI got into behavioral science experimentation by making an effort to apply the Scanlon plan to the company in 1952. The basic idea behind the Scanlon Plan is that people not only can be productive but want to be.

As John F. Donnelly has pointed out:

We are indebted to Rensis Likert of the Institute for Social Research at the University of Michigan for the concept of interlocking work teams. It has been very useful. To show you how the teams work, let's first contrast them with the usual organization setup. Here's a traditional organization chart. You see these groups of men all responsible to a single boss. There are many potential teams on this chart, but in spite of much management rhetoric, few of these groups operate as teams. Even when a group meets with its boss, there is a tendency to have a series of one-to-one dialogues with him. Likert found that some groups in companies do act as teams and consequently show consistently higher performance. There's good teamwork when the head of one team is also part of a successfully working team at the next higher level and becomes a sort of "linking pin" between the two teams.[11]

The Failure of the Non-Linear Systems Experiment

One of the most interesting experiments in participative management was carried out at Non-Linear Systems, Inc., of Del Mar, California. NLS was established in the early 1950s and the president and owner, Andrew F. Kay, built the company up from almost nothing to a peak of several hundred employees in the 1960s. The experiment in "participative management" was begun in 1960–61. The aim was to develop more of a human relations orientation toward work. Hourly rates of pay were done away with and no deductions were made for arriving late or leaving early. Project teams were introduced, and an eight-member Executive Council provided the operating policy, plan, and coordination. Introduction programs were introduced to familiarize new people with this novel setup. Sophisticated training procedures were introduced. Time clocks were removed because they were held to be dehumanizing. Further, following the view that keeping performance records vitiates self-control, the company operated on a "put nothing in writing basis."

In 1965, the experiment was terminated. Basically, the company was not performing in terms of either productivity or profitability. Why did the experiment fail? Partly because changing business conditions had made nonsense of the sales forecasts. The experiment had been a great success while the company was enjoying dramatic growth in sales, but in 1965 sales dropped sharply. Furthermore the company was unable to get the productive efficiency needed, when it began to experience severe competition. In the face of financial pressure, NLS moved back to conventional management. Thus the company probably avoided bankruptcy and is now operating with a small profit.

The General Foods Failure at Topeka

Some ten years ago General Foods Corporation opened a dog food plant in Topeka, Kansas, designed to run with semiautonomous work groups. Shopfloor people were allowed to make job assignments, schedule breaks, interview job applicants, and even decide on pay raises. The experiment was widely regarded as a model. At first the experiment was very successful, but GF is now discouraging publicity about the Topeka project. Economically the project was a success, but it became too threatening to many people in the company.

The project system came up against the GF bureaucracy. The project system was designed by Richard G. Walton, Professor of Business Administration at Harvard University, and eliminated a number of hierarchical levels. Employees worked under the direction of a "coach" rather than a supervisor, and rotated between dreary and meaningful jobs. While the system was a great success, it ran against corporate policies. Some of the Topeka managers careers were adversely affected because of their connection with this experiment. A number of managers involved left the company.

Costs and Requisites

The costs of introducing semiautonomous work groups can be high, both in material and personnel terms. As Panagiotis N. Fotilas points out:

> The best layouts are usually found in new plants, such as the Volvo automotive assembly plant in Kalmar, Sweden, which was designed specifically for semi-autonomous work groups.
>
> (Volvo, incidentally, calculates that the investment cost for the Kalmar plant was 10 percent higher than it would have been for a normal car assembly plant. Fiat, on the other hand, reports that its Rivalta plant, which was also designed for semi-autonomous work groups, cost four times as much.)
>
> Many companies with semi-autonomous work groups have also introduced new wage systems that are consistent with the new work organization. These systems are usually based on the performance of individual work groups as an incentive to increase cooperation among group members.[12]

Fotilas also specifies these requisites for success:

> Before semi-autonomous work groups can be introduced, it is imperative that the entire organization be made more democratic and that employees at all

levels be prepared to support efforts at innovation. It seems absolutely essential that decision making be decentralized and authority delegated throughout the organization.

If companies are to attack new problems, it is sensible that they mobilize all the knowledge and experience available in the organization. This is not a new principle, of course. But it is one that is still applied principally to managers and staff specialists. No value is placed on the knowledge, experience and skills of those who actually do the work every day with the existing equipment and on specific production processes. But their possible contributions to redesigning the workplace, planning the work, and solving problems must be regarded as at least as useful as that of the managers and experts.

Workers must be encouraged to accept responsibility and to participate in what are traditionally management decisions. But effective participation requires training for employees at all levels, from shopfloor workers to management. Participative groups with representatives from different levels should be formed to discuss and contribute ideas on the introduction of semi-autonomous work groups.

Supervisors should be involved from the start in planning and setting up of work groups. Studies indicate that whenever supervisors are included in the introduction of work groups, work groups have succeeded. And when they are not, work groups have failed.

Moreover, top management must be committed to any efforts to develop such groups. An unhindered flow of information from the top level to the bottom and vice versa is essential to the success of such work groups.

Investigations show clearly that the most successful workplace innovations and employee participation programs are found in companies where top management is completely committed and deeply involved in their application.[13]

The Effects of QWL Programs

What has been accomplished by these Quality of Work Life programs? Many have focused on job enrichment, which has prompted the redesigning of jobs and the development of work teams. These programs have improved productivity and restored dignity and equity to employees. All this leads to improved worker motivation and development of mutual trust between management and labor. Many contemporary personnel managers are familiar with the QWL programs at General Foods' Topeka pet food plant and Donnelly Mirrors in Holland, Michigan. No doubt a learning process has resulted from a knowledge of these programs.

A CONTINGENCY VIEW OF PARTICIPATION

The Demand for Participative Management

How do young American executives view the management of the system? The answer is that an increasing number of them expect to play a larger part in the critical decision making that affects the way they work. What they are looking for is participative management. Many young managers are sure they know better how to run the business than the boss, and even the unions are feeling this demand for participation. Further, a growing number of companies in the United States and Western Europe are offering the worker the chance to participate in management. In the United States this upsurge of interest in participative management has arisen because employers are worried about high absenteeism and low productivity.

Many managers and theorists start from the proposition that what is desirable in management is maximum feasible participation. The basic idea is that everybody should participate in everything. However, careful reading of political theorists suggests that this view may be misleading. It is wrong to presume that participation and classical democracy are synonymous. Political theorists have argued that in every society and organization an elite must rule, and that democracy is mainly involved in selecting the elite. And of course Michels in his famous "Iron Law of Oligarchy"[14] argues that an elite must emerge to take control even of a democracy. According to Michels, all organizations end up as dictatorships, with the top dogs running everything.

Problems with Participative Management

Unfortunately, participative management is more easily said than done. In spite of all discussions of self-fulfillment through job enrichment, a large number of American shop-floor people would still opt for more money rather than more enrichment. A program that will appeal to most will be more money, not too much challenge, and the four-day week.

There is also abundant evidence from studies of executives' attitudes that junior and middle management lack confidence in the decision-making skills of their superiors. Nevertheless, many young Americans taking employment as managers since the late 1970s, having grown up in the democratic 1960s, have developed a preference for participative work styles and are preoccupied with improving the quality of working life.

Toward a More Democratic Structure

If progress is to be made, then, it will be necessary also to make structural changes: i.e., to encourage the managers to participate in the process of deciding their own destiny, which means some kind of industrial democracy. In this area of managerial involvement in policy making North Americans could learn much, for once, from their cousins on the other side of the Atlantic.

As we have noted, in Western Europe a widespread effort is under way to beat the blues among not only blue-collar but also white-collar workers, including managers, which goes well beyond the current North American efforts to combat administrative authoritarianism through MBO, T-groups, the managerial grid, OD, intervention strategy, and process consultation. Such change strategies in Europe were part of a broad-based movement toward industrial democracy that sought to involve managers as elected representatives in the policy-making process. Such elected managers, besides pursuing their own interest, could become a check-and-balance restraining the shop-floor people from going too far.

The Chief Executive as Philosopher-King

What many of the studies cited in this chapter had in common was senior executive leadership which was a particular blend of the genius of a philosopher-king and an outstanding behavioral scientist. By providing leadership in getting managers involved in deciding their own destiny it is not what you do, but how you do it, and (perhaps even more important) whom you do it with, that counts.

But in any case, "winning the hearts and minds" of middle managers involves not only changes in management attitudes but also structural changes. The role of the Chief Executive in orchestrating the structural and attitude changes is vital. Perhaps we need another ancient mariner in the form of an executive steward (or perhaps only a representative) to tell our Chief Executive: "No man is an Island, entire of itself. . . . Any man's death diminishes me, because I am involved in Mankind; and therefore, never send to know for whom the bell tolls; it tolls for thee."

REFERENCES

1. *New York Times,* July 5, 1981. © 1981 by The New York Times Company. Reprinted by permission.
2. Michael R. Cooper, Brian S. Morgan, Patricia Mortensen Foley, and Leon B. Kaplan, "Changing Employee Values: Deepening Discontent?," *Harvard Business Review,* Jan.-Feb. 1979, pp. 124-25. Reprinted by permission. Copyright © 1979 by the President and Fellows of Harvard College. All rights reserved.
3. Frank Gibney, *Japan, the Fragile Superpower* (New York: Norton, 1975).
4. James O'Toole, *Making America Work: Productivity and Responsibility* (New York: Continuum, 1981).
5. "MOTOROLA A World Leader in Electronics—Quality and productivity through employee participation in Management." © 1982 Motorola Inc.

6. Paraphrased from William G. Ouchi, *Theory Z* (Reading, Mass.: Addison-Wesley Publishing Co., 1981).

7. J. M. Juran, "Product Quality—a Prescription for the West," *Management Review*, p. 9, (June 1981): © 1981 by J. M. Juran, 866 United Nations Plaza, New York, 10017. All rights reserved.

8. *New York Times*, June 20, 1976. © 1976 by The New York Times Company. Reprinted by permission.

9. Elliott Jacques, *Changing Culture of a Factory* (Tavistock Publications, London, 1951).

10. Robert H. Guest, "Quality of Work Life—Learning from Tarrytown," *Harvard Business Review*, July–Aug. 1979, p. 77. Reprinted by permission. Copyright © 1979 by the President and Fellows of Harvard College. All rights reserved.

11. An interview with John F. Donnelly, "Participative Management at Work," *Harvard Business Review*, Jan.–Feb. 1977, p. 120. Reprinted by permission. Copyright © 1977 by the President and Fellows of Harvard College. All rights reserved.

12. Excerpted, by permission of the publisher, from Panagiotis N. Fotilas, "Semi-autonomous Work Groups: An Alternative in Organizing Production Work?," *Management Review*, July 1981. © 1981 by AMACOM, a division of American Management Associations. All rights reserved.

13. *Ibid.*

14. Robert Michels *Political Parties*. Glencoe, Ill.: Free Press, 1958.

40. A Structure for Successful Worker Problem-Solving Groups

Gervase Bushe, Drew P. Danko, and Kathleen J. Long*

The advent in the United States of Quality Control Circles (or Quality Circles, as they are more often referred to) and their increasing popularity have encouraged the re-emergence of employee participation and participative management systems in manufacturing locations. Early attempts at participative management tended to focus on the values, attitudes, and supervisory styles of managers, while ignoring the attributes of individual subordinates (Argyris, 1957; McGregor, 1960). Later research discovered that not all employees wanted to participate, and that those who did wanted to participate to varying degrees (Vroom, 1960). With the focus on management style, little attention was paid to organizational structures and procedures that supported or inhibited involving greater numbers of people in the organization's decision-making processes.

Quality Circles have approached employee participation from the other end. They emphasize giving employees training and skills in problem analysis and problem solving, so that they will be more effective participants. Quality Circles also offer a particular set of structural and procedural elements to facilitate

Gervase Bushe is currently completing his doctoral degree in organizational behavior at Case Western Reserve University in Cleveland, Ohio. As a freelance consultant, he helped develop the materials for General Motors' Employee Participation Groups Program. He was subsequently hired on a part-time basis by GM's Organizational Research and Development Department to study and develop the EPG process. His research and consulting interests are in small group dynamics, organizational change, organization design, and sociotechnical systems.

Drew Danko has 15 years of experience in General Motors working as a researcher and consultant in the areas of job satisfaction, survey methodology, absenteeism, job design, organizational change, and quality of work life. He currently is focusing his efforts on employee participation groups and how they can impact the total organization. He has been with the Organizational Research and Development Department of General Motors since 1969.

Kathy Long was hired full time after an initial consulting assignment with the Education and Training Department of General Motors. Her work with their Research and Development group involved responsibility for the identification of corporate training needs and the design/development of special corporate management programs. She was the project manager for the development of GM's Employee Participation Group training materials. She is now with the Organizational Research and Development Department of GM; one of her primary responsibilities is to consult with manufacturing and staff groups on the implementation, maintenance, and diffusion of Employee Participation Groups.

employee participation. In this way, Quality Circles deal with some of the deficiencies of early participative management schemes and offer much more concrete guidelines for implementing and operating such a system.

In our experiences with over 20 manufacturing locations using some form of Quality Circles, we have found that only those organizations that have experimented with much more complex, home-grown Quality Circle programs establish viable long-term efforts. Those that strictly followed some of the early Quality Control Circle approaches available on the market had less success in surviving. In our attempts to understand the differences between those efforts that succeeded and those that failed, we have developed a model for worker problem-solving groups that encompasses much more of the total organization.

In this chapter we will share that model and its underlying rationale. We will begin by examining the deficiencies with the most popular Quality Circle programs currently on the market, particularly within the context of the typical structure and processes of manufacturing organizations. Then we will share the structural model we have developed and discuss how it deals with the deficiencies of currently marketed Quality Circle programs. Finally, we will discuss the usefulness of co-ordinators in implementing and diffusing the process.

QUALITY CONTROL CIRCLES: PROGRAM VERSUS INTERVENTION

One can hardly pick up a business journal or popular magazine today without finding numerous references to Japanese management and specifically to the Quality Control Circle. A quick review of business advertising shows the current "get-rich-quick" bucks are in Quality Circles. Many of these sales pitches are for warmed-over packages with little understanding of the Japanese approach and even less attention to the reality of the American industrial setting. In fact, the selling of Quality Circle programs parallels very nicely with what Krell (1981) describes as the marketing of Organization Development (OD), at least the "Mainline OD" version that aims unashamedly at satisfying the client's need for certainty and quick results. Krell states (p. 320):

Increasing one's business in this side of the field, for newcomers and established firms alike, will probably require more of an ability to package and promote OD technology than the ability to accomplish change or learning. Product design will continue to be more inclined to buy results than process, so emphasis should be on implementation rather than research or experiment

We sense that a marketing model of Quality Circles is being adopted by internal and external practitioners alike at the expense of change and learning. To better understand the implications of this argument, it is necessary to briefly look at what Quality Circles are and how they are typically implemented.

The Quality Circle process functions by allowing small groups of workers who are organized on a voluntary basis to meet regularly for the purpose of solving problems. These groups are typically composed of 8 to 12 people and generally meet weekly for an hour in or near their natural work environment. The supervisor is often a member and may or may not be the group's leader. The leader and members meet to discuss common problems, devise ways to collect and analyze factual data, develop work improvement plans, and make recommendations to management. Each member, including the leader, participates in the group's activities and shares in the problem-solving process. What sets the Quality Control Circle process apart from earlier quality improvement attempts (e.g., Zero Defects, Value Engineering, etc.) is the involvement of employees and supervisors in a setting where they meet as peers rather than as members of a hierarchy. This implicit power equalization, as well as the additional authority of the group to *identify* and *study* problems, sets the Quality Control Circle process apart.

We feel the marketing approach to Quality Circles falls short in three critical areas: where the focus is within the process (almost exclusively on creating Circles, with little attention to organizational readiness); where consultants enter and impact the organization (at the top and bottom only), and whom they exclude; and finally, how Circles are expected to operate within the organization (Circles as a means of organizational control versus employee influence).

Focus and Readiness

The first of these deficiencies concerns the focus within the process. Many consultants focus on the rapid creation of Quality Circles, whether or not the organization is ready for them.

As outlined in Table 40-1, typical Quality Circle implementation uses a consultant to train selected organizational members in quality control analysis and problem-solving techniques. These people then become leaders and train members who volunteer to join these groups. As membership enlarges, groups split in two and continue. In this way the system is expected to diffuse. A major selling point to managers is that Circles only recommend solutions to management. Management retains decision-making authority over the implementation or rejection of Circle ideas, thereby still retaining organizational control and power.

Although lip service is given to the necessity for "organizational readiness" for a Quality Circle process, it is not a major component of the prevalent Quality Circle model, where the primary focus is on delivering the necessary training for the groups to begin functioning. Unfortunately, this marketing orientation overlooks a significant reality of most traditional manufacturing organizations, where operations are based on rationalizing work into simplified tasks and on heavy dependence on inspectors for quality control. This type of

Table 40-1. The Prevalent QC Implementation Model.

- OUTSIDE CONSULTANT AND MANAGEMENT ENGAGE
- MANAGEMENT STEERING COMMITTEE IS FORMED
- CONSULTANT TRAINS A GROUP OF PEOPLE IN QC MATERIALS
- THESE PEOPLE BECOME GROUP LEADERS
- HOURLY EMPLOYEES ARE ASKED TO VOLUNTEER FOR GROUPS
- VOLUNTEERS ARE PLACED INTO GROUPS AND RECEIVE TRAINING IN QC MATERIALS FROM LEADERS
- GROUPS MEET FROM ONCE A WEEK TO ONCE A MONTH ON COMPANY TIME
- GROUPS' IDEAS ARE PRESENTED TO MANAGEMENT STEERING COMMITTEE
- STEERING COMMITTEE DECIDES WHETHER OR NOT TO IMPLEMENT IDEAS

system removes the major portion of quality responsibility from production workers and discourages the use of their intellectual capabilities. Because of this, production workers—if they talk to management at all—are accustomed to reporting *difficulties* up the chain of command, rather than *proposed solutions*. It is this basic false separation of problem solvers from non–problem solvers that needs to be changed, to have real worker participation in problem solving. To do this, an organization must have the cooperation of workers, union representatives, and supervisors *prior* to the implementation of a participative problem-solving process. Organizations vary as to how alienated workers are from supervision, and how alienated supervisors and managers are from each other. Within the prevalent Quality Control Circle implementation model, little or no attention is given to the problems of organizational alienation.

Recognizing that there are different types of organizations with their own structures, procedures, competencies, and crunch points, we want to make clear that we are dealing here with a particular kind of organization: manufacturing organizations that utilize production workers for the mass production of goods, as for example automobile plants, steel foundaries, toy factories, etc.

Numerous studies of these types of organizations have consistently found similarities in their structures and processes, regardless of the actual product being produced. Henry Mintzberg (1979) has recently coined the term "machine bureaucracy" to describe these kinds of organizations. After reviewing decades of research and writing, Mintzberg gives the following description of a machine bureaucracy: "highly specialized, routine operating tasks, very formalized procedures in the operating core, a proliferation of rules, regulations and formalized communication throughout the organization, large-sized units at the operating level, reliance on the functional basis for grouping tasks, relatively centralized power for decision-making, and an elaborate administrative structure with a sharp distinction between line and staff" (p. 315). This is the type of organization that is most concerned with eliminating uncertainty from the operating core and having smooth, uninterrupted production (Thompson, 1967). The main way in which the organization tries to make sure that people do what they must for smooth production is to control the way in which tasks are performed. Jobs are programmed and routinized, often tying the worker to a machine and/or an assembly line that, in a sense, controls his behavior.

These organizations employ engineers and other experts to design these work processes. Generally, the overriding concern is with maximizing efficiency. Little attention is paid to the psychosocial impact of these work designs. Close supervision is employed to make sure workers are doing their jobs properly. Wage rates and standards are usually outside the supervisor's control, so he or she has few incentives to offer as motivational tools. When necessary, the supervisor resorts to threats and punishment to enforce worker compliance. Furthermore, workers are actively discouraged from tampering with the work processes of experts because they have no "expertise." When offered, their opinions about how a job could be done are often ignored.

Supervisors often find themselves having to respond quickly to "emergencies" and to keep the "machine" running smoothly. This constant "firefighting" creates psychological pressure, reduces time available for decision making, and requires that subordinates respond immediately and without question to supervisory directives. Thus rigid, close supervision tends to prevail not only at the worker level but at all levels of the production hierarchy. Decision making tends to follow the formal chain of command, and subordinates are discouraged from challenging the views of superiors or experts.

The effect of all this on workers is that they usually feel treated as little more than tools. Often, they respond to this through passive aggression, telling supervisors, "I'm not paid to think" and refusing to do any more than the strict definition of their jobs. In the worst cases, this builds up to frustration and resentment, leading to incidents of open aggression such as property destruction and sabotage. This reinforces supervisors' beliefs that workers are lazy and that close supervision is required to maintain control over production processes. Indeed, Mintzberg describes machine bureaucracies as "organizations obsessed with control." The guiding assumption of work design is to try to make it "idiot-proof." Maintenance of a strong, authoritarian chain of command is reinforced through visible symbols of status and differentiation like privileged parking, different cafeterias, and restricted washrooms. Some of the relevant assumptions, norms and procedures of machine bureaucracies are outlined in Table 40-2.

Clearly, readiness for allowing workers to participate in problem solving depends on how closely an organization approximates this picture of a rigid, control-oriented bureaucracy. But Quality Circles are generally marketed as being applicable no matter what the organization's present state, with the rapid creation of Circles as the primary agenda.

Table 40-2. Characteristics of Typical Machine Bureaucracies.

Assumptions	Workers naturally dislike work. Workers are motivated by threats and money. Systems must be made idiot-proof.
Norms	Subordinates should only be seen working and not heard. There are experts for everything and only they know.
Procedures	Break down tasks into simple, repetitive actions. One person, one job. One person supervises many. Decisions are made by authorities. Obedience is rewarded.

Entry, Impact, and Exclusion

Just as minimal attention is given to the organization's state of readiness, almost no recognition is given to the evolutionary nature of a Quality Circle process and the changes it can stimulate within the organization during and after implementation. Simply stated, marketing-oriented practitioners and consultants view Quality Circles as an end state rather than as a transition state in the development of an organization. This "end state" mentality leads directly to the second deficiency concerning where consultants enter or impact the organization.

Most Quality Circle approaches today are being sold with the expectation that only minimal changes will be needed within the organization's existing structure for the process to work. The assumption is that an organization based on the management of individuals has the necessary framework to support a participative group-based system. Most Quality Control Circle approaches are primarily seen as floor-level programs attachable to any existing organization, the only top-level involvement being Steering Committee membership. Thus Quality Circles are generally not linked to existing internal structures such as the maintenance function or the reward or appraisal systems. The result is that most Quality Control Circle programs are seen as something management can simply add to the present operation to increase floor-level involvement.

This "top and bottom only" entry into an organization ignores the reality of most industrial settings. The impression is given that the union can be bypassed in implementing Quality Circle activity. Thus Quality Circles are seen as a management decision outside of joint union-management consideration. The impression is also given that it is not necessary to involve the second and third levels of supervision. This has the unfortunate result of leaving most Quality Circles without any real linkage into the organization's decision-making system, aside from its Steering Committee. The top/bottom approach leaves the union and middle management in limbo without any clear role, whereas upper management's role swells and exacerbates the problem of role overload. Not surprisingly, union representatives and middle managers often become the source of primary resistance to Quality Control Circles.

A critical aspect of most manufacturing organizations is that their work force is unionized. The presence of a strong union creates a number of problems and opportunities for "machine bureaucracies" considering greater involvement of their employees. Because management and unions are bound together through legal and contractual relations, interventions requiring fundamental change in the nature of work, roles, and relations require the consent and involvement of both parties. Typically, this requires an atmosphere of mutual trust and respect—one rarely found in unionized manufacturing facilities. Without some level of involvement and trust, unions are likely to perceive any kind of worker participation program as an attempt to get more work for less pay and/or an attempt to co-opt workers, thereby reducing the union's influence. Unless the local union feels sufficiently involved in the process, it will probably be successful in stonewalling or destroying it. With the introduction of Quality Circle projects, strong local unions will be very cautious about compromising themselves or their members. If there is a history of distrust and hostility, the union will probably distrust management's motives. Our experience is that projects are more likely to be successful by developing joint commitment to the project from union and management leadership at its inception.

A related aspect of union-management relations that bears directly on the efficacy of Quality Circles concerns what we refer to as the "shop-floor alienation cycle" (see Figure 40-1). The cycle was probably caused by the abuse of management control to maximize production at the expense of treating workers like replaceable cogs. But even where that is no longer the case, a vicious circle was started that probably still exists. The cycle has three groups of actors: supervision, hourly workers, and shop-floor union representatives. In organizations with high levels of distrust and frustration, many workers are angry at management and want to "get back at the system." One of the ways they do this is to apply pressure on their elected representative, the union official, to do so. The union representative, being an elected official, must perform for his constituency in visible ways, if he is to be re-elected. One of the most visible ways in which he can appease the demands of his constituents is to challenge the most visible power symbol of the system, the supervisor. The target of this challenge, the supervisor, naturally wants to defend him/herself and so retaliates. One of the supervisor's most accessible and effective strategies is to hassle and discipline the very workers this union representative is supposed to protect. Thus workers who must bear the brunt of the supervisor's anger are reinforced in their desire to get back at management, and the cycle continues on under its own fuel. Obviously, the Quality Circles must find some way to break this cycle or they will be torn up in it.

Aside from the dynamics of union resistance, the resistance of supervisors

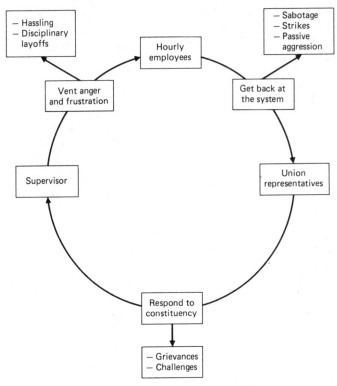

Fig. 40-1. The shop-floor alienation cycle.

and middle managers is not difficult to understand, if we recognize how different the assumptions, norms, and procedures underlying Quality Circles are from those in a traditional manufacturing organization. For instance, Quality Circles are founded on the assumption that workers want to do a good day's work, and with recognition and support will be motivated to produce a quality product. This is very different from traditional assumptions described earlier.

Table 40-3 outlines the differences between the assumptions, norms, and procedures of machine bureaucracies and Quality Circles. The radical cultural shift required in a manufacturing organization to make a Quality Circle process work is not emphasized in current Quality Circle packages, if understood at all by those who sell them. Consider for a moment how an uninvolved, uninformed foreman or middle manager with 20 years of experience in a machine bureaucracy would feel, upon learning that employee opinions and ideas are now to be actively solicited by management, that production employees are to be given training and to be paid for their meeting time, and that their ideas are to be seriously considered for implementation. After years of supporting the fractionization of work, middle managers must now accept that effective

Table 40-3. Comparison of Assumptions, Norms, and Procedures of Machine Bureaucracies and Quality Circles.

MACHINE BUREAUCRACY	QUALITY CIRCLES
ASSUMPTIONS	
Workers naturally dislike work.	Workers naturally want to be involved in their work.
Workers are motivated by threats and money.	Workers are motivated by recognition, praise, and the chance to make a difference.
Systems must be made idiot-proof.	Systems must allow for individual creative contribution.
NORMS	
Subordinates should only be seen working and not heard.	Subordinates should be encouraged to give their ideas and opinions.
There are experts for everything and only they know.	Everyone knows something about his or her work.
PROCEDURES	
Break down tasks into simple, repetitive actions.	Grapple with whole tasks.
One person, one job.	Whole group, whole task.
One person supervises many.	The group can develop its own internal leadership.
Decisions are made by authorities.	Decisions are made by group consensus.
Obedience is rewarded.	Innovation is rewarded.

group problem solving requires workers to understand whole tasks and how their parts fit in. These managers must suddenly adapt to the idea that Quality Circle programs, in effect, encourage workers to tamper with the designs of experts and to creatively contribute to the organization. The learning shock will logically create resistance, and the implementation process must aoknowledge that shock before it can deal with it. Exclusion of these managers from the implementation process only exacerbates the problem.

Quality Circles as a Means of Organizational Control

A third deficiency in the major Quality Circle packages on the market concerns how they define the operation of the Quality Circles within the organization. The functioning of the Quality Circle process in Japan is mainly centered on quality control or product quality. This is also true of the typical applications in this country and is reflected in the training, which focuses heavily on statistical quality-control analysis and technical problem-solving skills. Little attention is given to the skills necessary for effective group problem solving. The result is a limitation on the type of problems a group is trained to address (primarily those concerning product quality), and a reduction in its efficiency owing to its inability to recognize and deal with "group process" problems. No attention is paid to issues of authority and influence, so that issues pertaining to worker influence in the organization are not understood or dealt with until it may be too late. Use of a facilitator is advocated within most Quality Circle approaches, to deal with these issues. Unfortunately, this is a labor-intensive solution and that depends heavily on the time and skills of a single individual to recognize and correctly intervene within a group or the organization.

The statistical quality control (SQC) techniques taught in most of the Quality Circle packages available are extremely useful for production organizations and do aid in developing worker participation. They aid participation by teaching workers and supervisors a common language with which to interact around problems. They also aid in power equalization, because what is right is determined by the facts, not by who is in authority.

However, SQC technology plays into the machine bureaucracy's obsession for control. Sometimes the development of participative work systems are instituted as a sophisticated management control system, as opposed to an attempt to grant real influence to members lower down in the organizational hierarchy. As Dickson (1981) has pointed out, participative groups can be used to create conformity to top-management decisions by limiting the things the group can think about or do. Such pseudo-participation can initially increase people's job satisfaction and productivity. A burst of energy usually takes place, when people who have felt left out of the organization's decision making perceive they now have a chance to get in. But our experience has been that where real influence over decisions regarding people's work is denied them and organizational control is maintained, Quality Circle programs do not last more than 18 to 20 months. After the initial burst of energy they fade out rather quickly.

Quality Circles as an Intervention

We have just reviewed what we see as deficiencies in the typical Quality Control Circle process in manufacturing organizations that grow out of a conscious or unconscious adoption of a marketing model. An alternate approach guiding our efforts is a developmental or OD model.

An OD model suggests several significant differences. First of all, we would expand a client's definition of success from the number of Circles implemented or problems solved to include the effectiveness of the organization's processes. The implementation of Quality Control Circles should enable the organization to adjust better through effective problem solving at several levels. The organization should be better able to adapt, to be flexible, and should resist changes less because the Circle training itself is an experience in change. Work groups within and across functions should become better co-ordinated in their goal accomplishment efforts. And there should be better sharing of relevant information and utilization of knowledge across organizational boundaries.

From our perspective, Quality Circles can also be seen as an attempt to optimize the social and technical systems of the traditional manufacturing organization. Following Carlson's idea of "quality of work life" (QWL) as a developmental process (Carlson, 1980), Quality Control Circles could be incorporated as a logical and effective precursor for the development of semi-autonomous or totally autonomous groups in the redesign of a facility. At a minimum, Quality Control Circles have the potential to fine-tune the technical system, while providing employees with psychological rewards.

This developmental notion implies the final significant difference. We would encourage clients to view Quality Circles as an intervention in an organizational system, not sell it as a packaged program. Clients need to see Quality Control Circles as having the potential to create a transitional state by unfreezing the organization and moving it toward a new state. What the organization does in response to this intervention depends heavily on its readiness to move from a traditional machine bureaucracy to some kind of alternate organizational form.

A STRUCTURAL MODEL FOR SUCCESSFUL WORKER PROBLEM-SOLVING PROJECTS

Elements of Success

Developing participative work systems requires attention to three major areas: commitment, skills, and access to resources. Put simply, for participation to work, people must be committed to the organization's goals, particularly as they apply to individuals' jobs in the form of work process, standards, and rewards. People must have the skills to do the job without close supervision and the skills to work co-operatively with others. Finally, they must have access to the resources necessary to do the work and implement their ideas. People have to want to do the job, know how to do the job, and be able to get what they need to do the job.

Team approaches to worker participation characterize the bulk of efforts undertaken, and there are a number of good reasons for this. The small group is likely to offer a wide enough distribution of the skills necessary to get the job done. Further, it is the small group, even in machine bureaucracies, that sets the standards for production and enforces member behavior to conform to these standards (Roethlisberger and Dickson, 1939; Homans, 1950). Finally, it is the small group to which individuals really commit themselves through identification (Srivastva, Obert, and Neilsen, 1977).

Research and writing on participative work systems has tended to focus on the attitudes, assumptions, and beliefs of organizational members that are necessary to make such systems work. As stated previously, little attention has been paid to the structural issues involved. The exception to this has been in the literature treating the use of sociotechnical systems (STS) to design new plants (Pasmore and Sherwood, 1978). While STS techniques appear to provide an exciting new approach to designing participative manufacturing facilities, this approach has been singularly unsuccessful in redesigning existing machine bureaucracies (Trist, 1981).

The reasons given for these failures usually fall into one of the following categories: resistance from middle or upper management who feel their jobs are threatened as workers take on more responsibility; resistance from a local union not fully involved in the redesign effort; resistance from other workers not involved in the redesign who feel there are inequities in the system (Goodman, 1979; Pasmore, 1982).

It is clear from these efforts that before sociotechnical redesign is undertaken, the organization has to develop greater cooperative, participative ways of working, and greater trust than is usually found in machine bureaucracies. STS redesign may be where our efforts at worker participation eventually take us, but is not the place to begin.

In reviewing the efforts of successful and unsuccessful attempts to improve the quality of work life through developing employee problem-solving groups, we have begun to suspect that a critical determinant of success is the way in which these groups are structured into the existing organization. While we are only now collecting the data necessary to support or invalidate our theories, we think that certain elements of this structure are more important to successful problem-solving group projects than the kind of training provided.

Further, we have been forced to change our definition of success. Prior to our investigations, our criteria of success (and, we suspect, the criteria of most others who implement Quality Circles) was the number of active groups, the stability of the system, the utility of the ideas emanating from these groups, and the improvement of members' quality of work life. Our research has caused us to recognize the fundamental instability of small problem-solving groups within machine bureaucracies. The norms, assumptions, and values of the two systems are too contradictory. We now view successful worker problem-solving group efforts as those that prepare the organization for a restructuring effort which will put operating systems more in line with worker participation. This will probably mean eventually structuring the organization around semiautonomous teams at both management as well as worker levels in a peculiar kind of matrix system.

Two elements of the structure of problem-solving groups seem particularly critical. First, there appears to be a relation between the number of supervisors and managers in problem-solving groups and the success of the project: the more managers involved, the more successful the project. Second, the more union representatives are structured into the same groups as managers, the more successful the project. We will deal with these two elements in turn.

Involving Managers: the Parallel Organization

Because we are abstracting general principles from a number of participative work projects with different names, training designs, and goals, we have had to develop terms that encompass them. By a *parallel group* we mean any group of hourly and/or salaried employees intended to be permanent, with a mandate to participate in some facet of the organization's problem-solving and/or decision-making processes. A second aspect of a parallel group is that the norms and procedures in the group promote collaborative nonhierarchical relations. Practically speaking, this means that at the very least, decisions in the group are made by consensus. *Parallel organization* is the term we use to describe the total network of parallel groups in any one organization. We further distinguish parallel organization by the levels of the management hierarchy included in parallel groups.

As pointed out earlier, conventional Quality Circle programs only place workers (with or without foremen) into problem-solving groups; the rest of supervision is left out. Where foremen are in groups, we refer to this as a first-level parallel organization. Where general foremen are also in parallel groups, with or without hourly workers (though hourly workers participate in other parallel groups), we refer to this as a second-level parallel organization. And so forth. Figure 40-2 provides a structural view of the first-level parallel organization.

Many of the problems first-level parallel organizations create have to do with resistance from managers and supervisors left out of the system. One common problem is that supervisors are not sure what these groups are doing and sus-

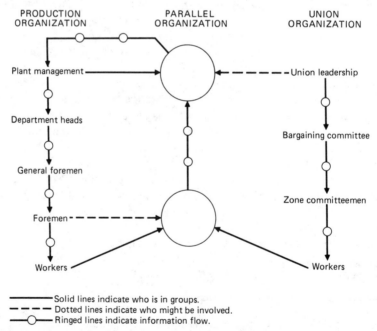

Fig. 40-2. A structural view of the prevalent quality circle model: first-level parallel organization.

pect that the groups are, at best, excuses for goofing off, and at worst, forums for subversive activity. Another common concern is that supervisors are not sure what their roles are in relation to these groups. Some fear that these groups are a direct threat to their authority. This is compounded when the structure calls for group recommendations to bypass the normal chain of command to a top-management steering committee.

Quite clearly, this "being left out" generates resistance not only among first-line supervisors, but at any level of the management hierarchy. In one location studied, first- and second-line supervisors were in parallel groups, but department heads were not. It was at the department head level that the greatest amount of resistance was evident.

One aspect of this is the "in-group, out-group" dynamics that are generated through membership in parallel groups. Where these groups are successful in developing collaborative norms, old status distinctions tend to break down. Foremen in parallel groups with workers begin identifying with the workers "against" the rest of management. New "we-they" lines are drawn that threaten the status quo of those not directly involved in the parallel organization.

The obvious conclusion, then, is that projects which attempt to involve workers through parallel group structures must involve supervisors and managers in parallel groups as well. We do not believe there is any one best way of structuring the involvement of supervisors and managers, although having overlapping membership in these groups (similar to the linking pin concept proposed by likert, 1967) does appear extremely effective in increasing communication flow and integrating group efforts. It does not even seem necessary that these parallel groups be called the same thing or be implemented as part of a total package. For example, one structure could have a department head, representatives from functional areas, and second-line supervisors in a parallel group called a "business team." Then second-line and first-line supervisors could be in a parallel group called a "coordination team," while workers with or without supervisors serve in "Quality Circles." The point is that this third-level parallel organization will be much more successful at reducing resistance to participative work systems than extensive training efforts or lower-level parallel organizations.

Structuring the Union In

A second consideration is the union's role in the parallel organization. More and more, research on implementing participative work systems is targeting union involvement and commitment as a key determinant of success (Goodman, 1979; Mansell, 1980). This holds true with our experience as well.

We have found the greatest success in improving quality of work life indicators at those locations where union representatives are structured into the parallel organization. Practically speaking, this means that union representa-

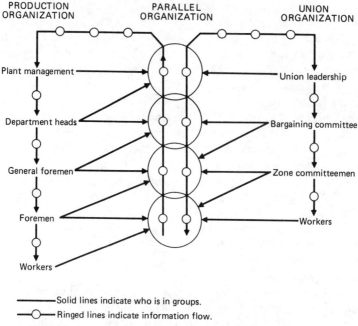

Fig. 40-3. A structural view of a totally integrated parallel organization.

tives should be members of the same parallel groups as supervisors, general supervisors, and department heads. This results in a totally integrated parallel organization (see Figure 40-3).

Perhaps the greatest pay-off here is in breaking down the shop-floor alienation cycle. As supervisors and union representatives work in parallel groups on common goals (such as improved QWL or job security through enhanced competitiveness), long-standing barriers begin to break down. Union representatives develop a greater appreciation of the pressures on supervisors and the complexity of the problems facing them. Supervisors develop a better appreciation of the skills and contributions union representatives provide in managing human affairs. Co-operative work relations fostered in the parallel group help each side realize that the other person is a decent, reasonable human being, and these changes in attitudes and behavior slip into day-to-day activities.

THE FIRST STRUCTURAL CHANGE: THE INTRODUCTION OF CO-ORDINATOR ROLES

Having people whose full-time responsibility is to implement and nurture a participative work system is a critical factor in its success. The introduction of these co-ordinator roles into the organization is typically the first change in the machine bureaucracy structure. In the organizations we have worked with, we have found that many individuals already employed have the appropriate skills and beliefs to do the job. All they need is a little extra outside training in the areas of group problem solving, team building, and organization development. The most desirable state of affairs is to begin with two full-time co-ordinators: one from the supervisory personnel, and one who is currently, or has been, a union representative. These should be two people who are intelligent, committed to worker participation, respected within the plant community, and able to develop into a team.

Other things also appear to be useful in facilitating their roles, though how critical these are we have not yet been able to determine. It is probably desirable that both these individuals report directly to the plant manager and the local bargaining committee. They should have some idea about the amount of money the plant has to spend on implementing worker participation. It may be useful for them to have some kind of secretarial backup, and they will need fairly regular access to adequate meeting rooms.

The co-ordinators' job is to develop implementation plans in consultation with others, to promote the concept, and to provide training to parallel groups. This training is probably most useful if done on site, in the context of a parallel group meeting.

CONCLUSION

The number of companies currently attempting Quality Circle projects under various names is growing at an exceedingly fast rate. Unfortunately, with the prevalent "marketing model" of Quality Circles, which pays little attention to the structures, processes, and management style of American industry, we fear that Quality Circles will go away as quickly as they have come. This is unfortunate for a number of reasons. First and foremost, involving greater numbers of employees at all levels in running the business can significantly contribute to organizational efficiency and effectiveness. We have avoided discussing the effects of work participation on such things as productivity, as this has been reviewed so many times before (see Locke and Schweiger, 1979). The consensus of these reviews has been that participation almost always increases job satisfaction, but not always productivity. Our experience is that those organizations which introduce higher-level parallel organizations show significant improvements on most conventional measures of manufacturing performance. It is not hard to understand why, if you accept Locke and Schweiger's argument that productive efficiency is a result of motivational and knowledge factors. Three aspects of the parallel organization contribute to organizational knowledge. The parallel organization creates mass amounts of information flow top down, bottom up, and laterally. It creates a time and place for problem solving to occur. Finally, it allows for more effective co-ordination of efforts.

Through the use of group problem-solving and consensus decision-making techniques, power equalization takes place, which in turn helps to build trust (Gamson, 1968). Trust is a critical factor in an organization's ability to adapt to changing environments. Further, it is an important aspect of employee motivation (Fox, 1974). Having influence over the conditions of day-to-day work allows people to have "a piece of the action" and motivates them to find more efficient, cost-saving means of producing a quality product.

Quality Circles have helped to point us in the right direction by emphasizing the need to gain the involvement of workers in the day-to-day running of the business. We have found that, by and large, once workers are convinced of the organization's sincerity in seeking their participation, their commitment is not a problem. More problematic is the commitment of supervisors and managers to manage in a participative manner. Part of this lack of commitment stems from being left out of the problem-solving groups, and part from not having the skills to manage participatively. In some situations supervisors do not trust their superiors to back them up. We have found that locations which have developed complex parallel organizations involving many supervisors and managers in problem-solving groups have been more successful at developing greater worker participation than those which have attempted to implement Quality Circles only at the work level. It seems that through this kind of structure supervisors develop their own interests in participative work systems and evolve greater commitment to worker participation as a whole. Greater trust between managerial levels is developed. Further, through working co-operatively with peers on problem-solving tasks, they gain greater skills in managing participatively.

We cannot overstress the need to fully involve the local union whereever one exists, both in cosponsoring the project and in participating in these parallel groups. This aids greatly in developing the trusting, co-operative environment necessary for participation to work, and helps to break down the destructive shop-floor alienation cycle.

In the final analysis, an effective organization seeking to involve and utilize all its people requires a structure that promotes co-operative relations. Traditional manufacturing organizations are often structured in a way that promotes conflict. The challenge ahead is to find new ways of organizing labor in the mass production of goods that will promote responsible autonomy for groups on the shop floor while maintaining productive efficiency.

REFERENCES

Argyris, C. *Personality and Organization.* New York: Harper and Row, 1957.

Carlson, H. C. "A Model of Quality of Work Life as a Developmental Process." In Burke, W., and Goodstein, L. D., eds., *Trends and Issues In OD: Current Theory and Practice,* pp. 83–123. San Diego, Calif.: University Associates, 1980.

Dickson, J. W. "Participation as a Means of Organizational Control." *Journal of Management Studies* 18 (1981): 159–76.

Fox, A. *Beyond Contract: Work, Power and Trust Relations.* London: Faber and Faber, 1974.

Gamson, W. A. *Power and Discontent.* Homewood, Ill.: Dorsey Press, 1968.

Goodman, P. S. *Assessing Organizational Change: The Rushton Quality of Work Experiment.* New York: Wiley-Interscience, 1979.

Homans, G. C. *The Human Group.* New York: Harcourt, Brace and Co., 1950.

Krell, T. C. "The Marketing of Organization Development: Past, Present, and Future." *Journal of Applied Behavioral Science* 17 (1981): 309–23.

Likert, R. *The Human Organization.* New York: McGraw-Hill, 1967.

Locke, E. A., and Schweiger, D. M. "Participation in Decision-Making: One More Look." In B. Staw, ed., *Research in Organizational Behavior,* pp. 265–339. Greenwich, Conn.: JAI Press, 1979.

Mansell, J. *Dealing with Some Obstacles to Innovation in the Work Place.* Toronto: Ontario Ministry of Labour; Quality of Working Life Centre, 1980.

McGregor, D. *The Human Side of Enterprise.* New York: McGraw-Hill, 1960.

Mintzberg, H. *The Structuring of Organizations.* Englewood Cliffs, N.J.: Prentice-Hall, 1979.

Pasmore, W. A. "Overcoming Road Blocks to Work Restructuring." *Organization Dynamics,* Spring 1982.

Pasmore, W. A., and Sherwood, J. J., eds. *Sociotechnical Systems: A Sourcebook.* La Jolla, Calif.: University Associates, 1978.

Roethlisberger, F. J., and Dickson, W. J. *Management and the Worker.* Cambridge, Mass.: Harvard University Press, 1939.

Srivastva, S., Obert, S. L., and Neilsen, E. H. "Organizational Analysis through Group Processes: A Theoretical Perspective for Organization Development." In C. Cooper, ed., *Organizational Development in the U.K. and the U.S.A.,* pp. 83–111. New York: MacMillan, 1977.

Thompson, J. D. *Organizations in Action.* New York: McGraw-Hill, 1967.

Trist, E. L. *The Evolution of Socio-technical Systems.* Toronto: Ontario Ministry of Labour; Quality of Working Life Centre, 1981.

Vroom, V. *Some Personality Determinants of the Effects of Participation.* Englewood Cliffs, N.J.: Prentice-Hall, 1960.

41. Project Teams: One Alternative to Enhance R & D Innovation

Joseph G. P. Paolillo*

INTRODUCTION

A project team is an integral part, a building block, in an organizational matrix design. This organizational unit is dedicated to the attainment of a specific goal, the successful accomplishment of which must be on time, within budget, and in conformity with predetermined specifications (Evan and Black, 1967).

The project team personnel are specialists from diverse fields; the number of experts and the disciplines represented vary with the mission of the team. The project team is typically organized by task (vertically) rather than by function (horizontally) (Evan and Black, 1967). The goal is to blend the advantages of the vertical structure, in which the control and performance associated with autonomous management are maintained for a given project, with the horizontal structure, in which continuity, flexibility, and the use of scarce talents may be achieved in a group. The pooling of various experts in the formation of project teams is often favored for activities where continuity, flexibility, and autonomy afforded by the project team are believed to contribute to creativity and innovation.

Much of the literature on matrix management and project teams is related to research and development divisions and/or organizations where there is a keen interest in creativity and innovation. Some examples would include firms involved in the aerospace industry and particularly the contractors working for the NASA and the Apollo programs. Descriptions of project teams are found in various other research and development settings including semiconductors, telecommunications, guidance systems, flight simulators, construction research, and toiletries. A good number of large multinational organizations have introduced company-wide matrix organizations and project teams. These

*Joseph G. P. Paolillo received a B.S. in chemistry from Ohio University, Athens, an MBA from the University of Delaware, Newark, and a Ph.D. from the Graduate School of Management and Business, University of Oregon, Eugene. He has worked for Inmone Corporation. At present, he is an Assistant Professor of Business Administration at the University of Wyoming, Laramie. His most recent works have appeared in *Research Management, IEEE Transactions on Egnineering Management, The Journal of Management Studies,* and *The Journal of Management.*

include TRW, Lockheed, ITT, and Dow-Corning. Most recently a wide variety of advertising agencies, entertainment companies, health service organizations, management consulting firms, educational institutions, and accounting firms have adopted project teams for activities where continuity, flexibility, and autonomy afforded by the project team are believed to contribute to creativity and innovation.

Social commentators have repeatedly admonished those individuals who are responsible for the design or redesign of organizations to devise organizations that are innovative, in preparation for a future that will be characterized by increasingly rapid social and technological change. Numerous problems arise in defining and analyzing organizational innovation because (1) the term has different meanings for different analysts, (2) there are different kinds of organizational innovations, and (3) there are a variety of ways in which the construct has been conceptualized and operationalized.

One difficulty is that most individuals expect an innovation to be of some positive value to the organization; that is, they expect it to result in a cost savings, or increased profit, or sales in new markets. There are unsuccessful innovations that do not become economically advantageous or that eventually fail because they are not accepted outside the organization. Additional problems arise if one attempts to differentiate between large and small improvements; these problems concern the degree of innovation, or how radical the innovation is. In the evaluation of various innovations, what criteria does one employ, and what is the proper time frame for evaluation?

These difficulties are multiplied when one considers the major types of innovations in organizational settings. Four categories of organizational innovation have been outlined:

1. *Product or service innovations:* the introduction of new products or services that the organization produces, sells, or gives away.
2. *Production-process innovations:* the introduction of new elements in the organization's task, production, or service operations, or advances in the company's technology.
3. *Organization-structure innovation:* the introduction of altered work assignments, authority relationships, communication systems, or formal reward systems into the organization. In addition, any other introduction that alters the interaction patterns among the employees in the organization.
4. *People innovation:* either of two alternatives that produce direct changes in the people within the organization: (a) altering the employees by dismissing old ones and/or hiring new ones, and (b) modifying the behavior or beliefs of the people in the organization via techniques such as education or psychoanalysis.

The innovations in each of these four categories could have either a positive or a negative impact on the viability or adaptability of the organization.

The problems arising from the fact that organizational innovation has been employed widely can best be demonstrated by enumerating the diverse conceptual and operational definitions of the construct. Conceptually, there seems to be little consensus with regard to the exact meaning of organizational innovation. Shephard (1967) considers an organization to have innovated when it learns to do something it did not know how to do before, whereas Mansfield (1963), Becker and Stafford (1967), and Mohr (1969) refer to innovation as the first use of a new product, process, or idea, and Evan and Black (1967) define innovation as the implementation of new procedures or ideas. Zaltman, Duncan and Holbeck (1973) adopt a different view by conceptually defining innovation as the propensity to adopt any idea, practice, or material artifact perceived to be new by the adopting unit. Generally, the numerous conceptual definitions of organizational innovation have tended to (1) use adoption and implementation interchangeably, and (2) confuse the actual adoption and the propensity to adopt a new product, process, or idea (adoption and propensity to adopt have entirely different meanings). The employment of the Zaltman et al conceptual definition circumvents the problem of evaluating the success and the degree of innovativeness of a particular innovation by emphasizing an organization's propensity to adopt an innovation, and not the question of market acceptance or evaluation. The designation of the R & D subsystem, rather than the entire organization, as the adopting unit allows one to concentrate on technical product-process innovations and to avoid the confusion inherent in analyzing multiple categories of innovation. A review of the literature regarding relationships between innovation and organizational characteristics reveals that organizational innovation is related to organizational size (Aiken and Hage, 1971; Mansfield, 1963; Mohr, 1969; Mytinger, 1968), structure (Aiken and Hage, 1971), and resources (Aiken and Hage, 1971; Mytinger, 1968). For purposes of this chapter, organizational size is the number of R & D employees; structure is measured by the number of formal supervisory levels and the size of the project team; and resources are the financial resources, operationalized as the size of the R & D budget in dollars.

When the previous research results are considered, several general factors can be noted. First, virtually all of the studies are industry or company specific (with the notable exception of the organizational climate study by Hall and Lawler, 1969), preventing cross-industrial or comparative analyses of organizational innovation. Second, the strength of the reported relationships between organizational characteristics and innovation may have been masked by utilizing the organization rather than the research and development subsystem as the unit of analysis. The present investigation seeks to provide data that will help to overcome these problems.

METHOD

Sample

This study was carried out with a sample of research scientists and professional engineers in six organizational R & D subsystems, representing the clothing, leisure time, electronic components, machinery, electronic instrumentation, and paper industries.

The sample consisted of six R & D directors, 20 laboratory supervisors, and 58 research scientists or engineers, for a total sample of 84 respondents. Ninety-six percent of the sample were male with mean tenure of ten years in their respective companies and five years in their present positions. Seventy-five of the 84 respondents (i.e., 89%) held advanced degrees.

Since the study was exploratory in nature, it was decided that the representativeness of the organizational R & D subsystems participating would not be as crucial as their heterogeneity with respect to industry. Therefore the primary determinant in selecting R & D subsystems was that they permit a meaningful test of the relationships being examined.

Data Analysis

The data were analyzed using Spearman-rank correlations, since this type of analysis makes no assumptions about the distribution of the data points and requires at least ordinal data.

Results

A summary of the findings are presented in Table 41-1. The number of R & D employees and of formal supervisory levels were significantly and negatively related to R & D subsystem innovation. Average size of research project teams

Table 41-1. Relationships Between R & D Subsystem
Characteristics and Perceived Innovativeness.

N = 84

INDEPENDENT VARIABLES	PERCEIVED INNOVATIVENESS
No. of R & D employees	−.37***
No. of formal supervisory levels	−.14*
Average size of research project teams	.20**
Size of R & D budget	−.05

* Significant at the .10 level (one-tailed test)
** Significant at the .05 level (one-tailed test)
*** Significant at the .001 level (one-tailed test)

was significantly and positively related to R & D subsystem innovativeness. The R & D budget was unrelated to (independent of) R & D subsystem innovation.

The present findings are based on a limited sample and suffer from all the limitations inherent in any nonprobability sampling method. One cannot claim that the sample is representative of a particular population, or that the statistics resulting from such a sample are unbiased and have known properties. And of course the Spearman-rank correlation coefficients should be interpreted as indices of association found within the context of the study.

Discussion

The relationship between size and R & D subsystem innovation was not in the direction indicated by prior research at the aggregate organizational level. Instead of the larger R & D subsystems, it was the smaller ones that were perceived by R & D personnel as being more closely associated with innovations. Apparently, large numbers of research scientists, engineers, and technicians were perceived as a hindrance to innovation. This may indicate that the quality, rather than quantity, of the R & D personnel above the technician level is more important in considering the innovativeness of R & D units. It might also indicate that the administrative (coordination) demands commensurate with large size may be a source of dissatisfaction; R & D personnel are known for their high autonomy needs. Such dissatisfaction could then be the basis for the relatively unfavorable view of innovativeness in larger R & D subsystems.

The structure of the R & D subsystems was shown to be correlated with perceived R & D subsystem innovativeness. The relationship was in the direction indicated by prior research at the organizational level. Essentially, R & D subsystems with few supervisory levels were perceived as being associated with innovation. In a related way we might extend this result to postulate that an R & D subsystem characterized by less supervision, including organizational informality and decentralization, is conducive to R & D subsystem innovativeness. This is consistent with the results of testing the independence of R & D subsystem and project team size with innovativeness. Small R & D subsystems with few formal supervisory levels are associated with R & D subsystem innovativeness.

The size of research project teams was shown to be strongly correlated with perceived R & D subsystem innovativeness. R & D subsystems utilizing large (five-man) research teams were perceived by R & D personnel to be more innovative than subsystems utilizing smaller (two-man) research teams. There are two apparent reasons for the correlation between the size of research project teams and innovativeness: (1) the greater the pool, the greater the probability that one will find a creative individual; (2) the collegial interaction and exchange of ideas afforded by relatively large research project teams were perceived by R & D personnel as being conducive to R & D subsystem innovativeness. Just how large the research team could be before size becomes dysfunctional is not answered here. Obviously, as team sizes become very large, the positive relationship may disappear.

The lack of an appreciable correlation between the size of an R & D unit's budget and its innovativeness appears to indicate that the nonfinancial resources available within the R & D subsystem—such as talented research personnel, a collegial atmosphere afforded by few supervisory levels, and utilization of project teams—may have a greater impact on R & D subsystem innovation than does the size of the underlying budget. This result draws attention to the need for R & D management to consider the allocation of nonfinancial as well as financial resources in their attempts to enhance R & D subsystem innovativeness.

Based on the results and preceding discussion, some provisional structural suggestions for the design of innovative R & D subsystems would include:

- a subsystem with a relatively small number of quality research scientists and engineers, or, in extremely large R & D divisions (e.g., Bell Laboratories), differentiated laboratories with each housing a relatively small number of research scientists and engineers
- a subsystem with a small number of formal supervisory levels
- a subsystem that utilizes research project teams of at least four or five people

The practical implications accruing from the results of this study are quite clear. Structural/design suggestions would include the elimination of superfluous hierarchical levels in an attempt to enhance subsystem innovativeness. This could be attempted through the expansion of supervisory spans of control and the decentralization of the formal hierarchy of authority within the R & D subsystem, or more simply by utilizing research project teams.

The results of this study provide some useful tentative recommendations for the design of innovative organizational R & D subsystems. The present findings could be made more conclusive through replication across different industries in a larger, heterogeneous sample of organizational R & D subsystems. Replication is essential with regard to the results that do not fit with prior aggregate organizational data, specifically (1) the relationship between R & D innovativeness and the independence of R & D budget size, and (2) the relationship between small R & D subsystem size and R & D subsystem innovativeness. Additional study could be accomplished through field experiments designed to control specific R & D subsystem characteristics, beyond the few addressed in this study.

More specifically, this exploratory study provides the foundation for a series of meaningful research questions. Given the results of this study, what is the optimum size for innovative R & D subsystems? Is the optimum range dependent upon the industrial setting and the type of research activity conducted in the R & D subsystem? If so, what is the direction and magnitude of these relationships? Similarly, what is the optimum number of formal supervisory levels to be found in innovative R & D subsystems, and what factors influence this range? In addition, what is the optimum size of research project teams, given our conclusion that utilizing project teams does indeed enhance innovativeness.

REFERENCES

Aiken, M., and Hage, J. "The Organic Organization and Innovation." *Sociology* 5 (1971): 63–82.

Becker, S., and Stafford, F. "Some Determinants of Organizational Success." *Journal of Business* 40 (1967): 511–518.

Cornbach, L. "Coefficient Alpha and the Internal Structure of Tests." *Psychometrika* 16 (1951): 279–316.

Evan, W., and Black, G. "Innovations in Business Organizations: Some Factors Associated with Success or Failure of Staff Proposals." *Journal of Business* 40 (1967): 519–30.

Gaddis, P. O. "The Project Manager." *Harvard Business Review,* May–June 1959, 89–97.

Hall, D., and Lawler, E. "Unused Potential in Research Development Organizations." *Research Management* 12 (1969): 339–54.

Mansfield, E. "Size of Firm, Market Structure and Innovation." *Journal of Political Economy* 71 (1963): 556–76.

Mohr, L. "Determinants of Innovation in Organizations." *American Political Science Review* 63 (1969): 111–26.

Mytinger, R. "Innovations in Local Health Services." Public Health Service Division of Medical Care Administration, U.S. Department of Health, Education and Welfare, Arlington, Va., 1968.

Rosner, M. "Administrative Controls and Innovativeness." *Behavioral Science* 13 (1968): 36–43.

Sapolsky, H. "Organizational Structure and Innovation." *Journal of Business* 40 (1967): 497–510.

Shephard, H. "Innovation Resisting and Innovation Producing Organizations." *Journal of Business* 40 (1967): 470–77.

Zaltman, G., Duncan, R., and Holbeck, J. *Innovations and Organizations.* New York: Wiley, 1973.

42. The Introduction of Matrix Management Into An Organization

David I. Cleland*

The decision of a large corporation to convert its management system to matrix management is generally made because it will provide an optimal management of resources. Such a conversion involves the organization's structure, decision processes, and prevailing cultural ambiance, as well as such complementary subsystems as those concerned with information, motivation, planning, and control. During the conversion process, effective communication is the key in developing the knowledge, skills, and attitudes of those associated with the change. Indeed, effective communication continues to play an important part in the ongoing success of any matrix management system.

The communication must go beyond the policy memoranda, letters, procedural instructions, and organizational charts that most managers use, for it is the role of interpersonal communication that is of paramount importance. This chapter describes the particular strategy used in the introduction of matrix management into a large transnational corporation and the part that interpersonal communication played during that conversion.**

THE SETTING

To make the most of emerging global market opportunities, a large transnational industrial corporation realigned its strategy to accommodate an increased effort in its international markets. For many years this corporation had been organized on a decentralized product structure, with basic profit-center responsibility at the departmental level. Managers were rewarded on an

---

*Dr. Cleland is a Professor of Engineering Management in the Industrial Engineering Department at the University of Pittsburgh, He is author or coauthor of ten books and has published numerous articles in leading national and internationally distributed technological, business management, and educational periodicals. Dr. Cleland has had extensive experience in management consultation, lecturing, seminars, and research.
**Certain material in this article has been paraphrased from David I. Cleland, "The Cultural Ambience of Matrix Management," *Management Review,* 1981, and "The Role of Communications in Effecting Change to a Matrix Management System," Project Management Institute 1980 Proceedings, Phoenix, Arizona.

individual incentive compensation system based on the financial performance of their organization. Everything counted at the profit center, everything was measured there, and key strategic and operational decisions were made by the profit center manager with the counsel of a functional staff.

The corporation had a long history of profit-center managers having equal authority and responsibility. In strategic matters these profit-center managers established objectives, goals, and strategies. Product planning was carried out at the profit-center level to include product technology, pricing, sourcing, configuration, and market support. Human resources, cash management, and investments also came under the profit-center manager's jurisdiction. In short, these profit center managers ruled fiefdoms where they called the strategic and operational moves. The senior managers of the corporation had all been appointed because of their ability as profit-center managers. It was within this cultural ambiance that a decision was made to reorganize into an international matrix management system.

The executives of this corporation faced a communication challenge: how to develop in key managerial personnel the knowledge, skills, and attitudes necessary for functioning in a successful international matrix management system. Three considerations were critical to the success of matrix management in this corporation: (1) dedication on the part of the key managers to make the matrix management system work; (2) the redesign of management processes and procedures to bring about a sharing of decisions, accountability, and financial rewards; and (3) basic attitudinal and cultural changes that emphasize team effort in the design and implementation of matrix organizational strategies such as participative management, consensus decision making, team management, and the like.

Four basic organizational designs were to be incorporated into this matrix management system: *functions, products, geography,* and *projects.* The already existing functional and product areas were to be augmented by geographic and project organizations. International project managers were appointed to manage projects in foreign locations. These projects entailed the management of resources across product divisions, functions, and geographic structures. The philosophical cornerstones of this international matrix management system were:

- Decision making, accountability, and financial results would be shared.
- The "country" would be the primary building block of the international organization.

A description of these cornerstones is rather simple, compared to the awesome task of getting key people intellectually and emotionally committed to support the matrix management system—which is where communication came into play. What follows is a description of the organizational change and communication used to implement a matrix management system in this transnational corporation.

THE ORGANIZATION CHANGE PROCESS

Six phases were identified in this change process. The change was designed to be evolutionary in nature, and to be accomplished over a two-to-three-year time period with maximum participation by the managers responsible for implementing the change. A key concern of the chief operating officer was how to design and implement the change, keeping in mind the necessity for an ongoing dialogue on why the change was needed and how it was to be effected.

Figure 42-1 depicts the strategic change process that was used to effect and communicate the changeover to the international matrix management system. In Phase 1 the chief operating officer, working with his staff, developed a tentative statement of what would be required to implement an international matrix management system. After this statement was developed, a task force

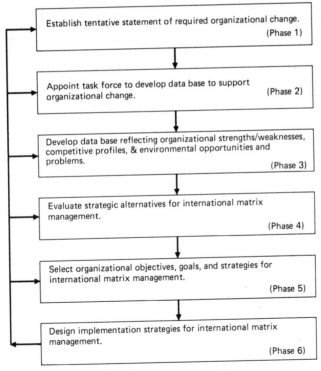

Fig. 42-1. Strategic change process.

was appointed to develop the data bases to support the organizational change (Phase 2). This task force, composed of approximately 45 people, represented different disciplines and different organizational levels. Over 300 key people were interviewed by the task force to develop data bases on everything from the strengths and weaknesses of the corporation in its international business, to a comprehensive analysis of the market opportunities and the nature and intensity of the competition (Phase 3).

Several key weaknesses in the existing organizational approach came to light during this phase:

- There was no effective organizational mechanism to pull together a team from several profit centers to bid on large projects in the international market.
- Several product divisions were attempting to sell to the same overseas customers without any corporate coordination that would present a common market position.
- An effective system for evaluating competitors' weaknesses, strengths, and probable strategies did not exist.
- The strong prerogatives of the profit-center manager in product planning—configuration, sourcing, pricing, and support services—often ignored key issues in geographic factors affecting market demand.
- Competitors were outperforming the corporation in virtually all markets.

In Phase 4 the task force assembled a mix of strategic alternatives for international matrix management, including an identification of alternative organizational approaches centered around various models of matrix management systems.

Phase 5 dealt with the strategic decision process of selecting organizational objectives, goals, and strategies for the international organization. An international organization was created on parity with the existing product-group structure. The matrix alignment of the international organization vis-á-vis the group-product structure is depicted in Figure 42-2. The sharing context implied in Figure 42-2 centered around the interfaces between product and geographic managers coming to focus in each country through the in-country subsidiary manager or product manager.

In Phase 6 the implementation strategies were designed and implementation was started. Figure 42-3 depicts this process. A first step was the reorganization of the corporation into a matrix structure to be complemented by support systems—information, accounting, personnel evaluation, budget, and so on. In Step 2 a consultant was engaged to conduct in-depth interviews with 30 key managers, to gain their perceptions of some of the relevant problems or opportunities to be encountered in implementing matrix management in the corpo-

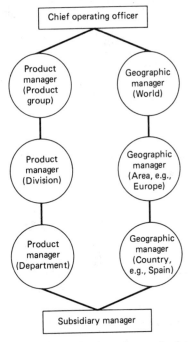

Fig. 42-2. International matrix organization.

ration. These inverviews, primarily nondirective in nature and conducted in the context of "diplomatic immunity," resulted in the development in Step 3 of a profile of key manager perceptions about matrix management consisting of the following:

- Corporate expected cultural changes
- Degree of knowledge—matrix management
- Required executive attitude change
- Change required—incentive systems
- Strategic planning in the matrix context
- Education required in matrix management
- What not to do in the educational effort
- Current executive experience in matrix management
- Expected modus operandi—matrix management
- Role of the new international organization
- Degree of commitment to new international matrix management system

The interviews provided a basis for the continued development of a matrix management system acceptable to those people who had to grapple with

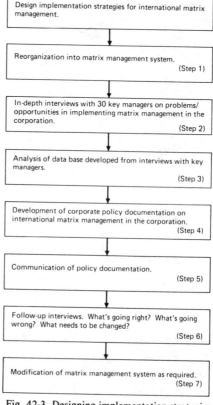

Fig. 42-3. Designing implementation strategies.

actually making it work. An evolutionary approach was selected to implement the system, starting first with the geographic area where the corporation had had the most success in the international marketplace.

In Step 4 corporate policy documentation was developed to portray the expected authority/responsibility/accountability patterns in the matrix system. Then, in Step 5, this policy documentation was disseminated through corporate seminars and discussion groups. Paralleling and supplementing this activity were formal educational programs that emphasized the theoretical aspects of how matrix management systems should be established, drawing on the best book and periodical literature available. The educational programs consisted of three-day training sessions for all key managers in the corporation. The tutorial strategy for the training sessions involved a bare minimum of lectures, depending instead on small five- or six-person discussion groups exploring the meaning of matrix management. Existing and potential matrix management situations were drawn from corporate experience and written up to use in the small group sessions. An important output of the small group discussions was a series of recommendations on how certain aspects of matrix management could be implemented in the corporation. Some examples of these included:

- Project management organizational design in the international context
- How to define an international project
- Identification of corporate policy needs to implement matrix management
- Development of a work breakdown structure for an international project
- How to use a linear responsibility chart to identify authority/responsibility / accountability patterns in matrix management
- Project manager candidate specifications
- How to select an international project manager
- Key problems/opportunities to be considered in making the matrix system work

The educational program, combined with the use of organizational examples, provided an effective means for communicating corporate policy on how the new matrix management system was expected to operate. All the key executives, including general managers and key functional and matrix managers, became involved at some point in the design and implementation of the new system. Each person so involved had ample opportunity to exert influence, to assist in making the new management system work. Many long days—sometimes extending into weeks and even months—were spent in making sure that corporate purposes and intents were effectively communicated throughout the corporate community. Key executives became zealous missionaries carrying the "gospel" of the new corporate matrix management system worldwide.

Step 6 was accomplished by a consultant who conducted follow-up interviews with key people to get answers to such questions as: What's going right? What's going wrong? What needs to be changed? And so on. These interviews helped to identify and attack some of the problems impeding the development of the system. Much had to be done to keep the system evolving, as depicted in Step 7 in Figure 42-3.

Several key characteristics of this corporation's approach facilitated the effective conversion to a matrix model of management. The common denominator of the organizational change was the candid and open communication that pervaded the organization from the top down while the change was under way. The openness of the organization was evident to people from the outside. Participative management, consensus decision making, the use of teams of executives, and the assistance of professionals to help in bringing about the change—all served to typify a cultural ambiance of free and open communication in the search for new organizational strategies.

THE COMMUNICATION PROCESS

A highly simplified model of the communication process consists of three basic elements: a sender, a receiver, and a message. When all three of these elements are present, communication may result. Communication takes place when a message is understood by the receiver as the sender intended it to be understood. Unfortunately, this doesn't always happen. Why not? Because "noise" can occur. Such noise can be anything from a physical interruption of the message to the experiences of the people who are trying to communicate. These experiences—shaped by language, values, customs, attitudes, behavior, and cultural setting—distort the way messages are perceived. This applies to both sender and receiver. In industrial organizations the cultural patterns that can inhibit or facilitate communication are established by tradition and by managerial strategies, policies, and procedures. When an organization undergoes a basic change in its objectives, goals, and strategies, such as the introduction of matrix management, the communication challenge is to recognize the cultural patterns in order to get the message to those managers who can make it happen. Therefore the introduction of a matrix management system into an organization affects communication processes within that organization.

Keeping in mind the basic model of the communication process that includes a sender, a receiver, and a message distorted by noise, the question can be asked: Did the manner in which this transnational corporation developed its matrix management system make sense from a communications viewpoint?

The sender in this case was senior corporate management, which wanted to communicate the design and implementation of a matrix management system throughout the corporation. The receiver was the body of key managers and professionals who could make the system work. The noise potentially consisted of a wide range of factors and forces: language, values, customs, attitudes, behavior, culture, organizational structure, traditional ways of doing things, and so on. This noise was very real. The objective of the organizational development was to change the culture from a traditional management system to one that was fully and effectively committed to matrix management. This objective was accomplished, but only through a lengthy process of interaction between senders and receivers working through an abundance of noise.

Had the communication factor of this organizational change not been considered, it is doubtful if the change would have occurred as effectively as it did. The size and complexity of this transnational corporation and the information overload on the managers were already awesome. The strategic shift to a matrix management system placed additional demands on the need to assimilate the flow of information. In the matrix management system, real authority in the corporation became increasingly dependent on those people who had access to information and could control which information was provided to the managers. The commitment of these people to matrix management, and their willingness to communicate in the matrix context were vital to the success of the new management system.

SUMMARY

Matrix management is hard to launch and challenging to operate. The more conventional the culture has been, the more challenges will emerge in moving to the matrix form. The following caveats are in order, for those who plan to initiate and use the matrix design.

- Patience is absolutely necessary. It takes time to change the system and people who make the matrix work.
- Promote by word and example an open and flexible attitude in the organization. Encourage the notion that change is inevitable, and that a free exchange of ideas is necessary to make project management work.
- Develop a scheme for organizational objectives, goals, and strategies that will provide the framework for an emerging matrix management culture.
- Accept the idea that some people may never be able to adjust to the unstructured, democratic ambiance of the matrix culture.
- Be mindful of the tremendous importance that team commitment plays in managing matrix activities. As in sports, commitment is an absolute prerequisite to becoming a championship team.
- Provide a forum whereby conflict can be resolved before it deteriorates into interpersonal strife.
- Realize that matrix management is not a panacea for organic organizational maladies. On the contrary, implementation of a matrix management system brings to light many organizational problems that have remained hidden in the conventional line and staff organization.
- Remember that the route an organization follows in its journey to the matrix design must evolve out of the existing culture.
- Recognize that senior management support and commitment are essential to success.
- Work for uninhibited, thorough, and complete communication within the company. Information requirements for matrix management need definition. Persons who have a need to know require access to the information to do their job. Those in key positions have to understand and use the project-generated information.
- Be aware that shifting to a matrix form is easier for the younger organization. For large, well-established companies with a rigid bureaucracy, the shift will be quite formidable.
- Institute a strong educational effort to acquaint key managers and professionals with the theory and practice of project management. Time should

be taken to do this right at the start, using the existing culture as a point of departure.

The real culture of matrix management refers to actual behavior—those things and events that really exist in the life of an organization. The introduction of matrix management into an existing culture will set into motion a "system of effects" that will make attitudes, values, beliefs, and management systems more participative and democratic. Thus a new cultural context for the sharing of decisions, results, rewards, and accountability will ultimately emerge, but only as a result of careful guidance, open lines of communication, and the patient understanding that neither human beings nor organizations can change without grasping the nature of the change.

Index

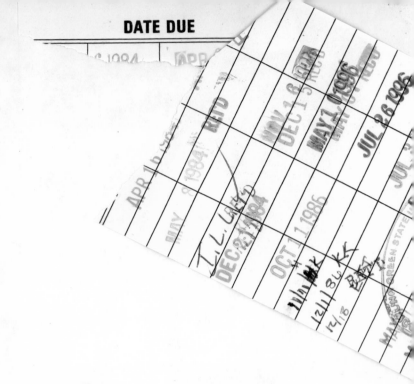